W9-APH-377

Numerical Methods
for Scientists
and Engineers

Numerical Methods for Scientists and Engineers

SECOND EDITION

R. W. HAMMING

Adjunct Professor of Computer Science,
Naval Postgraduate School

Dover Publications, Inc., New York

Copyright © 1962, 1973 by R. W. Hamming.
All rights reserved under Pan American and International Copyright Conventions.

Published in Canada by General Publishing Company, Ltd., 30 Lesmill Road, Don Mills, Toronto, Ontario.
Published in the United Kingdom by Constable and Company, Ltd.

This Dover edition, first published in 1986, is an unabridged republication of the second edition (1973) of the work first published by McGraw-Hill, Inc., New York, in 1962.

Manufactured in the United States of America
Dover Publications, Inc., 31 East 2nd Street, Mineola, N.Y. 11501

Library of Congress Cataloging-in-Publication Data

Hamming, R. W. (Richard Wesley), 1915-
    Numerical methods for scientists and engineers.

    Reprint. Originally published: New York: McGraw-Hill, 1973.
    Bibliography: p.
    Includes index.
    1. Numerical analysis—Data processing.   I. Title.
[QA297.H28   1987]   519.4   86-16226
ISBN 0-486-65241-6

THE PURPOSE OF COMPUTING
IS INSIGHT, NOT NUMBERS

To study, and when the occasion arises to put what one has
learned into practice—is that not deeply satisfying?
Confucius, Analects 1.1.1

CONTENTS

* Starred sections may be omitted.

* Starred sections may be omitted.

PREFACE

There has been much progress in the 10 years since the first edition was written, but of the many books that have appeared on the topic none has put the emphasis on the frequency approach and its use in the solution of problems. For these reasons, a second edition seems necessary.

The material has been extensively rearranged, rewritten, and added to, so that in some respects it is a new book; however, the main aims, style, and motto have not changed.

As always, the author is greatly indebted to others for much that is in the book. Most important are his management and colleagues at the Bell Telephone Laboratories. Professor Roger Pinkham has over the years been a constant source of stimulation and inspiration. It would take a list of at least 100 names to thank all who have contributed to some extent, and at the top of this list would be M. P. Epstein. My thanks also go to all the unmentioned people on the list and to A. Ralston for many helpful suggestions. Thanks also to Mrs. Jeannie Waddel for typing and helping to organize the manuscript.

R. W. HAMMING

PART I

Fundamentals and Algorithms

1

AN ESSAY ON NUMERICAL METHODS

1.1 THE FIVE MAIN IDEAS

Numerical methods use numbers to simulate mathematical processes, which in turn usually simulate real-world situations. This implies that there is a *purpose* behind the computing. To cite the motto of the book, The Purpose of Computing Is Insight, Not Numbers. This motto is often thought to mean that the numbers from a computing machine should be read and used, but there is much more to the motto. The choice of the particular formula, or algorithm, influences not only the computing but also how we are to understand the results when they are obtained. The way the computing progresses, the number of iterations it requires, or the spacing used by a formula, often sheds light on the problem. Finally, the same computation can be viewed as coming from different models, and these different views often shed further light on the problem. *Thus computing is, or at least should be, intimately bound up with both the source of the problem and the use that is going to be made of the answers—it is not a step to be taken in isolation from reality.*

Much of the knowledge necessary to meet this goal comes from the field of application and therefore lies outside a general treatment of numerical methods. About all that can be done is to supply a rich assortment of methods and to comment on their relevance in general situations.   This art of connecting the specific problem with the computing is important, but it is best taught in connection with a field of application.

The second main idea is a consequence of the first.   If the purpose of computing is insight, not numbers, as the motto states, then it is necessary to *study families and to relate one family to another when possible,* and to avoid isolated formulas and isolated algorithms.   In this way a sensible choice can be made among the alternate ways of doing the problem, and once the computation is done, alternate ways of viewing the results can be developed.   Thus, hopefully, the insight can arise.   For these reasons we tend to concentrate on systematic methods for finding formulas and avoid the isolated, cute result.   It is somewhat more difficult to systematize algorithms, but a unifying principle has been found.

This is perhaps the place to discuss some of the differences between numerical methods and numerical analysis (as judged by the corresponding textbooks). Numerical analysis seems to be the study in depth of a few, somewhat arbitrarily selected, topics and is carried out in a formal mathematical way devoid of relevance to the real world.   Numerical methods, on the other hand, try to meet the need for methods to cope with the potentially infinite variety of problems that can arise in practice.   The methods given are generally chosen for their wide applicability in creating formulas and algorithms as well as for the particular result being found at that point.

The third major idea is *roundoff error.*   This effect arises from the finite nature of the computing machine which can only deal with finitely represented numbers.   But the machine is used to simulate the mathematician's number system which uses infinitely long representations.   In the machine the fraction $\frac{1}{3}$ becomes the terminated decimal 0.333...3 with the obvious roundoff effect.   At first this approximation does not seem to be very severe since usually a minimum of eight decimal places are carried at every step, but the incredible number of arithmetic operations that can occur in a problem lasting only a few seconds is the reason that roundoff plays an important role.   *The greatest loss of significance in the numbers occurs when two numbers of about the same size are subtracted so that most of the leading digits cancel out, and unless care is taken in advance, this can happen almost any place in a long computation.*

Most books on computing stress the estimation of roundoff, especially the bounding of roundoff, but we shall concentrate *on the avoidance of roundoff.*   It seems better to avoid roundoff than to estimate what did not have to occur if common sense and a few simple rules had been followed *before* the problem was put on the machine.

The fourth main idea is again connected with the finite nature of the machine, namely that many of the processes of mathematics, such as differentiation and integration, imply the use of a limit which is an infinite process.   The machine has finite speed and can only do a finite number of operations in a finite length of time. This effect gives rise to the *truncation error* of a process.

We shall generally first give an exact expression for the truncation error and deduce from it various bounds.   A moment's thought should reveal that if we had an exact expression, then it would be practically useless because to know the exact error is to know the exact answer.   However, the exact-error expression is very useful in studying families of formulas, and it provides a starting point for a variety of error bounds.

The fifth main idea is *feedback*, which means, as its name implies, that numbers produced at one stage are fed back into the computer to be processed again and again; the program has a loop which uses the output of one cycle as the input for the next cycle.   This feedback situation is very common in computing, as it is a very powerful tool for solving many problems.

Feedback leads immediately to the associated idea of *stability* of the feedback loop—will a small error grow or decay through the successive iterations?   The answer may be given loosely in two equivalent ways: first, if the feedback of the error is too strong and is in the direction to eliminate the error (technically, negative feedback), then the system will break into an oscillation that grows with time; second and equivalently, if the feedback is delayed too long, the same thing will happen.

A simple example that illustrates feedback instability is the common home shower.   Typically the shower begins with the water being too cold, and the user turns up the hot water to get the temperature he wants.   If the adjustment is too strong (he turns the knob too far), he will soon find that the shower is too hot, whereupon he rapidly turns back to cold and soon finds it is too cold.   If the reactions are too strong, or alternately the total system (pipes, valve, and human) is too slow, there will result a " hunting " that grows more and more violent as time goes on.   Another familiar example is the beginning automobile driver who over-reacts while steering and swings from side to side of the street.   This same kind of behavior can happen for the same reasons in feedback computing situations, and therefore the stability of a feedback system needs to be studied before it is put on the computer.

1.2   SECOND-LEVEL IDEAS

Below the main ideas in Sec. 1.1 are about 50 second-level ideas which are involved in both theoretical and practical work.   Some of these are now discussed.

At the foundation of all numerical computing are the actual numbers themselves. The floating-point number system used in most scientific and engineering computations is significantly different from the mathematician's usual number system. The floating-point numbers are not equally spaced, and the numbers do not occur with equal frequency. For example, it is well known that a table of physical constants will have about 60 percent of the numbers with a leading digit of 1, 2, or 3, and the other digits—4, 5, 6, 7, 8, and 9—comprise only 40 percent.

Although this number system lies at the foundation of most of computing, it is rarely investigated with any care. People tend to start computing, and only after having frequent trouble do they begin to look at the system that is causing it.

Immediately above the number system is the apparently simple matter of evaluating functions accurately. Again people tend to think that they know how to do it, and it takes a lot of painful experience to teach them to examine the processes they use before putting them on a computer.

These two mundane, pedestrian topics need to be examined with the care they deserve *before* going on to more advanced matters; otherwise they will continually intrude in later developments.

Perhaps the simplest problem in computing is that of finding the zeros of a function. In the evaluation of a function near a zero there is almost exact cancellation of the positive and negative parts, and the two topics we just discussed, roundoff of the numbers and function evaluation, are basic, since if we do not compute the function accurately, there can be little meaning to the zeros we find. Because of the discrete structure of the computer's number system it is very unlikely that there will be a number x which will make the function $y = f(x)$ exactly zero. Instead, we generally find a small interval in which the function changes sign. The size of the interval we can use is related to the size of the argument x, since for large x the number system has a coarse spacing and for x small (in size) it has a fine spacing. This is one of the reasons that the idea of the *relative error*

$$\text{Relative error} = \left| \frac{\text{true} - \text{calculated}}{\text{true}} \right|$$

plays such a leading role in scientific and engineering computations. Classical mathematics uses the *absolute error*

$$\text{Absolute error} = |\text{true} - \text{calculated}|$$

most of the time, and it requires a positive effort to unlearn the habits acquired in the conventional mathematics courses. The relative error has trouble near places where the true value is approximately zero, and in such cases it is customary to use as the denominator

$$\max\{|x|, |f(x)|\}$$

where $f(x)$ is the function computed at x.

The problem of finding the complex zeros of an analytic function occurs so often in practice that it cannot be ignored in a course on numerical methods, though it is almost never mentioned in numerical analysis. A simple method resembling one used to find the real zeros is very effective in practice.

In the special case of finding all the zeros of a polynomial the fact that the number of zeros (as well as other special characteristics) is known in advance makes the problem easier than for the general analytic function. One of the best methods for finding them is an adaptation of the usual Newton's method for finding real zeros, and this discussion is used to extend, as well as to analyse further, Newton's method. It is only in situations in which a careful analysis can be made that Newton's method is useful in practice; otherwise its well-known defects outweigh its virtues.

What makes the problem of finding the zeros of a polynomial especially important, besides its frequency, is the use made of the zeros found. The method is a good example of the difference between the mathematical approach and the engineering approach. The first merely tries to find some numbers which make the function close to zero, while the second recognizes that a pair of "close" zeros will give rise to severe roundoff troubles when used at a later stage. In isolation the problem of finding the zeros is not a realistic problem since the zeros are to be used, not merely admired in a vacuum. Thus what is wanted in most practice is the finding of the multiple zeros as multiple zeros, not as close, separate ones. Similarly, zeros which are purely imaginary are to be preferred to ones with a small real part and a large imaginary part, *provided* the difference can reasonably be attributed to uncertainties in the underlying model.

Another standard algorithmic problem both in mathematics and in the use of computation to solve problems is the solution of simultaneous linear equations. Unfortunately much of what is commonly taught is usually not relevant to the problem as it occurs in practice; nor is any completely satisfactory method of solution known at present. Because the solution of simultaneous linear equations is so often a standard library package supplied by the computing center and because the corresponding description is so often misleading, it is necessary to discuss the limitations (and often the plain foolishness) of the method used by the package. Thus it is necessary to examine carefully the obvious flaws and limitations, rather than pretending they do not exist.

The various algorithms for finding zeros, solving simultaneous linear equations, and inverting matrices are the classic algorithms of numerical analysis. Each is usually developed as a special trick, with no effort to show any underlying principles. The idea of an *invariant algorithm* provides one common idea linking, or excluding, various methods. An invariant algorithm is one that in a very real sense attacks the problem rather than the particular representation supplied to the computer. The idea of an invariant algorithm is actually fairly simple and obvious once understood. In many kinds of problems there are one or more classes of transformations that will transform one representation of the equations into another of the same form. For example, given a polynomial

$$P(x) = a_n x^n + a_{n-1} x^{n-1} + \cdots + a_0 = 0$$

the transformation of multiplying the equation by any nonzero constant does not really change the problem. Similarly, when $a_0 \neq 0$, replacing x by $1/x$ while also multiplying the equation by x^n merely reverses coefficients. These transformations form a group (provided we recognize the finite limitations of computing), and it is natural to ask for algorithms that are invariant with respect to this group, where invariant means that if the problem is transformed to some equivalent form, then the algorithm uses, at all stages, the equivalent numbers (within roundoff, of course). In a sense the invariance is like dimensional analysis—the scaling of the problem should scale the algorithm in exactly the same way. It is more than dimensional analysis since, as in the example of the polynomial, some of the transformations to be used in the problem may involve more than simple scaling.

It is surprising how many common algorithms do not satisfy this criterion. The principle does more than merely reject some methods; it also, like dimensional analysis, points the way to proper ones by indicating possible forms that might be tried.

1.3 THE FINITE DIFFERENCE CALCULUS

After examining the simpler algorithms, it is necessary to develop more general tools if we are to go further. The *finite difference calculus* provides both the notation and the framework of ideas for many computations. The finite difference calculus is analogous to the usual infinitesimal calculus. There are the difference calculus, the summation calculus, and difference equations. Each has slight variations from the corresponding infinitesimal calculus because instead of going to the limit, the finite calculus stops at a fixed step size. This reveals why the finite calculus is relevant to many applications of computing: in a sense it *undoes* the limiting process of the usual calculus. It should be evident that if a limit process cannot be undone, then there is a very real question as to the soundness of the original derivation, because it is usually based on constructing a believable finite approximation and then going to the limit.

The finite difference calculus provides a tool for estimating the roundoff effects that appear in a table of numbers *regardless* of how the table was computed. This tool is of broad and useful application because instead of carefully studying each particular computation, we can apply this general method without regard to the details of the computation. Of course, such a general method is not as powerful as special methods hand-tailored to the problem, but for much of computation it saves both trouble and time.

The summation calculus provides a natural tool for approaching the very common (and often neglected) problem of the summation of infinite series, which is the simplest of the limiting processes (since the index n of the number of terms taken runs through the integers only).

The solution of finite difference equations is analogous to the solution of differential equations, especially the very common case of linear difference equations with constant coefficients, which is a valuable tool for the study of feedback loops and their stability. Thus finite difference equations have both a practical and a theoretical value in computing.

1.4 ON FINDING FORMULAS

Once past the easier algorithms and tools for doing simple things in computing, it is natural to attack one of the central problems of numerical methods, namely, the approximation of infinite operations (operators) by finite methods. Interpolation is the simplest case. In interpolation we are given some samples of the function, say, $y(-1)$, $y(0)$, and $y(1)$, and we are asked to *guess* at the missing values—to read between the lines of a table. While it is true that because of the finite nature of the number system used there are only a finite number of values to be found, nevertheless this number is so high that it might as well be infinite. Thus interpolation is an infinite operator to be approximated.

There is no sense to the question of interpolation unless some additional assumptions are made. The *classical assumption* is that given $n + 1$ samples of the function, these samples determine a unique polynomial of degree n, and this polynomial is to be used to give the interpolated values. With the above data consisting of three points, the quadratic through these points is

$$P(x) = \frac{x(x - 1)}{2} y(-1) + (1 - x^2)y(0) + \frac{x(x + 1)}{2} y(1)$$

We are to use this polynomial $P(x)$ as if it were the function. This method is known as the *exact matching* of the function to the data.

The error of this interpolation can be expressed as the $(n + 1)$st derivative (of the original function) evaluated at some generally unknown point 0 in the interval. Unfortunately in practice it is rare to have any idea of the size of this derivative.

For samples of the function we may use not only function values $y(x)$ but also values of the derivatives $y'(x)$, $y''(x)$, etc., at various points.   For example, the cubic exactly matching the data $y(0)$, $y(1)$, $y'(0)$, and $y'(1)$ is

$$P(x) = (1 - 3x^2 + 2x^3)y(0) + (3x^2 - 2x^3)y(1) + (x - 2x^2 + x^3)y'(0) \\ + (x^3 - x^2)y'(1)$$

It is important to use analytically found derivatives when possible.   Then we can usually get a higher order of approximation at little extra cost since generally once the function values are computed, the derivatives are relatively easy to compute. No new radicals, logs, exponentials, etc., arise, and these are the time-consuming parts of most function evaluation.   Of course a sine goes into a cosine when differentiated, but this is about the only new term needed for the higher derivatives. Even the higher transcendental functions, like the Bessel functions, satisfy a second-order linear differential equation, and once both the function and the first derivative are found, the higher derivatives can be computed from the differential equation and its derivatives (which are easy to compute).   Thus we shall emphasize the use of derivatives as well as function values for our samples.

Although a wide variety of function and derivative values may be used to determine the interpolating polynomial, there are some sets, rather naturally occurring, for which $n + 1$ data samples do not determine an nth-degree polynomial. Perhaps the best example is the data $y(-1)$, $y(0)$, $y(1)$, $y''(-1)$, $y''(0)$, and $y''(1)$ which do not determine a fifth-degree polynomial—the positions and accelerations at three equally spaced points do not determine a quintic in general.

The classic method for finding formulas for other infinite operators, such as integration and differentiation, is to use the interpolating polynomial as if it were the function and then to apply the infinite operator to the polynomial.   For example, if we wish to find the integral of a function from -1 to $+1$, given the values $y(-1)$, $y(0)$, and $y(1)$, we find the interpolating quadratic as above and integrate it to get the classical Simpson's formula:

$$\int_{-1}^{1} y(x)\, dx = \tfrac{1}{3}y(-1) + \tfrac{4}{3}y(0) + \tfrac{1}{3}y(1)$$

This process is called *analytic substitution*; in place of the function we could not handle we take some samples, exactly match a polynomial to the data, and finally analytically operate on this polynomial.   This is the classical method for finding formulas.   It is a two-step method: find the interpolating function and then apply the operator to this function.

There is another direct method that is *almost* equivalent to the analytic-substitution method.   In this method we make the formula true for a sequence of functions $y(x) = 1, x, x^2, x^3, \ldots, x^m$. For example, to derive Simpson's formula by this method we assume the form

$$\int_{-1}^{1} y(x)\, dx = a_{-1} y(-1) + a_0\, y(0) + a_1 y(1)$$

and substitute the sequence of functions 1, x, and x^2. The three resulting equations determine the three unknown coefficients a_i, and the resulting formula is exactly the same (in this case). The two methods differ in the case where there is no interpolating polynomial; it may be that there is a formula even if there is no interpolating polynomial. For example, we have the formula

$$\int_{-1}^{1} y(x)\, dx = \tfrac{1}{21}[5y(-1) + 32y(0) + 5y(1)] - \tfrac{1}{315}[y''(-1) - 32y''(0) + y''(1)]$$

which is exact for sixth-degree polynomials when, as we have noted above, there is in general no interpolating polynomial of fifth degree.

It would seem as if the two methods were equivalent, for if there were an interpolating polynomial, then the formula would surely be true for the corresponding powers of x; and conversely, if it were true for the individual powers, then it would be true for any linear combination, namely a polynomial. The difference lies in the words *if there is an interpolating polynomial, then* .... It can happen that the two-stage process fails on the first step, but the one-step direct method will work.

There are two main advantages of the direct method. First the derivations are much easier, and second the direct method provides a basis for extensive generalizations. The importance of this method is hard to overestimate. It means that we can find a very wide range of formulas, all within a common framework of ideas and methods, and that we will therefore be able to compare one formula to another and then decide which one to use. Perhaps more important, it means that we can decide the kind of formula we want and then with this single method and its generalizations find almost any formula we want—we can fit the formula to the problem rather than fit the problem to the formula. Thus we can try to achieve the insight that is the main goal of the book, as stated in our motto, The Purpose of Computing Is Insight, Not Numbers.

With a general method for finding polynomial approximation formulas it is necessary to have a corresponding method for finding the error of the formula. The general method of finding the error is somewhat difficult to understand the first time. It is based on the use of a Taylor series with the integral remainder, and by substituting this into the formula for which we want the error (and manipulating the results a bit) we get the desired error formula. Once we have the exact-error term, it can be transformed in various ways to get suitable practical-error estimates.

The method for finding the truncation error term for polynomial approximation unfortunately gives it in the form of a derivative, much as the interpolation method did. This, as noted before, is unfortunate because the high-order derivatives are seldom available.

This brings up a central dilemma. Should one use a high-order formula (error term has a high-order derivative) or use the repetition of a low-order formula—the composite formula? The answer is simple in principle. It depends on the size of the high-order derivative as compared to the other lower-order derivative. This is, of course, almost no answer at all, because we seldom can decide which is better, and the basis of the choice depends on the location in the complex plane of the singularities of the function being integrated—something we seldom know.

1.5   CLASSICAL NUMERICAL ANALYSIS

Much of classical numerical analysis, as we have indicated, is based on polynomial approximation for the infinite operations of differentiation, integration, and interpolation. The polynomial approximation is also used in the numerical integration of ordinary differential equations. The most widely used methods for this are the predictor-corrector methods. A polynomial is fitted to some of the data at past points and is used to extrapolate to the next point—to *predict*. The predicted value is used in the differential equation to get the predicted slope. This slope along with past data is used to find another polynomial which produces the *corrected value*, and the corrected value of the slope is found. If the predicted and corrected values are sufficiently close, then the step is accepted as accurate enough, and if not, the step size of integration may be halved (doubled if the two values are too close).

There are so many possible predictor-corrector methods of the same order of accuracy that it is necessary to have a general theory to compare the various formulas within a common framework. Otherwise chaos and prejudice would reign.

So far we have discussed the exact-matching interpolating polynomial, and this is the more usual method. There are other methods, more or less classical, for the selection of the approximating polynomial. One method is the minimum sum of squares of the residuals between the formula and the data given. Another more modern method picks the polynomial with the minimum maximum error—the minimax, or Chebyshev, approximation. Still other criteria could be used if desired, though the labor of finding the polynomial may be fairly high in some cases.

1.6 MODERN NUMERICAL METHODS— FOURIER APPROXIMATION

The difficulty with polynomial approximation in practice is that it is in the nature of polynomials to "wiggle" and to go to infinity for large absolute values of the argument x. Physically occurring functions tend to wiggle much less than polynomials and to remain bounded for large values of the argument. Thus polynomials are a poor basis for approximation, even though they are easy to compute and to think about. The fact that the Weierstrass approximation theorem states that any continuous function can be uniformly approximated in a closed interval by a polynomial is irrelevant for two reasons. First, the degree of the Weierstrass polynomial is generally very high for even a low degree of approximation; second, we are not finding the polynomial the way the theorem states it can be found. Indeed, it is "well known"[1] that for the simple function $y(x) = 1/(1 + x^2)$ in the interval $|x| \leq 3.63 \ldots$ the sequence of polynomials that exactly matches the function at a set of equally spaced points does not approach the function uniformly as the number of points increases indefinitely—the function and the polynomial differ by arbitrarily large amounts, and the sequence of approximating polynomials fails to converge.

Since polynomials are rather poor functions to use for approximating many functions that occur in practice, it is natural to look for other sets of functions. Among the many sets that are known to be complete (meaning that they can approximate any continuous function in a closed interval) the functions $\sin nx$ and $\cos nx$ $(n = 0, 1, \ldots)$ have been the most studied and are the most useful. Approximation in terms of them is usually called Fourier approximation because J. B. J. Fourier (1768–1830) used them extensively in his work.

In the simplest case of approximating a function in an interval we are given a periodic function of period, say 2π, and are asked to approximate the function $y(x)$ by a form

$$y(x) = \frac{a_0}{2} + \sum_{k=1}^{\infty} (a_k \cos kx + b_k \sin kx)$$

It is easy to show that since the functions are orthogonal, that is,

$$\int_0^{2\pi} \cos kx \cos mx \, dx = \begin{cases} 0 & k \neq m \\ \pi & k = m \neq 0 \\ 2\pi & k = m = 0 \end{cases}$$

$$\int_0^{2\pi} \sin kx \cos mx \, dx = 0$$

$$\int_0^{2\pi} \sin kx \sin mx \, dx = \begin{cases} 0 & k \neq m \\ \pi & k = m \neq 0 \end{cases}$$

---

[1] Meaning that it can be found in the literature.

the coefficients in the expansion are given by

$$a_k = \frac{1}{\pi} \int_0^{2\pi} y(x) \cos kx \, dx \qquad b_k = \frac{1}{\pi} \int_0^{2\pi} y(x) \sin kx \, dx$$

The Fourier functions have a number of interesting properties beyond merely remaining bounded for all values. The error of an approximation that uses only a finite number of terms can be expressed in terms of the function rather than, as in the polynomial case, some high-order derivative. Furthermore, the rate of convergence, that is, how fast the finite series approaches the function as we take more and more terms, can be estimated easily from the discontinuities of the function and its derivatives.

Perhaps most important is the simple fact that the effect of taking equally spaced samples of the continuous function can be easily understood. The higher frequencies (speeds of rotation) appear as if they were lower frequencies. This effect, called *aliasing* because one frequency goes under the name of another, is a familiar phenomenon to the watchers of TV and movie westerns. As the stage coach starts up, the wheels start going faster and faster, but then they gradually slow down, stop, go backwards, slow down, stop, go forward, etc. This effect is due *solely* to the sampling the picture makes of the real scene. Figures 1.6.1 and 1.6.2 should make the effect clear. Once we know the sampling rate, we know exactly what frequencies will go into what frequencies. The highest frequency that is correct is called the *Nyquist*, or *folding*, *frequency*. In the polynomial situation we have no such simple understanding of the effect of sampling.

It might appear that to calculate all the coefficients of a Fourier expansion would involve a great deal of computing, but the recently discovered fast Fourier transform (FFT) method requires about $N \log N$ operations to fit N data points. This discovery has greatly increased the importance of Fourier approximation.

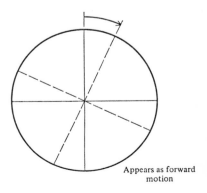

Appears as forward motion

FIGURE 1.6.1

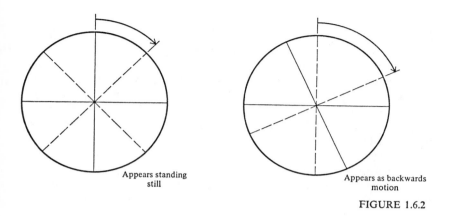

Appears standing
still

Appears as backwards
motion

FIGURE 1.6.2

Most of our functions are not periodic, and so the Fourier series is of limited importance. For the general nonperiodic function there is a corresponding Fourier integral

$$f(t) = \int_{-\infty}^{\infty} F(\sigma)e^{2\pi i \sigma t}\, d\sigma$$

$$F(\sigma) = \int_{-\infty}^{\infty} f(t)e^{-2\pi i \sigma t}\, dt$$

The two functions $f(t)$ and the *transform* $F(\sigma)$ have the same information: one describes the function in the space of the variable t, while the other describes it in the space of frequencies σ. These two equivalent views are part of the reason that the Fourier integral gives such a useful approach to many problems.

When we resort to using a finite number of sample points, we again face aliasing, exactly the same effect as before. The folding frequency now provides a more natural barrier and leads to the concept of a "band-limited function," meaning that the frequencies in the function are confined to a band.

The idea of a band-limited function is mirrored very closely in real problems. A hi-fi system handles all the frequencies in a band and cuts off above and below rather sharply; the better the hi-fi system, the wider the band. The idea of a band of frequencies applies to most information transmission systems, servomechanisms, and feedback control situations. This means, among other things, that from the physics of the problem we can estimate the spacing we will need for our samples, and vice versa, that from the spacing we can estimate the frequency content of the solution.

With the aid of the Fourier integral, which closely parallels the Fourier series with its nice mathematical properties, we can show the effect of taking a finite slice of a function from a potentially infinitely long function. For example, the light from a pulsar or Cepheid variable star shines for many years; yet we observe it for a night or two and from that limited record try to estimate what the star is doing. It is often important to know the effect of the length of the observation on what we can hope to learn about the star. Similarly in other control problems, the length of the observation affects what we can see, and the Fourier integral enables us to understand this limitation.

Using this new tool of Fourier approximation, we can then look back at the polynomial approximation methods and see them in a new way. Once this new way becomes familiar, we can understand much more clearly what we were doing before. Exactly the same computations can be viewed in a new, more revealing light. This is especially true for the communications and control problems that tend to dominate our technological age.

Once this new way of looking at computing becomes familiar, it is natural to begin designing formulas to meet new criteria. In place of the discrete set of integers that were the exponents of the variable x in polynomial approximation, we now have a continuous band of frequencies to use and examine. We can pick the formula that minimizes some property of this continuous error curve (in the frequency space to be sure). One popular method is to make the error curve have a minimax (Chebyshev) property.

This new approach is relevant to many design problems. For example, in designing a simulator for humans to use while training for airplane or space travel, we are interested in building the simulator so that it "feels right" to the human who is being trained. This to a first approximation means that the Fourier transform (more generally, the Laplace transform) of the simulator should be close to the transform of the real thing. It is less important that the simulator and real vehicle go exactly the same place than it is to "feel right." Thus we are no longer to judge the method of integrating a system of differential equations that describe the simulator by how well the solutions agree, but rather by how well the transforms agree—which is not the same thing! This new method of design is known as *the method of zeros and poles*, and unfortunately it involves classical network theory.

The role of digital communications systems is steadily increasing; more and more we are sampling the continuous analog signal that occurs naturally in the real world and converting it to a sequence of discrete digits. Thus we face not only sampling but *quantization effects* of the digitizing of the analog signal. This effect is like that of roundoff in many ways, but it is usually much more severe, and because it occurs at the beginning, it again limits us to what we can hope to see of the original underlying physical phenomena. Without a good understanding of these limitations we are not likely to understand what the numbers coming out of the computer mean and do not mean.

This leads gradually to the design of digital filters, which filter digital signals much as the old analog filters were used in processing analog signals, with radio and television being a couple of familiar examples. The differences from the continuous signals are significant in the digital case, and it is the digital case that digital computers must use.

1.7 OTHER CLASSES OF FUNCTIONS USED IN APPROXIMATIONS

After polynomial and Fourier approximations, the exponential set of functions is most used. The three sets, polynomials, Fourier, and exponentials (together with combinations of the three), are *invariant* under a translation of the origin. This is an important property because in many situations there is no natural origin, and without this property the choice of the origin would affect the answer. For these three classes the answer is independent of the origin, though the particular set of coefficients used in the approximation will differ as the origin is chosen differently.

For the exponential functions there is the Laplace transform corresponding to the Fourier transform, but it has much more difficult properties from the computing point of view.

1.8 MISCELLANEOUS

When a problem has a singularity, as many practical problems do (often because of the mathematical idealization), then the structure of the singularity indicates the class of approximating functions to use, and the position gives the natural origin. The methods used in this case are similar to those of the three usual cases.

Sometimes the problem has a natural set of functions to use, and again the methods we have developed can be applied, though the details may get a bit messy in many cases.

Optimization occurs frequently in practice, and the person practicing numerical methods needs to know something about this rapidly growing field. Most simulations imply an optimization in the background—the simulation is being done to optimize some aspect of the situation.

A central idea of mathematics is *linear independence*. In computing it is natural that this idea, which involves a yes no situation, becomes more vague. Clearly in computing there will be some degree of linear independence, and various forms for representing the same information will have varying degrees of linear independence.   The idea is still in its infancy and needs a great deal more development, but it is clearly a central idea in computing.

One of the more difficult problems requiring an algorithm is finding the eigenvalues and eigenvectors of a matrix.   Unfortunately, there is a great lack of understanding of what the problem actually is and of what the answers are to be used for, and there are no widely accepted methods for the general case.   For the particular case of a symmetric (also for a Hermitian) matrix reasonably effective methods are known.

1.9   REFERENCES

The problem of supplying further references is a vexing one.   The literature is rapidly changing when compared to the lifetime of a book, and as a result most references would soon be out of date and misleading.   Furthermore, in a text like this where most chapters can, and some have, been expanded into whole books, there is little point in giving a lot of isolated references which will probably be ignored by most readers.   We shall assume that a few standard textbooks are available and usually refer the reader to them for further information.   The occasional reference to the literature is to amplify a point that is not in standard textbooks.

2
NUMBERS

2.1 INTRODUCTION

Numbers are the basis of numerical methods. Thus logically they belong at the beginning of a course on numerical methods. On the other hand, psychologically they occur rather late in the development.

The situation in numerical methods is very like that in the calculus course where the real number system is basic to the limit process. The calculus course, therefore, often begins with a detailed, fairly rigorous discussion of the real number system. Unfortunately, at this point in his education the student has little reason to care about the topic, and it always turns out to be the most difficult part of the entire course. Furthermore, the topic generally repels the student, and this attitude carries over to the rest of the course.

The history of mathematics further shows that the real number system was very late in developing. The discoverers and developers of the calculus ignored the niceties of the number system for many years. The biological principle "ontogeny recapitulates phylogeny" means that "the development of the individual tends to repeat the development of the species." This is very relevant to teaching; the history of a subject gives important clues as to the ordering and relative difficulties of the material being taught.

History likewise shows that for a long time the number system used in computing was essentially ignored and taken for granted. Putting the topic first, therefore, requires justification, because we are asking the beginner to learn material whose importance he is not psychologically prepared to accept. The justification is the same as that for the calculus course. If we are to make rapid progress and not to have to repeat some material several times, then it is necessary to start with a firm foundation. The author's own experience was that only after many years of computing did he come to understand how the number system used by the machines affected what was obtained and how at times it led him astray.

Thus we are asking the beginner to accept on faith that the material in this chapter is basic and to put aside his natural psychological prejudices in favor of the logical approach. Of course he wants to get on to solving real problems and not to fuss with apparently trivial, irrelevant details of the number system used by machines, which he thinks he understands anyway. In compensation we will try to make the material a bit more dramatic than usual in order to sustain his interest through this desert of logical presentation. Probably he should plan to review this chapter later several times until he becomes thoroughly familiar with many of the various peculiar features of the number system used by the machine. In a sense it is the *real* number system, since they are the only numbers that can occur in the computation; there are no other numbers, and the mathematician's "real" number system is purely fictitious.

2.2   THE THREE SYSTEMS OF NUMBERS

There are three systems of numbers in the usual computing machine. First there are the *counting numbers* 0, 1, 2, . . . , 32,767 (or some other finite, rather small number). Note that this system begins with 0 rather than 1. Unless this fact is thoroughly learned, when a loop is written in a program, the loop will not be able to be done *no* times (meaning that it cannot be skipped). This number system is usually intimately connected with the index registers of the machine, and the range of the numbers is thereby determined. This is the familiar counting system, and little more need be said beyond the fact that it is definitely bounded and does not extend to infinity.

The second system of numbers is the familiar *fixed-point number* system. Typical numbers are:

$$3.141592654$$
$$0.012345678$$
$$-123.4567890$$

These numbers all have a fixed length, usually the word length of the machine (or some simple multiple of it), and the differences between successive numbers are the same. They are the familiar numbers of hand computing. Almost all the tables of numbers that the beginner has used (trigonometric, logarithmic, etc.) are in fixed-point notation. The chief difference between the machine's system and hand calculation is that the human often changes the number of digits he carries as circumstances seem to warrant, while the machine generally keeps the same number of digits (it may occasionally shift from single to double or even multiple precision).

The third system of numbers is the *floating-point number* system, which is closely related to the so-called " scientific notation " used in many parts of science. The system is designed to handle both very large and very small numbers. Typical numbers are:

$$.31415927 \times 10^1$$
$$.12345678 \times 10^{-1}$$
$$-.\underbrace{12345678}_{\text{mantissa}} \times 10^{4}\underset{\text{exponent}}{\nwarrow}$$

The first block of eight digits in these above numbers is called the *mantissa* (in analogy with logarithms), and the last digit is called the *exponent* (often ranging from at least -38 to $+38$).

Usually the mantissa and the exponent are stored in the same word of the machine, and as a result the mantissa of a floating-point number is usually shorter than the corresponding fixed-point number.

For convenience we shall use in the examples in the book a three-digit mantissa and a one-digit exponent for our floating-point number system. Not only does this save space, but it also makes the examples much easier to follow. Occasionally we will use a three-digit fixed-point number system. Examples of our floating-point numbers are:

$$\pi = \quad .314 \times 10^1$$
$$\frac{1}{\sqrt{2}} = \quad .707 \times 10^0$$
$$\frac{-\pi}{1,000} = -.314 \times 10^{-2}$$
$$0 = \quad .000 \times 10^{-9}$$

2.3 FLOATING-POINT NUMBERS

The floating-point number system has a number of unfamiliar properties, and the rest of this chapter is devoted to examining some of them.

First, the zero

$$0 = .000 \times 10^{-9}$$

is relatively isolated from the adjacent numbers since the next two positive numbers

$$.100 \times 10^{-9} \quad \text{and} \quad .101 \times 10^{-9}$$

are 10^{-12} apart. No other number has a mantissa beginning with a zero digit.

Second, each decade has exactly the same number of numbers, 900 in all, running

$$\text{from } .100 \times 10^{a} \text{ to } .999 \times 10^{a} \quad a = -9, -8, \ldots, 0, \ldots, 9$$

Within a decade the numbers are equally spaced, but the spacing increases each decade. Thus the spacing is a fixed, arithmetic spacing for 900 numbers, followed by a "geometric spacing jump" and then another block of 900 arithmetically spaced numbers, etc.

The number 0 and the number -0 (namely, $-.000 \times 10^{-9}$) are logically the same, and some machines make them the same, but some do not. Usually there is no infinity corresponding to zero.

In mathematics there is a single, unique zero, while in computing there are two kinds of zeros. The first, as in mathematics, occurs in expressions like

$$a \cdot 0 = 0$$

and behaves like a proper zero. But the zero that occurs in expressions like

$$a - a = 0$$

can come from a form like

$$1 - (1 - \varepsilon)(1 + \varepsilon) = 0$$

whenever ε^2 is less than $\frac{1}{2} 10^{-3}$ (3 being the number of decimal places carried in the mantissa). Clearly this zero differs from the mathematical zero. It is this zero that arises in the subtraction of two apparently equal-sized numbers which causes so much trouble when the finite arithmetic of the machine is equated to the infinite arithmetic of mathematics.

All these remarks appear to be obvious, but their consequences continually affect how we compute and the results we get from the machine.

It is natural in a floating-point number system to measure the error of a number by the size of the difference *relative to* the correct number, and thus

$$\text{Relative error} = \left| \frac{\text{true} - \text{calculated}}{\text{true}} \right|$$

This is distinctly different from the conventional *absolute error* used in much of mathematics, where

$$\text{Absolute error} = |\text{true} - \text{calculated}|$$

The idea of relative error fits most physical situations very well, since it is *scale-free*; that is, a change in the size of the units of measurement does not change the size of the relative error, while it does change the absolute error.

The use of relative error fails, however, when the true value is zero, or close to it. For example, in computing sin π we would have

$$\text{Relative error} = \left| \frac{\sin \pi - \sin(.314 \times 10^1)}{\sin \pi} \right|$$

and since $\sin(.314 \times 10^1)$ is not likely to be exactly zero, the relative error would be infinite. For this reason it is often better in computing the relative error to use

$$\max\{|x|, \ |f(x)|\}$$

as the denominator rather than $f(x)$.

As a result of the discrete spacing of the numbers, there are numbers that cannot come out of some calculations. For example, consider evaluating tan x. The slope of the function is always greater than 1; therefore as we go through adjacent numbers in x, the corresponding values of tan x (see Fig. 2.3.1) must

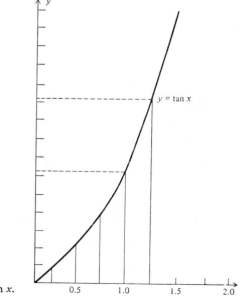

FIGURE 2.3.1
Graph of $y = \tan x$.

occasionally skip over some of the possible numbers (since both x and y start at the origin and y has gotten further in its range than has x in its range).

When searching for gaps in the function values for consecutive values of the argument, a somewhat crude rule to use is that the spacing is (roughly) proportional to the size of the number. This produces the spacing in the y values (roughly) proportional to $x(dy/dx)$, and if the spacing in the number system around y is finer, then there will be gaps, that is, if

$$|y| < \left| x \frac{dy}{dx} \right| \quad \text{or} \quad |y/x| < \left| \frac{dy}{dx} \right|$$

This can be translated into words as follows: when the slope of the line from the origin to the current point on the curve is less than the derivative at that point, there are likely to be gaps between consecutive function values.

It might be thought that if the slope is less than 1, as for the sine function in the first quadrant, then there would be no missed values, but this neglects the fine structure of the number system. Consider the table (in radians)

x	$\sin x$
1.00	.841
1.01	.847
1.02	.852
1.03	.857
1.04	.862

We see that x has shifted to a new decade and hence to a coarser spacing than that of the function values, thus producing the gaps in the sequence of function values.

PROBLEMS

2.3.1 Show that our number system has exactly 34,201 different numbers.

2.3.2 Discuss the gaps in the y values of $y = \sqrt{x}$.

2.3.3 Discuss the gaps in the y values of $y = x^2$.

2.3.4 Discuss the gaps in $y = \cos x$ for the first quadrant.

2.4 HOW NUMBERS COMBINE

Having looked at the individual numbers, let us now look very briefly at how numbers combine in the four arithmetic operations. We assume familiarity with conventional arithmetical practice.

   The product of two three-digit numbers in conventional arithmetic is a five-
or six-digit number, but in our number system there are only three-digit numbers!
Which number shall we take as the product? Common sense suggests taking the
three-digit number closest to the product. The following is a mechanism for
doing this. If the leading digit of the full six-digit product is a zero, then shift the
mantissa one position to the left and at the same time decrease the exponent (which
is the sum of the exponents of the two factors) by 1 to compensate for the shift.
Neglecting the sign of the product, add a 5 in the fourth position. Provided an
overflow on the left does not occur, the first three digits of what remains are taken
as the product. If an overflow does occur, then further shifting and adjusting of
the exponent are required.

$$
\begin{array}{r}
.512 \times 10^1 \\
\times \ .106 \times 10^2 \\
\hline
3072 \\
0000 \\
0512 \\
\hline
\end{array}
$$

$.054272 \times 10^3$
$.542720 \times 10^2$ shift left
$+ \quad 5$ round product
_____
$.543 \quad \times 10^2 =$ product

   This process of rounding produces a very slight *bias* because the ambiguous
case of 500 in the last three places of the mantissa should be rounded up half the
time and rounded down half the time, but the mechanism just described *always*
rounds up. The effect is too slight to worry about in practice; it is chiefly of
theoretical interest.
   Division is somewhat more complicated, but the effect of rounding is much
the same: we select the three-digit number closest to the mathematically correct
quotient, again with a slight bias.
   Addition and subtraction require first comparing the exponents of the two
numbers and, if necessary, shifting one mantissa with respect to the other before
combining. The final shifting for roundoff is much the same as for multiplication.
Subtraction can produce many leading zeros (or even all zeros, in which case the
machine produces $.000 \times 10^{-9}$), and these need to be shifted off before the rounding occurs. (Beware of using $.000 \times 10^0$ as zero.)

Compute $.314 \times 10^1 + .419 \times 10^{-1}$.

$$
\begin{array}{rl}
 & .314 \quad \times 10^1 \\
+ & .00419 \times 10^1 \\
\hline
 & .31819 \times 10^1 \\
+ & \quad 5 \qquad\qquad \text{\scriptsize round} \\
\hline
 & .318 \quad\; \times 10^1 = {\scriptsize \text{sum}}
\end{array}
$$

Compute $.315 \times 10^2 - .314 \times 10^2$.

$$
\begin{array}{rl}
 & .315 \quad \times 10^2 \\
- & .314 \quad \times 10^2 \\
\hline
 & .001 \quad \times 10^2 \\
 & .10000 \times 10^0 \quad \text{\scriptsize shift left} \\
+ & \quad 5 \qquad\qquad \text{\scriptsize round} \\
\hline
 & .100 \quad\; \times 10^0 = {\scriptsize \text{difference}}
\end{array}
$$

Compute $.749 \times 10^2 + .436 \times 10^2$.

$$
\begin{array}{rl}
 & .749 \quad \times 10^2 \\
+ & .436 \quad \times 10^2 \\
\hline
 & 1.185 \quad \times 10^2 \\
 & .1185 \times 10^3 \quad \text{\scriptsize shift right} \\
+ & \quad 5 \qquad\qquad \text{\scriptsize round} \\
\hline
 & .119 \quad\; \times 10^3 = {\scriptsize \text{sum}}
\end{array}
$$

This briefly describes what ideally happens in roundoff. In practice there are many minor variations and some not so minor. For example, since this process of rounding is expensive in both hardware and machine speed, some machines merely *chop off* or drop the last digits once the initial shifting has lined up the leading digits. At first glance chopping may seem to be only moderately more severe than rounding, but unfortunately in some problems chopping can produce an " epidemic " in the sense that in cycle after cycle of computing the effect will be in the same direction and the cumulative effect will grow almost linearly. (See Sec. 23.4.)

PROBLEMS

2.4.1 Show that a shift due to roundoff can occur.

2.4.2 Discuss using $.000 \times 10^a$ as zero.

2.5   THE RELATIONSHIP TO MATHEMATICS AND STATISTICS

By now it should be clear that the mathematician's number system is significantly different from the floating-point number system of the machine, and the latter one is used in most scientific and engineering calculations.   What relationship have they to each other?

Up to now we have adopted the harsh view that the numbers in the machine are the only numbers there are and that the mathematician's numbers are purely artificial.   This is very useful for many purposes, especially when trying to debug a program by accounting for every last digit.   It is also necessary if we are to understand the limitations of what can be done on the machine.

On the other hand, usually the machine is used to simulate the mathematician's system, and the machine's number system is sometimes a poor approximation of what is needed.   The approximations in the initial numbers plus the continual roundoffs in almost every arithmetic operation produce differences that can be serious.   This is especially true when a subtraction produces a large number of leading zeros which are then shifted off (and the exponent is correspondingly adjusted).

To understand roundoff *in the large*, it is customary to regard it as a *random process* in spite of the obvious fact that the same program run repeatedly will produce the same—not random—results (assuming that the machine is not defective). This assumption of random behavior is not essentially different from what is done in many real-world applications of statistics.   In the statistical mechanics of gas molecules we simply do not want to know the detailed behavior of the 6.02×10^{23} molecules in a mole of gas; we merely want the effects of the average behavior. Similarly in practice we do not want to know the exact roundoff at all stages of a computation, rather we want to know their average effect.

Thus statistics continually enters into numerical methods because of the random roundoff effects.   Furthermore, we usually deal only with samples of the functions and not with the functions themselves.   As a result, statistics lies in the background of much of what we do, and occasionally it steps up into the foreground.   It is necessary, therefore, whether we like the topic or not, to give serious attention to the statistical effects in computing.

2.6   THE STATISTICS OF ROUNDOFF

Let us look first at the distribution of the roundoff of a single number.   In the mathematician's number system all numbers in a short interval occur with equal frequency.   All numbers in the interval $x_0 - \frac{1}{2} \le x_0 < x_0 + \frac{1}{2}$ (measured in units of the last digit) go into the number x_0 (where $x_0 > 0$ is a number in the machine). Thus the roundoff has a uniform probability distribution in the last digit (Fig. 2.6.1).

$$p(x) = \begin{cases} 1 & (x_0 - \tfrac{1}{2}, x_0 + \tfrac{1}{2}) \\ 0 & \text{otherwise} \end{cases}$$

$$\int_{-\infty}^{\infty} p(x)\, dx = 1$$

The two most commonly used measures of a distribution are the mean (average) and the variance. The mean is the *first moment* of the distribution $p(x)$,

$$\text{Av}\{p(x)\} = \int_{x_0 - 1/2}^{x_0 + 1/2} x p(x)\, dx = \int_{x_0 - 1/2}^{x_0 + 1/2} x\, dx = \frac{(x_0 - \tfrac{1}{2})^2 - (x_0 - \tfrac{1}{2})^2}{2}$$

$$= x_0$$

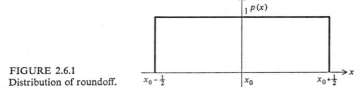

FIGURE 2.6.1
Distribution of roundoff.

while the variance is the *second moment about the mean* (measures the square of the "spread")

$$\text{Var}\{p(x)\} \equiv \sigma^2 = \int_{x_0 - 1/2}^{x_0 + 1/2} (x - x_0)^2 p(x)\, dx = \int_{-1/2}^{1/2} t^2\, dt = \tfrac{1}{12}$$

Experimental tests of the uniform distribution of roundoff seem to show that it is a reasonable model.

How shall we represent this roundoff? It is conventional to let x be the true (mathematician's) number and $x + \varepsilon$ be the computer number, where ε is the additive roundoff. This notation is suitable for the fixed-point number system, but for floating-point numbers it is better to use

$$x(1 + \varepsilon) \qquad |\varepsilon| \leq \tfrac{1}{2} \times 10^{-2} \qquad (2.6.1)$$

We have chosen to let x be the mathematician's number rather than the computer's number because in this book we are concerned with computing more from the user's point of view than from the machine's. The difference in notation between the two approaches is minor in principle, but it has the effect of focusing the attention on one aspect or another in *every* formula in which the roundoff occurs.

Roundoff, once started, propagates through subsequent operations. For example, in the multiplication of two numbers

$$x_1(1 + \varepsilon_1)x_2(1 + \varepsilon_2) = x_1 x_2 (1 + \varepsilon_1 + \varepsilon_2 + \varepsilon_1 \varepsilon_2)$$

Usually $\varepsilon_1 \varepsilon_2$ can be neglected, but we need to add the roundoff ε from the present operation to get the total roundoff

$$x_3(1 + \varepsilon_3) = x_1 x_2 (1 + \varepsilon_1 + \varepsilon_2 + \varepsilon) \qquad |\varepsilon| < \tfrac{1}{2} \times 10^{-2}$$

$$\varepsilon_3 = \varepsilon_1 + \varepsilon_2 + \varepsilon$$

Similarly for the other operations.

PROBLEMS

2.6.1 Calculate the mean and variance for chopping (see Sec. 2.4).

2.6.2 Examine the propagation of roundoff through division.

2.6.3 Show that for roundoff the third moment about the mean is 0 and the fourth is 1/80.

2.7 THE BINARY REPRESENTATION OF NUMBERS

The various features of the floating-point number system have been discussed in terms of the familiar decimal representation. However, most computing is done on machines that use the binary form for representing numbers. The binary representation system is increasingly familiar these days, and so only a few details will be given. As a general rule, if you have trouble with the binary system, then probably it is because you do not really understand the decimal system, and the way out of your trouble is to think about the similar situation in decimals. For example, a decimal number means

1,414.214 =

$$1 \times 10^3 + 4 \times 10^2 + 1 \times 10^1 + 4 \times 10^0 + 2 \times 10^{-1} + 1 \times 10^{-2} + 4 \times 10^{-3}$$

Similarly the binary number

1011.101 =

$$1 \times 2^3 + 0 \times 2^2 + 1 \times 2^1 + 1 \times 2^0 + 1 \times 2^{-1} + 0 \times 2^{-2} + 1 \times 2^{-3}$$

It is important to recognize the difference between the *kind* of number system used (counting, fixed, or floating) and the *form* of the representation of the number

(binary, octal, decimal, hexadecimal, etc.).   Regardless of the form of the representation, the number is the same; thus *3* in decimal and *11* in binary *are the same
number.*   It is conventional to speak of the binary number or the decimal number
to save words, but in careful thinking it is necessary to differentiate between the
number system and the form of the representation.

Perhaps the main stumbling block with the binary system is converting from
decimal to binary and back.   For example, consider writing the number 417 in the
binary representation.   That is, we wish to write 417 as a sum of powers of 2 with
coefficients of either 0 or 1

$$417 = 1 \cdot 2^n + a_{n-1} 2^{n-1} + a_{n-2} 2^{n-2} + \cdots + a_0 2^0$$

If we divide both sides by 2, the remainder is a_0.   If we divide the quotient by 2,
the remainder is a_1, and so on.

$$
\begin{array}{r|l}
2 & 417 \\ \hline
2 & 208 + 1 = a_0 \\ \hline
2 & 104 + 0 = a_1 \\ \hline
2 & 52 + 0 = a_2 \\ \hline
2 & 26 + 0 = a_3 \\ \hline
2 & 13 + 0 = a_4 \\ \hline
2 & 6 + 1 = a_5 \\ \hline
2 & 3 + 0 = a_6 \\ \hline
2 & 1 + 1 = a_7 \\ \hline
2 & 0 + 1 = a_8
\end{array}
$$

$$417 = 110 \quad 100 \quad 001$$

To convert back we reverse the process.   Multiply a_8 by 2 and add a_7;
multiply the sum by 2 and add a_6; etc.

$$
\begin{array}{l}
1 \\
\underline{2} \\
2 + 1 = 3 \\
\quad\quad \underline{\times\, 2} \\
\quad\quad 6 + 0 = 6 \\
\quad\quad\quad\quad \underline{\times\, 2} \\
\quad\quad\quad\quad 12 + 1 = 13 \\
\quad\quad\quad\quad\quad\quad \underline{\times\, 2} \\
\quad\quad\quad\quad\quad\quad 26 + 0 = 26
\end{array}
$$

etc.

The above process works for integers. For the fractional part we use a similar trick of doubling and using the overflow on the left. For example,

$$.762 = a_{-1}2^{-1} + a_{-2}2^{-2} + a_{-3}2^{-3} + \ldots$$

double

$$1.524 = a_{-1} + a_{-2}2^{-1} + a_{-3}2^{-2} + \ldots$$

and so $a_{-1} = 1$.

In full

$$
\begin{array}{r}
.762 \\
2 \\
\hline
1 \mid 524 \\
2 \\
\hline
1 \mid 048 \\
2 \\
\hline
0 \mid 096 \\
2 \\
\hline
0 \mid 192 \\
2 \\
\hline
0 \mid 384 \\
2 \\
\hline
0 \mid 768 \\
2 \\
\hline
1 \mid 536
\end{array}
$$

Hence

$$.762 = .110\ 000\ 1 \ldots$$

Reversing the process gets from the binary representation to the decimal.

Previously prepared tables for conversion purposes are another way of converting from one form of representing a given number to another form of representing the same number.

The conversions are customarily done by the computing machine, and since the machine works in binary arithmetic and since the above discussion has used decimal arithmetic, there are differences in exactly what happens in the machine. It is not hard to take the machine's point of view and to deduce how to convert its way.

Some words of caution are needed however. The terminating decimal

$$.1 = \tfrac{1}{10} = .0001\ 1001\ 1001\ 1 \dots$$

does not terminate in binary, and as a result adding 1/10 to itself 10 times will not produce exactly 1 (just as in conventional decimals $1/3 + 1/3 + 1/3 = .333 + .333 + .333 = .999 \neq 1$). Thus using the floating-point representation of numbers for counting or for logical control is inviting trouble.

Another point of warning is needed. The conversion routines cannot always read in a decimal number, convert it to binary and back, and give the result that is the same as the original number—at least not when the number of digits in the two forms is reasonably balanced. Take, for example, a three decimal to ten binary digit (binary digit is usually abbreviated *bit*) conversion. The 100 decimal numbers

$$100 \text{ numbers}\begin{cases}9.00 \to 1001.000000 \\ 9.01 \to 1001.xxxxxx \\ 9.02 \to 1001.xxxxxx \\ \dots\dots\dots\dots\dots \\ 9.99 \to \underbrace{1001.xxxxxx}\end{cases}$$

64 distinct numbers

must go into 64 binary representations, so that some distinct decimals *must go into the same* binary representation and cannot be distinguished in the reconversion process. Thus in spite of the fact that $10^3 < 2^{10} = 1{,}024$, there will be times when a decimal number is read in and comes back slightly, and annoyingly, different.

PROBLEMS

Convert to binary:
2.7.1   1,728
2.7.2   1,972
2.7.3   1,066
2.7.4   0,345
2.7.5   0.592
2.7.6   1/12
2.7.7   1/16

Convert to decimal:
2.7.8   101 001
2.7.9   111 111

2.7.10 100 001
2.7.11 .111 111
2.7.12 .100 001
2.7.13 Show that the above argument for the nonunique conversion also applies to the conversion from 8 decimals to 27 binary digits.
2.7.14 Describe the terminating decimals that also terminate in binary.

2.8 THE FREQUENCY DISTRIBUTION OF MANTISSAS

Although the mantissas of floating-point numbers are equally spaced, and hence occur equally frequently in the form of representation, *they are not equally frequent in practice.* Instead, the probability of getting the leading digit 1, 2, or 3 in a decimal number is about 60 percent. For example, consider the 50 physical constants whose leading digits are given in Table 2.8.1. In Sec. 2.9 we shall show why we care about this phenomenon beyond mere curiosity.

This effect can be explained in terms of the mathematician's smooth number system since it is a characteristic of the numbers themselves and not peculiar to the finite representation that the machine necessarily uses. We shall show that it is

Table 2.8.1 THE DISTRIBUTION OF THE LEADING
DIGITS OF 50 PHYSICAL CONSTANTS*

Leading digit N	Number of cases observed	Theoretical number Eq. (2.8.3)	Difference
1	16	15	1
2	11	9	2
3	2	6	−4
4	5	5	0
5	6	4	2
6	4	3	1
7	2	3	−1
8	1	3	−2
9	3	2	1
	50	50	

*From Handbook of Mathematical Functions, *AMS 55*, National Bureau of Standards, 1964 and Dover Publications, Inc.

reasonable to adopt the model for the distribution that the *probability density* for observing the number x in the base b is

$$r(x) = \frac{1}{x \ln b} \qquad \frac{1}{b} \le x < 1 \qquad (2.8.1)$$

This is called the *reciprocal distribution* (Fig. 2.8.1) for obvious reasons.

The cumulative probability distribution is defined as

$$R(x) = \int_{1/b}^{x} r(t)\, dt = \frac{\ln x + \ln b}{\ln b}$$

$$R\!\left(\frac{1}{b}\right) = 0 \qquad \text{and} \qquad R(1) = 1 \qquad (2.8.2)$$

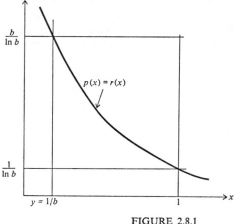

FIGURE 2.8.1
Reciprocal distribution.

Thus the probability of observing the leading digit N is

$$R\!\left(\frac{N+1}{b}\right) - R\!\left(\frac{N}{b}\right) = \frac{\ln (N+1) - \ln N}{\ln b} \qquad (2.8.3)$$

which is what the table confirms.

We examine first the way multiplication transforms various distributions. Let x come from the probability (density) distribution $f(x)$, let y come from $g(y)$, and let the product z have the distribution $h(z)$.   Further, let their cumulative distributions be, respectively, $F(x)$, $G(y)$, and $H(z)$, the measure of the set of points for which $xy \le z$.

A study of Fig. 2.8.2 shows that

$$H(z) = \int_{1/b}^{z} \int_{1/b}^{z/(bx)} f(x)g(y)dy\,dx + \int_{1/b}^{z} \int_{1/(bx)}^{1} f(x)g(y)\,dy\,dx$$
$$+ \int_{z}^{1} \int_{1/(bx)}^{z/x} f(x)g(y)\,dy\,dx$$
$$= \int_{1/b}^{z} f(x)\left[G\left(\frac{z}{bx}\right) - G\left(\frac{1}{b}\right) + G(1) - G\left(\frac{1}{bx}\right) \right] dx$$
$$+ \int_{z}^{1} f(x)\left[G\left(\frac{z}{x}\right) - G\left(\frac{1}{bx}\right) \right] dx$$

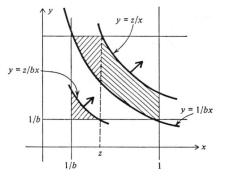

FIGURE 2.8.2
The cumulative probability distribution
for the product $z = xy$.

Differentiating $H(z)$ with respect to z gives the density distribution

$$h(z) = f(z)\left[G\left(\frac{1}{b}\right) - G\left(\frac{1}{b}\right) + G(1) - G\left(\frac{1}{bz}\right) - G(1) + G\left(\frac{1}{bz}\right) \right]$$
$$+ \int_{1/b}^{z} f(x)g\left(\frac{z}{bx}\right) \frac{1}{bx}\,dx + \int_{z}^{1} f(x)g\left(\frac{z}{x}\right) \frac{1}{x}\,dx$$
$$= \frac{1}{b} \int_{1/b}^{z} \frac{f(x)}{x} g\left(\frac{z}{bx}\right) dx + \int_{z}^{1} \frac{f(x)}{x} g\left(\frac{z}{x}\right) dx \qquad (2.8.4)$$

This is the basic formula which describes how the process of multiplication combines the distributions of mantissas of two numbers x and y to give the distribution $h(z)$ of the mantissa of the product.

Suppose, now, that one of the two factors, say y, has the reciprocal distribution; that is,

$$g(y) = \frac{1}{y \ln b}$$

Putting this in Eq. (2.8.4), we get

$$h(z) = \frac{1}{b} \int_{1/b}^{z} \frac{f(x)}{x} \frac{bx}{z \ln b} \, dx + \int_{z}^{1} \frac{f(x)}{x} \frac{x}{z \ln b} \, dx$$

$$= \frac{1}{z \ln b} \left(\int_{1/b}^{z} f(x) \, dx + \int_{z}^{1} f(x) \, dx \right) = \frac{1}{z \ln b} \qquad (2.8.5)$$

Obviously the same applies if we assume $f(x)$ is the reciprocal distribution. Thus we have shown that if one of the factors of a product comes from the reciprocal distribution, then regardless of the distribution of the other factor, the product has the reciprocal distribution. This may be called the *persistence of the reciprocal distribution* once it is established. In a long sequence of multiplications if at least *one* factor has the reciprocal distribution, then the result has the reciprocal distribution.

Next we show how the reciprocal distribution can arise. We need a measure of how close a distribution is to the reciprocal distribution. After some tries we measure the distance of $h(z)$ from the reciprocal distribution $r(z)$ by

$$\max_{1/b \leq z \leq 1} \left\{ \left| \frac{h(z) - r(z)}{r(z)} \right| \right\} \equiv D\{h(z)\} = D\{h\} \qquad (2.8.6)$$

This measures the maximum relative error (which is natural for floating-point numbers).

We just showed that

$$r(z) = \frac{1}{b} \int_{1/b}^{z} \frac{f(x)}{x} r\left(\frac{z}{bx}\right) \, dx + \int_{z}^{1} \frac{f(x)}{x} r\left(\frac{z}{x}\right) \, dx$$

Subtract this from Eq. (2.8.4) and divide by $r(z)$ to get the form for the distance function

$$\frac{h(z) - r(z)}{r(z)} = \frac{1}{b} \int_{1/b}^{z} \frac{f(x)}{x} \left\{ \frac{g[z/(bx)] - r[z/(bx)]}{r(z)} \right\} \, dx$$

$$+ \int_{z}^{1} \frac{f(x)}{x} \left[\frac{g(z/x) - r(z/x)}{r(z)} \right] \, dx$$

But

$$bxr(z) = \frac{bx}{z \ln b} = r\left(\frac{z}{bx}\right)$$

$$xr(z) = \frac{x}{z \ln b} = r\left(\frac{z}{x}\right)$$

and we have

$$\frac{h(z) - r(z)}{r(z)} = \int_{1/b}^{z} f(x) \left\{ \frac{g[z/(bx)] - r[z/(bx)]}{r[z/(bx)]} \right\} dx$$

$$+ \int_{z}^{1} f(x) \left[\frac{g(z/x) - r(z/x)}{r(z/x)} \right] dx$$

Since $f(x) \geq 0$ in the two intervals

$$\left| \frac{h(z) - r(z)}{r(z)} \right| \leq \int_{1/b}^{z} f(x)D\{g\}\, dx + \int_{z}^{1} f(x)D\{g\}\, dx$$

$$\leq D\{g\}$$

for all z, it follows that

$$D\{h\} \leq D\{g\} \qquad (2.8.7)$$

How fast does the distribution approach the limiting distribution? Remembering that for any distribution $g(x)$

$$\int_{1/b}^{1} [g(x) - r(x)]\, dx = 0$$

we form a guess at the increase that must occur in replacing the square brackets by their maximum $D\{g\}$. We can also calculate how the distance decreases for a continued product of a sequence of numbers taken from a flat (uniform) distribution. See Table 2.8.2

Similar results can be obtained for division. For example, Eq. (2.8.4) becomes

$$h(z) = \frac{1}{z^2} \int_{1/b}^{z} xf(x)g\left(\frac{x}{z}\right) dx + \frac{1}{bz^2} \int_{z}^{1} xf(x)g\left(\frac{x}{bz}\right) dx$$

and the rest follows to produce the corresponding results for division.[1]

Table 2.8.2 DISTANCE FROM THE FLAT DISTRIBUTION TO THE LIMITING DISTRIBUTION

Number of factors	Distance	Percentage of original distance
1	1.558	100.0
2	0.3454	22.2
3	0.0980	6.29
4	0.0289	1.85

[1] For further results see R. W. Hamming, On the Distribution of Numbers, *Bell Systems Technical Journal*, vol. 49, no. 8, pp. 1609–1625, October, 1970.

PROBLEMS

2.8.1 Derive the corresponding results for division.

2.8.2 If $f(x) = f(y) = b/(b-1)$, show that for multiplication

$$h(z) = \frac{b}{(b-1)^2} [\ln b - (b-1)\ln z]$$

2.8.3 If $f(x) = g(y) = b/(b-1)$, show that for division

$$h(z) = \frac{1}{2(b-1)} \left(b + \frac{1}{z^2}\right)$$

2.9 THE IMPORTANCE OF THE RECIPROCAL DISTRIBUTION

The reciprocal distribution has many applications. For example, consider the placing of the decimal or binary point before rather than after the first nonzero digit. Scientific convention places it after, while computing convention places it before, the digit. The difference first arose in the earliest electronic computers where the choice in fixed-point notation meant that the product would not produce an overflow on the left. What justification can we *now* give? If it is placed before, we run the risk in floating point of having a leading zero and of requiring time to shift this off and adjust the exponent accordingly. On the other hand if it is after the digit, we run the risk of getting two digits before the point and of requiring an exponent adjustment to get to standard form. What are the probabilities of these two (complementary) events? If $xy \le 1/b$, then the probability of a shift when both factors are from the reciprocal distribution is (Fig. 2.9.1)

$$p = \int_{1/b}^1 \int_{1/b}^{1/(bx)} \frac{1}{x \ln b} \frac{1}{y \ln b} \, dy \, dx$$

$$= \int_{1/b}^1 \frac{1}{\ln^2 b} \left[\frac{\ln 1/(bx) - \ln 1/b}{x}\right] dx = \int_{1/b}^1 \frac{1}{\ln^2 b} \left(-\frac{\ln x}{x}\right) dx$$

$$= \frac{1}{\ln^2 b} \left(-\frac{\ln^2 x}{2}\right)\Big]_{1/b}^1 = \frac{1}{2}$$

And so it is a matter of indifference where the point is placed in this model. For numbers from the flat uniform distribution the probability depends on the base b and for $b = 2, p \approx 0.38$.

In the design of optimal library routines it is necessary to know the distribution of the input data. In the cases of square root and exponential routines it is the distribution of the mantissa that is most important, but for other routines such as the sine it is necessary to know something about the distribution of the ex-

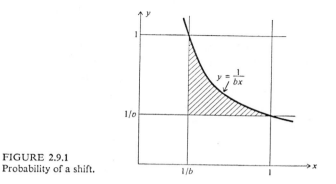

FIGURE 2.9.1
Probability of a shift.

ponents. Unfortunately almost nothing at this time is actually known, nor are there any theories available to suggest what might be found under some suitable conditions.

We will later give some other applications of the reciprocal distribution.

PROBLEMS

2.9.1 In a base 16 system written as groups of four binary digits, show that the numbers of the form

$$.1xxx \ldots$$
$$.01xx \ldots$$
$$.001x \ldots$$
$$.0001 \ldots$$

are equilikely for the reciprocal distribution.

2.9.2 In information theory the information measure of a distribution is

$$I = -\int_{-\infty}^{\infty} p(x)\ln[p(x)]\,dx$$

Apply this to the reciprocal distribution and compute the information loss from the uniform distribution for base b.

2.10 HAND CALCULATION

Hand calculations are used to see how a computation might go *before* programming it for a machine and to check the results after a machine run. We no longer need to do extensive hand calculations, and as a result it is a dying art—one that is best left to die quietly.   A whole book could be devoted to the topic, and the knowledge of its contents would be of some use sometimes, but it is simply not worth the effort to learn in detail when there are so many other things more worth knowing.   Besides, the topic is dull and disorganized.

The main thing to note is the types of error that are most common in hand calculation.   The two most prominent errors are both in copying numbers.   The first is to reverse a pair of digits—87 becomes 78—and the second is to double the wrong number in a triple of numbers—776 becomes 766.   They are the same errors that are common in dialing phone numbers.

FUNCTION EVALUATION

3.1 INTRODUCTION

Once we understand the number system, we can examine how numbers combine when we evaluate functions. Mathematically equivalent formulas can have very different roundoff characteristics. If the way we evaluate a function produces a great deal of roundoff, then we cannot use it for practical computing. We need, therefore, to see how to *avoid* roundoff when evaluating functions; we need to learn how to evaluate functions for machine computation.

3.2 THE EXAMPLE OF THE QUADRATIC EQUATION

The formula for finding the zeros of the quadratic equation

$$ax^2 + bx + c = 0$$

is given in textbooks as

$$x = \frac{-b \pm \sqrt{b^2 - 4ac}}{2a}$$

An examination of the formula suggests that when $4ac$ is small with respect to b^2, that is,

$$|4ac| \ll b^2$$

then

$$-b + \sqrt{b^2 - 4ac} \qquad \text{for } b > 0$$

$$-b - \sqrt{b^2 - 4ac} \qquad \text{for } b < 0$$

will result in severe cancellation for one root.   Consider, for example, the particular case:

$$x^2 - 80x + 1 = 0 \qquad \begin{cases} a = & 0.100 \times 10^1 \\ b = & -0.800 \times 10^2 \\ c = & 0.100 \times 10^1 \end{cases}$$

$$x = \frac{0.800 \times 10^2 \pm \sqrt{0.640 \times 10^4 - 0.400 \times 10^1}}{0.200 \times 10^1}$$

$$= \frac{0.800 \times 10^2 \pm 0.800 \times 10^2}{0.200 \times 10'} \qquad \qquad \text{Approximate true values}$$

$$= \begin{cases} 0.800 \times 10^2 = 80 & \qquad 80 - \frac{1}{80} = 80\left(1 - \frac{1}{80^2}\right) \\ \\ 0.000 \times 10^{-9} = 0 & \qquad \frac{1}{80} + \frac{1}{80^3} = \frac{1}{80}\left(1 + \frac{1}{80^2}\right) \end{cases}$$

How can we avoid this cancellation?
Since

$$a(x - x_1)(x - x_2) = ax^2 - a(x_1 + x_2)x + ax_1 x_2$$

it follows that the product of the two zeros is c/a.   If we use the case without the cancellation to find x_1 and then find x_2 from

$$x_2 = \frac{c}{ax_1}$$

we will get

$$x_1 = 0.800 \times 10^2$$

$$x_2 = \frac{1}{0.800 \times 10^2} = 0.125 \times 10^{-1}$$

We need, therefore, to rearrange the formula for the zeros to pick out the one without the cancellation. One way is

$$x_1 = \frac{-b \pm \sqrt{b^2 - 4ac}}{2a} \qquad \text{Use opposite sign of } b$$

$$x_2 = \frac{c}{ax_1}$$

How good are these answers? One way of answering is to try to reconstruct the polynomial from the zeros.

$$(x - 0.800 \times 10^2)(x - 0.125 \times 10^{-1}) = x^2 - x(0.800 \times 10^2) + 0.100 \times 10^1$$

The reconstruction is exact; therefore, in some sense, the answers must be exact since we appear to have lost no information around the whole loop.

This is not the complete answer on how to evaluate the formula; we still need to worry about (1) underflow, (2) overflow, and (3) $b^2 - 4ac < 0$, but these are not relevant here.

PROBLEM

3.2.1 Find the formula for the roots of $ax^2 + 2bx + c = 0$ and note the savings in arithmetic.

3.3 REARRANGEMENT OF FORMULAS

It would appear that there are an unlimited number of tricks for rearranging formulas to avoid severe cancellation (in some region of the argument x), but they are often the *same* tricks that were used in the calculus course to rearrange the expressions that arose in the delta process of formally taking the derivative of a function.

EXAMPLE 3.1   Evaluate $\sqrt{x + 1} - \sqrt{x}$ for large x. As in the calculus we rationalize the numerator by

$$(\sqrt{x + 1} - \sqrt{x})\left(\frac{\sqrt{x + 1} + \sqrt{x}}{\sqrt{x + 1} + \sqrt{x}}\right) = \frac{1}{\sqrt{x + 1} + \sqrt{x}}$$

which has no cancellation.   ////

EXAMPLE 3.2 Evaluate $\dfrac{\sin x}{x}$ for small x.

In floating point there is no trouble except for $x = 0$. As a special case, if $x = 10^{-2}$, then

$$\sin x = 10^{-2}$$

and

$$\frac{\sin x}{x} = 1$$

as it should. In general since for small x

$$\sin x = x - \frac{x^3}{6} + \cdots$$

then

$$\frac{\sin x}{x} = 1 - \frac{x^2}{6} + \frac{x^4}{120} - \cdots$$

is accurately evaluated by the $\sin x$ routine followed by division by x. The trouble occurs in the fixed-point number system (which we do not use). ////

EXAMPLE 3.3 Evaluate $\sin(x + \varepsilon) - \sin x$ for small ε.
Using the trigonometric identity

$$\sin a - \sin b = 2 \cos \frac{a + b}{2} \sin \frac{a - b}{2}$$

we have

$$= 2 \cos \left(x + \frac{\varepsilon}{2}\right) \sin \frac{\varepsilon}{2}$$

as a suitable form. ////

EXAMPLE 3.4 Evaluate $\dfrac{1 - \cos x}{\sin x}$ for small x.
We have the identities

$$\frac{1 - \cos x}{\sin x} = \frac{\sin x}{1 + \cos x} = \tan \frac{x}{2} \qquad ////$$

EXAMPLE 3.5 Evaluate

$$\int_N^{N+1} \frac{dx}{1 + x^2} = \arctan(N + 1) - \arctan N$$

for large N.

In Fig. 3.3.1 for large N the area is the shaded region and is expressed as the difference between two large areas.   We use the identity

$$\arctan a - \arctan b = \arctan \frac{a - b}{1 + ab}$$

to get

$$\arctan \frac{1}{1 + N(N + 1)}$$

as a suitable form.   Note we also have avoided one arctan evaluation which is the time-consuming part of the computation.   ////

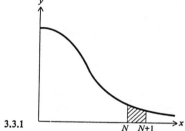

FIGURE 3.3.1

EXAMPLE 3.6   For large x evaluate

$$\frac{1}{\sqrt{x}} - \frac{1}{\sqrt{x + 1}}$$

We write it as

$$\frac{\sqrt{x + 1} - \sqrt{x}}{(\sqrt{x + 1})(\sqrt{x})} = \frac{1}{\sqrt{x + 1}\sqrt{x}(\sqrt{x + 1} + \sqrt{x})} = \frac{1}{(x + 1)\sqrt{x} + x\sqrt{x + 1}} \qquad ////$$

PROBLEMS

For large x, or ε small with respect to x, rearrange for evaluation:

3.3.1 $\dfrac{1}{x + 1} - \dfrac{1}{x}$

3.3.2 $\tan(x + \varepsilon) - \tan x$ *Ans.* $\dfrac{\sin \varepsilon}{\cos x \cos(x + \varepsilon)}$

3.3.3 $\dfrac{1}{x+1} - \dfrac{2}{x} + \dfrac{1}{x-1}$ *Ans.* $\dfrac{2}{x(x^2-1)}$

3.3.4 $\sqrt[3]{x+1} - \sqrt[3]{x}$

3.3.5 $\cos(x + \varepsilon) - \cos x$

3.3.6 $\displaystyle\int_N^{N+1} \dfrac{dx}{x} = \ln(N+1) - \ln N$ N large

3.3.7 $e^{\varepsilon} - 2 + e^{-\varepsilon}$ *Ans.* $4 \sinh^2(\varepsilon/2)$

3.4 SERIES EXPANSIONS

Sometimes rearrangements cannot be found to remove the cancellation, and some other device must be used.   One of the more effective tricks from the calculus course is the expansion of the functions into a Taylor series about some suitable point.

EXAMPLE 3.7   Evaluate $e^x - 1$ for small x.
Expanding e^x about $x = 0$, we get

$$e^x - 1 = 1 + x + \frac{x^2}{2} + \frac{x^3}{6} + \cdots - 1$$

$$= x\left(1 + \frac{x}{2} + \frac{x^2}{6} + \cdots\right) \qquad ////$$

EXAMPLE 3.8   Evaluate for large N

$$\int_N^{N+1} \ln x \, dx = (x \ln x - x) \Big|_N^{N+1}$$

$$= (N+1)\ln(N+1) - N \ln N - 1$$

which has severe roundoff trouble in this form.   See Fig. 3.4.1.
There are many ways of going about this problem.   One way is to write it as

$$N[\ln(N+1) - \ln N] + \ln(N+1) - 1 = N \ln\left(1 + \frac{1}{N}\right) + \ln(N+1) - 1$$

and use

$$\ln(1+x) = x - \frac{x^2}{2} + \frac{x^3}{3} - \frac{x^4}{4} + \cdots \qquad |x| \text{ small}$$

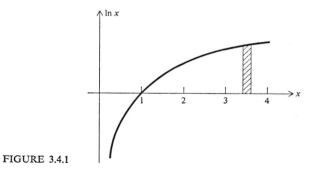

FIGURE 3.4.1

to get

$$= N\left(\frac{1}{N} - \frac{1}{2N^2} + \frac{1}{3N^3} - \frac{1}{4N^4} + \cdots\right) + \ln(N+1) - 1$$

$$= \ln(N+1) - \frac{1}{2N} + \frac{1}{3N^2} - \frac{1}{4N^3} + \cdots \qquad ////$$

EXAMPLE 3.9   Evaluate $\dfrac{1}{x} - \operatorname{ctn} x$ for small x.

We have

$$\frac{1}{x} - \operatorname{ctn} x = \frac{1}{x} - \frac{\cos x}{\sin x}$$

$$= \frac{\sin x - x \cos x}{x \sin x}$$

$$= \frac{\left(x - \dfrac{x^3}{3!} + \dfrac{x^5}{5!} - \cdots\right) - x\left(1 - \dfrac{x^2}{2!} + \dfrac{x^4}{4!} - \cdots\right)}{x\left(x - \dfrac{x^3}{3!} + \dfrac{x^5}{5!} - \cdots\right)}$$

$$= \frac{\dfrac{x^3}{3}\left(1 - \dfrac{x^2}{10} + \cdots\right)}{x^2\left(1 - \dfrac{x^2}{6} + \cdots\right)} = \frac{x}{3}\left(1 + \frac{x^2}{15} + \cdots\right) \qquad ////$$

EXAMPLE 3.10   For small x evaluate

$$\frac{\ln(1-x)}{\ln(1+x)} = \frac{-\left(x + \frac{x^2}{2} + \frac{x^3}{3} + \frac{x^4}{4} + \cdots\right)}{x - \frac{x^2}{2} + \frac{x^3}{3} - \frac{x^4}{4} + \cdots}$$

Divide out the two series formally to get

$$\frac{\ln(1-x)}{\ln(1+x)} = -\left(1 + x + \frac{x^2}{2} + \frac{5x^3}{12} + \cdots\right)$$ ////

If we are given an ε that is small, then when we evaluate

$$\ln(1 + \varepsilon)$$

by computing first $1 + \varepsilon$, we lose most of the accuracy in this first addition.   For this reason many math package libraries include the function:

given ε, compute $\ln(1 + \varepsilon)$ directly

PROBLEMS

For large x, or ε small with respect to x, compute:

3.4.1 $\dfrac{1 - \cos \varepsilon}{\varepsilon^2}$

3.4.2 $\dfrac{\varepsilon(e^{(a+b)\varepsilon} - 1)}{(e^{a\varepsilon} - 1)(e^{b\varepsilon} - 1)}$ *One ans.* $\dfrac{a+b}{ab}$

3.4.3 $\ln \dfrac{1 - \varepsilon}{1 + \varepsilon}$ *Ans.* $-2\left(\varepsilon + \dfrac{\varepsilon^3}{3} + \dfrac{\varepsilon^5}{5} + \cdots\right)$

3.4.4 $\sqrt{\dfrac{e^{2\varepsilon} - 1}{e^\varepsilon - 1}}$

3.4.5 $\dfrac{\varepsilon - \sin \varepsilon}{\varepsilon - \tan \varepsilon}$

3.4.6 $(x + 1)^{1/n} - x^{1/n}$

3.5   USE OF MACHINE TO DECIDE

The beginner is inclined to believe that he must personally analyze each problem and program the machine accordingly.   Only after a while does it occur to him that the machine can make the choices, provided the program is written properly.

For example, consider evaluating

$$e^{ax} - 1$$

for a range of x and a set of parameter values a. When $|ax|$ is small, we need to use the series approximation; otherwise we can use the direct evaluation and subtract. If we do not mind the loss of one bit in the direct evaluation, then we need only use the series for $|ax| < \frac{1}{10}$. (*Note:* $e^{0.1} = 1.105....$)

To evaluate $e^{ax} - 1$, see Fig. 3.5.1.

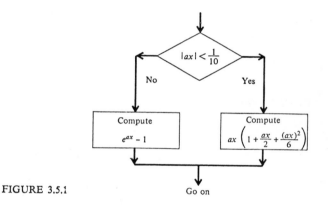

FIGURE 3.5.1

The term neglected is of relative size

$$\left| \frac{(ax)^3}{24} \right| < 10^{-3}$$

so that even if $ax < 0$, the relative error is less than $\frac{1}{2} \times 10^{-3}$, and there is no loss in accuracy in the power series approximation.

3.6 THE MEAN VALUE THEOREM

The mean value theorem is the third item from the calculus course that is widely used in computing, especially in the more theoretical parts of deriving errors of formulas.

The mean value theorem   Let $y = f(x)$ be continuous for $a \leq x \leq b$ and possess a derivative at each x for $a < x < b$. Then there is at least one number θ between a and b (see Fig. 3.6.1) such that

$$f(b) - f(a) = (b - a)f'(\theta) \qquad a < \theta < b \qquad (3.6.1)$$

This is equivalent to stating that at the position θ the slope of the derivative is parallel to the chord joining the two endpoints.

Equation (3.6.1) can sometimes be used to avoid roundoff (and sometimes a lot of machine time). For example,

$$\ln(N + a) - \ln N = a\frac{1}{\theta}$$

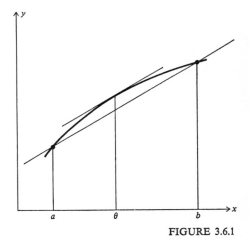

FIGURE 3.6.1

and it is reasonable to pick θ as the midvalue

$$\ln(N + a) - \ln N \approx \frac{2a}{2N + a}$$

Let $N = 100$ and $a = 1$; then

$$\ln 101 - \ln 100 = 4.61512 - 4.60517 = 0.00995$$

Again,

$$\frac{2}{200 + 1} = \frac{1}{100.5} = 0.00995$$

and no logs were evaluated in the computation.

PROBLEMS

Apply the mean value theorem and check the result if numbers are given.

3.6.1 $\sqrt{82} - \sqrt{81}$

3.6.2 $\sin(x + \varepsilon) - \sin x$

3.6.3 $\arctan(N + 1) - \arctan N \qquad N = 10$

3.6.4 $\displaystyle\int_{N}^{N+1} \ln x \, dx \qquad N = 10$

3.7 SYNTHETIC DIVISION

Polynomials play a central role in computing. Not only do they occur naturally, but everything that the four arithmetic operations (addition, subtraction, multiplication, and division) can produce is the equivalent of one polynomial divided by another. It is important, therefore, to examine with some care the problem of evaluating a polynomial for a given value of the argument x.

The polynomial

$$P(x) = a_N x^N + a_{N-1} x^{N-1} + a_{N-2} x^{N-2} + \cdots + a_1 x + a_0$$

can be written in the "chain" (nested) form

$$\{\cdots [(a_N x + a_{N-1})x + a_{N-2}]x + \cdots a_1\}x + a_0$$

which involves N additions and N multiplications.

The chain method is the same that occurs in the synthetic division process of dividing a polynomial by a linear factor $x - a$. For example, the particular polynomial

$$P(x) = x^4 + 6x^2 - 7x + 8$$

divided by $x - 2$ leads to

$$
\begin{array}{r}
x^3 + \ 2x^2 + 10x + 13 = \text{quotient} \\
\text{Divisor} = x - 2\ \overline{\smash{\big)}\ x^4 + 0x^3 + \ 6x^2 - \ 7x + \ 8} \\
\underline{x^4 - 2x^3} \\
2x^3 + \ 6x^2 \\
\underline{2x^3 - \ 4x^2} \\
10x^2 - \ 7x \\
\underline{10x^2 - 20x} \\
13x + \ 8 \\
\underline{13x - 26} \\
+ \ 34 = \text{remainder}
\end{array}
$$

which can be written as

$$\frac{P(x)}{x - 2} = \frac{x^4 + 6x^2 - 7x + 8}{x - 2} = x^3 + 2x^2 + 10x + 13 + \frac{34}{x - 2}$$

or as

$$P(x) = \underbrace{x^4 + 6x^2 - 7x + 8}_{\text{polynomial}} = \overset{\uparrow}{(x - 2)}\underbrace{(x^3 + 2x^2 + 10x + 13)}_{\text{quotient}} + \overset{\uparrow}{34}$$

When $x = 2$, we get

$$P(2) = 34 = \text{remainder}$$

This is a special case of the remainder theorem.

The remainder theorem   If a polynomial $P(x)$ is divided by $x - a$, we get

$$P(x) = (x - a)Q(x) + R$$

where $Q(x)$ is the quotient and R is the remainder, and we have

$$P(a) = R$$

If $R = 0$, then a is a zero of $P(x)$.

The division process can be simplified by noting that the powers of x are place holders and need not be written *provided* zero coefficients are supplied for the missing terms.   Furthermore, the quotient need not be written on the top line since it is given by the numbers at the bottom of each column.   Finally, we need write not $x - a$, but a, and then add, instead of subtract, in the body of the form. As a result of all these changes we get

In this form it is clearly the chain method of evaluating a polynomial $P(a)$.

If the quotient is again divided by the same linear factor, we will get another quotient and another remainder r_1.   Dividing this new quotient again by the same factor, etc., we find that we are performing exactly the same process that was used to represent a number as a sum of powers of a number base (Sec. 2.7 where we used the base 2), and correspondingly we are representing a polynomial as a sum of powers of the linear factor we were dividing by.   Thus we will get

$$P(x) = r_0 + r_1(x - a) + r_2(x - a)^2 + \cdots + r_N(x - a)^N$$

Using the Taylor-series expansion of a function

$$f(x) = f(a) + (x - a)f'(a) + (x - a)^2 \frac{f''(a)}{2!} + \cdots$$

we see that

$$r_0 = P(a)$$
$$r_1 = P'(a)$$
$$r_2 = \frac{P''(a)}{2!}$$
$$r_3 = \frac{P'''(a)}{3!}$$

$$\cdots\cdots\cdots\cdots$$

Using our earlier example,

$$
\begin{array}{r|rrrr}
2 & 1 & 0 & 6 & -7 & 8 \\
  &   & 2 & 4 & 20 & 26 \\
\hline
2 & 1 & 2 & 10 & 13 & \boxed{34} = P(2) \\
  &   & 2 & 8 & 36 \\
\hline
2 & 1 & 4 & 18 & \boxed{49} = P'(2) \\
  &   & 2 & 12 \\
\hline
2 & 1 & 6 & \boxed{30} = \dfrac{P''(2)}{2} \\
  &   & 2 \\
\hline
  & 1 & \boxed{8} = \dfrac{P'''(2)}{6}
\end{array}
$$

and we have

$$P(x) = 34 + 49(x - 2) + 30(x - 2)^2 + 8(x - 2)^3 + (x - 2)^4$$

3.8 ROUNDOFF EFFECTS

The preceding is a purely mathematical result. How does it work out in the machine's floating-point number system with its ever-present roundoff? In particular, how large a remainder R can be considered as being zero?
We start by assuming that the coefficients a_i of the polynomial have roundoff errors in the floating-point form

$$a_i(1 + \varepsilon_i)$$

and we shall assume that the number a at which we want to evaluate the polynomial is exactly known.   Since double-precision arithmetic is very widely available and usually costs little more to use than single precision,[1] it is widely used in polynomial evaluation.   In this case the roundoff of the evaluation process contributes almost nothing to the error in the result—almost all the error arises from the initial coefficients.   Since the coefficients occur linearly in the division process, the result is the sum of two divisions, one on the true coefficients a_i and the other on the errors $a_i \varepsilon_i$.   If we assume that the ε_i are bounded,

$$|\varepsilon_i| \le \varepsilon$$

then carrying out the following division process

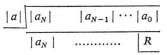

gives $R\varepsilon$ as the error bound on the remainder.   This is an error bound due to the error in the initial coefficients.

If single-precision arithmetic is used in the evaluation process, then we must also consider the effects of the rounding of the individual products and sums. Each arithmetic operation will include a factor

$$1 + \varepsilon^*$$

If we examine the process of synthetic division, we observe that the first coefficient a_N goes through $2N$ arithmetic operations and so picks up $2N$ factors of this form. The next coefficient a_{N-1} will have $2N - 1$ operations done on it, the next a_{N-2} will have $2N - 3$ operations, etc., down to an a_0 which will have 1 operation.

We are going to use the method of "backward analysis," meaning that the roundoff errors will be regarded as equivalent to small changes in the initial coefficients—we will answer the question "what problem has exactly the calculated answer?"   The multiplication produces a factor $1 + \varepsilon$ on each of the terms that enter into the multiplicand.   But the addition process may shift one number with respect to the other before the addition and roundoff take place.   We are supplying a term $1 + \varepsilon$ to both numbers, thus putting in more roundoff than there should be, but the more the relative shift done, the less this excess.   The typical term is of the form

$$a_k(1 + \varepsilon_k)a^k(1 + \varepsilon_1^*) \cdots (1 + \varepsilon_{2k+1}^*)$$

and the error in the polynomial evaluation from this term will be

$$a_k a^k [(1 + \varepsilon_k)(1 + \varepsilon_1^*) \cdots (1 + \varepsilon_{2k+1}^*) - 1]$$

---

[1] This is not true for the special-function evaluation of e^x, $\ln x$, $\sin x$, etc.

which is, neglecting products of ε_i's in comparison to ε_i's

$$a_k a^k[\varepsilon_k + \varepsilon_1^* + \varepsilon_2^* + \cdots + \varepsilon_{2k+1}^*]$$

For the purpose of bounding the ultimate error, all the ε^* are

$$|\varepsilon_j^*| \leq (\tfrac{1}{2})2^{-k}$$

for a k-bit machine. Thus if in place of a_k in the polynomial we wrote

$$a_k\left[1 + \varepsilon_k + \left(\frac{2k+1}{2}\right)2^{-k}\right]$$

we would have more than allowed for all the roundoff of the single-precision arithmetic done in the evaluation process, and we could proceed as before, using these modified coefficients to obtain a bound.

3.9 COMPLEX NUMBERS—QUADRATIC FACTORS

It frequently happens that a function with real coefficients is to be evaluated for a complex argument $x + iy$. For example, given

$$w = f(z) = \frac{e^z + e^{-z}}{2}$$

evaluate this for $z = x + iy$. We have

$$f(z) = \frac{e^x \cos y + e^{-x} \cos y}{2} + \frac{ie^x \sin y - e^{-x} \sin y}{2}$$

$$= \frac{e^x + e^{-x}}{2}\cos y + i\left(\frac{e^x - e^{-x}}{2}\right)\sin y$$

It is also very common to be asked to find

$$f(z) + f(\bar{z})$$

where the bar means the conjugate $x - iy$. In the above example we will have

$$f(z) + f(\bar{z}) = (e^x + e^{-x})\cos y$$

The particular case of a polynomial is of importance, both because it occurs frequently and because it has special properties. The direct evaluation using complex arithmetic is unnecessarily expensive in machine time. We use the fact that evaluating the polynomial at a point $a + ib$ is the same as finding the remainder when dividing by the corresponding linear factor. Noting that corresponding to the linear factor we can construct the real quadratic factor

$$[z - (a + ib)][z - (a - ib)] = z^2 - 2az + a^2 + b^2$$

we divide the polynomial by this quadratic factor (using the obvious extension of synthetic division to quadratic factors) to get the remainder $r_1 z + r_0$.

Consider the special case of computing $P(2 - 3i)$ where, as before,

$$P(z) = z^4 + 6z^2 - 7z + 8$$

The quadratic factor is

$$[z - (2 - 3i)][z - (2 + 3i)] = z^2 - 4z + 13$$

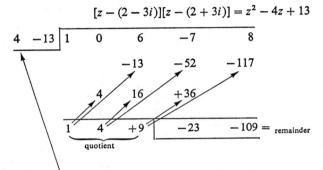

Thus

$$P(z) = (z^2 - 4z + 13)(z^2 + 4z + 9) + (-23z - 109)$$

At $z = 2 - 3i$ we have $z^2 - 4z + 13 = 0$

so that

$$P(2 - 3i) = -23(2 - 3i) - 109$$
$$= -155 + 69i$$

Note how much complex arithmetic we have avoided in this process. This is a very useful trick: we replace the evaluation of a polynomial at a complex number by a division process using only real numbers and at the last moment let the complex number enter (only linearly).

Error bounds follow along the lines of the linear-factor theory, and need not be developed here in detail.

PROBLEMS

Evaluate the given polynomial for the argument value given.

3.9.1 $P(x) = x^6 + 6x^5 + 15x^4 + 20x^3 + 15x^2 + 6x + 1$; for $x = -1 + i$.

3.9.2 $P(x) = x^4 - 2x^3 + 5x^2 + 4x + 7$; for $x = \dfrac{-1 + i\sqrt{3}}{2}$.

3.9.3 $P(x) = x^4 - 2x^3 + 5x^2 + 4x + 7$; for $x = \dfrac{1 + \sqrt{5}}{2}$.

3.9.4 Discuss the evaluation of $P'(x + iy)$.

3.10 REPEATED EVALUATIONS

Often we want to evaluate a function not once, but at a series of equally spaced points (arguments). If machine time and costs are important, then significant savings can be made by taking advantage of various features of the function.

EXAMPLES

$$e^{a + (n+1)h} = e^{a + nh} e^h$$

$$\begin{cases} \sin[a + (n+1)h] = \sin(a + nh) \cos h + \cos(a + nh) \sin h \\ \cos[a + (n+1)h] = \cos(a + nh) \cos h - \sin(a + nh) \sin h \end{cases}$$

$$\ln[a + (n+1)h] = \ln(a + nh) + \frac{h}{a + nh} - \frac{1}{2}\left(\frac{h}{a + nh}\right)^2 + \cdots \qquad h \ll a + nh$$

$$////$$

For polynomials see Sec. 9.5.

The need for frequent evaluations of a polynomial even when the arguments are not equally spaced is a common occurrence. For example, most library routines for the special functions have some polynomial-evaluation steps. It comes as a surprise to many people to learn that for polynomials of degree six or higher there are arrangements that can save in the number of arithmetic operations used versus those used by the obvious chain method (see Ralston [49], sec. 7.2, for more details). Note that (1) often these shorter methods lead to severe roundoff troubles, and (2) the conversion to the shorter form can be programmed on a machine once and for all.

PROBLEMS

3.10.1 Discuss using the addition formula for $\tan x$ for finding successive, equally spaced values of $\tan x$. Include a discussion of the probable error trouble.

3.10.2 Note that the routine for successive values of the sine requires the corresponding cosine values; develop a method that uses only the last two sine values.

3.10.3 Examine the roundoff error propagation in the sine routine and the effect of writing $\cos h$ as $1 - \phi(h)$ for h small.

3.11 OVERFLOW AND UNDERFLOW

We have ignored one of the realities of computing, namely that not only are the mantissas limited, but so also are the exponents. This limitation gives rise to both overflow and underflow. It may seem strange at first that with the usual wide range of at least 10^{-38} to 10^{38} there should be trouble with exponents. But this ignores a number of facts. First, it is quite common to have adequate theories, with analytical results, at both ends of some range, and the computation is to cover the ground between where certain effects (terms) are completely negligible to where certain other effects are similarly to be ignored. Thus it is natural that in some cases being run there will be terms that are very small or very large.

Secondly, even for modest ranges some very common functions such as the exponential will have very large (small) values. For example, $e^{25} = 0.720 \times 10^{11}$.

Thirdly, many mathematical theories about the real world include singularities, and it is sometimes necessary to compute very close to the singularity in order to discover the nature of it. (See Chap. 42.)

When underflow occurs, it is almost always satisfactory to replace the number by the machine zero, 0.000×10^{-9}. Usually there is no comparable infinity to use when overflow occurs, and at best the overflow number is replaced by some very large number such as 0.900×10^9 (to leave room for some further increase without again tripping the overflow alarm).

Because of the lack of symmetry in the machine hardware and corresponding number system, it is preferable to arrange computations to have an underflow rather than an overflow. Usually this is not hard to do. For example, to compute

$$\frac{e^x}{e^x - 1} \qquad \text{for } 1 \le x \le 10$$

we write

$$\frac{e^x}{e^x - 1} = \frac{1}{1 - e^{-x}}$$

It should be obvious how to do this in many cases; what is necessary is to think about it *before* programming the details for a machine.

4

REAL ZEROS

4.1 INTRODUCTION

The problem of finding the real zeros of a continuous function occurs frequently in science and engineering and has therefore received extensive treatment. We shall give only a few of the simpler methods that are easy to understand and use. We will look at the problem of finding complex zeros in Chap. 5, and in Chap. 6 we will examine the important special case of polynomials.

It is in the nature of the problem of finding the real zeros of a function that the positive and negative terms of the function almost exactly cancel. We therefore face roundoff, and all that was said in Chap. 3 about how to evaluate a function to avoid unnecessary roundoff clearly applies to the problem of finding real zeros of a function.

The floating-point number system puts further restraints on what can be expected. We can at best hope to keep the relative error under control, and we cannot expect to find zeros far from the origin with great absolute accuracy. The roundoff also means that we cannot expect to find a number that makes the function exactly zero. It is for this reason that generally we do not try to find a zero; rather we try to find an interval in which the function changes sign. We then measure the accuracy by the length of the interval relative to the maximum of the function value and the argument.   But even worse, we can expect in practice to find a sequence of values, some with plus and some with minus signs!

4.2   GRAPHICAL SOLUTION

One of the easiest methods for humans to use is the graphical method of drawing a curve and noting where it crosses the axis, or sometimes a pair of curves whose mutual crossings indicate the zeros.   Unfortunately this method is particularly bad for unaided machines, and so we will look at it only briefly to set the stage for further methods that are more suitable for machines.   For machines with graphical output and with a human examining the output, it can be very useful and easy to use.

EXAMPLE 4.1   Consider the problem of finding the real zeros of the function

$$y = e^{ax} - x^2$$

We picture in our minds the plot of $y = e^{ax}$ (see Fig. 4.2.1) and the plot of $y = x^2$ and look for their crossings.   If a were small and positive, it is clear that the x^2 curve will rise quickly enough to cross the exponential curve at least once, and since ultimately the exponential grows faster than any power of x, there will be a second positive crossing.

When we plot the curves, we get a picture that confirms what we expected. It is natural to ask for the value of a beyond which there is only the negative zero and no crossing on the positive side.   To find this we note that as a increases, the two positive crossings will approach each other and finally merge into a double zero.   At this point both the function and the derivative will have a zero at the same place.

$$y = e^{ax} - x^2 = 0$$
$$y' = ae^{ax} - 2x = 0$$

Eliminate e^{ax} to get

$$ax^2 - 2x = 0$$

or

$$\begin{cases} x = 0 & \text{(spurious solution)} \\ x = \dfrac{2}{a} \end{cases}$$

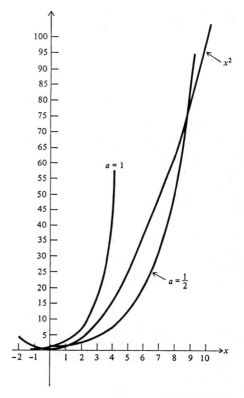

FIGURE 4.2.1

Putting this in the second equation, we get

$$ae^2 - \frac{4}{a} = 0$$

or for $a > 0$,

$$\begin{cases} a = \dfrac{2}{e} \\ x = e \\ y = 0 \end{cases}$$ ////

PROBLEMS

4.2.1   Find graphically the first two positive zeros of $\tan x - x = 0$ to one decimal place.

4.2.2   Find graphically the maximum of $y = -x \ln x (0 \le x \le 1)$.   Check analytically.

4.2.3   Find the value of a for which $\sin x - ax = 0$ has a double zero.

4.2.4   Show that there are no real zeros for $y = e^x - \ln x - 1$.

4.3   THE BISECTION METHOD

One of the best, most effective methods for finding the real zeros of a continuous function is the bisection method.   The bisection method begins with an interval (x_1, x_2) in which the function changes sign, which is measured in the machine by

$$f(x_1)f(x_2) < 0$$

We then pick the midpoint x_3 (bisect the interval) and evaluate the function there.   If

$$f(x_1)f(x_3) \begin{cases} < 0 & \text{then there is a sign change in } (x_1, x_3) \\ > 0 & \text{then there is a sign change in } (x_3, x_2) \\ = 0 & \text{then } x_3 \text{ is a zero} \end{cases}$$

Repeated bisections reduce the interval containing the sign change to an arbitrarily small length—or until we meet the granularity of the number system around x and the machine computes the midpoint as one of the two end values. Ten bisections reduce the interval length by more than a factor of $\frac{1}{1,000}$—three decimal place improvement!   Thus in practice it is rare that the bisection method will be used for 20 steps.

The method is *robust* in the sense that small roundoff errors will not prevent the method from giving an interval with a sign change, and if roundoff is misleading you, it is not the fault of the method but of the program that evaluates the function.   There is one danger, namely that what you implicitly thought was a continuous function had a pole and that you end up straddling an odd-order pole, but this can be checked for easily.

The bisection method supposed that an interval had been found which had a sign change in it, and the problem is usually stated as that of finding all the real zeros in some given interval.   This requires a search for intervals of sign change. If we search with a large step size h, then we run the risk of stepping over a pair of zeros or of getting an interval with an odd number of zeros and finding only one of them in the end.   If we use a small step size h, then we will spend most of the time looking where there are no zeros.   *It is an engineering judgment to resolve this dilemma.*

The problem of multiple zeros can be troublesome. One way to cope with it is to search for sign changes in both the function (which indicates all odd-order zeros) and the first derivative (which indicates all even-order zeros), but without a lot of elaborate analysis the machine cannot be expected to solve the general problem of the multiplicity of the zeros it isolates.

EXAMPLE 4.2   Find $\sqrt{2}$.
   This is a zero of $y = x^2 - 2$.   Try $h = \frac{1}{2}$, starting at $x = 0$, for the search.

$$y(0) = -2$$
$$y(\tfrac{1}{2}) = -\tfrac{7}{4}$$
$$y(1) = -1$$
$$y(\tfrac{3}{2}) = \tfrac{1}{4}$$

We have the interval

$$1 < x < 1.5$$

Bisect and try $x = 1.25$; $y(1.25) = -0.4375$.   We have the interval

$$1.25 < x < 1.5$$

Bisect and try $x = 1.37$; $y(1.37) = -0.1231$.   We have the interval

$$1.37 < x < 1.5$$

Bisect and try $x = 1.43$; $y(1.43) = 0.0449 \ldots > 0$.   And so forth.   ////

As a practical matter, given the original interval, it is usually best to decide on the *number* of subintervals to try in the search method and on the number of bisections to use in case a sign change is found, rather than on interval sizes.

PROBLEMS

Find to one decimal place the real zero of:
4.3.1 $he^h = 1$
4.3.2 $\cos x = x$
4.3.3 $x \ln x = 1$
Using the bisection method, compute:

4.3.4 $\dfrac{1 + \sqrt{5}}{2}$ *Hint:*   Find a quadratic of which it is a zero.

4.3.5 $\sqrt[3]{2}$
4.3.6   Elaborate on the last paragraph of this section.

4.4   THE METHOD OF FALSE POSITION

The method of false position (*regula falsi*) is a very ancient method, and it attempts to do better than the obviously slow bisection method.   The method of false position again starts with two points at which the function has opposite signs; that is, we have an interval in which there is a zero, and we wish to decrease the interval. The straight line through the two points is used in place of the function, and the zero of the straight line is used in place of the midpoint of the bisection method. The main weakness of the method—its slow approach from one side—is apparent from the picture.   Further obvious faults tend to exclude the method from practical use on machines.

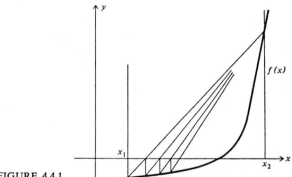

FIGURE 4.4.1

In more detail, given the two points $(x_1, f(x_1))$ and $(x_2, f(x_2))$ (see Fig. 4.4.1), the line through them is

$$y(x) = y(x_1) + \frac{y(x_2) - y(x_1)}{x_2 - x_1} (x - x_1)$$

whose zero is given by

$$x = x_1 - y(x_1) \frac{x_2 - x_1}{y(x_2) - y(x_1)}$$

$$= \frac{x_1 y(x_2) - x_2 y(x_1)}{y(x_2) - y(x_1)}$$

Note (1) the symmetry of the formula and (2) that $y(x_2) - y(x_1)$ is in fact an addition because we assumed $y(x_1) y(x_2) < 0$.

4.5 MODIFIED FALSE POSITION

A simple modification of the false-position method eliminates its worst feature, the one-sided approach to the zero with the resulting large interval. The modification of the false-position method consists in dividing by 2 during each cycle of the computation the function value at the end that is kept. Figure 4.5.1 shows how this modification improves the method.

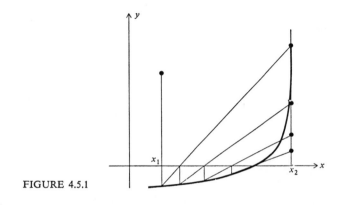

FIGURE 4.5.1

The method is usually (but not always) faster than the bisection method and has the defect that we cannot tell in advance how many steps will be required to obtain an interval length less than the preassigned accuracy needed.

In more detail the algorithm is the following. Given a continuous function $f(x)$ for an interval $a \leq x \leq b$ with $f(a)f(b) < 0$,

1 Compute $x = \dfrac{af(b) - bf(a)}{f(b) - f(a)}$.

2 Compute $f(x)$.

3 Compute $f(a)f(x)$. If $f(a)f(x)$ $\begin{cases} < 0 & \text{then} \begin{cases} x & \text{becomes new } b \\ f(x) \text{ becomes new } f(b) \\ \dfrac{f(a)}{2} \text{ becomes new } f(a) \end{cases} \\ = 0 & \text{then you are done} \\ > 0 & \text{then} \begin{cases} x & \text{becomes new } a \\ f(x) \text{ becomes new } f(a) \\ \dfrac{f(b)}{2} \text{ becomes new } f(b) \end{cases} \end{cases}$

EXAMPLE 4.3   Find the zero of $y = he^h - 1$ (using two-decimal arithmetic to make it easily followed).

It is easy to see that

$$f(0) = -1 < 0$$

$$f(1) = e - 1 > 0$$

hence

$$a = 0 \qquad f(a) = -1$$

$$b = 1 \qquad f(b) = 1.72$$

$$x = \frac{af(b) - bf(a)}{f(b) - f(a)} = \frac{0 \times 1.72 - 1(-1)}{1.72 - (-1)} = \frac{1}{2.72} = 0.37$$

$$f(x) = 0.37 \times 1.45 - 1 = 0.54 - 1 = -0.46$$

next step

$$a = 0.37 \qquad f(a) = -0.46$$

$$b = 1 \qquad \frac{f(b)}{2} = 0.86$$

$$x = \frac{0.37(0.86) - 1(-0.46)}{0.86 + 0.46} = \frac{0.778}{1.32} = 0.59$$

$$f(x) = 0.59 \times 1.80 - 1 = 0.06$$

next step

$$a = 0.37 \qquad \frac{f(a)}{2} = -0.23$$

$$b = 0.59 \qquad f(b) = 0.06$$

$$x = \frac{0.37(0.06) - 0.59(-0.23)}{0.06 + 0.23} = \frac{0.16}{0.25} = 0.55$$

$$f(x) = 0.55 \times 1.73 - 1 = -0.05$$

next step

$$a = 0.55 \qquad f(a) = -0.05$$

$$b = 0.59 \qquad \frac{f(b)}{2} = 0.03$$

$$x = \frac{0.55(0.03) - 0.59(0.05)}{0.03 + 0.05} = \frac{0.046}{0.08} = 0.58$$

$$f(x) = 0.038$$

$$a = 0.55 \qquad \frac{f(a)}{2} = -0.025$$

$$b = 0.58 \qquad \frac{f(b)}{2} = 0.038$$

$$x \approx 0.565$$

End.   ////

An alternate version of the modified false-position method halves the ordinate kept *only* when it stays on the same side of the zeros as on the previous step. This modification is believed to speed up the convergence at the cost of a slight amount of extra programming and testing. Unfortunately the method has not been adequately analyzed, and experimental tests of selected functions are not completely convincing.

PROBLEMS

Using the modified false-position method for three steps, find the zero of:

4.5.1 $y = \tan x - \dfrac{1}{1 + x^2} \qquad 0 \le x < \dfrac{\pi}{2}$

4.5.2 $y = x^2 - 2 \qquad 0 \le x \le 2$

4.5.3 $y = x \ln x - 1 \qquad 1 \le x \le 2$

4.5.4 $y = \cos x - x \qquad 0 \le x \le \dfrac{\pi}{2}$

4.5.5 $y = x^3 - 2 \qquad 0 \le x \le 2$

4.5.6 The choice of halving the kept function value was arbitrary. Discuss other possible choices and when to use them.

4.5.7 Draw a flow diagram of the modified false-position method.

4.6   NEWTON'S METHOD

Newton's method for finding the real zeros of a function $y = f(x)$ is usually taught in the calculus course.   The idea behind the method is the "analytic substitution" of the local tangent line for the function and then the use of the zero of this line as the next approximation to the zero of the function.

In mathematical notation the tangent line at x_k (Fig. 4.6.1) is

$$y(x) = f(x_k) + f'(x_k)(x - x_k)$$

and hence solving for $x = x_{k+1}$, the next approximation is

$$x_{k+1} = x_k - \frac{f(x_k)}{f'(x_k)}$$

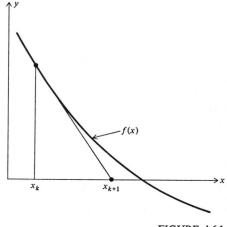

FIGURE 4.6.1

When it works, Newton's method is fine, but it has three obvious faults as can be seen in Fig. 4.6.2.   Thus, unless the local structure of the function is known in some detail, the unmodified Newton's method is to be avoided.

The worst features of the method can be partially compensated for by a simple device.   Let us introduce a *metric*, or distance, to measure how well we are doing.   We will use $|f(x)|$ as this metric.   The beginning of the next step in the iteration process involves the evaluation of $f(x)$ at the new point.   If this new

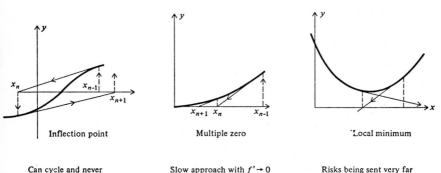

Inflection point	Multiple zero	'Local minimum
Can cycle and never converge	Slow approach with $f' \to 0$ and trouble in division step	Risks being sent very far away for next approximation

FIGURE 4.6.2

value is not smaller in size than the previous value, we will not accept the step, but instead we will go back and halve the step

$$x_{k+1} - x_k = \Delta x_k = \frac{-f(x_k)}{f'(x_k)}$$

using instead

$$\frac{\Delta x_k}{2}$$

If this does not decrease the distance, we repeat the halving process until it does, because we *know* that the local slope of the curve is in the direction to decrease the distance and that a sufficiently small step will work (until the granularity of the number system gets in the way). If this modification is used, it is well to add another, namely that no new step is more than twice the preceding step (to prevent the tedious shortening-up of the step size cycle after cycle). In this way the troubles from inflection points are eliminated, and the method will creep up on multiple zeros. Local minima, which are not always easily recognized, can cause trouble in a machine.

How shall we end the iteration in Newton's method? The approach to the zero becomes one-sided, and we do not have an interval in which the zero lies. If we use the size of the function at the point as the criterion, we will get different answers depending on whether we use $f(x) = 0$ or $kf(x) = 0$ as our original equation. If we use the step size $\Delta x = x_{k+1} - x_k$ as a measure, then when we quit, we may be far from the zero and be only slowly creeping up on it. There is no completely satisfactory answer, and we will suggest using the size of the step *relative* to the zero being found as a measure.   As will be shown in the Sec. 4.7, the error tends to be squared each step when we are finally close (in some sense), and this fact can be used by comparing three consecutive step sizes,

$$\frac{\Delta x_{k+1}}{(\Delta x_k)^2} \approx \frac{\Delta x_k}{(\Delta x_{k-1})^2} \approx \frac{-f''(x_{k+1})}{2f'(x_{k+1})}$$

4.7   THE CONVERGENCE OF NEWTON'S METHOD

Newton's method is highly favored theoretically because when it finally starts to converge, the convergence is quite rapid.   To show this we need a couple of important mathematical results.

The first result we need is the Taylor series with an integral remainder.   For our immediate use this can be found by integrating by parts the obvious identity

$$y(x) = y(a) + \int_a^x y'(s)\, ds$$

Setting

$$u = y'(s) \qquad du = y''(s)\, ds$$
$$dv = ds \qquad v = -(x - s)$$

we have

$$y(x) = y(a) - [(x - s)y'(s)]\Big|_a^x + \int_a^x y''(s)(x - s)\, ds$$

$$= y(a) + (x - a)y'(a) + \int_a^x y''(s)(x - s)\, ds$$

The second mathematical result we need is that if $f(x)$ and $g(x)$ are continuous and if $f(x) \geq 0$ $(a \leq x \leq b)$, then

$$\int_a^b f(x)g(x)\, dx = g(\theta) \int_a^b f(x)\, dx \qquad \text{for some } \theta \qquad a < \theta < b$$

To prove this, write g as the minimum of $g(x)$ and write G as the maximum (in the interval).   Thus

$$g \leq g(x) \leq G$$

Multiply this inequality by the nonnegative quantity $f(x)$

$$gf(x) \leq f(x)g(x) \leq Gf(x)$$

and then integrate (sum) to get

$$g \int_a^b f(x)\, dx \leq \int_a^b f(x)g(x)\, dx \leq G \int_a^b f(x)\, dx$$

Now consider the quantity, as a function of t,

$$\phi(t) = g(t) \int_a^b f(x)\, dx - \int_a^b f(x)g(x)\, dx$$

At the point where

$$g(t) = g, \text{ then } \phi(t) \le 0$$
$$g(t) = G, \text{ then } \phi(t) \ge 0$$

Hence, since $\phi(t)$ is continuous, there is a value $t = \theta$ such that

$$\phi(\theta) = 0$$

or

$$\int_a^b f(x)g(x)\, dx = g(\theta) \int_a^b f(x)\, dx$$

We use these results as follows: write $f(x)$ in the Taylor-series form where a is the current guess x_k and x is the zero of the function.

$$f(x) = 0 = f(x_k) + (x - x_k)f'(x_k) + \int_{x_k}^x (x - s)f''(s)\, ds$$

The next guess x_{k+1} is given by Newton's formula

$$0 = f(x_k) + (x_{k+1} - x_k)f'(x_k)$$

Subtracting, we get

$$(x - x_{k+1})f'(x_k) + \int_{x_k}^x (x - s)f''(s)\, ds = 0$$

Using the second theorem, since $x - s$ is of constant sign,

$$(x - x_{k+1})f'(x_k) + f''(\theta) \int_{x_k}^x (x - s)\, ds = 0$$

$$(x - x_{k+1})f'(x_k) + f''(\theta) \frac{(x - x_k)^2}{2} = 0$$

But

$$x - x_{k+1} = \varepsilon_{k+1} = \text{next error}$$
$$x - x_k \quad = \varepsilon_k \quad = \text{current error}$$

hence

$$\varepsilon_{k+1} = \frac{-f''(\theta)}{2f'(x_k)}\, \varepsilon_k{}^2$$

is an exact formula for the new error in terms of the current error; we square the error each step (almost).   This is often described by the remark that with Newton's method you double the number of correct digits each step (assuming that $f''(0)/[2f'(x_k)]$ is around 1 in size).

But let us not be deceived by this result.   Normally it is not the final rate of convergence that controls the number of iterations; it is the initial rate of convergence.   Here the bisection method shines—it starts out well in comparison with many other methods.

PROBLEMS

Apply Newton's method, using the starting value given to:

4.7.1 $y = xe^x - 1$ $x = 1/2$
4.7.2 $y = \arctan x - 1$ $x = 1$
4.7.3 $y = \ln x - 3$ $x = 10$
4.7.4 Using $y = x^n - 1$, show how for some values of x_k the local convergence of Newton's method can be very slow.

4.8 INVARIANT ALGORITHMS

The idea of an invariant algorithm is simple but important.   The importance arises from being one of the few ideas that provide some unity in the field of algorithms with its many varied methods and tricks.

The idea is easily illustrated by applying it to Newton's method.   We instinctively feel that it is the same problem whether we find the zeros of $y = f(x)$ or of $y = cf(x)$.   A moment's examination of Newton's method shows that at the step where we compute $f(x)/f'(x)$ the multiplicative factor c is automatically removed.   We also feel that if in place of x we were to use $c_1 x$ as the argument of the function, then we still have essentially the same problem.   *If* we scale the starting value by the corresponding amount, then we again see that the successive values of the estimates x_k also scale properly; the iteration equation is homogeneous in c_1.   Due to the nature of the floating-point number system a translation of the origin *does not* leave the problem unchanged.

What about the stopping rule?   Should it not also be invariant under these two classes of transformations?   The use of the change in the estimate $x_{k+1} - x_k$ *relative* to the estimate x_k clearly scales properly.   The use of the step size, the use of the size of the function, and many other possible stopping rules do not scale properly and are not appropriate for an invariant algorithm.

In general, an algorithm is invariant with respect to some class of transformations.   The class is found by considering the source of the problem, the mathematical structure of the problem, and the use that is going to be made of the results.   The class usually forms a *group*, provided the finite limitations of the machine are recognized.   The importance of invariance with respect to a group of transformations has long been recognized in mathematics.   That which is invariant is usually recognized as being more fundamental than the chance form of the particular representation of the problem; in Euclidean geometry only things invariant with respect to translation and rotation are regarded as geometric entities.   Similarly, an algorithm which transforms properly with respect to a class of transformations is more basic than one that does not.   In a sense the invariant algorithm attacks the problem and not the particular representation used (though clearly it uses a particular representation in the computation done).   The slight differences in roundoff that can occur are not basic to the idea, but come from the nature of the finite structure of the number system used.   On the other hand, the roundoff differences from different representations will *usually* not be large, except for ill-posed problems.

Invariance is something like dimensional analysis.   We realize instinctively that invariance should apply to proper equations and proper algorithms, and if it does not, then this is a warning sign to look a good deal closer before going on.   And like dimensional analysis, invariance in an algorithm can point to forms that are suitable and ones that are not, and often delimit the acceptable forms so as to practically give the proper one (see, for example, the stopping rule for Newton's method).   But invariance may include transformations that are more than mere linear scaling, and so the idea of invariant algorithms includes more than dimensional analysis.

The larger the class of transformations under which the algorithm is to be invariant, the more restricted the algorithm and (like much of mathematics) the more simple and powerful the result when finally found.   In the search for an algorithm the invariance requirement will block off many fruitless paths.

It is easy to see that two classes of transformations, multiplying the function by a constant and stretching the x axis, are reasonable to be used on the problem of finding zeros of functions.   Applying this invariance to the bisection method, we see that if the initial interval is properly scaled, then it is a matter of scaling the stopping rule; and if the stopping rule is not so scaled, then there is probably something wrong.

Similarly the false-position and modified false-position methods are invariant *provided* the stopping rules are picked properly.   In particular, the method of iterating the bisection method a fixed number of times, as well as the interval length relative to the size of the zero being found, is invariant.   Stopping rules, like the size of the function or the length of the interval, do not scale properly and should be avoided in practice.

In the future we will use this idea of invariance as a tool for finding and examining algorithms, and we will find that some of the classic methods are not invariant.

4.9   REMARKS ON COMPARING ALGORITHMS

Having given a number of algorithms for finding real zeros, how are we to compare them and others yet to be discovered so that we can pick one in a reasonable fashion when necessary to do a particular problem?

Evidently the bisection method is powerful, but slow in the final steps when compared to Newton's method.   But it is quite likely that the bisection method is much faster in the early stages.   And in practice it is usually the initial rates of convergence, not the final ones, that matter because we rarely want high accuracy in the zeros we find—in mathematical problems yes, but in engineering and science usually we don't know the other parts sufficiently well to justify a many-digit answer.

The modified false position is so much better than the simple false position that it is the second choice to consider.   When and in what respects does it compare favorably with the bisection method?   Evidently it is likely to be faster on reasonable functions, but it can be slower on some functions.   Nothing is really known at this time about the distribution of functions that occur in practice whose zeros we want to find, and so little can be said.

Newton's method is the classic one for finding zeros.   It requires finding and coding the derivative.   Some methods of rating algorithms count the effort to evaluate the derivative the same as to evaluate the function—thus counting one step as the equal of two in other methods—but this is *not* reasonable.   For most functions, once the parts of the function have been found, there are very few new parts that are expensive to compute when the derivative is found.   No new radicals, logs, or exponentials can arise; at worst a sine can go into a cosine (or the reverse), and the special functions are the time-consuming parts of most function evaluations.   All the straight arithmetic is usually small in time when compared to the special-function-evaluation routines..

Two methods which are sometimes discussed, but which we have so far ignored, should be mentioned.   The *secant method* resembles the false-position method, *except* that it uses the plausible argument that the last two values, rather than a pair that lies one on each side, should be kept.   The danger in the method is especially clear when roundoff is considered, since by chance a locally horizontal tangent can cause great trouble.

The other method is called *successive substitutions*. Given one or more equations, they are each solved explicitly for a different variable, even if it also appears on the other side, and the old values are substituted into one side to find updated new values of the unknowns. This method is powerful when done by hand, but it is hard to use on machines in situations that have not been preanalyzed. Evidently its success depends on arranging matters so that large errors in an unknown tend to produce smaller ones in the next cycle of iteration. This technique is related to the idea of stability which will be discussed in more detail in later chapters.

EXAMPLE 4.4 Consider again the example in Sec. 4.2 of finding the zeros of

$$e^{x/2} - x^2 = 0$$

this time using $a = \frac{1}{2}$.
We have the two curves (Fig. 4.9.1)

$$y = x^2$$
$$y = e^{x/2}$$

Solving the second equation for x, we have

$$y = x^2$$
$$x = 2 \ln y$$

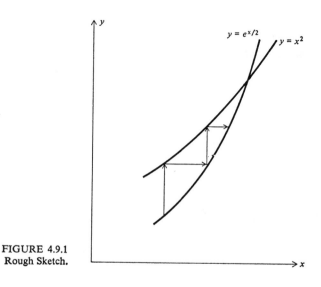

FIGURE 4.9.1
Rough Sketch.

If we start with $x = 5$, we get in turn (using two decimals)

$$x = 5$$
$$y = x^2 = 25$$
$$x = 2 \ln y = 6.4$$
$$y = 41$$
$$x = 7.4$$
$$y = 55$$
$$x = 8.0$$
$$y = 64$$
$$x = 8.3$$
$$y = 69$$
$$x = 8.5$$
$$y = 72$$
$$x = 8.5$$

(which is a poor answer), and we approach the upper zero from below.   If we were to reverse the substitution loop, we would approach the lower zero from above.

////

It is easy to draw sketches to show what will happen in particular cases, but this is exactly the kind of thing that machines have a lot of trouble doing. Thus the successive-substitution method is suitable only for carefully analyzed situations (of which there are many), but the details are easily worked out in any particular situation, and nothing more need be said here.

PROBLEMS

4.9.1   Make sketches showing the dangers of the secant method.

4.9.2   Show that the proposed modification of Newton's method is invariant.

4.10   TRACKING ZEROS

The problem is often not just to find the real zeros of a function, but rather to find the zeros as a function of some parameter in the equation.   In this situation it is clearly foolish to solve each parameter value as if it were a new problem.   Instead, from the values of the zeros for the previous parameter value(s) we can form a good guess at where those for the current parameter value probably lie. Methods for tracking the zeros as functions of the parameter can be developed from the predictor-corrector methods discussed in Chap. 23. It is a complex, difficult task to design a foolproof method of tracking zeros since sooner or later almost every possible trouble will occur.

In tracking zeros one should use a local coordinate system (like the moving trihedral in differential geometry) to avoid artifacts of the problem's coordinate system such as vertical and horizontal tangents. This remark applies to almost all graphical work.

For this chapter probably the best, easily available references are Ralston [49] and Traub [59] (see References in the back of this book).

5

COMPLEX ZEROS

5.1 INTRODUCTION

The problem of finding the complex zeros of a function in some region occurs frequently in practice, though not as frequently as finding real zeros. It is curious that the real-zero problem has been extensively investigated while the complex-zero problem has generally been ignored. Evidently it is a field ripe for further research. We shall give only two simple methods which are based to a great extent on the bisection method.

The functions we shall consider are called "analytic in the region of investigation," meaning that at every point in the region the function has a convergent Taylor series. For convenience only, we shall assume the region is a rectangle in the complex plane.

While finding real zeros, we used the notation $y = f(x)$; in the complex plane it is customary to use the notation $z = x + iy$ as the independent variable and $w(z) = u(x, y) + iv(x, y)$ as the dependent variable. Thus the single condition that $w(z) = 0$ is equivalent to the two simultaneous conditions

$$u(x, y) = 0$$
$$v(x, y) = 0$$

where $u(x, y)$ is called *the real part* and $v(x, y)$ is called *the imaginary part* of $w(z)$.

Very frequently the function we are examining has only real values for $w(z)$ when z takes on real values; that is, $w(x + i0)$ has only real values. In this important case we have the well-known result that if $x + iy$ is a zero of $w(z)$, then $x - iy$ is also a zero.

The theorem is important because it means that in this very common case we need to examine only the upper (or lower) half of the complex plane for zeros.

As an exercise in dealing with complex numbers we shall prove this theorem. The letter i can occur in the function definition as well as in the argument $z = x + iy$; for example,

$$w(z) = \sin z = \frac{e^{iz} - e^{-iz}}{2i}$$

and we need to distinguish these two appearances. Given $w = f(z)$, we shall use a long bar over both the function and the argument to mean the conjugate values, that is, when i is replaced by $-i$. A short bar over the function means that only the i's in the function are changed, while the bar over the argument means that only in the argument are the i's changed to $-i$'s. The assumption that the function takes on real values for real arguments means that, as in the above case of $\sin x$,

$$w = w(x) = \bar{w}(x)$$

The statement that $x + iy$ is a zero means that

$$w(x + iy) = 0$$

We need to show also that $w(x - iy) = 0$. We have in turn

$$w(x + iy) = 0 = \overline{w(x + iy)} = \bar{w}(x - iy) = w(x - iy)$$

and we have proved the theorem.

PROBLEMS

For practice in handling complex functions, find u and v

5.1.1 If $w = z^3$

5.1.2 If $w = \tan z$

5.1.3 If $w = \ln z$

5.1.4 If $w = \sqrt{z}$

5.2 THE CRUDE METHOD

The bisection method for finding real zeros is both very easy to understand and very robust, and it is natural to try to extend the method to finding complex zeros. This extension depends on finding the right way of looking at the bisection method in order to generalize it to complex zeros (keeping in mind the invariance principle).

One approach is to regard the search method for locating the zero as recording plus or minus at each point along the real axis according to the sign of the function at the point. In the complex plane for each point $x + iy$ we record

the quadrant number in which the function value falls. Thus at each point, we record 1, 2, 3, or 4, and we record a 0 whenever either $u(x, y) = 0$ or $v(x, y) = 0$ or both. This produces a picture of a region with numbers attached to each mesh point (see Fig. 5.2.1).

We can now take colored pencils and color in each quadrant. We will find that typically four quadrants meet at a point, each having an angle of about 90° at the point. At this point there is evidently a zero, since it corresponds to a point in the w plane where both $u(x, y)$ and $v(x, y)$ are zero.

The $u(x, y) = 0$ curves divide quadrants 1 and 2, and 3 and 4, while the $v(x, y) = 0$ curves divide quadrants 1 and 4, and 2 and 3.

As we shall later show (Sec. 5.4), where four quadrants meet at a point, it is a simple zero; where eight quadrants meet (each having an angle of about 45°), it is a double zero; etc.

Having located the general region of the zero, we may clearly enlarge the region as much as we please by plotting our mesh points at finer and finer spacing to get further accuracy until we run into either roundoff or the granularity of the number system.

This crude method is easy to understand, reliable, and accurate, but its fault is that it requires unnecessary computing (and also implies human intervention in the process of finding the zeros). We will later refine the method, but even in the crude form it is very useful and effective.

PROBLEMS

5.2.1 Prove that for functions such that $w = \bar{w}$, the x axis of the plot is a line of zeros corresponding to $v = 0$.

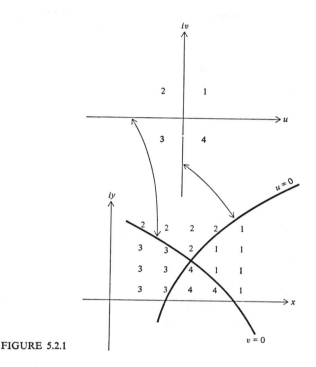

FIGURE 5.2.1

5.3   AN EXAMPLE USING THE CRUDE METHOD

As an example of how the crude method works, consider the problem of finding the complex zeros of the function

$$w = w(z) = e^z - z^2$$

which lie near the origin.   For this function we have $w(x) = \bar{w}(x)$, and so we need only explore the upper half-plane.   We try the rectangular region

$$-\pi \leq x \leq 2\pi$$
$$0 \leq y \leq 2\pi$$

The quadrant numbers are easily computed on a machine and are plotted in Fig. 5.3.1.   The x axis is a line of 0s, as it should be (there is one other number which is almost zero which we have marked as a 0).   In drawing the curves $u = 0$

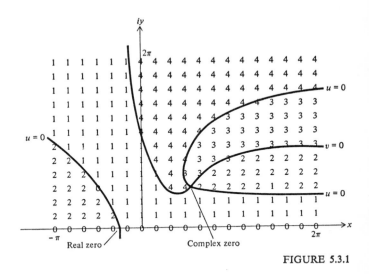

FIGURE 5.3.1

and $v = 0$, we remember that the curves in the lower half-plane are the mirror images (in the x axis) of those in the upper half, but that the quadrant designations interchange 1 and 4 as well as 2 and 3.

Does this picture seem to be reasonable? The real zero on the negative real axis seems to be about right because we know that as x goes from 0 through negative values, e^x decreases from 1 toward 0, whereas z^2 goes from 0 to large positive values. Thus, e^x and z^2 must be equal at some place (and we easily see that this happens before we reach -1).

The complex zero we found is likely to be one of a family of zeros with the next one appearing in the band

$$2\pi \leq y \leq 4\pi$$

An examination of the picture shows that it is a reasonably convincing display of the approximate location of the complex zero. We could easily refine the particular region if we wished by simply placing our points in a closer mesh.

5.4   THE CURVES $u = 0$ AND $v = 0$ AT A ZERO

At any point $z = z_0$ (Fig. 5.4.1) the Taylor expansion of a function has the form

$$f(z) = f(z_0) + f'(z_0)\frac{z - z_0}{1!} + f''(z_0)\frac{(z - z_0)^2}{2!} + f'''(z_0)\frac{(z - z_0)^3}{3!} + \cdots$$

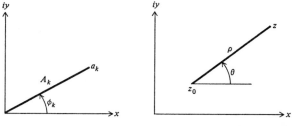

FIGURE 5.4.1

or

$$f(z) = a_0 + a_1(z - z_0) + a_2(z - z_0)^2 + a_3(z - z_0)^3 + \cdots$$

For each k we set

$$a_k = A_k e^{i\phi_k} \qquad A_k \text{ real}$$

and we also set

$$z - z_0 = \rho e^{i\theta}$$

Therefore, the Taylor expansion has the form

$$f(z) = A_0 e^{i\phi_0} + A_1 \rho e^{i(\phi_1 + \theta)} + A_2 \rho^2 e^{i(\phi_2 + 2\theta)} + \cdots$$

At a simple zero $f(z_0) = 0$, then $A_0 = 0$, and for small ρ the immediate neighborhood of z_0 $f(z)$ "looks like"

$$A_1 \rho e^{i(\phi_1 + \theta)}$$

or

$$f(z) \approx A_1 \rho [\cos(\phi_1 + \theta) + i \sin(\phi_1 + \theta)]$$

The $u = 0$ curves (Fig. 5.4.2) are approximately given by

$$A_1 \rho \cos(\phi_1 + \theta) = 0$$

or

$$\theta = -\phi_1 + \frac{\pi}{2} + k\pi \qquad k = 0, 1$$

and the $v = 0$ curves are approximately given by

$$A_1 \rho \sin(\phi_1 + \theta) = 0$$

or

$$\theta = -\phi_1 + k\pi \qquad k = 0, 1$$

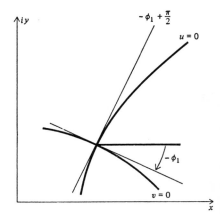

FIGURE 5.4.2

We see that the $u = 0$ and $v = 0$ curves intersect at right angles, and hence the quadrants we plan to color each have an angle of approximately 90° at the zero. The picture we color is therefore easy to interpret at a simple zero. Note that this happened at both zeros of the example in Sec. 5.3.

At a double zero both $f(z_0)$ and $f'(z_0)$ are 0, so that the Taylor series looks like

$$f(z) = A_2 \rho^2 e^{i(\phi_2 + 2\theta)} + A_3 \rho^3 e^{i(\phi_3 + 3\theta)} + \cdots$$

We have for small ρ

$$u \approx A_2 \rho^2 \cos(\phi_2 + 2\theta)$$
$$v \approx A_2 \rho^2 \sin(\phi_2 + 2\theta)$$
$$u = 0 \qquad \theta = -\frac{\phi_2}{2} + \frac{\pi}{4} + \frac{k\pi}{2}$$
$$v = 0 \qquad \theta = -\frac{\phi_2}{2} + \frac{k\pi}{2}$$

and the angles of the colored quadrants are approximately 45° (see Fig. 5.4.3). It is easy to see that for a triple zero the colored quadrant angles will be approximately 30°, and in general for a multiplicity of m, we shall have the $u = 0$ and $v = 0$ curves meeting at approximately $\pi/(2m)$ radians (or 90°/m).

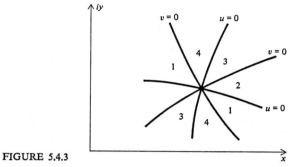

FIGURE 5.4.3

As an example of a double zero, consider Example 4.1 in Sec. 4.2 where we asked for a double zero of

$$w(z) = e^{az} - z^2$$

and found that $a = 2/e$. What does the picture of this look like? We should have the u and v curves crossing at 45° angles for this transcendental function. We have plotted the quadrant numbers only where quadrant changes occur, which is all that is needed. (See Fig. 5.4.4.)

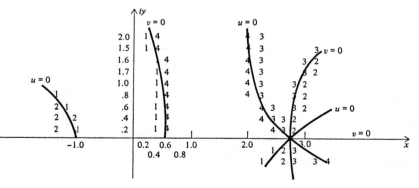

FIGURE 5.4.4

PROBLEMS

5.4.1   Sketch the $u = 0$ and $v = 0$ curves for $w = z^2 - x + \frac{1}{2}$ using the lattice points

$$x = \frac{k}{4} \qquad k = 0, 1, 2, 3, 4$$

$$y = \frac{m}{4} \qquad m = 0, 1, 2, 3, 4$$

and check by using the quadratic equation formula.

5.4.2   Prove that for a zero of order m, the quadrant angles are $\pi/(2m)$ radians.

5.5   A PAIR OF EXAMPLES OF $u = 0$ AND $v = 0$ CURVES

The following pair of simple polynomials illustrate how the $u = 0$ and $v = 0$ curves behave at zeros and elsewhere in the plane.

The first example is a simple cubic with zeros at -1, 0, and 1. The polynomial is

$$\begin{aligned} w = w(z) &= (z + 1)z(z - 1) = z^3 - z \\ &= (x + iy)^3 - (x + iy) \\ &= (x^3 - 3xy^2 - x) + i(3x^2y - y^3 - y) \end{aligned}$$

The real curves are defined by

$$u = x^3 - 3xy^2 - x = 0$$

or

$$x(x^2 - 3y^2 - 1) = 0$$

This is equivalent to two equations

$$x = 0$$
$$x^2 - 3y^2 - 1 = 0$$

The latter is a hyperbola whose asymptotes are

$$y = \pm \frac{x}{\sqrt{3}}$$

The imaginary curves are defined by

$$v = 3x^2y - y^3 - y = 0$$

This again is equivalent to two equations

$$y = 0$$
$$3x^2 - y^2 = 1$$

The latter is again a hyperbola, but this time with asymptotes

$$y = \pm\sqrt{3}x$$

The figure we have drawn (Fig. 5.5.1) looks reasonable. In the first place, at each simple zero the real and imaginary curves cross at right angles as they should according to theory. Secondly, far out, that is, going around a circle of large radius, the curves look as if they were from a triple zero; and the farther out we go, the more the local effects of the exact location of the zeros tend to fade out.

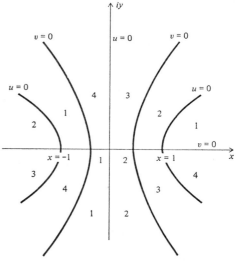

FIGURE 5.5.1

For the second example, we move the zero from $z = -1$ to $z = 0$, which makes a double zero at $z = 0$. The polynomial is

$$\begin{aligned}
w = z^2(z - 1) &= z^3 - z^2 \\
&= (x + iy)^3 - (x + iy)^2 \\
&= x^3 - 3xy^2 - x^2 + y^2 + i(3x^2y - y^3 - 2xy)
\end{aligned}$$

The real curve is

$$u = x^3 - 3xy^2 - x^2 + y^2 = 0$$

Solving for y^2, we have

$$y^2 = \frac{x^2(x-1)}{3x-1}$$

which is easily plotted as it has a pole at $x = \frac{1}{3}$, zeros at 0 and 1, and symmetry about the x axis.   As we expect, the asymptotes are parallel to

$$y = \pm \frac{x}{\sqrt{3}}$$

The imaginary curve is

$$v = 3x^2y - y^3 - 2xy = 0$$

which is

$$y(3x^2 - y^2 - 2x) = 0$$

or

$$y = 0$$

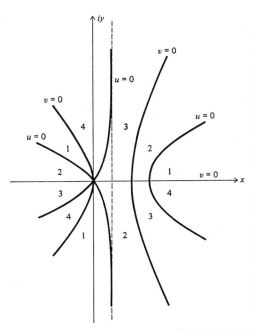

FIGURE 5.5.2

and

$$3(x - \tfrac{1}{3})^2 - y^2 = \tfrac{1}{3}$$

The last curve has asymptotes

$$y = \pm\sqrt{3}(x - \tfrac{1}{3})$$

which is what we expect.

Sketching these curves (Fig. 5.5.2), we see that far out in the complex plane they are the same as in the previous example. At the double zero there are two real curves and two imaginary curves alternating and crossing at 45° angles. The rest of the curves look as if the real and imaginary lines were being forced together by the movement of the zero from $z = -1$ to $z = 0$, and as if they tend to repel each other strongly.

Note that in the example in Sec. 5.3 the infinite sequence of zeros must be considered in judging the reasonableness of the shapes of the curves.

5.6 GENERAL RULES FOR THE $u = 0$ AND $v = 0$ CURVES

We cite without proof *the principle of the argument* which comes from complex-variable theory. This principle states that as you go around any contour, rectangular or not, in a counterclockwise direction, you will get a progression of quadrant numbers like

$$1, 1, 1, 2, 2, 2, 3, 3, 4, 4, 1, \ldots$$

with as many complete cycles 1, 2, 3, 4 as there are zeros inside (we are assuming that there are no poles in the region in which we are searching). There may at times be retrogressions in the sequence of quadrant numbers such as

$$1, 1, 1, 2, 2, 3, 3, 2, 3, 3, 4, 4, 4, 1, \ldots$$

for some suitably shaped contour, but the total number of cycles completed is exactly the number of zeros inside.

We have glossed over the instances of jumping over a quadrant number (say a 1 to a 3), and we will take that up later. We have assumed that the 0s that may occur are simply neglected, as they do not influence the total number of complete cycles.

To understand this principle of the argument, the reader can try drawing various closed contours in the previous examples. No matter how involved he draws them, he will find that he will have the correct number of zeros inside when he counts either $+1$ if he circles the zero in the counterclockwise direction or -1 if he circles the zero in the clockwise direction. Double zeros count twice, of course.

In the general analytic function, the $u = 0$ and $v = 0$ curves can be tilted at an angle to the coordinate system. They can be somewhat distorted and involved, but they must obey the three following restraints:

1 At a zero the curves must cross alternately and be spaced according to the multiplicity of the zero.

2 Far away from any zeros, the local placement of the zeros must tend to fade out and present the pattern of an isolated multiple zero having the number of all the zeros inside (with their multiplicities).

3 The number of cycles of 1, 2, 3, 4 going counterclockwise along *any* closed contour must equal the number of zeros inside that contour (when counted properly).

These three conditions so restrict the behavior of the curves we are following that many pathological situations are eliminated and the problem is thus made tractable.

PROBLEMS

5.6.1 Sketch the curves for the polynomial having zeros at

$$z = i \qquad z = i \qquad z = 0$$

5.6.2 Sketch the curves for

$$w = z^4 + 1$$

5.6.3 Sketch the curves for

$$w = z^4 + 2z^2 + 1$$

5.7 THE PLAN FOR AN IMPROVED SEARCH METHOD

One of the main faults of the crude method is that it wastes a great deal of machine time in calculating the function values (and corresponding quadrant numbers) at points which lie far from the $u = 0$ and $v = 0$ curves and hence give relatively little information.

Instead of filling in the whole area of points as we did, we propose to trace out only the $u = 0$ curves and to mark where they cross the $v = 0$ curves which gives, of course, the desired zeros. See the example at the end of Sec. 5.3 for how this might appear if we track *both* the $u = 0$ *and* the $v = 0$ curves.

The basic search pattern is to go counterclockwise around the area we are examining and look for a $u = 0$ curve, which will be indicated by a change from quadrant number 1 to 2 (or 2 to 1) or else from 3 to 4 (or 4 to 3).   See Figs. 5.7.1 and 5.7.2.

When we find such a curve, we will track it until we meet a $v = 0$ curve, which is indicated by the appearance of a new quadrant number other than the two we were using to track the $u = 0$ curve.

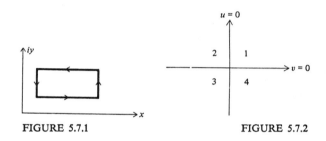

FIGURE 5.7.1   FIGURE 5.7.2

Having found the general location of a zero, we will pause to refine it, but we will need to pass over the zero finally to continue tracking our $u = 0$ curve until it goes outside the area where we are searching for zeros.

We need to know if this plan will probably locate all the zeros ("probably" depending on the step size we are using and not on the basic theory behind the plan).   By the principle of the argument, the number of cycles we find is the number of zeros inside the region.   Thus, the $u = 0$ and $v = 0$ curves we care about *must* cross the boundary of the region—they cannot be confined within the region—and our search along the boundary will indeed locate all the curves we are looking for (unless the step size of the search is too large) (Fig. 5.7.3).

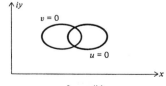

FIGURE 5.7.3   Impossible

It may happen that occasionally there is a jump in the quadrant numbers, say from 1 to 3.   We can easily see that we have in one step crossed both a $u = 0$ and a $v = 0$ curve.   Just where these two curves cross is, of course, not known, though probably it is near the edge (Fig. 5.7.4).   If we want to find this zero, then we assume that it is a 1, 2, 3; whereas, if we wish to ignore it, we assume that it is a 1, 4, 3.   We have to modify our search plan accordingly, but this is a small detail that is not worth going into at this point.

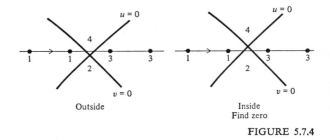

Outside   Inside
                               Find zero

FIGURE 5.7.4

5.8   TRACKING A $u = 0$ CURVE

How shall we track a $u = 0$ curve?   For convenience we start at the lower left-hand corner of the rectangular region we are examining for zeros and go counterclockwise, step by step, looking for a change in quadrant numbers that will indicate that we have crossed a $u = 0$ curve (Fig. 5.8.1).   When we find such a change, we construct a square[1] inside our region, using the search interval as one side.

The $u = 0$ curve must exit from the square so that a second side of the square will have the same change in quadrant numbers.   We continue in this manner,

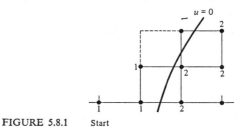

FIGURE 5.8.1   Start

[1] Squares only if x and y are comparable in size or importance (or both); otherwise, suitably shaped rectangles to maintain relative accuracy.

each time erecting a square on the side that has the quadrant number change, until we find that a different quadrant number appears.

When this occurs, we know that we have crossed a $v = 0$ curve (Fig. 5.8.2) and that we are therefore near a complex zero of the function we are examining.

By eye it is easy to make the new square, but it is more difficulty to write the details of a program that properly chooses the two points of the next square to be examined. It is also necessary at each stage to check whether the curve we are tracking has led us out of the region we are searching, and if so, we must mark the exit (Fig. 5.8.3) so that when we come to it at a later time (while we are going around the contour), we do not track this same $u = 0$ curve again, this time in the reverse direction of course.

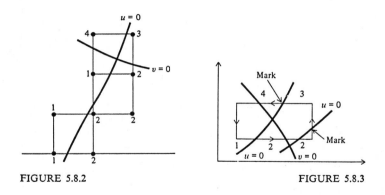

FIGURE 5.8.2   FIGURE 5.8.3

PROBLEMS

5.8.1   In the example of Sec. 5.3, apply the tracking method to the complex zero in the first quadrant.

5.8.2   Find the complex zeros as in the example in Sec. 5.3, except in the region $0 \le x \le 2\pi$, $2\pi \le y \le 4\pi$.

5.9 THE REFINEMENT PROCESS

Having located a square during our tracking of a $u = 0$ curve which has three distinct quadrant numbers, we have the clue that we are near a zero. We need, therefore, to refine our search pattern and locate the zero more accurately. The simplest way to do this is to bisect the starting side of the square that first produced the three different quadrant numbers.

If we use this smaller size, the $u = 0$ curve crosses one of the two halves, and selecting this half, we erect a square (of the new size). We may or may not find a third quadrant number. If we do not, then we continue the search with the new step size.

Usually within three steps (Fig. 5.9.1) we again have a square with three distinct quadrant numbers. We again halve and repeat the process, continuing until we have as small a square as we please (or run into either roundoff or the quantum size of the machine's number system).

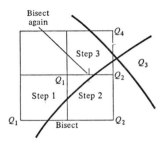

FIGURE 5.9.1

Evidently when we stop, we know that the zero is either in the square we have isolated or, at worst, in an adjacent square (again assuming that our step sizes are so small that the curves we are dealing with are reasonably smooth and well behaved).

What to do when you meet a zero value while tracking one of the curves is a small coding nuisance and is not a basic difficulty.

PROBLEMS

5.9.1   Discuss how to handle a zero value in the refinement process.

5.9.2   Discuss the use of other methods than bisection for the refinement process.

5.10 MULTIPLE ZEROS IN TRACKING

So far we have tacitly assumed that while we were tracking a $u = 0$ curve, we would come to a simple, isolated zero. But what happens when there is a double (or two very close—close for the step size we are using at the moment) or higher-order zero?

At a double zero we may find that the two quadrant numbers on the far side from the initial side of the square have the same quadrant numbers but are reversed (see Fig. 5.10.1). Thus we need, in fact, not only to check whether we have found a new quadrant number but also to check *at each step* that the same numbers are not in diagonally opposite corners, indicating two zeros or a double zero.

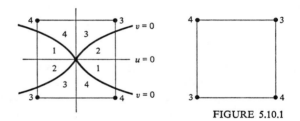

FIGURE 5.10.1

Once we sense that we are near a multiple zero, we are reasonably well off, because we can then afford relatively elaborate computing to clarify the matter. In principle it is only necessary to go around the suspected location with a sufficiently fine mesh of points to find the change in the argument and hence the

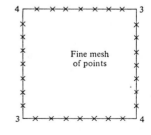

FIGURE 5.10.2

number of zeros inside (Fig. 5.10.2). Once we know this, we can proceed as the situation suggests. One way is to regard it as a new problem with a much finer search size and make an "enlargement" of the region.

Another method, which somewhat simplifies the problem of multiple zeros, is not to search one curve at a time but to go across the bottom side and find all the intervals that contain a $u = 0$ curve (Fig. 5.10.3). Then we push the calculations

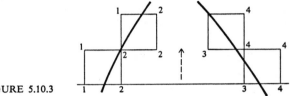

FIGURE 5.10.3

up one square, regardless of how many squares sidewise they go (and watch the sides for new $u = 0$ curves).

It is only when two or more $u = 0$ curves meet that we need to worry about a multiple zero. Thus, unless the second $u = 0$ curve is parallel to the bottom side of the rectangle, or so nearly so as to come as a total surprise, we have a very simple clue as to when to suspect a multiple zero. It may be that the two are meeting head on and are the ends of a single curve, which is not hard to detect; otherwise, we are reasonably well off in making an estimate of the number of $u = 0$ curves that are coming into the local region and hence of the multiplicity of the zero.

This method of pushing up one row at a time has the added advantage that it eliminates a lot of testing to see if the curve we are pursuing is leading us out of the region we are searching.

But let us be clear about one thing: We are vulnerable to being fooled by having chosen too large a search step, just as in the bisection method, and so the method cannot be made absolutely foolproof. What we want is a reasonably economical and safe method. All we need is a warning that there is a complicated situation at or near the location, and then we can simply apply our microscope via the enlarging process to isolate what is going on; we can repeat this enlargement process until we run into the granularity of the number system of the computing machine.

5.11 FUNCTIONS OF TWO VARIABLES

The problem of finding the real simultaneous solutions of

$$g(x, y) = 0$$
$$h(x, y) = 0$$

can be partially mapped onto the problem of finding the complex zeros of $f(x + iy) = 0$ by assigning the quadrant numbers in the obvious way:

If	Quadrant
$g > 0, h > 0$	1
$g > 0, h < 0$	2
$g < 0, h < 0$	3
$g < 0, h > 0$	4

The crude method will work, though we are not sure that the zero curves will meet at right angles at a simple zero.   If we try the tracking method, we are not sure (since the principle of the argument need not apply) that we shall find all the curves as we trace the boundary.   Still, the method will sometimes give useful results.   Fig. 5.7.3 is now possible.

PROBLEM

5.11.1 Discuss the motion of the zeros of $w = e^{ax} - x^2 = 0$ as functions of the parameter a.

6

*ZEROS OF POLYNOMIALS

6.1 WHY STUDY THIS SPECIAL CASE?

We have just discussed the location of both real and complex zeros of a function, and it is natural to ask why we should examine the special case of polynomials,

$$w = P(z) = a_N z^N + a_{N-1} z^{N-1} + \cdots + a_0$$

This can be answered in a number of different ways.

1. The problem occurs frequently, and special methods can save a great deal of machine time and trouble.
2. For polynomials we can apply the invariance principle very effectively because we have a large class of transformations that leave the problem essentially unchanged.
3. A polynomial of degree N has exactly N zeros, and we therefore know when we have found all the zeros.
4. The factor theorem shows that in the polynomial case, zeros and factors are equivalent, and we can use the divisibility as a tool in finding the zeros.

5   When we find a zero, we can "deflate" the polynomial by dividing out this factor and thus obtain a simpler problem to solve.

6   In applications it is often of crucial importance to identify multiple zeros as multiple zeros and not as close ones.

7   In applications it is often true that the zeros are known to be real and/or pure imaginary and those that fall off the two axes were moved by roundoff.   Therefore, the polynomial routine can "nudge" such zeros back onto their respective axes *provided* the difference is attributable to roundoff effects.

8   In applications it appears that often the user is more interested in the zeros as a self-consistent set rather than as a set of individually accurate, but unrelated, zeros.

9   From the factors we find, we can reconstruct the polynomial (approximately) and compare these coefficients with the original to form some measure of the loss of information around the whole loop.   Since the reconstruction is fairly direct, presumably most of the loss occurs while finding the factors.   (Compare Sec. 3.2 where we discussed quadratics.)

The point about multiple zeros needs elaboration.   If a double zero is reported as a pair of close zeros, then when they are used it is highly likely that there will be great trouble with roundoff.   We have adopted the attitude that we prefer to avoid roundoff rather than bound it by an elaborate rule, and this means that we must make special efforts to identify multiple zeros so that later the roundoff will not cause trouble.

As an example of the effect of a double zero being reported as a pair of close ones, consider the problem of solving the simple linear differential equation

$$y'' + 2y' + y = 0 \qquad y(0) = 1 \qquad \text{and} \qquad y'(0) = 0$$

The characteristic equation is

$$m^2 + 2m + 1 = 0$$
$$(m + 1)^2 = 0$$
$$m = -1, -1$$

The solution is, therefore,

$$y = e^{-x}(C_1 + C_2 x)$$
$$= e^{-x}(1 + x)$$

But if due to roundoff the roots were given as

$$m_1 = -1 + \varepsilon$$
$$m_2 = -1 - \varepsilon \qquad \varepsilon \text{ small}$$

then the general solution would be

$$y = C_1 e^{-(1-\varepsilon)x} + C_2 e^{-(1+\varepsilon)x}$$

Applying the initial conditions we get

$$C_1 + C_2 = 1$$
$$-(1-\varepsilon)C_1 - (1+\varepsilon)C_2 = 0$$

and

$$y = \frac{1}{2}\left[\left(1 + \frac{1}{\varepsilon}\right)e^{-(1-\varepsilon)x} - \left(\frac{1}{\varepsilon} - 1\right)e^{-(1+\varepsilon)x}\right]$$

Note the roundoff trouble, especially near $x = 0$.

As a second example of this important effect consider the integration of rational functions. If the zeros of the denominator were multiple but were reported as close, then instead of

$$\int_0^\infty \frac{dx}{(a^2 + x^2)^2} = \int_0^{\pi/2} \frac{a\sec^2 t}{a^4 \sec^4 t}\,dt = \frac{1}{a^3}\int_0^{\pi/2}\cos^2 t\,dt = \frac{\pi}{4a^3}$$

we would have the integral (with b close to a)

$$\int_0^\infty \frac{dx}{(x^2 + a^2)(x^2 + b^2)} = \frac{1}{b^2 - a^2}\int_0^\infty \left(\frac{1}{x^2 + a^2} - \frac{1}{x^2 + b^2}\right)dx$$
$$= \frac{1}{b^2 - a^2}\left(\frac{1}{a}\arctan\frac{x}{a} - \frac{1}{b}\arctan\frac{x}{b}\right)\Big|_0^\infty$$
$$= \frac{\pi}{2}\frac{1}{b^2 - a^2}\left(\frac{1}{a} - \frac{1}{b}\right)$$

This expression can be reworked, using the methods of Chap. 3, to avoid the obvious heavy cancellation. In this simple example the trouble is obvious and typical of the general case of integration of rational functions when a multiple zero occurs in the denominator.

PROBLEMS

6.1.1   Devise another example to demonstrate the effect of close zeros on subsequent computations.

6.1.2   Discuss the case

$$\int_0^a \frac{dx}{(a+x)^3} = \frac{3}{8a^2} \qquad a > 0$$

Hint: use $-a$, $-(a-\varepsilon)$, $-(a+\varepsilon)$ as zeros.

6.2 INVARIANCE PRINCIPLE

Polynomials provide a good example of the use of the invariance principle because we have three different transformations that leave the polynomial essentially unchanged:

1 The transformation of $P(z)$ into $cP(z)$

2 The transformation of $P(z)$ into $P(cz)$

3 The transformation of $P(z)$ into $z^N P\left(\dfrac{1}{z}\right)$

The special case under (2), where $c = -1$, reverses the direction of the x axis and can be used, if we wish, to confine the search to finding positive zeros only.

The reciprocal transformation (3) has the effect of reversing the coefficients of the polynomial. This is mirrored in the synthetic division process. We have written the polynomial (a linear combination of successive powers of z) as if $|z| > 1$ and have put the highest power of z first, much as we would when writing numbers. But if we think of z as being less than 1 in size, then we would naturally write it as

$$a_0 + a_1 z + a_2 z^2 + \cdots + a_N z^N$$

and in the division process divide by (for a quadratic factor)

$$1 - p'z - q'z^2$$

The remainder would be

$$r'_{N-1} z^{N-1} + r'_N z^N$$

If in the usual process the divisor were a factor of the polynomial, then the remainders r_1 and r_0 would both be zero; in the second case it would be r'_{N-1} and r'_N that would be zero. A little thought shows that these two are equivalent and that the two quotients *must be the same*. Thus we have two alternate ways of using the third transformation: either we can reverse the coefficients of the polynomial (which carries the region $1 \le |z| < \infty$ into $1 \ge |z| > 0$) or else equivalently we can do our synthetic division in the reverse way, from constant to higher powers rather than the more conventional order of higher to constant powers.

We shall check regularly that the methods we propose to use obey the invariance principle with respect to these three transformations.

PROBLEMS

6.2.1 The product of a number of factors $(x - c_1)(x - c_2) \cdots (x - c_N)$ leads to a polynomial whose coefficients are the elementary symmetric functions

$$E_1 = \sum c_i, \; E_2 = \sum c_i c_j, \ldots, E_N = c_1 c_2 \cdots c_N$$

Thus for

$$x^N + a_{N-1}x^{N-1} + \cdots + a_0$$

with zeros $c_1, c_2, \ldots, c_N$,

$$a_{N-k} = (-1)^k E_k$$

Discuss the roundoff errors in constructing this polynomial.

6.2.2   Devise a measure of closeness of two polynomials using only the coefficients. Defend your choice.

6.3   THE PLAN

One of the consequences of the complex zeros of a real function occurring in pairs is that by the fundamental theorem of algebra a polynomial with real coefficients can be written as a product of real linear and real quadratic factors. This approach has the advantage, as shown in Sec. 3.9, of working only with real numbers and of avoiding special programming for complex numbers (also saving machine time). The approach to the problem of finding the factors has the further advantages of ease of thought and of finding what is most often needed in the next step, namely, the factors rather than the actual zeros.

Therefore, after examining the given coefficients to see what kind of polynomial we have and making the obvious possible simplifications, we will first find all the real linear factors, watching carefully for multiple factors, and deflate the polynomial each time we find one. Then we will search for the real quadratic factors, again watching for multiple factors and deflating each time we find a factor. Further, each time we find a quadratic factor, we will check to see if the zeros of the factor might have been on one of the axes and had been moved off due to roundoff.

The methods we use will be guided by the requirements of the invariance principle.

6.4   PREPROCESSING THE POLYNOMIAL

Often the polynomial which we are given is said to be of degree N, but is not. For example, the leading coefficient a_N may be zero (perhaps for a particular value of a parameter in the equation), which means that the polynomial has a zero at infinity. Further it may be that the last coefficient a_0 is zero, which means that there is a zero at the origin. In both cases these zeros should be removed before starting. To do this we can test for

$$a_0 a_N = 0$$

and if the relation holds, we must find out which of the factors is zero and remove the corresponding zero from the polynomial. This test applied recursively will remove all these trivial zeros.

Next we need to find out if it is a polynomial in some power of z, say z^2 or z^3, and if it is, we should make the proper substitution to reduce the effective degree of the polynomial. For example, the polynomial

$$x^{12} - 6x^8 + 4x^4 + 1 \equiv (x^4)^3 - 6(x^4)^2 + 4(x^4) + 1$$

is actually a cubic in x^4 and should be treated as a cubic and not as a polynomial of twelfth degree.

The factor k that we want is the common factor of all the exponents of terms with nonzero coefficients. To find this factor we start with the constant term $a_0 \neq 0$ and look for the next higher term with a nonzero coefficient. If it is a_1, we are done—$k = 1$; but if it is some higher power, we find the factors of the exponent which are also factors of the degree N. Each factor that divides N is a potential candidate for further search to see if all terms in the polynomial have exponents that are multiples of it. In this way we can find the highest common factor of the exponents those terms present in the polynomial.

Suppose we have found such a number $k > 1$. Making the obvious substitution,

$$z^k = z'$$

we have a polynomial in z' of much lower degree to solve, one having no common factors of the exponents of the terms present.

This transformation must, of course, be undone when we finally produce the answers. For the real linear factors

$$z' = a = z^k$$

Write a in the polar form

$$a = \rho e^{i\theta} \qquad \text{where} \qquad \rho = |a| \qquad \theta = \begin{cases} 0 & \text{for } a > 0 \\ \pi & \text{for } a < 0 \end{cases}$$

$$z^k = \rho e^{i(\theta + 2\pi m)} \qquad m = 0, 1, \ldots, k - 1$$

Taking the kth root, we have the k zeros

$$z_m = \rho^{1/k} e^{i[(\theta + 2\pi m)/k]} = \rho^{1/k}\left(\cos\frac{\theta + 2\pi m}{k} + i\sin\frac{\theta + 2\pi m}{k}\right)$$

$$m = 0, 1, \ldots, k - 1$$

For a quadratic factor

$$(z')^2 - pz' - q = z^{2k} - pz^k - q \qquad q < 0$$

write it in the form

$$(z^k - \rho e^{i\theta})(z^k - \rho e^{-i\theta}) = z^{2k} - (2\rho \cos \theta)z^k + \rho^2$$

so that

$$\rho^2 = -q \qquad\qquad \rho = \sqrt{-q}$$
$$\text{or}$$
$$2\rho \cos \theta = p \qquad\qquad \theta = \arccos \frac{p}{2\rho}$$

We therefore have the k real quadratic factors

$$(z - \rho^{1/k} e^{i[(\theta + 2\pi m)/k]})(z - \rho^{1/k} e^{-i[(\theta + 2m\pi)/k]})$$

$$\equiv z^2 - 2\rho^{1/k} \cos\left(\frac{\theta + 2m\pi}{k}\right)z + \rho^{2/k} \qquad m = 0, 1, \ldots, k - 1$$

or

$$\begin{cases} p_m = 2\rho^{1/k} \cos \dfrac{\theta + 2m\pi}{k} \\ q_m = -\rho^{2/k} \end{cases} \qquad m = 0, 1, \ldots, k - 1$$

6.5   THE REAL ZEROS

We first remove the real zeros.   For this purpose we can use either the bisection or the modified false-position method.   In both cases we make an initial search for intervals with a sign change.   When we find a sign change, we can refine the interval as much as we wish, and then we deflate the polynomial.

Now the product of all the zeros is

$$(-1)^N \frac{a_0}{a_N}$$

so that the geometric mean of all the zeros

$$r = \sqrt[N]{\left|\frac{a_0}{a_N}\right|}$$

provides a natural unit of length.   The invariance principle tells us that since the reciprocal transformation and the reversal of the negative and positive axes leave the polynomial unchanged, we must treat the four ranges

$$0 \leq z \leq r$$

$$0 \geq z \geq -r$$

$$0 \leq \frac{1}{z} \leq \frac{1}{r} \qquad \text{where } z \text{ is a real number}$$

$$0 \geq \frac{1}{z} \leq -\frac{1}{r}$$

in essentially the same way. As a matter of engineering judgement we will search each range, using $2N$ equally sized steps, and take one step in each range before going to the next step. We start searching at zero because we prefer the risk of an underflow to an overflow and because we instinctively feel that the roundoff propagation will be less. Thus, except for trivial details the process of searching is invariant under the three transformations,[1] since the refinement method was invariant under the first two and the third is covered by breaking the problem into four ranges.

We need, of course, to watch for multiple zeros. What do we mean by a zero of multiplicity k? Since a polynomial of degree N can be factored into N factors (not necessarily distinct), it is conventional (because the remainder theorem of Sec. 3.7 connects the zeros and factors of a polynomial) to identify *each* factor with a zero and to say that a polynomial of degree N has exactly N zeros. This is apparently how the idea of multiple zeros arose.

The Taylor-series expansion of a function gives a second approach. The number of successive coefficients (of the now possibly infinite sequence of coefficients) that vanish at a point is the multiplicity of the zero. In this fashion the idea of multiplicity, defined originally for polynomials, is extended to analytic functions. At this point we do not need the further extension to functions such as the very common

$$y = \sqrt{a^2 - x^2}$$

which could be said to have a real zero of order $\frac{1}{2}$ at $x = \pm a$.

The repeated synthetic division process in Sec. 3.7 identified $k!$ times the successive remainders with the kth derivative. While we are searching for a zero by either the bisection or modified false position, we will evaluate the successive derivatives together with the estimates of the roundoff noise to find which is the highest-order derivative that can be regarded as vanishing. Using the next remainder in the sequence (which is therefore not zero), we make the preceding

---

[1] It is easy to devise an algorithm which uniquely tells in which order the four ranges are to be searched, but it does not appear worth using in practice.

one as small as we can (we could use Newton's method here if we pleased). For a possible double zero, we are finding the place where the derivative is zero, for a triple zero, the local inflection point, etc., which seems to be quite a reasonable way to place the multiple zero. We need, of course, to check that the original zero is still within roundoff, and if it is not, we then have guessed at too high a multiplicity and must go back and try again.

In our search process we watch not only for sign changes in the function values, but also to be safe we watch for sign changes in the first derivative. Together these isolate both the odd- and even-order zeros (probably).

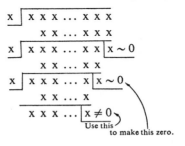

PROBLEMS

6.5.1   Prove that even in the presence of roundoff when we end the search process, we will have a polynomial of even degree.

6.5.2   Devise a reasonable algorithm to carry out the remark in footnote 1 in Sec. 6.5.

6.5.3   Show that

$$\frac{1}{\rho(\cos\theta + i\sin\theta)} = \frac{\cos\theta - i\sin\theta}{\rho}$$

or

$$\frac{1}{x + iy} = \frac{x - iy}{x^2 + y^2}$$

enables us to undo the reciprocal transformation.

6.6   PLAN FOR FINDING COMPLEX ZEROS

Having (probably) removed all the real zeros, we will have left a polynomial of even degree

$$w(z) = P_{2M}(z) = a_{2M}z^{2M} + a_{2M-1}z^{2M-1} + \cdots + a_0$$

Searching for real quadratic factors instead of complex zeros means that we are not searching the complex x, iy plane but the pq plane of the coefficients of the quadratic factor (see Fig. 6.6.1). These two planes are connected by

$$Q(z) = z^2 - pz - q = 0$$

This means that for both a point in the upper half of the complex plane and its conjugate point there is one point in the pq plane. The line of the x-axis goes into the curve of multiple zeros, namely when the radical is zero

$$\sqrt{p^2 + 4q} = 0$$

or more conveniently,

$$p^2 \equiv -4q$$

Inside this parabola the zeros of the quadratic factor are complex, along the curve they are multiple, and above the curve they are a pair of real distinct zeros (which

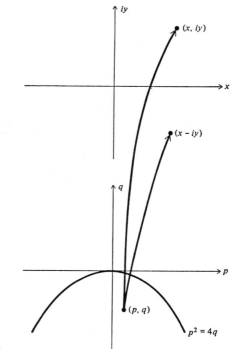

FIGURE 6.6.1

we have removed from the polynomial).   Since we have deliberately searched for
multiple zeros and have removed all the real zeros, we are definitely *inside* the
parabola and not " near " it.

We next examine the basic Bairstow method that we will use in finding the
real quadratic factors.

6.7   BAIRSTOW'S METHOD

When we divide a polynomial of even degree $P_{2M}(z)$ by a quadratic factor $Q(z) = z^2 - pz - q$, we get a quotient and a remainder

$$r_1 z + r_0$$

As in Sec. 3.7, repeated divisions of the quotient by $Q(z)$ produce an expansion in
powers of $Q(z)$, namely,

$$P_{2M}(z) = a_{2M} Q^M(z)$$
$$+ (r_{2M-1} z + r_{2M-2}) Q^{M-1}(z) + \cdots + (r_3 z + r_2) Q(z) + (r_1 z + r_0)$$

We know that $Q(z)$ is a factor if and only if $r_1 = r_0 = 0$.

Since the remainders

$$r_1 = r_1(p, q)$$
$$r_0 = r_0(p, q)$$

are analytic functions in p and q, we may expand them in a Taylor series about the
current point (p, q).   The value at any nearby point (p^*, q^*) in the pq plane is
given by

$$r_1(p^*, q^*) = r_1(p, q) + \frac{\partial r_1}{\partial p} \Delta p + \frac{\partial r_1}{\partial q} \Delta q + \cdots$$

$$r_0(p^*, q^*) = r_0(p, q) + \frac{\partial r_0}{\partial p} \Delta p + \frac{\partial r_0}{\partial q} \Delta q + \cdots$$

where

$$\Delta p = p^* - p$$
$$\Delta q = q^* - q$$

As a first approximation we drop all the terms beyond the linear ones.
Next we pick (p^*, q^*) so that

$$r_1(p^*, q^*) = 0$$
$$r_0(p^*, q^*) = 0$$

will improve our guesses at p and q. Note that this is essentially Newton's method in two variables.

To get the required partial derivatives, we observe that $P_{2M}(z)$ does not depend on p and q, and we differentiate

$$\frac{\partial P_{2M}}{\partial p} \equiv \{\cdots\}Q(z) + (r_3 z + r_2)\frac{\partial Q}{\partial p} + \frac{\partial r_1}{\partial p} z + \frac{\partial r_0}{\partial p} = 0$$

$$\frac{\partial P_{2M}}{\partial q} \equiv \{\cdots\}Q(z) + (r_3 z + r_2)\frac{\partial Q}{\partial q} + \frac{\partial r_1}{\partial q} z + \frac{\partial r_0}{\partial q} = 0$$

These equations are *identities* in z, hence are true for all values of z and in particular for those which make $Q(z) = 0$. Using one of these values, we have

$$(r_3 z + r_2)(-z) + \frac{\partial r_1}{\partial p} z + \frac{\partial r_0}{\partial p} = 0$$

$$(r_3 z + r_2)(-1) + \frac{\partial r_1}{\partial q} z + \frac{\partial r_0}{\partial q} = 0$$

But since $z^2 = pz + q$ and we are operating formally, we regard z as a complex number and set the linear terms and (separately) the constant terms equal to zero.[1] Thus we get the four equations

$$r_3 p + r_2 = \frac{\partial r_1}{\partial p}$$

$$r_3 q = \frac{\partial r_0}{\partial p}$$

$$r_3 = \frac{\partial r_1}{\partial q}$$

$$r_2 = \frac{\partial r_0}{\partial q}$$

Using these, the Newton approximations (the truncated Taylor-series expansions) become the usual Bairstow equations

$$(r_3 p + r_2)\,\Delta p + r_3\,\Delta q = -r_1$$
$$(r_3 q)\,\Delta p + r_2\,\Delta q = -r_0$$

The determinant of these equations is

$$D = r_2{}^2 + r_2 r_3 p - r_3{}^2 q$$

$$= r_3{}^2\left[\left(\frac{r_2}{r_3}\right)^2 + p\frac{r_2}{r_3} - q\right]$$

---

[1] We are using the fact that if $az + b = 0$, with a and b real and z complex, then $a = b = 0$.

From the last form (valid only if $r_3 \neq 0$), we see that the ratio r_2/r_3 is a zero of $z^2 + pz - q = 0$ (note the plus sign on the z term), and if the original quadratic $Q(z) = z^2 - pz - q$ has complex zeros, then $D \neq 0$. If $r_2 = r_3 = 0$, then we are "near" a possible multiple zero which we will investigate later. The possibility that $q = 0$ and $r_2 = 0$ is also unlikely [though we shall start with $q = 0$ as our starting approximation for $Q(z)$] since we have removed all zeros at the origin, which implies $q \neq 0$.

The modification of Newton's method (Sec. 4.6) uses a metric $|P(z)|$. We shall at this point use the equivalent, but more convenient, metric

$$|P(z)|^2$$

evaluated at the point where $Q(z) = z^2 - pz - q = 0$. Thus

$$|P(z)|^2 = (r_1 z + r_0)(r_1 \bar{z} + r_0)$$
$$= r_1^2 (-q) + r_1 r_0 p + r_0^2$$

Experience shows that using this modification of Newton's method, Bairstow's method will converge *even* for multiple factors and cannot wander or fail to converge.

When we find a quadratic factor, we need to check that its zeros are not near the imaginary axis (we have in the removal of the real zeros effectively checked that they are not near the real axis.) If p^2 is very much less than $-4q$ (which is a positive number), then we need to consider whether or not the quadratic factor $z^2 - q$ would be acceptable within roundoff, and if it is, then we prefer it.

6.8   CONVERGENCE OF BAIRSTOW'S METHOD

The unmodified Bairstow method is famous for not converging, but the simple modification of Newton's method removes most of this trouble. It is worth examining further to see why we do not get into trouble in this application of Newton's method in two variables—in particular why a *local* minimum does not trap us.

In the theory of functions of a complex variable there is a minimum modulus theorem (much like the maximum modulus theorem) which says that if at a point z_k in the z plane the function $f(z) \neq 0$ (and $f(z)$ is not a constant), then there are values of $f(z)$ in the immediate neighborhood of z_k such that

$$|f(z)| < |f(z_k)|$$

We are, of course, not in the z plane, but as discussed in Sec. 6.6, we are operating in the pq plane inside the parabola and not "near" the edge. Therefore there is a local one-to-one mapping of points in one plane onto points in the other, and there are points in the pq plane where the function is lower—in this problem there can be no local minima other than zero. When we solve the Bairstow equations for the next step (except in the fantastically unlikely case of exact cancellation of roundoff effects), we will find that there are terms in the polynomial that will produce a step, possibly small, that takes us away from any local saddle points (the two-dimensional equivalent of a local minimum trouble in one variable), and we will not get stuck, unable to go to a lower point. It was partly for this reason that we reduced the polynomial to one which had no common factor of the exponents of the nonzero terms.

6.9 MULTIPLE ZEROS

We have developed the theory for the linear and quadratic factors so that they parallel each other. Thus there is little more to be said about finding multiple quadratic factors except to say we do the analogous steps.

We should not become too excited about how accurate in fact are various multiple or close zeros. Consider, for example, the two functions

$$f_1 = (z + 1)^8 = z^8 + 8z^7 + 28z^6 + 56z^5 + 70z^4 + 56z^3 + 28z^2 + 8z + 1$$
$$f_2 = (z + 1)^8 + (\tfrac{1}{10})^8 = z^8 + \cdots + 1 + (\tfrac{1}{10})^8$$

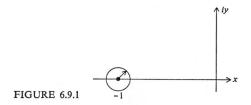

FIGURE 6.9.1

To eight decimal places they are the same; yet the first has an eightfold zero at $z = -1$, while the other has its zeros on a circle of radius $1/10$ about $z = -1$. Evidently near multiple zeros it is difficult to know what is really happening—but often in practice it does not matter. If it does, then probably the problem was posed wrong! But for computing purposes we must report them as multiple zeros if we are to avoid subsequent roundoff troubles, and the multiple zero can be located reasonably accurately.

7

LINEAR EQUATIONS AND MATRIX INVERSION

7.1 INTRODUCTION

The problems of simultaneous linear equations and matrix inversion have been exhaustively studied, and yet our knowledge is not all that we wish it were. Indeed, the very volume of papers indicates the difficulties of the two problems. We shall begin by presenting the simpler, rather widely accepted views of the matter and gradually point out the deficiencies in our theories that arise in some circumstances. In spite of the many objections that we shall raise for which there are at present only a few partial answers, the present methods often provide satisfactory results. Many of the points we wish to make will be illustrated by simple examples (it does not take a big machine to study and do research in numerical analysis), but very little is known as to how typical the examples are in practice.

7.2 GAUSSIAN[1] ELIMINATION—SIMPLIFIED VERSION

The solution of a set of simultaneous linear equations by the successive elimination of the unknowns is taught in beginning algebra courses and is known as *Gaussian elimination*. We consider first a simple concrete example:

$$3x - 4y + z = 1$$
$$2x + y + 2z = 3$$
$$x + 2y - z = 5$$

First divide the top equation by 3:

$$x - \tfrac{4}{3}y + \tfrac{1}{3}z = \tfrac{1}{3} \qquad (7.2.1)$$

and then subtract 2 times Eq. (7.2.1) from the second and also subtract Eq. (7.2.1) from the third, thus eliminating x from the last two equations.

$$x - \tfrac{4}{3}y + \tfrac{1}{3}z = \tfrac{1}{3} \qquad (7.2.2)$$
$$\tfrac{11}{3}y + \tfrac{4}{3}z = \tfrac{7}{3}$$
$$\tfrac{10}{3}y - \tfrac{4}{3}z = \tfrac{14}{3}$$

Next divide the second equation by $\tfrac{11}{3}$

$$y + \frac{4}{11}z = \frac{7}{11} \qquad (7.2.3)$$

and subtract 10/3 of this from the third equation to get

$$-\frac{84}{33}z = \frac{84}{33}$$

Thus

$$z = -1$$

Substitute $z = -1$ in Eq. (7.2.3) to get

$$y = \frac{7}{11} - \frac{4}{11}z = 1$$

and substitute the values of y and z in Eq. (7.2.2) to get

$$x = \tfrac{1}{3} + \tfrac{4}{3}y - \tfrac{1}{3}z = \tfrac{1}{3} + \tfrac{4}{3} + \tfrac{1}{3} = 2$$

Let us generalize this familiar process. Given N equations in N unknowns $x_1, x_2, \ldots, x_N$,

$$a_{1,1}x_1 + a_{1,2}x_2 + a_{1,3}x_3 + \cdots + a_{1,N}x_N = b_1$$
$$a_{2,1}x_1 + a_{2,2}x_2 + a_{2,3}x_3 + \cdots + a_{2,N}x_N = b_2$$
$$\cdots\cdots\cdots\cdots\cdots\cdots\cdots\cdots\cdots\cdots\cdots\cdots\cdots\cdots\cdots\cdots$$
$$a_{N,1}x_1 + a_{N,2}x_2 + a_{N,3}x_3 + \cdots + a_{N,N}x_N = b_N$$

[1] (Johann) Carl Friedrich Gauss (1777–1855).

we divide the first equation by $a_{1,1}$ and then subtract $a_{2,1}$ times this first result from the second equation, $a_{3,1}$ times the initial result from the third, etc., until we have $N - 1$ equations in the $N - 1$ variables $x_2, x_3, \ldots, x_N$. Using these $N - 1$ equations, we eliminate x_2 in the same way, leaving $N - 2$ equations in $x_3, \ldots, x_N$. Repeating this process a total of $N - 1$ times, we finally come down to one equation in the variable x_N.

In the *back solution* we solve the last equation for x_N, put this in the next-to-last equation to get x_{N-1}, put these two into the third from the last equation to get $x_{N-2}, \ldots$, until we have all the unknowns (determined in the order $x_N, x_{N-1}, \ldots, x_1$).

We now have our solution.   The amount of arithmetic is roughly[1]

$$N^2 + (N - 1)^2 + (N - 2)^2 + (N - 3)^2 + \cdots + 1 = \frac{N(N + 1)}{2} \frac{2N + 1}{3}$$

additions and the same number of multiplications—a total of approximately $2N^3/3$ operations in the forward part of the solution.   The back solution has of the order of N^2 operations, and for large N this is not significant.

Sometimes the value of the determinant of the system of equations is needed. When we divide an equation by its leading coefficient, we alter the value of the determinant by this amount.   But when we multiply an equation by a constant *and* add it to another equation, we do not alter the value of the determinant.   Thus the value of the determinant is the product of the N divisors we use.

PROBLEMS

Solve and find the determinant value:

7.2.1 $3x + 6y + 9z = 39$
        $2x + 5y - 2z = 3$
        $x + 3y - z = 2$

7.2.2 $x + y + z = 6$
        $2x + 3y + z = 1$
        $x - y + z = 3$

7.2.3 $x - y + z = 1$
        $2x + 2y - z = 3$
        $3x - 3y + z = 1$

7.2.4 $ax + by = e$
        $cx + dy = f$

---

[1] See Sec. 11.3 for the summation formula.

7.3 PIVOTING

If the solution of simultaneous linear equations is this simple, then why give all this attention to the problem? The reason is simply that in the above description we have overlooked many important details.

In the first place we ignored the possibility that one of the N coefficients that we were dividing by might have been zero. The answer is, of course, to pick another nonzero coefficient, but which one? The customary description says to pick as our *pivot* (as it is called) the largest in the column of the unknown you are eliminating. This is called *partial pivoting*.

This in turn raises the question of in which order the unknowns should be eliminated. Why the order $x_1, x_2, \ldots, x_N$? If instead of looking only in the next column we look at the array of all the coefficients $a_{i,j}$ and pick the largest as the pivot, then this is called *complete pivoting*. It is generally agreed that while complete pivoting is not worth the very large effort necessary to find the largest of all those left at each stage, partial pivoting is worth the effort.

Partial pivoting can be regarded as a reordering of the equations, though this should not be done in the machine since it is apt to confuse the user of the results if he ever wants to see how the elimination went in his problem. Complete pivoting can be regarded as reordering both the equations *and* the subscripts of the unknowns.

The conventional argument for using the largest element as the pivot goes as follows. Selecting the largest element means that at each stage when it is combined with another equation, the pivoting equation will be multiplied by a number less than 1 in size, suggesting that the roundoff will therefore be less. But this is not true in a very real sense. Let L_i be the ith equation and c be the multiplier. If we eliminate one way, we get (symbolically)

$$cL_i + L_j = L_j' = \text{the new equation}$$

while if we eliminate using the *same* two equations but in the reverse way, we get

$$L_i + \frac{1}{c} L_j = \frac{1}{c} L_j' = \text{the new equation}$$

It should be obvious that these two equations have the same relative floating-point errors but different fixed-point errors. Thus the widely used argument that we should use large pivots *because* they reduce the roundoff is not sound for floating-point computation.

PROBLEMS

Solve, using pivoting:

7.3.1 $x + y + z = 6$
$2x - y + z = 3$
$3x + 2y - z = 4$

7.3.2   Prob. 7.2.3

7.4   GAUSS-JORDAN ELIMINATION

If instead of the back solution we use the last equation to eliminate x_N in the top $N - 1$ equations and then use x_{N-1} in the next-to-last equation to eliminate all the x_{N-1}'s, etc., we will come to a diagonal system of equations (Fig. 7.4.1) with

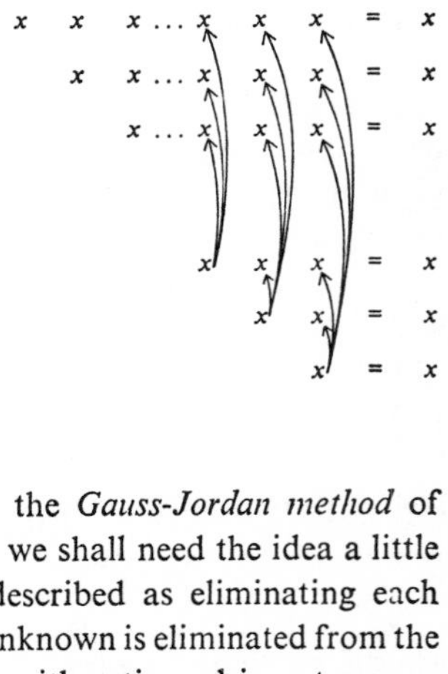

FIGURE 7.4.1

the solution explicitly given.   This is known as the *Gauss-Jordan method* of solution.   It is not recommended in practice, but we shall need the idea a little later.   Sometimes the Gauss-Jordan method is described as eliminating each unknown from the earlier equations just after the unknown is eliminated from the equations below it.   This, clearly, requires more arithmetic and is not recommended.

7.5   SCALING

The method of pivoting used words like "pick the largest element," but if instead of writing down an equation we wrote down c times the equation, we would quite possibly get a different pivot.   Clearly the algorithm is not invariant as it stands; what we do will depend in many problems on the haphazard way the equations were written down.   This suggests that some initial scaling is necessary before we begin the solution process.   Scaling can be regarded as trying to make the algorithm invariant.

We can clearly multiply the ith equation by any nonzero constant r_i we please and mathematically still have the same set of equations, and we can replace the jth variable by $c_j x_j$, which effectively multiplies the jth column by c_j.   And we could, if we pleased, multiply the right-hand sides b_i by a constant, as well as multiply all the numbers in the problem by a constant M.   These give the transformations that we intuitively feel leave the system of equations the same and for which we wish our algorithm to be invariant.

If we were to pick the same pivots each time (as judged by the subscripts i, j), then the algorithm would be invariant under all transformations of the same problem, though it will take a little thought on the reader's part to follow the elimination process and the back solution processes through to convince himself of this.   The r_i merely affect the multipliers used to do the elimination, the c_j change all the numbers in the jth column by the same factor, the right-hand sides similarly are merely scaled by a multiplication constant, and the multiplier M has no real effect.

We face, therefore, the problem of making some sense of the words "pick the largest element in the next column as the next pivot."   It is conventional to say that the transformations should be used to scale the rows so that the maximum in each row is around 1 in size and then we scale the columns similarly (or else scale the columns and then the rows).   Sometimes it is observed that we should not actually perform the scaling, as this will cause unnecessary roundoff, but merely store away the scaling factors and use them when we come to "pick the largest . . . ."

As we shall later show, scaling by rows and then columns, or scaling by columns and then rows, does not lead to the same result; indeed, one way may lead to disaster while the other does not.   Thus the usual advice about scaling is both superficial and misleading.

If this is so, then what about the whole method of Gaussian elimination? Before we panic, it should be observed that many times the problem comes in a form where the equations have been "naturally scaled" by the way they came about, especially in equations that arise in mathematical problems.   But there are times when the equations do indeed need some preliminary scaling before starting the elimination process.   Section 7.6 discusses one possible approach to scaling that produces a unique result and has the property that any time you alter the system by any of the transformations:

Interchange of rows
Multiplication of a row by a nonzero constant r_i

Interchange of columns

Multiplication of a column (including the right-hand side) by a nonzero constant c_i

Multiplication of everything by a constant M

the scaling method will, within roundoff, remove the transformation effects, thus producing the basis for an invariant algorithm.

7.6 INVARIANT SCALING—ANALYSIS OF VARIANCE

The method of analysis of variance in statistics provides a natural tool for the invariant scaling of a set of simultaneous linear equations.

We suppose, first, that all the $a_{i,j} \neq 0$ and set

$$|a_{i,j}| = 2^{d_{i,j}}$$

We imagine multiplying each row by 2^{r_i} ($i = 1, 2, \ldots, N$), each column (including the right-hand side) by 2^{c_j} ($j = 1, 2, \ldots, N + 1$), and everything by 2^M. The exponent in position i, j is now

$$d_{i,j} + r_i + c_j + M$$

If we now minimize the sum of the squares of all the exponents, we will find values such that, in the least-squares sense, the exponents will be as close to zero as they can be made to be; that is, the variance of the exponents is minimized.

Therefore we minimize

$$m = \sum_{j=1}^{N+1} \sum_{i=1}^{N} (d_{i,j} + r_i + c_j + M)^2$$

Differentiating with respect to each of the unknowns r_i and c_j, we get

$$\frac{\partial m}{\partial r_i} = 2 \sum_{j=1}^{N+1} (d_{i,j} + r_i + c_j + M) = 0 \qquad i = 1, 2, \ldots, N$$

$$\frac{\partial m}{\partial c_j} = 2 \sum_{i=1}^{N} (d_{i,j} + r_i + c_j + M) = 0 \qquad j = 1, 2, \ldots, N + 1$$

We have more than enough parameters, and so we arbitrarily set

$$M = \frac{-1}{N(N+1)} \sum_{j=1}^{N+1} \sum_{i=1}^{N} d_{i,j}$$

The solution to these equations is

$$r_i = \frac{-1}{N+1} \sum_{j=1}^{N+1} (d_{i,j} + M) \qquad \text{hence } \sum_{i=1}^{N} r_i = 0$$

$$c_j = \frac{-1}{N} \sum_{i=1}^{N} (d_{i,j} + M) \qquad \text{hence } \sum_{j=1}^{N+1} c_j = 0$$

as can be seen by substituting these into the derivative equations

$$\sum_{j=1}^{N+1} [d_{i,j} - (d_{i,j} + M) + 0 + M] = 0$$

$$\sum_{i=1}^{N} [d_{i,j} + 0 - (d_{i,j} + M) + M] = 0$$

Thus r_i and c_j are the appropriate negative averages of

$$d_{i,j} + M$$

and M is the negative of the average over all the $d_{i,j}$.

This approach minimizes the variance of the scaled exponents, which seems like a good thing to do. However, we are unable to connect this property directly with the subsequent elimination process, or at present with any other method of solution, so that the method is merely suggestive and plausible.

If any of the coefficients of the original system of equations were zero, then the corresponding value of $d_{i,j}$ would be minus infinity, and we exclude them as "missing data." With data missing the solution of the system is harder to find. We refer the reader who wishes to pursue this unconventional scaling method to the standard books on the topic of analysis of variance. We note, however, that we are merely looking for a reasonable solution to these equations to get some way of reasonably scaling the original system. Note that we do not need to get integral solutions to minimize the roundoff due to scaling, because we intend not to actually scale the equations but only to use these numbers to pick the pivots in the original system of equations

7.7 RANK

During Gaussian elimination on a system of linear equations it may happen that at some point all the rest of the coefficients in a column are zero. This means that the rank of the matrix of coefficients is not N but lower. Indeed, in principle we may find at several stages that we are blocked by a column of zeros. Equally likely we may find a row, or rows, of zeros blocking us.

In the case of finding a row of zeros this clearly means that the linear combination of the earlier rows used is equal to the row of all zeros; it means that there is a linear dependence and that the rank is not N.

Most of the time the proposer of the problem believed that the system of equations was of full rank and that it would determine a unique solution. The fact that the system is not of full rank does not end the problem, since the proposer needs to know what the dependence is in order to go back and reformulate his problem. How can we find the dependence? We could follow the elimination process to find the linear combination that produced it. Alternatively, we can drop one of the earlier rows and try again; if at the same point we again find all zeros, then we know that the dropped row is not part of the dependence. But if we do not find all zeros, then it is part of the dependence. Dropping one row at a time, we can find the minimal linearly dependent set. Then it is a matter of finding the actual coefficients of these equations, which can be done by keeping track of the coefficients used in the elimination process.

It is a central theorem of linear algebra that the row and column ranks are the same, which means that if we find a row dependence, then there must be a corresponding column dependence, and vice versa. Evidently we can operate on the columns as we did the rows to find a column dependence.

Mathematics often does not make distinctions that some applications require. For example, Euclidean geometry does not distinguish between left-handed and right-handed triangles: given three sides of one triangle equal to three sides of another triangle, the triangles are congruent (see **Fig. 7.7.1**). But for many applications (such as wearing gloves, studying some chemical compounds, etc.) it is necessary to distinguish between the two different orientations. For these applications Euclidean geometry must be altered a bit.

Similarly for simultaneous equations: though in principle the row and column ranks are the same, in practice the meaning to be attributed to one may be different from the other. In practice a simple row dependence, say two rows are the same, is quite different from two columns being the same.

Row dependence	*Column dependence*

$$2x + 4y + 6z = 10 \qquad 2x + 4y + 6z = 10$$
$$x + 2y + 3z = 5 \qquad\quad x + 2y - z = 4$$
$$\text{pivot}\quad x - y - z = -7 \qquad \text{pivot}\quad x + 2y + 7z = 6$$
$$x + 2y + 3z = 5 \qquad\quad x + 2y + 3z = 5$$
$$0y + 0z = 0 \qquad\qquad 0y - 4z = -1$$
$$-3y - 4z = -12 \qquad\quad 0y + 4z = 1$$

row of zeros   column of zeros

FIGURE 7.7.1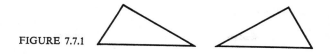

This difference means that in partial pivoting when we find a zero element, we should at that point check to see if the whole row is zeros (we will automatically find the column of zeros if that is the case). When we find a dependence, we should isolate it as best we can and exhibit it to the user. We should, probably, continue to search for further dependencies before stopping.

Sometimes it is necessary to carry out the solution of a system of linear equations when the rank is less than N. We cite without proof the main result for this. If the rank of the system is $r < N$, then some $N - r$ of the variables may be transferred to the right-hand side and the system solved in terms of them as if they were parameters. If the rank of the augmented matrix is the same as the rank of the matrix, then there is an $(N - r)$-parameter solution; otherwise the equations are inconsistent.

PROBLEMS

7.7.1 Find the rank and linear dependence.
$$x + y + z = 2$$
$$2x + 4y - 6z = 8$$
$$2x + 3y - 2z = 6$$

7.7.2 Solve the system
$$x + y = a$$
$$2x + 2y = 2a$$

7.7.3 Find the linear dependence, both row and column, of:
$$x + 3y - z = 5$$
$$2x - y - 2z = 7$$
$$4x + 5y - 4z = 17$$

7.8 ILL-CONDITIONED SYSTEMS

All the above has been carried out in terms of exact numbers. The floating-point numbers we use will generally not produce rows or columns of exact zeros, and we will have trouble determining the rank. Indeed, it may well be that the row and column ranks are not the same as judged by our tests for rank. How to resolve this kind of trouble is not very well understood at present.

Given the system

$$x + \tfrac{1}{3}y = 1$$
$$2x + \tfrac{2}{3}y = 2$$

The rank is 1.
Put the system in a computer:

Note coefficients

$$0.100 \times 10^1 x + 0.333 \times 10^0 y = 0.100 \times 10^1$$
$$0.200 \times 10^1 x + 0.667 \times 10^0 y = 0.200 \times 10^1$$

Now the rank is 2, and we can solve the system uniquely!

$$0.100 \times 10^{-2} y = 0$$
$$\begin{cases} y = 0 \\ x = 1 \end{cases}$$

Our trouble with determining the rank plus our trouble with initial scaling to give pivoting some meaning indicate that we need to find something to replace these two ideas. This needed idea is "ill-conditioning," and as used in the literature it is ill-defined. Ill-conditioned means, vaguely, that small changes in the given numbers will produce large changes in the results.

The ill-conditioning may arise from three distinct sources. First, it may be the original problem itself. Thus a pencil balanced on its point (Fig. 7.8.1) presents us with a situation in which small changes in the initial position will result in large changes in the subsequent position in a short time. This is usually referred to as an *unstable system*.

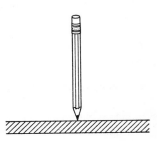

FIGURE 7.8.1

Second, the trouble may be not in the problem but in the representation used. For example, if we solve the differential equation

$$\frac{d^2y}{dx^2} = y \qquad y(0) = 1 \qquad y(0) = -1$$

by using the hyperbolic functions as our basis for representation, then the solution is

$$y = \cosh x - \sinh x$$

and we will indeed see that small changes give rise to large differences later (Fig. 7.8.2). But if we choose the exponential functions (which are mathematically completely equivalent) as a basis, we get

$$y = e^{-x}$$

For large values of x there is no trouble with this representation.

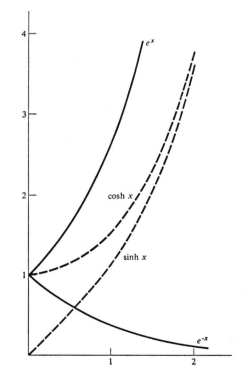

FIGURE 7.8.2

Third, the source of the ill-conditioning may be in the method of solution. Consider, for example, the following symmetric system of equations which appear to be reasonably scaled.

$$
\begin{aligned}
3x + 2y + z &= 3 + 3\varepsilon \\
2x + 2\varepsilon y + 2\varepsilon z &= 6\varepsilon \qquad\qquad \varepsilon \text{ small} \\
x + 2\varepsilon y - \varepsilon z &= 2\varepsilon
\end{aligned}
$$

$$
\begin{aligned}
x + \tfrac{2}{3}y + \phantom{(-\tfrac{4}{3}+2\varepsilon)}\tfrac{1}{3}z &= 1 + \varepsilon \\
(-\tfrac{4}{3} + 2\varepsilon)y + (-\tfrac{2}{3} + 2\varepsilon)z &= -2 + 4\varepsilon \\
(-\tfrac{2}{3} + 2\varepsilon)y + (-\tfrac{1}{3} - \varepsilon)z &= -1 + \varepsilon
\end{aligned}
$$

For small ε these are ill-conditioned and are almost linearly dependent.   *But* if we first eliminate x between the original second and third equations, we get

$$\varepsilon y - 2\varepsilon z = -\varepsilon$$

In floating-point arithmetic this is equivalent to

$$y - 2z = -1$$

and we can now solve, using the first two equations with this equation to get

$$y = z = 1$$

Put these in the first equation and we will have roundoff trouble, but in either of the other two we get accurately

$$x = \varepsilon$$

The fact that the system may be well conditioned and apparently reasonably scaled and that still the method of Gaussian elimination appears ill-conditioned is apparently not widely known nor its importance appreciated.

The conventional "scale by rows and then by columns, or vice versa," may cause ill-conditioning.   Given the matrix of coefficients

$$
\begin{pmatrix}
3 & \dfrac{1}{\varepsilon} & \dfrac{1}{\varepsilon} \\[2mm]
1 & 2\varepsilon & \varepsilon \\[2mm]
2 & \varepsilon & \varepsilon
\end{pmatrix}
$$

where ε is small and due to roundoff effects we equate

$$1 \pm \varepsilon^2 = 1$$

Scale by rows first   Scale by columns first

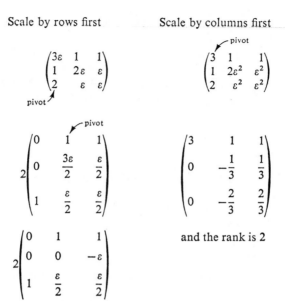

and the back
substitution is easy   and the rank is 2

PROBLEMS

7.8.1 Devise another example of apparently scaled systems which are ill-conditioned by
straight Gaussian elimination but are in fact not ill-conditioned when solved
properly.

7.8.2 Scale the "row and column scaling" example by the method of Sec. 7.6.

7.9 THE RIGHT-HAND SIDES CAN
CAUSE ILL-CONDITIONING

It is widely believed that ill-conditioning depends *only* on the coefficients of the
unknowns, but this is false as the following example shows.   Take the same
symmetric system as in Sec. 7.8 and change the right-hand sides to form the
system

$$3x + 2y + z = 6$$
$$2x + 2\varepsilon y + 2\varepsilon z = 2 + 4\varepsilon$$
$$x + 2\varepsilon y - \varepsilon z = 1 + \varepsilon$$

The solution is $x = y = z = 1$, and the system is indeed ill-conditioned because we are unable to solve the system so that the factor ε can be *divided out* as we did before *without* depending on the exact cancellation of the integer parts on the right-hand side (which, of course, will produce a large shift in the floating point addition) and hence depend on the rounded-off parts of the ε terms.   As a result the answers are sensitive to relatively small changes in the coefficients of x.

PROBLEMS

7.9.1   Explain " why " the above example works.   Devise another example showing this effect.

7.9.2   It is often thought that it is the angle between the lines that makes ill-conditioning. For $N = 2$ show that this statement is false.

7.10   A DISCUSSION OF GAUSSIAN ELIMINATION

Gaussian elimination is the main tool for solving simultaneous linear equations. In spite of all the faults and misconceptions about the method it is often reasonably effective.   How can this be ?   As we observed earlier, many systems are naturally scaled by the source from which they come.

Consider pivoting next.   Once the system is scaled, the value of the determinant is fixed, and the larger the early pivots are, the smaller the later ones must be. It is not correct that big pivots are good and small pivots are *automatically* bad; rather, large numbers become large *only* through multiplication by large numbers, while small numbers can be small because of multiplications by small numbers *or else* by heavy cancellation.   Thus a large number is comparatively safe, while a small one is suspect.   By picking the large numbers as pivots we tend to avoid the numbers that were formed by heavy cancellation.

But we may doubt the wisdom of the basic method.   Given a system of equations and the general processes used in eliminating one variable, if we ask how to produce the maximum amount of linear dependence in the next stage, then we would be tempted to use one equation to eliminate from all the others so that at the next stage everywhere we look we would see this same one equation—that is, we would do something like pivoting.   It is possible instead to eliminate a variable using adjacent equations thus avoiding the same equation in every derived equation.   We can clearly see a mechanism that produces ill-conditioning in Gaussian elimination if we imagine a system of equations scaled and with the pivot in the top equation, but also with the largest numbers in two other columns being in the top equation (Fig. 7.10.1). When we use the pivot equation, these largest two numbers tend to dominate all the derived equations, thus producing a linear (column) dependence. Indeed, looking at some of the earlier examples, we can see this is the mechanism behind why the examples "worked" (i.e., failed).

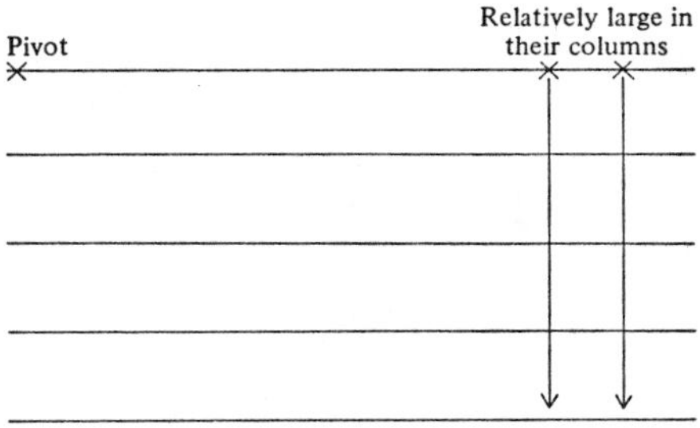

FIGURE 7.10.1

The user should be wary of the use of bounds as a measure of the quality of the results and especially as the basis for a comparison of two methods. Bounds tell only part of the story. The answer may be very much better than the careful bounds indicate. Backward analysis which pushes the errors committed back onto the original coefficients and states that the numbers obtained are a solution to a nearby system of equations is of little use *unless* the zero coefficients are kept as zero in the nearby system. (We should use the floating-point concept of relative error and not the fixed-point idea of additive error—a point that is rarely noticed in backward analysis.)

In many places in the literature the condition of a sytem of equations is given in terms of the singular values of the coefficient matrix. The singular values of a matrix A are the eigenvalues of $A^T A$ (see Chap. 45). Of course if A is symmetric, then the singular values are just the squares of the eigenvalues. But Sec. 7.9 shows that this cannot be the whole story.

If the coefficient matrix is symmetric, then it can be written as the sum of its principal idempotents (projections), each multiplied by an eigenvalue. The accuracy of the solution can then be related to this expression, as is well summarized by Richard Rosanoff[1]:

[1] Rosanoff and Nishimoto, Zeus on the Leus, Space Division, North American Rockwell Corp.

> Idempotents summed are the
> Array baby,
> When truncated they create
> Dismay baby,
> You cannot invert them
> Any way baby,
> Gauss can't give you anything
> But love.

7.11   MATRIX INVERSION 1

The problem of finding the inverse of a square matrix $A = (a_{i,j})$ is very common. Since matrices are not in general commutative, there is the idea of a left inverse and a right inverse

$$A_L^{-1}A = I$$
$$AA_R^{-1} = I$$

It is an important mathematical result that these two inverses are the same.   The proof follows easily from the associativity.

$$A_L^{-1}AA_R^{-1} = \begin{cases} (A_L^{-1})(AA_R^{-1}) = (A_L^{-1}I) = (A_L^{-1}) \\ (A_L^{-1}A)(A_R^{-1}) = (IA_R^{-1}) = (A_R^{-1}) \end{cases}$$

However, this may be very far from true for some computed matrices.   An inverse may be close when multiplied on one side (in the sense that the off-diagonal elements are small) but very far when multiplied on the other side.   For example, let

$$A = \begin{pmatrix} 1.00 & 1.00 \\ 1.00 & 0.99 \end{pmatrix} \qquad C = \begin{pmatrix} -89 & 100 \\ 90 & -100 \end{pmatrix}$$

then

$$AC = \begin{pmatrix} 1.0 & 0.0 \\ 0.1 & 1.0 \end{pmatrix} \qquad CA = \begin{pmatrix} 11 & 10 \\ -10 & -9 \end{pmatrix}$$

Given the coefficients of a square matrix $A = (a_{i,j})$, if we were to solve the systems with the right-hand sides

first	$1, 0, 0, \ldots, 0$
second	$0, 1, 0, \ldots, 0$
third	$0, 0, 1, \ldots, 0$
etc.	

and arrange the solutions in a matrix form with the solutions as columns, then we would have the inverse. To see why this is so consider what you would do if you had these special fundamental solutions and were asked to solve the same system with the right-hand sides $b_1, b_2, b_3, \ldots, b_N$. You would multiply these fundamental solutions in turn by $b_1, b_2, \ldots, b_N$ and add the results.

Thus, given

$$Ax = b$$

we have

$$x = A^{-1}b$$

which is what you did to get the answer—you multiplied the (b_i) by a matrix.

PROBLEMS

7.11.1 Find an example when AA^{-1} and $A^{-1}A$ are very different.

7.11.2 Find the inverse of

$$\begin{pmatrix} 1 & 0 & 1 \\ 0 & 1 & 0 \\ -1 & 0 & 1 \end{pmatrix}$$

7.11.3 Find the inverse of

$$\begin{pmatrix} 0 & 2 & 3 \\ 3 & 0 & 2 \\ 3 & 2 & 0 \end{pmatrix}$$

7.12   MATRIX INVERSION 2

A completely equivalent method of finding the inverse is to write the $2N \times N$ matrix

$$\begin{pmatrix} a_{1,1}a_{1,2} & \cdots & a_{1,N} & 1 & 0 & \cdots & 0 \\ a_{2,1}a_{2,2} & \cdots & a_{2,N} & 0 & 1 & \cdots & 0 \\ \cdots\cdots\cdots\cdots\cdots\cdots\cdots\cdots\cdots\cdots\cdots\cdots \\ a_{N,1}a_{N,2} & \cdots & a_{N,N} & 0 & 0 & \cdots & 1 \end{pmatrix}$$

and to use the steps of the Gauss-Jordan elimination to transform the first N columns into the identity matrix. We will find that the last N columns are the inverse of the first N columns. To see this we must first recognize that each step we used can be written as multiplication on the left by some matrix. When we are done, we can, by the associative law of multiplication, gather all the matrices we used into one single matrix, say C.   We have

$$CA = I$$

hence

$$C = A^{-1}$$

But the matrix C also operated on the last N columns (the identity matrix), so when we are done, we have A^{-1} in the last N columns.

EXAMPLE 7.1   Given the matrix

$$\begin{pmatrix} 1 & 1 & 4 \\ -1 & 0 & 1 \\ 1 & -1 & -5 \end{pmatrix}$$

form

$$\begin{pmatrix} 1 & 1 & 4 & 1 & 0 & 0 \\ -1 & 0 & 1 & 0 & 1 & 0 \\ 1 & -1 & -5 & 0 & 0 & 1 \end{pmatrix}$$

$$\begin{pmatrix} 1 & 1 & 4 & 1 & 0 & 0 \\ 0 & 1 & 5 & 1 & 1 & 0 \\ 0 & -2 & -9 & -1 & 0 & 1 \end{pmatrix}$$

$$\begin{pmatrix} 1 & 1 & 4 & 1 & 0 & 0 \\ 0 & 1 & 5 & 1 & 1 & 0 \\ 0 & 0 & 1 & 1 & 2 & 1 \end{pmatrix}$$

$$\begin{pmatrix} 1 & 1 & 0 & -3 & -8 & -4 \\ 0 & 1 & 0 & -4 & -9 & -5 \\ 0 & 0 & 1 & 1 & 2 & 1 \end{pmatrix}$$

$$\begin{pmatrix} 1 & 0 & 0 & 1 & 1 & 1 \\ 0 & 1 & 0 & -4 & -9 & -5 \\ 0 & 0 & 1 & 1 & 2 & 1 \end{pmatrix}$$

The inverse is

$$\begin{pmatrix} 1 & 1 & 1 \\ -4 & -9 & -5 \\ 1 & 2 & 1 \end{pmatrix}$$ ////

PROBLEMS

Using the method of Sec. 7.12, find the inverse of the following matrices and check the results.

7.12.1 $\begin{pmatrix} a & b \\ c & d \end{pmatrix}$

7.12.2 $\begin{pmatrix} 1 & 0 & -1 \\ 1 & 1 & 0 \\ 0 & 1 & 1 \end{pmatrix}$

7.12.3 $\begin{pmatrix} a & 0 & 0 \\ 0 & b & 0 \\ 0 & 0 & c \end{pmatrix}$

7.12.4 $\begin{pmatrix} 2 & 3 & 5 \\ 7 & 11 & -13 \\ 17 & -19 & 23 \end{pmatrix}$

REFERENCES

Probably the two best references to this highly developed field are Forsythe [11] and Noble [46].

8

*RANDOM NUMBERS

8.1 WHY RANDOM NUMBERS?

Randomness is traditionally looked upon as something to be avoided if possible. The idea that randomness may be used constructively strikes many people as strange. However, many parts of modern computing practice use random numbers, and we shall give a few examples in this chapter.

In spite of formal axioms, probability is an intuitive idea, just as in Euclidean geometry the ideas of point, line, and plane are intuitive. Randomness is closely connected with probability. As an example of how the modern, complex idea of a random process can arise, consider the simple act of "tossing a well-balanced coin." If we record a 1 for a head and a 0 for a tail, we have a simple random variable. If next we consider not a single toss but a sequence of tosses, then we have a sequence of 0s and 1s, which we may regard as a binary number. Thus we are led by easy stages into gradually considering an indefinite number of tosses and then to regarding the resulting sequence of 0s and 1s as the binary representation of a number between 0 and 1. As a result we have for each of the nondenumerable numbers in $0 \leq x < 1$ a corresponding sequence of tosses of a

simple coin.   This is the idea of a random, or stochastic, process; corresponding
to some index λ we have a family of functions

$$f_\lambda(t)$$

where the variable t is often regarded as time, either continuous or at discrete
intervals.

   With our finite machine we cannot hope to actually carry out the mathematician's random process; instead we will approximate it by some finite sequence
of finite numbers.   Furthermore, as we shall show, in most cases we will not deal
with what is intuitively random numbers, but rather we will generate a perfectly
definite sequence that is completely predictable.   This approach violates the view
of random that is usually equated with nonpredictable.   However, in practice
we do not care if the numbers are theoretically predictable.   We care only that
with regard to their use they are random.

   The idea of a definite random process is almost a contradiction, and so we
must approach the matter slowly.   First we examine a couple of very simple
examples to see how random numbers might be used in many applications and
why we are not alarmed at their actual structure in these cases.   Evidently we
would not use a predictable sequence of random numbers in such applications as
lottery selections and other processes that interact directly with humans; we use
them only with machines and then *only* with care.

   After looking at the simple uses, we will look both at possible sources of
uniformly distributed random numbers and at why we prefer a predictable source.
Only then will we look at how to generate them and prove that they have certain
properties.   Since it is a difficult topic, we will give a few moments' consideration
to the complex matter of testing random numbers.

   We then look at various distributions other than the flat one and at ways of
using various tricks from statistics to aid in saving machine time.   As is so often
the case, what we cover in one short chapter can be, and has been, expanded into
a whole book.

8.2   SOME USES OF RANDOM NUMBERS

The most obvious use of random numbers is in the simulation of random processes such as nuclear disintegrations, people coming at random for a particular
service, telephone traffic studies, and making random choices in some decision
process, say playing a game.

   Less obvious are the so-called "Monte Carlo" applications in which a
definite process is replaced by a random process that arrives at the same result.

The most famous example is probably the Buffon[1] needle.   In 1773 Buffon observed that if a needle of length $L \leq 1$ were tossed at random onto a horizontal surface ruled with equally spaced lines, say at unit spacing, then the probability of the needle crossing a line is

$$\text{Probability} = \frac{2L}{\pi}$$

He reasoned, therefore, that he could experimentally determine the value of π by making repeated trials.

Let us derive this result (see Fig. 8.2.1).   The position of the needle's center is within 1/2 unit of some line, and the words " tossed at random " mean that the probability is uniform in the interval $0 \leq x \leq \frac{1}{2}$.   The angle θ that the needle

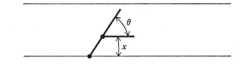

FIGURE 8.2.1

makes with the direction of the lines varies in the interval $0 \leq \theta < \pi$, and again the words " tossed at random " are taken to mean that the probability distribution is uniform in the interval.   A crossing will occur if and only if

$$x \leq \frac{L}{2} \sin \theta$$

From Fig. 8.2.2 we see that this probability is the ratio of the shaded area to the area of the whole rectangle

$$\text{Probability} = \frac{\int_0^\pi (L/2)\sin \theta \, d\theta}{\int_0^\pi \frac{1}{2} \, d\theta} = \frac{2L}{\pi}$$

If the value of π were inaccurately known, then this would be a good method for finding that it is around 3.   With some care in drawing the lines, measuring the needle and the crossings, and tossing at random, we could get perhaps 3.1 with some reliability; but to get 3.14 would tax our abilities and patience.

Many generalizations of this famous needle problem have been used.   One,

[1] Georges Louis Leclerc, Comte de Buffon (1707–1788).

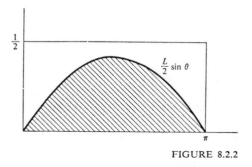

FIGURE 8.2.2

for example, estimates the amount of "grain boundary" in microphotographs of magnetic materials.

A similar Monte Carlo method,[1] this time estimating e, is based on the observation that if $2k$ numbers x_i are drawn in sequence from a random equilikely source, then the probability that they are all in ascending order $x_1, x_2 x_3 \cdots x_{2k}$ is

$$\frac{1}{(2k)!}$$

This probability is easily found from the observation that the $2k$ numbers x_i can have $(2k)!$ possible orderings and that only one ordering will be in ascending order. The probability that a sequence of trials yielding an increasing sequence of x_i's will fail on the odd trial $2k + 1$ is the difference

$$\frac{1}{(2k)!} - \frac{1}{(2k + 1)!}$$

Thus the total probability that a sequence of drawings of random numbers from an equilikely source will produce a rising sequence that ends with an even number of numbers is

$$\sum_{k=1}^{\infty} \left[\frac{1}{(2k)!} - \frac{1}{(2k + 1)!} \right] = \sum_{k=0}^{\infty} \left[\frac{1}{(2k)!} - \frac{1}{(2k + 1)!} \right] = \frac{1}{e}$$

An experiment using 252 runs gave $1/e = 0.381$ (a 3.5 percent error).

In both of these examples the experiment is a test more of the source of the random numbers used than of the result. Indeed, in the second case it provided a nice, simple check on the derivation of the formula—the test would have caught the elusive factor of 2, the slip in the minus sign, and many other typical human errors in mathematics. It is a common research practice to "experimentally

---

[1] Due to Roger Pinkham, (1929———).

verify some result " by some simple numerical tests before building further on the result, and Monte Carlo techniques are one more tool for doing this.

Another use of random numbers, which are most often random decimal digits (base 10) or random bits (base 2), is in game playing on a machine.   When things are equally favorable, then often it is desirable to make a random choice— this, for example, will tend to defeat the opponent who tries to play the same opening repeatedly until he finds a weakness in the formulas used by the machine.

The uncritical use of randomness to solve complicated problems should be tempered with a dose of reality.   It may well be true that by using random trials one can expect to find the solution, but the size of the task may be like the famous case of the monkeys and the typewriters.   In this story a number of monkeys are set down at typewriters, and we wait until they type out in sequence all the books in the British Museum in the card-catalog order.   There is a finite chance that the first letter will be typed correctly, and indeed, in the infinitude of time this will happen infinitely often.   Among this infinite number of cases there is a positive probability that the second correct letter will by typed next, and among all these infinitely many cases of two letters . . .   Often grandiose proposals to use a random search are this situation in some disguise!   A random search should not be started unless some careful estimates of the running time are made—otherwise you are apt to get a very large computing bill and waste a lot of time in the preparation of the program with little hope of getting a result.

PROBLEMS

8.2.1   Using the integral $-\int_0^\infty e^{-x} \ln x \, dx = \gamma$, set $e^{-x} = t$ and estimate γ, using 10 random points.

8.3   SOURCES OF RANDOM NUMBERS

The first source of random numbers likely to come to mind is some natural, physical phenomena such as cosmic rays, nuclear disintegrations, or perhaps sampling at random an alternating frequency.   Such devices have been built in the past, and they do have their uses.   Their main defect is that if they are used and something interesting happens in a computation, then it is practically impossible to repeat the run to find out how the effect came about.   This foolishness is repeatedly done; a clever programmer may realize that by consulting the clock time at widely separated intervals and using the last digit of the time, he has what is probably a source of random digits, but he will fail to realize the consequences for the research because he thereby removes the repeatability of the computation.

Tables of random digits have been published, the most famous being the Rand Table of One Million Random Digits and One Hundred Thousand Normal Deviates.[1] In principle such tables could be stored on a tape and read into a machine as needed, but this is rarely done because of the slowness of the input and because in large problems with random numbers 1 million digits is hardly enough. It is possible to use a table to create a new table, for example by adding adjacent numbers and dropping the carry, but this does not seem to be popular, although this device was used by Rand to make their numbers more random.

The third source is mathematical formulas. Many sources have been proposed, such as the old "in a k-digit machine, square the last random number and keep the middle k digits of the product as the new random number," but few are kept in use for long. The ones most often used are still multiplicative congruence generators of the form

$$x_{n+1} \equiv x_n \rho (\mathrm{mod}\ 2^k)$$

where we take the last k digits of the fixed-point product (on a k-digit binary machine). In words, "x_{n+1} is congruent to x_n modulo 2^k" means the difference between x_{n+1} and x_n is divisible by 2^k.

As an illustration of this kind of a random-number generator consider using 000 001 as x_0 and 100 101 as the multiplier on a six-bit machine. We get Table 8.3 of products. As we can easily see, all the numbers are of the form $4k + 1$.

[1] Free Press, Chicago, 1955.

Table 8.3

Binary number	Decimal equivalent	Drop last two digits
000 001	1	0
100 101	37	9
011 001	25	6
011 101	29	7
110 001	49	12
010 101	21	5
001 001	9	2
001 101	13	3
100 001	33	8
000 101	5	1
111 001	57	14
111 101	61	15
010 001	17	4
110 101	53	13
101 001	41	10
101 101	45	11
000 001	1	0

The last binary digit is hardly random, nor is the next to the last, but dropping both of them, we have the decimal representations of the numbers $0, 1, \ldots, 15$ in what looks like a random order. However, each number occurs only once in the cycle of 16 numbers, and that is definitely suspicious.

As we will prove in Sec. 8.4, using any $(k \geq 3)$-bit machine will have the same effects. All the numbers, once the last two digits are dropped, are represented exactly once, but otherwise the numbers seem to be in a chaotic order. The periodicity is for the typical 35-bit machine

$$2^{k-2} = 2^{33} \approx 8.59 \times 10^9$$

and seems to be adequately long for most simple applications.

PROBLEMS

8.3.1 Make a random-number generator using a 5-bit word length with $\rho = 13$ and $x_0 = 1$.

8.3.2 Find the period of a 6-bit random-number generator using $\rho = 33$ and $x_0 = 1$.

8.3.3 Using the generator in the text, remove the last 3 bits and study the probability of one number being followed by another, i.e., the correlation of adjacent numbers.

8.3.4 Repeat the random-number generator in the text except start with $x_0 = 3$. Compare the two runs.

8.4 THE RANDOM-NUMBER GENERATOR

We now study random-number generators of the form

$$x_n \equiv \rho x_{n-1} (\text{mod } 2^k)$$

where mod 2^k means "keep the remainder after dividing the product by 2^k," and this, in a binary machine, is simply the last k digits of the product, the lower part of the accumulator. How do we pick the multiplier ρ and the starting value, and what can we expect for the resulting sequence of random numbers?

If 2 divides x_0, then all the products will have at least one 0 at the end, and we are wasting one bit of the machine capacity. Hence we take x_0 as an odd number.

Similarly, if ρ is even, then the products will gradually accumulate 0s on the right-hand end until

$$x_{k+1} = \rho^{k+1} x_0 = 0 (\text{mod } 2^k)$$

and all numbers past this point will be zero.

Therefore ρ is taken as an odd number.

Now all odd numbers can be written in one of the forms

$$8t + 3 \qquad 8t + 1 \qquad 8t - 1 \qquad 8t - 3$$

for some integer t. How shall we pick t, and which of the four forms should we use?

Theorem 8.4.1 If $\rho = 8t \pm 3$, then the period of the sequence is 2^{k-2}.

PROOF If the period is to be 2^{k-2}, then the theorem states that

$$\rho^{2^{k-3}} \not\equiv 1 \pmod{2^k}$$
$$\rho^{2^{k-2}} \equiv 1 \pmod{2^k}$$

which is equivalent to saying 2^k divides $(\rho^{2^{k-2}} - 1)$ but does not divide $(\rho^{2^{k-3}} - 1)$.

We begin by observing that using

$$a^2 - 1 = (a + 1)(a - 1)$$

repeatedly we can write

$$(\rho^{2^{k-2}} - 1) \equiv (\rho^{2^{k-3}} + 1)(\rho^{2^{k-4}} + 1) \cdots (\rho + 1)(\rho - 1)$$

For each factor $(i \geq 1)$

$$\rho^{2^i} + 1 = 1 + (3 \pm 8t)^{2^i}$$
$$= 1 + 3^{2^i} + \sum_{k=1}^{2^i} (\pm 1)^k C(2^i, k)(8t)^k 3^{2^i - k}$$

But

$$1 + 3^{2^i} = 1 + (4 - 1)^{2^i} = 1 + (1 - 4)^{2^i}$$
$$= 1 + 1 + \sum_{k=1}^{2^i} (-4)^k C(2^i, k)$$

so that

$$\rho^{2^i} + 1 = 2 + \sum_{k=1}^{2^i} C(2^i, k)[(\pm 1)^k 3^{2^i - k} t^k 8^k + (-4)^k]$$

and 2 divides $\rho^{2^i} + 1$, but 4 does not.

Returning to the factored form, we use this in each of the first 2^{k-3} terms and conclude that

$$2^{k-3} \text{ divides } (\rho^{2^{k-3}} + 1)(\rho^{2^{k-4}} + 1) \cdots (\rho^2 + 1)$$

but 2^{k-2} does not divide it.

We have left the two factors

$$(\rho + 1)(\rho - 1) = \rho^2 - 1 = (8t \pm 3)^2 - 1$$
$$= 64t^2 \pm 48t + 9 - 1$$
$$= 8(8t^2 \pm 6t + 1)$$

and 2^3 divides it, but 2^4 does not.   Thus we have

$$2^{k-3} \cdot 2^3 = 2^k$$

divides the original

$$\rho^{2^{k-2}} - 1$$

but no higher power does.

Rearranged, we have

$$\rho^{2^{k-3}} - 1 \not\equiv 0 (\mathrm{mod}\ 2^k)$$
$$\rho^{2^{k-2}} - 1 \equiv 0 (\mathrm{mod}\ 2^k) \qquad\qquad ////$$

Theorem 8.4.2. If $\rho = 8t - 3$, then the sequence $x_0, x_1, \ldots, x_{2^{k-2}-1}$ generated by the formula is some permutation of

$$1, 5, 9, \ldots, 2^k - 3 \qquad \text{if} \qquad x_0 \equiv 1(\mathrm{mod}\ 4)$$
$$3, 7, 11, \ldots, 2^k - 1 \qquad \text{if} \qquad x_0 \equiv 3(\mathrm{mod}\ 4)$$

PROOF   Consider the numbers

$$\rho^n \qquad n = 0, 1, \ldots, 2^{k-2} - 1$$

The difference between consecutive terms

$$x_0 \rho^{n+1} - x_0 \rho^n = x_0 \rho^n(\rho - 1)$$
$$= x_0 \rho^n(8t - 4)$$

is divisible by 4.   We know that the period is 2^{k-2}; hence we have 2^{k-2} distinct numbers whose differences are all divisible by 4, and the theorem follows from this.   ////

How shall we pick the value of t?   For example, if $t = 1$, then $\rho = 5$ and whenever a small x_i occurs then for some time the succeeding x_i will increase— definite structure in the supposedly random sequence.   Similarly for ρ near 1.0....   It is customary to pick a t so that ρ is near 1/2, say $\rho = 0.10....$

If in the sequence x_i of random numbers we drop the last two bits, then we will have all possible (2^{k-2})-bit numbers in some order.   In practice it is customary to avoid more than the last two bits but to recognize that dropping many more does not help very much.

PROBLEMS

8.4.1 Examine the periods for $\rho = 8t \pm 1$ and $\rho = 8t \pm 3$.

8.4.2 Discuss the cycles that all the 2^k numbers fall into.

8.4.3 In picking a multiplier note that *both* ends need to be considered. A multiplier ending in ...01 will reproduce exactly the last two digits, and one in ...11 will cause the last digit to alternate. Extend this analysis to numbers ending in ...001 and others.

8.5 TESTING A RANDOM-NUMBER GENERATOR

Now that we know what we are going to use as a source of random numbers it is necessary to consider how well it meets our intuitive notions of what a random sequence of numbers should be.

In the first place we have implied that the numbers should be equilikely, that is, we were talking about a uniform distribution in an interval, which is conventionally taken as $0 \leq x < 1$ by putting a binary point in front of the first digit. It is easy to see from Theorem 8.4.2 that until we get close to the granularity of the number system, we have an *exactly* uniform distribution in the full period. This bothers us a bit since it is too uniform. We are sampling without replacing until we exhaust the whole set of numbers we are going to use, and *random* intuitively means that there is a chance, small though it may be, that the same number can come up twice in succession. Indeed there is a reasonable probability that in the long sequence there would be one such case. But this defect is not serious for most applications, and if we drop more digits on the right, we can get sequences in which such numbers occur a fixed number of times. Again this is too uniform and is not compatible with our intuitive feelings.

Taking the numbers two at a time, we can study the pattern of number pairs, which we again expect to be uniform. It is possible to show theoretically that the number pairs are not bad, but carried far enough by studying triples, etc., we will be in trouble. In practice the effect is seldom serious.

There are almost an infinite number of tests one can apply to a random-number sequence, run tests, poker-hand draws, power spectral tests, etc., and a large computer could be run full-time testing various generators. Randomness is essentially a negative property and cannot be proved in a mathematical sense by any finite amount of testing of a finite length of data. Every finite string of random numbers will be found to have some peculiar property (even if it is only that it does not have a peculiar property!). The topic of testing fascinates some people, and if they wish, they can follow the topic further in the references at the end of the chapter.

Experience, which is not an infallible guide, indicates that with care congruential generators give useful results, but carelessness regularly trips up the unwary. The use of an additional operation of adding a constant to the product before taking the last k digits has not been proved worth the time and trouble.

In the final analysis, randomness, like beauty, is in the eye of the beholder, and how the numbers are to be used determines their randomness.

8.6 OTHER DISTRIBUTIONS

We now have a source of random numbers with a flat distribution. In many applications other distributions are needed. In principle the following device works. We equate the cumulative distribution $F(y)$ of the distribution $f(y)$ that we want to the cumulative distribution x of the flat distribution and solve for the new one

$$\int_0^x 1 \, dx = x = \int_0^y f(y) \, dy = F(y)$$

Applying the inverse operation F^{-1}, we have

$$F^{-1}\{F(y)\} = y = F^{-1}(x)$$

For example, suppose we want the exponential distribution e^{-y}. Then

$$f(y) = e^{-y}$$

$$F(y) = \int_0^y e^{-y} \, dy = 1 - e^{-y}$$

$$e^{-y} = 1 - x$$

$$y = -\ln(1 - x)$$

and using the sequence of random numbers x_i from a flat distribution, we get

$$y_i = -\ln(1 - x_i)$$

or equally good,

$$y_i = -\ln x_i$$

This has been used many times with excellent success.

If this method of inverting the cumulative function is used on the normal distribution

$$\frac{e^{-y^2}}{\sqrt{2\pi}}$$

then the inverse function has to be approximated somehow—perhaps with the aid of auxiliary tables.

An alternate method for getting a normal distribution is to appeal to the central-limit theorem and simply add several random numbers from a flat (or any other) distribution. Since the numbers in the source are independent and have a variance,

$$\sigma^2 = \int_0^1 (x - \tfrac{1}{2})^2 \, dx = \frac{(x - \tfrac{1}{2})^3}{3} \bigg|_0^1 = \frac{1}{12}$$

The sum of 12 of them will have a variance of 1. Thus the common rule is to "add 12 numbers from our generator and subtract the constant 6 from the sum to get a normal distribution with mean zero and unit variance."

How good is this distribution? Clearly it is exactly zero outside of the interval $\pm 6\sigma$. The probability of a value being outside is approximately 2×10^{-9} and hardly has a serious effect.

This brings up a philosophical point. Do we really want genuine random numbers, or do we want a set of homogenized, guaranteed, and certified numbers whose effect is random but at the same time we do not run the risk of the fluctuations of a truly random source? When we look at how we are using the numbers, we usually find that we want to get the security of a large sample by taking as small a sample as we can. We do not want to run the risk of wild fluctuations. Thus our too-flat random generator and the normal distribution which eliminates the tails far out are closer to what we often want than the real thing is!

Many other distributions can be found by various methods described in the two references at the end of the chapter.

8.7 RANDOM MANTISSAS

One of the many uses of random numbers is in the simulation of random computations, or as a source of random data to test a routine. As we showed in Sec. 2.8, the distribution of mantissas is not flat but rather has the reciprocal distribution

$$p(x) = \frac{1}{x \ln b}$$

This distribution also applies to random physical data or random data for testing a mathematical library routine for average running time, etc. Thus we need a source of random mantissas.

One of the results proved in Sec. 2.8 is that a continued product of numbers from any reasonable distribution has a mantissa that rapidly approaches the reciprocal distribution. We gave a short table showing the rapidity of approach for numbers from our flat distribution.

A specific algorithm based on this observation uses the x_i from the random-number generator and forms

$$Y_n = Y_{n-1}x_n \qquad \text{(shifted)}$$

where *shifted* means that after the multiplication the possible leading zeros are shifted off to form a legitimate floating-point mantissa.

How well does this method work? Experimental verification is given by 8,192 trials. Counting the number of mantissas falling in each of N categories, we get the results shown in Table 8.7. The last two columns give the number of sign changes in the residuals between the observed and theoretical distributions. The expected number of sign changes might be thought to be $(N - 1)/2$, but since for $N = 2$ it is clear that one sign change must occur (because the mean of the residuals is zero), we have used $N/2$ as the expected number. A comparison of these two columns shows that there are no systematic errors in some regions. The chi-square test in the second column shows that the size of the deviations is approximately right, neither too small nor too large. Thus the generator appears to be safe to use.

It is interesting to note that probably this generator has a significantly longer period than the period of the source x_i.

PROBLEMS

8.7.1   Apply this generator to the 6-bit random numbers of Sec. 8.3.

Table 8.7   DISTRIBUTION OF 8,192 RANDOM MANTISSAS

N	χ^2	Degrees of freedom	Residuals Sign changes	Expected
64	61.392	63	30	32
32	22.804	31	14	16
16	11.150	15	8	8
8	7.724	7	5	4
4	3.261	3	2	2
2	1.467	1	1	1

8.8 SWINDLES

Problems using random numbers are often both tricky and inclined to be expensive in machine time; therefore it is wise to have a competent practicing statistician handy to discuss the computation before you begin.

A number of methods have been devised to make Monte Carlo computations and other computations using random numbers much more efficient, sometimes by factors of 1,000 or more. These are known by the colorful name of *swindles* to go along with the name *Monte Carlo*. One of the methods is called "antithetic variables," meaning that when one case occurs, then it is negatively correlated in a known way with another case. For example, in the Buffon needle case a cross is used, instead of a needle, so that if one arm is likely to cross a line, then the other is not.

Many other statisticians' tricks can be used, such as stratified sampling. Again, perhaps an analytical solution can be found for the bulk of the cases and the random sampling used only for the small difference. There are no systematic methods in the field as yet; the low cunning of trickery seems to be necessary for success, and this art lies outside the scope of a general textbook.

8.9 NOISE SIMULATION

Probably the greatest use of random numbers is in the simulation of noise. Rarely (beyond simple roundoff) does the noise resemble the flat distribution of our generator. For *independent* samples the transformation to other distributions is easy in principle, and since it does not need to be extremely accurate in most cases, it is fairly easy in practice. But most simulations of noise require that the samples be dependent from step to step and that they have various internal correlations. This topic is much too difficult to take up here, and often the proper choices depend on knowledge of the field of application. The proposer of a problem may not realize that he does not have independent samples of noise at the step size to be used in the computation, and care is necessary to avoid elegantly solving the wrong problem.

REFERENCES

The three best references are Jansson [22], Knuth [28], and Shreider [51], with Knuth being the more complete and latest one.

9

THE DIFFERENCE CALCULUS

9.1 INTRODUCTION

We have examined some algorithms for solving a few of the more common problems of numerical computations. Before examining the next major group of problems, we need to develop a few mathematical tools. The main tool needed is the *finite difference calculus*, which corresponds to the usual infinitesimal calculus *except* that we do not go the the limit but instead stop at a definite, finite step size. One reason for the importance of the finite difference calculus is that most of the time we have data (samples) of our functions at a sequence of equally spaced points x_k $(k = 0, 1, \ldots, N)$,

$$f(a), f(a + h), f(a + 2h), \ldots, f(a + Nh)$$

An example is the sequence of partial sums of a series

$$s(n) = \sum_{k=0}^{n} a_k \qquad n = 0, 1, 2, \ldots, N$$

A second reason is that when the difference is divided by the step size h, it provides an estimate for the derivative (which is a limit process and cannot be carried out by a finite machine).

If there is difficulty in understanding the finite difference calculus, then it is probable that the corresponding part of the infinitesimal calculus is not well understood and should be reviewed.

Because we are dealing with a finite number of steps and a finite number of operations, we do not get into the area of existence theorems or other difficult questions in mathematics; it is usually clear that we can carry out the proposed operations and that a number will result. Its relation to what is wanted may be obscure, but that is not the point at the moment. Because of the finite nature of our computations we can operate formally without regard to the usual questions of rigor.

It is worth recalling that differential equations, for example, are usually derived from some physical situation by first finding a finite approximation and then taking the limit. If the reverse step cannot be taken (when small steps are used), then there is some doubt about the validity of the differential equation. Thus it is reasonable to assume that the finite difference approximations we make are likely to be fairly accurate. Later we will examine the errors of such approximations. In this chapter and the next four, we will be dealing with finite, equally spaced data in a formal manner.

If the data are at the points

$$x = a, a + h, a + 2h, \ldots$$

we can reduce the situation to a standard form by the substitution

$$t_k = \frac{x_k - a}{h} = k \qquad k = 0, 1, 2, \ldots$$

Much of the time we will assume that this has been done. It is often convenient to write

$$f(0) = f_0$$
$$f(1) = f_1$$
$$\cdots\cdots\cdots\cdots$$
$$f(k) = f_k$$

In using these results we will have to either reduce the problem to the standard form or transform our formulas to the spacing of the original problem. Often it is a matter of indifference which we do, though once in a while it does make a significant difference in the amount of arithmetic that must be done. We can, as in analytic geometry, regard the transformation as an "alias" which keeps the points fixed and moves the coordinate system, thus giving the points new names; or we can regard the transformation as an " alibi " which keeps the coordinate system fixed and regard the problem as having been moved.   For example, the transformation

$$x' = x - a$$
$$y' = y - b$$

can be regarded as either shifting the coordinates of the circle

$$(x - a)^2 + (y - b)^2 = r^2$$

so that it is now labeled

$$x'^2 + y'^2 = r^2$$

or else as moving the circle to the origin.   Whichever way we view it, we must keep that view to avoid unnecessary confusion.

PROBLEMS

9.1.1   If x takes the values $11, 9, 7, \ldots, -11$, find the transformation that reduces this to the standard form $0, 1, \ldots, 11$.   *Ans.* $t = (11 - x)/2$

9.1.2   Reduce $3.0, 3.5, 4.0, 4.5, \ldots, 10$ to standard form.   *Ans.* $t = 2(x - 3)$

9.1.3   Reduce to standard form: $x = a, a - h, a - 2h, \ldots, a - (n - 1)h$.

Ans. $t = \dfrac{a - x}{h}$

9.1.4   Transform Simpson's formula

$$\int_{-1}^{1} f(x)\, dx = \tfrac{1}{3}[f(-1) + 4f(0) + f(1)]$$

to the interval $a \le x \le b$.

Ans. $\displaystyle \int_a^b f(x)\, dx = \frac{b - a}{6}\left[f(a) + 4f\left(\frac{a + b}{2}\right) + f(b)\right]$

9.1.5   Transform the interval $a \le x \le b$ to the interval $c \le t \le d$, assuming a, b, c, d finite.

Ans. $t = \dfrac{(c - d)x + ad - bc}{a - b}$

9.1.6   Transform the interval $0 \le x \le \infty$ to the interval $-1 \le t \le 1$.

Ans. $t = \dfrac{x - 1}{x + 1}$ or $x = \dfrac{1 + t}{1 - t}$

9.2 THE DIFFERENCE OPERATOR

The basic operation of the finite difference calculus is the difference operator Δ defined by

$$\Delta f(x) \equiv f(x + h) - f(x)$$

This operation is familiar from the calculus where it is used in the process of finding a derivative.

We can imagine the operator Δ as separated from the function on which it operates, just as we imagine the derivative operator d/dx operating on a function $f(x)$ or the integration operator $\int \ldots dx$ acting on a function $f(x)$.

The difference operator is linear, as are differentiation and integration. By this we mean that if a and b are constants, then

$$\Delta (af(x) + bg(x)) = a \, \Delta f(x) + b \, \Delta g(x)$$

This property of being linear is very important and makes the operator Δ very convenient to use. Finding zeros of a function is not a linear operation.

As an example of the linearity and application of the Δ operator, consider applying it to the quadratic

$$y = ax^2 + bx + c$$

We get, using the linearity of Δ and simple algebra,

$$\begin{aligned}
\Delta y = \Delta(ax^2 + bx + c) &= a \, \Delta(x^2) + b \, \Delta x + c \, \Delta(1) \\
&= a[(x + h)^2 - x^2] + b[(x + h) - x] + c[1 - 1] \\
&= 2axh + ah^2 + bh
\end{aligned}$$

The difference operator applied to a product gives

$$\begin{aligned}
\Delta[f(x)g(x)] &= f(x + h)g(x + h) - f(x)g(x) \\
&= f(x + h)g(x + h) - f(x + h)g(x) + f(x + h)(gx) - f(x)g(x) \\
&= f(x + h) \, \Delta g(x) + g(x) \, \Delta f(x)
\end{aligned}$$

Since $f(x)$ and $g(x)$ are interchangeable, this is equivalent to the alternate form

$$\Delta[f(x)g(x)] = f(x) \, \Delta g(x) + g(x + h) \, \Delta f(x)$$

Note that the $x + h$ appears in *only* one argument in either case. Also note how closely the formula resembles that of the calculus

$$\frac{d}{dx}[f(x)g(x)] = f(x)\frac{d}{dx}[g(x)] + g(x)\frac{d}{dx}[f(x)]$$

Similarly for the quotient we get

$$
\Delta\left[\frac{f(x)}{g(x)}\right] = \frac{f(x+h)}{g(x+h)} - \frac{f(x)}{g(x)}
$$

$$
= \frac{f(x+h)g(x) - g(x+h)f(x)}{g(x)g(x+h)}
$$

$$
= \frac{f(x+h)g(x) - f(x)g(x) + f(x)g(x) - g(x+h)f(x)}{g(x)g(x+h)}
$$

$$
= \frac{g(x)\,\Delta f(x) - f(x)\,\Delta g(x)}{g(x)g(x+h)}
$$

which again has one argument at $x + h$ and closely resembles the corresponding formula from the calculus

$$
\frac{d}{dx}[f(x)] = \frac{g(x)\dfrac{d}{dx}[f(x)] - f(x)\dfrac{d}{dx}[g(x)]}{g^2(x)}
$$

Calculus books usually give a long list of formulas for derivatives, both for the direct use and for later inversion to get a table of integrals. Similarly, in the difference calculus we need a list both for present use and for the later inversion into a summation table. We will, however, give only a short list.

$$
\Delta \sin(ax+b) = 2\sin\frac{ah}{2}\cos\left[a\left(x + \frac{h}{2}\right) + b\right]
$$

$$
\Delta \cos(ax+b) = -2\sin\frac{ah}{2}\sin\left[a\left(x + \frac{h}{2}\right) + b\right]
$$

$$
\Delta \tan(ax+b) = \sin ah \,\sec(ax+b)\sec[a(x+h)+b]
$$

$$
\Delta \arctan(ax+b) = \frac{ah}{1 + [a(x+h)+b](ax+b)}
$$

$$
\Delta\, a^x = a^x(a^h - 1)
$$

$$
\Delta \ln x = \ln\left(1 + \frac{h}{x}\right)
$$

In some respects the role of the number e in the calculus is played by the number 2 in the finite difference calculus. For example, if $a^h = 2$, then

$$
\Delta\, a^x = a^x
$$

In the common case of unit spacing ($h = 1$) we have $a = 2$ and

$$
\Delta\, 2^x = 2^x
$$

PROBLEMS

9.2.1 Verify the above list of results.

9.2.2 Show that

$$\Delta\left(\frac{1}{1+x^2}\right) = \frac{-h(2x+h)}{(1+x^2)[1+(x+h)^2]}$$

9.2.3 Show that

$$\Delta \sinh x = \frac{(e^h - 1)(e^x + e^{-(x+h)})}{2}$$

9.2.4 Find

$$\Delta(x \sin x)$$

9.2.5 Find

$$\Delta[\sec^2(ax + b)]$$

9.2.6 Show that

$$\Delta\left(\frac{1}{\sqrt{x}}\right) = \frac{-h}{\sqrt{x}\sqrt{x+h}(\sqrt{x} + \sqrt{x+h})}$$

9.2.7 Show that
(a)
$$\Delta(x^4) = \Delta(x^2 \cdot x^2)$$
(b)
$$\Delta(x^n) = \Delta(x^{n-1} \cdot x)$$

9.2.8 Show that

(a) $\Delta \sin(ax + b) = 2 \sin \dfrac{ah}{2} \sin\left(ax + b + \dfrac{h}{2} + \dfrac{\pi}{2}\right)$

(b) $\Delta \cos(ax + b) = 2 \sin \dfrac{ah}{2} \cos\left(ax + b + \dfrac{h}{2} + \dfrac{\pi}{2}\right)$

9.2.9 Find

$$\Delta\left(\frac{ax+b}{cx+d}\right)$$

9.2.10 Find

$$\Delta(2^x \sin x)$$

9.3 REPEATED DIFFERENCES

Since $\Delta f(x)$ is a function of x, we can apply the Δ operator again to obtain the second difference

$$\Delta[\Delta f(x)] \equiv \Delta^2 f(x)$$

This notation corresponds to the notation for the second derivative in the ordinary calculus

$$\frac{d}{dx}\left\{\frac{d[f(x)]}{dx}\right\} = \frac{d^2}{dx^2}[f(x)]$$

Repeated use of the Δ operation gives

$$\Delta[\Delta^{r-1}f(x)] = \Delta^r f(x)$$

In the example of the quadratic (Sec. 9.2)

$$y(x) = ax^2 + bx + c$$

we got

$$\Delta y = 2ahx + ah^2 + bh$$

The second difference is

$$\Delta^2 y = 2ah(x + h) - 2ahx = 2ah^2$$

while the third difference is, clearly,

$$\Delta^3 y = 0$$

It is not an accident that the third difference is zero; rather it is the consequence of the following important theorem.

The fundamental theorem of the difference calculus   The Nth difference of a polynomial of degree N

$$y(x) = a_N x^N + a_{N-1} x^{N-1} + \cdots + a_0$$

is a constant $a_N N! h^N$, and the $(N + 1)$st difference is zero.   The proof is very much like that found in the calculus for the corresponding result.   We first prove the lemma.

LEMMA   If $y(x)$ is a polynomial of degree N, then $\Delta y(x)$ is a polynomial of exactly degree $N - 1$.

PROOF OF LEMMA   For the special function $y(x) = x^N$ we find, using the binomial theorem,

$$\Delta y = (x + h)^N - x^N = \sum_{k=0}^{N} C(N, k) x^{N-k} h^k - x^N$$

$$= Nhx^{N-1} + \frac{N(N - 1)}{2} h^2 x^{N-2} + \cdots + h^N$$

Thus, when the Δ operator is applied, the term x^N becomes a polynomial of exactly degree $N - 1$ with leading coefficient Nh.   Using the linearity property of the operator decreases each term of a polynomial by one degree, and the term in x^{N-1} cannot cancel out. Thus the lemma is proved.   ////

PROOF   We simply use the lemma N times and find that only the term from x^N remains and that it has the coefficient $N! h^N$.   The next difference makes this constant cancel, and the theorem is proved.

////

This theorem is basic to much of the classical parts of numerical analysis; in one way or another an appeal is regularly made to it.

PROBLEMS

9.3.1 Prove the theorem by mathematical induction.

9.3.2 Using $h = 1$, compute the second and fourth differences of $y = x^4 - 4x^3 + 6x^2 - 4x + 1$.

$$Ans. \quad \Delta^2 y = 12(x - 1)^2, \Delta^4 x = 24$$

9.3.3 Using $h = 1$, find all the differences of $x(x - 1)(x - 2)(x - 3)$.

9.3.4 Find

(a) $\Delta^k \left(\dfrac{1}{x} \right)$

(b) $\Delta^k (\ln x)$

9.3.5 Find

$$\Delta^k [\sin (a + bx)]$$
$$\Delta^k [\cos(a + bx)]$$

Hint: Prob. 9.2.8

9.3.6 Find $\Delta^k (a^x)$.

9.4 THE DIFFERENCE TABLE

When using higher differences, it is useful to imagine that the numbers are arranged in the form of a difference table (Table 9.4.1), although the differences are probably not stored in a machine in this way.

Table 9.4.1 DIFFERENCE TABLE

x	$y(x)$	$\Delta y(x)$	$\Delta^2 y(x)$	$\Delta^3 y(x)$
0	$y(0)$			
		$\Delta y(0)$		
1	$y(1)$		$\Delta^2 y(0)$	
		$\Delta y(1)$		$\Delta^3 y(0)$
2	$y(2)$		$\Delta^2 y(1)$	
		$\Delta y(2)$		$\Delta^3 y(1)$
3	$y(3)$		$\Delta^2 y(2)$	
		$\Delta y(3)$		$\Delta^3 y(2)$
4	$y(4)$		$\Delta^2 y(3)$	
		$\Delta y(4)$		
5	$y(5)$			
.	.			
.	.			
.	.			

As an example of a difference table consider the values of the sine-integral function as shown in Table 9.4.2 (we have used the common convention and have written the differences as if the decimal point were at the right-hand end of the numbers).   Higher differences are usually carried until the numbers either tend to oscillate or are mainly zero.

It is easy to check a difference table for accuracy (when computed by hand). If we sum any column of differences and add the sum to the top number in the preceding column, we must get the bottom number in the preceding column.   For example, in Table 9.4.2 the sum of the Δ^3 is -26.   This added to the -2 of $\Delta^2(0)$ gives $-2 - 26 = -28$, the bottom value of the Δ^2 column, namely $\Delta^2(0.8)$.

Most tables are given in fixed-point notation, at least locally, though over various ranges differing numbers of figures may be printed.   This means that it is often possible to reconstruct the table from the top diagonal line of differences plus the last column of differences.   In a sense the differences contain the same information that the function does.   The only exception to this is when the function, or some column of differences, changes sign and the difference (which is now a sum of two numbers) in the next column requires a shift and possible loss of information due to the carry beyond the extreme left.

Table 9.4.2   DIFFERENCE TABLE OF
$$\text{Si}(x) = \int_0^x \frac{\sin t}{t}\, dt$$

x	$\text{Si}(x)$	Δ	Δ^2	Δ^3	Δ^4
0.0	0.0000				
		999			
0.1	0.0999		-2		
		997		-6	
0.2	0.1996		-8		5
		989		-1	
0.3	0.2985		-9		-4
		980		-5	
0.4	0.3965		-14		3
		966		-2	
0.5	0.4931		-16		-1
		950		-3	
0.6	0.5881		-19		0
		931		-3	
0.7	0.6812		-22		0
		909		-3	
0.8	0.7721		-25		0
		884		-3	
0.9	0.8605		-28		
		856			
1.0	0.9461				

Although we have used the tables as if they started at zero, it is frequently useful to imagine that the table extends indefinitely in both directions and we have only a sample of the entire function.   At other times it is useful to shift the origin to the place in which we are currently interested.   Thus we will be quite vague about the actual origin of a table and occasionally shift without mentioning it specifically.

PROBLEMS

9.4.1   Make a difference table for the sine function at $10°$ spacing, using a 5-place table $(0 \leq x \leq 90°)$.   Check your arithmetic by the above mentioned method.

9.4.2   Make a difference table for $y = x^4$.

9.4.3   Describe a difference table for the function $y = a^x$.

9.4.4   Describe a difference table for the function $y = \sin(\pi x/3)$.

9.5   TABULATING A POLYNOMIAL AT A REGULAR SPACING

As observed in Sec. 3.10, we frequently wish to evaluate a function at a sequence of equally spaced points.   This is especially true in evaluating polynomials.   For example, it is customary in processing data to first remove a polynomial trend. This means that the polynomial must be evaluated at every point so that it can be subtracted from the data.

The difference table gives a powerful way of constructing the left-hand column from the top line of differences plus one column of high-order differences. For a polynomial we know from the fundamental theorem that the Nth differences are a constant, and so the top line of differences alone will enable us to reconstruct the whole table.   The process is as follows and is based on the obvious relation: From

$$\Delta f(k) = f(k + 1) - f(k)$$

we get

$$f(k + 1) = f(k) + \Delta f(k)$$

Consider the specific example of

$$y(x) = 3x^2 - 6x + 9 \qquad y(0) = 9$$
$$\Delta y(x) = \qquad 6x - 3 \qquad \Delta y(0) = -3$$
$$\Delta^2 y(x) = 6 \qquad\qquad \Delta^2 y(0) = 6$$

Table 9.5 shows the method of construction from the differences.

Even when a function is not exactly a polynomial, this still provides a powerful method of constructing many values, before making a small adjustment due to the fact that the differences are not exactly a constant.

However, roundoff is a problem, as always.   In the case of a polynomial if the calculation of the differences is exact, then the rest of the arithmetic can usually be kept under reasonable control.   But if there is a small roundoff error in the differences, this error will be greatly magnified in the function.   For example, consider the table with all zeros to be true, but suppose that the kth differences are in error by a small fixed amount ε at each location.   What does the function become?

The preceding column becomes $n\varepsilon$, the next $[n(n-1)/2]\varepsilon, \ldots$, etc.   Roughly the function looks like

$$\frac{n^k}{k!}\,\varepsilon$$

which in time becomes very large.   Thus it is customary, when using this method of computing a function, to periodically correct the accumulated errors.

Table 9.5 $y(x) = 3x^2 - 6x + 9$

x	y	Δy	$\Delta^2 y$
0	9		
		−3	
1	6		6
		3	
2	9		6
		9	
3	18		6
		15	
4	33		6
		21	
5	54		6
		27	
6	81		6
		33	
7	114		6
		39	
8	153		6
		45	
9	198		6
		51	
10	249		

Check: $y(10) = 300 - 60 + 9 = 249$

9.6   THE FACTORIAL NOTATION

In the calculus x^n plays an important role. For example, in the Taylor series

$$f(x) = a_0 + a_1 x + a_2 x^2 + \cdots = \sum_{k=0}^{\infty} a_k x^k$$

an arbitrary function is expanded in powers of x. The main reason that x^n plays this leading role is that

$$\frac{d}{dx} x^n = n x^{n-1} \qquad n \geq 1$$

Thus the coefficients of the Taylor series are determined by differentiating and evaluating the derivatives at the origin

$$f(0) = a_0$$
$$f'(x) = a_1 + 2a_2 x + 3a_3 x^2 + \cdots$$
$$f'(0) = a_1$$
$$f''(x) = 2a_2 + 6a_3 x + \cdots$$
$$f''(0) = 2a_2$$

and in general

$$f^{(k)}(0) = k! a_k$$

When differences are used instead of derivatives, the functions corresponding to x^n are

$$g_n(x) \equiv x(x-1) \cdots (x - n + 1)$$

where there are n factors.
We have

$$\Delta g_n(x) = (x+1)x(x-1) \cdots (x - n + 2) - x(x-1) \cdots (x - n + 1)$$
$$= [(x+1) - (x - n + 1)]x(x-1) \cdots (x - n + 2)$$
$$= nx(x-1) \cdots (x - n + 2)$$

For convenience we will write

$$g_n(x) = x^{(n)} \equiv x(x-1)(x-2) \cdots (x - n + 1) \qquad n = 1, 2, \ldots$$

and we have just shown that

$$\Delta x^{(n)} = n x^{(n-1)}$$

We need to extend the range of definition to all integers (much as were the powers of x in algebra). As a basis we use the identity

$$x^{(n)} = x^{(m)}(x - m)^{(n-m)} \qquad n > m \geq 1$$

If we formally set $m = 0$, we get

$$x^{(n)} = x^{(0)}x^{(n)}$$

which suggests that we define

$$x^{(0)} \equiv 1$$

much as we made $0! \equiv 1$.

For negative exponents we use the same identity and set $n = 0$

$$1 = x^{(0))} = x^{(m)}(x - m)^{(-m)}$$

or

$$(x - m)^{(-m)} = \frac{1}{x^{(m)}}$$

Set $x - m = y$ and we have the more convenient form

$$y^{(-m)} = \frac{1}{(y + m)^{(m)}} = \frac{1}{(y + m)(y + m - 1) \cdots (y + 1)}$$

Note that

$$x^{(m)}x^{(-m)} \neq x^{(0)} \qquad m \neq 0$$

Corresponding to the Taylor-series expansion in powers of x, we have a formal expansion in factorials

$$f(x) = b_0 + b_1 x^{(1)} + b_2 x^{(2)} + \cdots = \sum_{k=0}^{\infty} b_k x^{(k)}$$

The coefficients are determined in exactly the same way, this time by differencing and evaluating at zero

$$f(0) = b_0$$
$$\Delta f(0) = b_1$$
$$\Delta^2 f(0) = 2b_2$$
$$\cdots\cdots\cdots\cdots$$
$$\Delta^k f(0) = k! b_k$$

Hence we have the *Newton expansion*

$$f(x) = \sum_{k=0}^{\infty} \frac{\Delta^k f(0)}{k!} x^{(k)} = \sum \Delta^k f(0) C(x, k)$$

where $C(x, k)$ is the usual binomial coefficient.

The conversion of a polynomial expressed in powers of x to a sum of factorials is straightforward and is based on the same idea used to convert: a number of a given base (Sec. 2.7); a polynomial to a sum of powers of $x - a$ (Sec. 3.7); or a polynomial to powers of a quadratic (Sec. 6.7). We divide the polynomial by x, the quotient by $x - 1$, that quotient by $x - 2$, etc. The successive remainders are the coefficients of the factorial representation. For example, consider the polynomial

$$P_4(x) = 2x^4 - 3x^3 + x^2 - 1$$

$$
\begin{array}{r|rrrr|r}
0 & 2 & -3 & 1 & 0 & -1 \\
  &   & 0 & 0 & 0 & 0 \\
\cline{1-6}
1 & 2 & -3 & 1 & 0 & \boxed{-1} \\
  &   & 2 & -1 & 0 \\
\cline{1-5}
2 & 2 & -1 & 0 & \boxed{0} \\
  &   & 4 & 6 \\
\cline{1-4}
3 & 2 & 3 & \boxed{6} \\
  &   & 6 \\
\cline{1-3}
  & 2 & \boxed{9}
\end{array}
$$

then

$$P_4(x) = 2x^{(4)} + 9x^{(3)} + 6x^{(2)} + 0x^{(1)} - 1$$

The first factor by zero need not be done, of course, but was included for logical presentation.

PROBLEMS

9.6.1 Using Newton's formula, find a polynomial which takes on the following values of $P(n)$.

n	0	1	2	3	4	5
$P(n)$	41	43	47	53	61	71

Ans. $P(n) = n^2 + n + 41$ [for $n < 41$ note that $P(n) = $ prime number]

9.6.2 Show that

$$1^{(-m)} = \frac{1}{(m+1)!}$$

9.6.3 Show that

$$n^{(n)} = n!$$

9.6.4 Compute $x^{(m)}x^{(-m)}$ where $x = m$. *Ans.* $\dfrac{(m!)^2}{(2m)!}$

9.6.5 Simplify $1/x^{(-m)}$.

9.6.6 Describe the difference table of $x^{(k)}$.

*9.7 STIRLING NUMBERS OF THE FIRST KIND

Although we have already shown how to go from a polynomial in powers of x to its factorial representation, the central role of powers of x in the continuous case and of factorials in the discrete case requires that we look more closely at the topic.

To express $x^{(n)}$ in powers of x we write

$$x^{(n)} = \sum_{k=0}^{n} s(n, k)x^k \qquad (9.7.1)$$

and set out to find the numbers $s(n, k)$. These numbers are called *Stirling*[1] *numbers of the first kind*. For $n = 1$ we have

$$x^{(1)} = x = s(1, 0) + s(1, 1)x$$

whence

$$s(1, 0) = 0 \qquad s(1, 1) = 1$$

For $n = 2$ we have

$$x(x - 1) = x^2 - x = s(2, 0) + s(2, 1)x + s(2, 2)x^2$$

whence

$$s(2, 0) = 0 \qquad s(2, 1) = -1 \qquad s(2, 2) = 1$$

It is easy to see that for all $n \geq 1$, $s(n, 0) = 0$, but $s(0, 0) = 1$.

Rather than continuing one power at a time we write

$$x^{(n+1)} = (x - n)x^{(n)}$$

and use Eq. (9.7.1) on both sides to get

$$\sum_{k=1}^{n+1} s(n + 1, k)x^k \equiv (x - n) \sum_{k=1}^{n} s(n, k)x^k$$

$$\equiv \sum_{k=1}^{n} s(n, k)x^{k+1} - n \sum_{k=1}^{n} s(n, k)x^k$$

$$\equiv \sum_{k=1}^{n} [s(n, k - 1) - ns(n, k)]x^k + s(n, n)x^{n+1}$$

Table 9.7

n \ k	1	2	3	4	5
1	1				
2	−1	1			
3	2	−3	1		
4	−6	11	−6	1	
5	24	−50	35	−10	1

[1] James Stirling (1692–1770).

Equating the highest power of x, we get

$$s(n + 1, n + 1) = s(n, n) = \cdots = s(1, 1) = 1 = s(0, 0)$$

For x^k we get

$$s(n + 1, k) = s(n, k - 1) - ns(n, k) \qquad k = 1, 2, \ldots, n$$

There is no simple formula for the Stirling numbers of the first kind. Table 9.7 gives a few of the numbers.

PROBLEMS

9.7.1 Extend Table 9.7 one line to $n = 6$.

9.7.2 Using Table 9.7, show that $x^{(3)} + 2x^{(2)} + x^{(1)} - 1 = (x^2 + 1)(x - 1)$.

9.7.3 Using Table 9.7, write $x^{(5)} + 10x^{(4)}$ as a polynomial.

*9.8 STIRLING NUMBERS OF THE SECOND KIND

The Stirling numbers of the second kind express x^n in terms of factorials:

$$x^n = \sum_{k=0}^{n} S(n, k)x^{(k)}$$

As in Sec. 9.7, we first compute a few cases and then find the general relation. For $n = 1$

$$x = S(1, 0) + S(1, 1)x$$

so that

$$S(1, 0) = 0 \qquad S(1, 1) = 1$$

For $n = 2$

$$x^2 = S(2, 0) + S(2, 1)x + S(2, 2)x(x - 1)$$

so that

$$S(2, 0) = 0 \qquad S(2, 1) = 1 \qquad S(2, 2) = 1$$

Table 9.8

n \ k	1	2	3	4	5
1	1				
2	1	1			
3	1	3	1		
4	1	7	6	1	
5	1	15	25	10	1

Again it is clear that

$$S(n, 0) = 0 \qquad n > 0$$

and

$$S(0, 0) = 1$$

The recurrence relation follows from

$$x^{n+1} = x \cdot x^n$$

$$\sum_{k=0}^{n=1} S(n + 1, k)x^{(k)} = x \sum_{k=0}^{n} S(n, k)x^{(n)} = \sum_{k=0}^{n} S(n, k)(x - k)x^{(n)} + \sum_{k=0}^{n} kS(n, k)x^{(k)}$$

$$= \sum_{k=0}^{n} S(n, k)x^{(k+1)} + \sum_{k=0}^{n} kS(n, k)x^{(k)}$$

$$= \sum_{k=0}^{n} [S(n, k - 1) + kS(n, k)]x^{(k)} + S(n, k)x^{(n+1)}$$

Equate like factorials

$$S(n + 1, k) = S(n, k - 1) + kS(n, k) \qquad k = 1, 2, \ldots, n$$

$$S(n + 1, n + 1) = S(n, n) = \cdots = S(1, 1) = 1 = S(0, 0)$$

Some of the numbers are given in Table 9.8.

PROBLEMS

9.8.1   Extend Table 9.8 to $n = 6$.

9.8.2   Write x^5 in factorial form using Table 9.8.   Check by synthetic division.

9.8.3   Write $x^3 + 7x^2 - 9x + 3$ in factorial form.

9.9   ALTERNATE NOTATIONS

The choice of the notation for the first difference

$$\Delta f(x) = f(x + h) - f(x)$$

was arbitrary and unfortunately not symmetric.   We might just as well have chosen the *backward* differences

$$\nabla f(x) = f(x) - f(x - h)$$

What would have happened if we had?   We would have found that the ascending factorials

$$^{(n)}x = x(x + h)(x + 2h) \cdots [x + (n - 1)h]$$

would give the relation we need for handling things, and we would have found a
backward Newton formula.   Similarly we would have found different Stirling
numbers $\sum(n, k)$ and $\sigma(n, k)$.   These are related to the old ones by

$$\sum(n, k) = (-1)^{n+k}s(n, k) = |s(n, k)|$$
$$\sigma(n, k) = (-1)^{n+k}S(n, k)$$

Thus, there is a completely analogous theory for the backward differences.   The
resulting formulas are occasionally useful, but in a first course they are a luxury.
Similarly *central differences* and *mean differences* are sometimes used.
Again they are merely a change in notation and a convenience in some situations
but not worth the trouble in an introductory course.
Many books make use of elaborate symbolic methods, for which the alternate notations are quite useful.   We have used only the operator Δ in a few
simple, finite cases and have so far avoided the more dubious symbolic operators.
For example, the Taylor series can be written

$$f(a + h) = f(a) + hf'(a) + h^2f''(a) + \cdots$$
$$= \sum_{k=0}^{\infty} \frac{h^k D^k}{k!} [f(a)]$$
$$= e^{hD}f(a)$$

Therefore

$$\Delta f(a) = (e^{hD} - 1)f(a)$$

or symbolically

$$\Delta = e^{hD} - 1$$
$$e^{hD} = 1 + \Delta$$
$$hD = \ln(1 + \Delta) = \Delta - \frac{\Delta^2}{2} + \frac{\Delta^3}{3} - \frac{\Delta^4}{4} + \cdots$$
$$D = \frac{1}{h}\left(\Delta - \frac{\Delta^2}{2} + \frac{\Delta^3}{3} - \cdots\right)$$

We will, however, occasionally do some formal manipulation with what
amounts to symbolic methods, because the methods lead easily and rapidly to
useful results.   The history of mathematics is full of processes that were once
condemned and are now considered legitimate—but also of ones that lead to false
results.   It is an art to be safe but not rigid.

9.10   AN EXAMPLE OF AN INTEGRAL EQUATION

With even this little bit of theory it is time to give an example of how it can be used to solve useful problems.   The illustration is one that occurred in solid-state physics.   The problem is to calculate

$$g(y) = \frac{d}{dy} \int_0^y \frac{f(x)}{\sqrt{y - x}} \, dx \qquad 0 \le y \le 1$$

when the data for $f(x)$ is given at $x = 0, 0.1, 0.2, \ldots, 1.0$.

Examining the problem, we immediately see that at the upper limit, when $x = y$, the integrand becomes infinite.   Furthermore, the result of the integration is going to be differentiated with respect to y, and when we differentiate under the integral sign with respect to y, we will get an integral that diverges.   Thus the problem requires some care.

To start our thinking, for what functions $f(x)$ could we solve the problem? It is natural to try x^n

$$g(y) = \frac{d}{dy} \int_0^y \frac{x^n \, dx}{\sqrt{y - x}}$$

This suggests the usual substitution

$$x = y \sin^2 \theta$$
$$dx = 2y \sin \theta \cos \theta \, d\theta$$

and

$$g(y) = \frac{d}{dy} \left(\frac{2y^{n+1}}{\sqrt{y}} \right) \int_0^{\pi/2} \sin^{2n+1} \theta \, d\theta = \frac{(2n + 1)y^n}{\sqrt{y}} \, W_{2n+1}$$

where W_{2n+1} is the Wallis[1] integral

$$W_{2n+1} = \int_0^{\pi/2} \sin^{2n+1} \theta \, d\theta = \frac{2 \cdot 4 \cdots (2n)}{1 \cdot 3 \cdot 5 \cdots (2n + 1)}$$

$$= \frac{2n}{2n + 1} \, W_{2n-1} \qquad W_1 = 1$$

Thus we can handle all the cases of x^n.   Since the operator

$$\frac{d}{dy} \int_0^y \frac{(\cdot)}{\sqrt{y - x}} \, dx$$

is linear, *if* we can write

$$f(x) = a_0 + a_1 x + \cdots + a_{10} x^{10} = \sum_{k=0}^{10} a_k x^k$$

---

[1] John Wallis (1616–1703).

we will have as the solution

$$g(y) = \frac{1}{\sqrt{y}} \sum_{n=0}^{10} (2n + 1) W_{2n+1} a_n y^n$$

To find the representation of $f(x)$ we use Newton's formula (with a spacing of $\frac{1}{10}$)

$$f(x) = f(0) + 10x \, \Delta f(0) + \frac{10x(10x - 1)}{2!} \Delta^2 f(0)$$

$$+ \cdots + (10x)^{(10)} \Delta^{10} f(0)$$

The Stirling numbers of the first kind will get us from the factorial expansion to the polynomial we want, and we are done.

As we will later see, it is often dangerous to use a polynomial of degree as high as 10, but in the case that inspired the example, where the data were crude and the answer was not expected to be much better, the results were satisfactory as judged by these facts:

1 The graph of the approximating polynomial we found from Newton's formula looked reasonable.

2 The results were compatible with other parts of the physical theory.

3 The results stimulated further laboratory measurements of the same effect.

As a final step in the problem, why did we succeed?   What other kinds of problems can we handle the same way?   Evidently if we can analytically integrate for some family of functions and can represent our function as a linear combination of the family, then we have a solution.

10

ROUNDOFF ESTIMATION

10.1 WHY ROUNDOFF AGAIN?

By now the reader has begun to suspect that numerical methods are much too concerned with roundoff. After all it is a small effect and should not bother the practical man who needs three or four figures at most. If single precision is not enough, then double precision surely will do the trick, and modern machines often have double precision built in so that it is hardly slower than single precision.

But consider the following situation. You are using the zeros of a polynomial, and the output from the library routine (unlike the one proposed in Chap. 6) produces, due to roundoff, a pair of close zeros instead of a double zero. As we discussed in Sec. 6.1, this will probably lead to severe cancellation in the subsequent computing. When the problem is repeated with double precision, the zeros reported by the polynomial routine are, of course, just that much closer, and the subsequent computation again loses almost all the accuracy there is. Double precision did not get you out of the trouble as it should have; it merely moved it around a bit and made it look a bit different.

It is often true, as many people say, that double precision is the answer to roundoff troubles. At least most of the time this is true *provided* you know that roundoff has caused the trouble.   But among other questions that naturally arise is, how are you to know *after* a computation is done that roundoff has caused inaccuracies?   By looking at the printed answers?   Just how are you going to do this and make any reasonable estimate of the roundoff present?   One of the goals of this chapter is to answer this question when the results are in the form of a table of numbers at equal spacing of some variable or parameter.

It is also reasonable to ask for an estimate of the roundoff *before* a computation is begun.   Of course this estimate will be less accurate than one made after the computation is done, but it also may be more valuable in saving machine and human time.   We have maintained that it is better to avoid serious roundoff by using forethought than to bound it after it happens.   It is further reasonable to ask how the machine can help us with the problem of roundoff *during* the computation.

Looking at the problem another way, there are (1) estimates of bounds, (2) guaranteed bounds, (3) statistical estimates of roundoff and its variance, and (4) even backward analysis (mentioned in Secs. 3.8 and 7.10), which gives answers to roundoff in the form " The answer you obtained is the exact answer to a problem that is within so much of the original problem."

This whole chapter is devoted to a few aspects of roundoff.   We will treat roundoff again from an entirely different point of view in Part III—a viewpoint which in many respects is more fundamental than this first careful look at a difficult, important topic.   Unfortunately for the user, all too little is known about roundoff.   When he faces his first large decision based on extensive numerical computation, the user will begin to see why he wishes he knew more about it.

10.2   RANGE ARITHMETIC (INTERVAL ARITHMETIC[1])

The basic idea of range, or interval, arithmetic is very simple.   In place of each number we carry two numbers, the largest (maximum) and the least (minimum) that it could be.   At the start of a computation these ranges are assigned, perhaps of zero length.   Each arithmetic operation is replaced by a pair of operations, one that finds the largest and rounds this up and the other that finds the smallest and rounds this down.   Thus, the result of each arithmetic operation is a pair of numbers marking the range in which it is *certain* that the result lies.

Range arithmetic is usually simulated by an interpretive system on a machine, though in principle it could be wired into a machine.   The details of the simulation are messy but not fundamentally difficult.

[1] See Moore [43].

In a short computation the resulting range is usually reasonable in size, but in a long computation the range may be so great that it is useless.   It may be thought that the true answer lies near the middle of the range, but this is clearly false since if the range of x were in the interval $0 \leq x \leq 1$ and the same were true for y, then the assumption that both lie near the middle leads to the product lying near the $1/4$ mark in the range, which is not the middle of its range.

Range arithmetic has another fault.   It is the essence of good, iterative computation that the iterations improve the result, but that the improvement takes place at no one step in the loop.   For example, in the trivial case of applying Newton's method for finding the square root of a number N, we are led to

$$y = x^2 - N$$
$$y' = 2x$$
$$x_{n+1} = x_n - \frac{x_n^2 - N}{2x_n}$$
$$= \frac{1}{2}\left(x_n + \frac{N}{x_n}\right)$$

In words, the new guess is the average of the current estimate x_n and the quotient N/x_n.   If one number is too large, then the other is automatically too small, and the average is a very good estimate.   How is range arithmetic to sense that this simple loop has in it this internal correlation between the two numbers?   Yet this is the essence of good computing.

All that range arithmetic can do is provide some bounds, probably very much too wide, within which the result must lie.   Sometimes this is good enough and the above objections are irrelevant, but sometimes they are very appropriate. Thus range arithmetic is a valuable (at times expensive) tool, but it is not the complete answer to roundoff.

PROBLEMS

10.2.1   Draw a flow diagram for the operation of multiplication in simulating range arithmetic.   (Note a range may cover 0.)

10.2.2   Do the same as in Prob. 10.2.1 but for division.

10.2.3   Discuss the characteristics of a loop in which the range will correctly contract as the iterations progress.

10.3 ERROR PROPAGATION IN A DIFFERENCE TABLE

The difference table, Table 9.4.1, is one of the basic tools of roundoff estimation. Suppose we have a table of equally spaced values of a function and there is one error and this is at the ith entry, $y_i(1 + \varepsilon_i)$. What happens in the difference table? The difference operator is linear, and so the resulting difference table can be regarded as the sum of the correct values plus the difference table of the single error $y_i\varepsilon_i$. This latter table is of the form shown in Table 10.3 (where we have suppressed the multiplicative factor $y_i\varepsilon_i$ from all the numbers).

The numbers in the table are the binomial coefficients. To prove this we merely write the operator equation, where E is the displacement operator, $Ef(x) = f(x + h)$. Since

$$\Delta[f(x)] = f(x + h) - f(x)$$

we have

$$\Delta = E - 1$$

and

$$\Delta^k = (E - 1)^k$$
$$= \sum (-1)^m C(k, m) E^{k-m}$$

Table 10.3

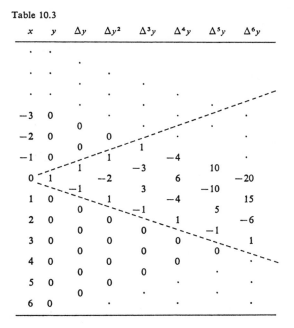

x	y	Δy	Δy^2	$\Delta^3 y$	$\Delta^4 y$	$\Delta^5 y$	$\Delta^6 y$
·	·						
		·					
·	·		·				
		·		·			
·	·		·		·		
		·		·		·	
−3	0		·		·		·
		0		·		·	
−2	0		0		·		·
		0		1		·	
−1	0		1		−4		·
		1		−3		10	
0	1		−2		6		−20
		−1		3		−10	
1	0		1		−4		15
		0		−1		5	
2	0		0		1		−6
		0		0		−1	
3	0		0		0		1
		0		0		0	
4	0		0		0		·
		0		0		·	
5	0		0		·		·
		0		·			
6	0		·		·		·

Suppose next that there is a small error $y_k \varepsilon_k$ in *each* entry of the table of function values. The result will be the correct difference table plus a sum of difference tables like the one with a single error. Thus the error will be the sum of the errors in the individual tables. To find this expression we apply the above formula to get

$$\text{Error in } \{\Delta^k y_0\} = \sum_{m=0}^{k} (-1)^m C(k, m) y_{k-m} \varepsilon_{k-m}$$

If the $|y_{k-m} \varepsilon_{k-m}|$ are bounded by ε, then the error is bounded by

$$\left| \sum_{m=0}^{k} C(k, m) \right| \varepsilon = (1 + 1)^k \varepsilon = 2^k \varepsilon$$

This bound can be attained if all the errors are the same size but with alternating sign, that is, if

$$y_{k-m} \varepsilon_{k-m} (-1)^m = \varepsilon$$

Most of the time the bound will not be attained and is very pessimistic.

10.4   THE STATISTICS OF ROUNDOFF

Since absolute bounds on the roundoff are apt to be very pessimistic, it is often necessary to resort to more realistic, but admittedly more dangerous, statistical estimates.

The roundoff of a single number y_i is distributed equilikely in the interval $-1/2 \leq x < 1/2$ of the last digit kept. In a continuing product we will have

$$y_1 (1 + \varepsilon_1) y_2 (1 + \varepsilon_2) \cdots y_n (1 + \varepsilon_n)$$

and neglecting the product of two or more ε_i, we find that this is

$$y_1 y_2 \cdots y_n [1 + (\varepsilon_1 + \varepsilon_2 \cdots + \varepsilon_n)]$$

By the central-limit theorem a sum of n independent random numbers approaches the normal distribution as n approaches infinity,

$$\frac{1}{\sqrt{2\pi}} e^{-x^2}$$

A similar effect applies for divisions except that we use

$$\frac{1}{1 + \varepsilon_i} = 1 - \varepsilon_i$$

and the corresponding ε_i appear with minus signs.

For addition and subtraction there is a shift before the mantissas are combined. It is more difficult to say exactly what happens, but the general effect is much the same. While the individual roundoffs are from a flat distribution, the roundoff effect of many arithmetic operations *tends* toward a normal distribution. This is readily verified experimentally.

In a difference table the bound we obtained was again very pessimistic. What is reasonable to expect?

To apply statistics in this case we need first to form a model of the situation we are estimating. The roundoff in a single table is perfectly definite, but we *imagine* that it is one table drawn from an *ensemble* of similar tables but with varying roundoffs. It is natural to suppose that negative and positive roundoffs are equilikely and that the roundoffs at different places in the table are independent. In mathematical notation these assumptions mean that

$$Av\{\varepsilon_i\} = 0$$
$$Av\{\varepsilon_i \varepsilon_j\} = 0 \qquad i \neq j$$

The assumption of independence of the roundoff at successive positions is sometimes dangerous, especially in recursive calculations where the computation for one value depends upon that of the previous value or values.

When we apply our statistical averages, we average the ε_i over the ensemble of tables, *not* over the index i, in the following formulas. We have

$$Av\{y_i \varepsilon_i\} = y_i \, Av\{\varepsilon_i\} = 0$$
$$\text{Variance}\{y_i \varepsilon_i\} = \text{Var}\{y_i \varepsilon_i\}$$
$$= Av\{y_i^2 \varepsilon_i^2\} = \sigma^2$$

where σ^2 is some number which we will later examine more closely and which is *assumed* to be independent of i.

For the kth differences in a table we have correspondingly

$$Av\{\Delta^k y_i \varepsilon_i\} = Av\left\{\sum_{m=0}^{k} (-1)^m C(k, m) y_{k-m} \varepsilon_{k-m}\right\}$$

$$= \sum_{m=0}^{k} (-1)^m C(k, m) y_{k-m} Av\{\varepsilon_{k-m}\} = 0$$

$$\text{Var}\{\Delta^k y_i \varepsilon_i\} = Av\left\{\sum_{m=0}^{k} (-1)^m C(k, m) y_{k-m} \varepsilon_{k-m}\right\}^2$$

$$= \sum_{r=0}^{k} \sum_{s=0}^{k} (-1)^{r+s} C(k, r) C(k, s) y_{k-r} y_{k-s} Av\{\varepsilon_{k-r} \varepsilon_{k-s}\}$$

Only when $r = s$ is the average not zero, hence

$$\text{Var}\{\Delta^k y_i \varepsilon_i\} = \sum_{r=0}^{k} C^2(k, r) \, \text{Av}\{y_{k-r} \varepsilon_{k-r}\}^2$$

$$= \sum C^2(k, r)\sigma^2$$

which defines σ^2 as the original "noise." Using the identity of Prob. 10.4.1,

$$\text{Var}\{\Delta^k y_i \varepsilon_0\} = C(2k, k)\sigma^2 = \frac{(2k)!}{(k!)^2}\sigma^2$$

The result we have is that the variance in the kth difference column is $C(2k, k)$ times the variance in the original table values—differencing *amplifies* the noise in a table by a factor $(2k)!/(k!)^2$. (See Table 10.4.)

Therefore, in order to estimate the variance (noise) in a table, we estimate the noise by finding the noise in the kth difference column and dividing this by $C(2k, k)$.

Sections 10.5 and 10.6 will show which value of k to use.

Table 10.4　ROUNDOFF AMPLIFICATION FACTORS

Order of difference	Variance $C(2k, k)$	Root-mean-square noise $\sqrt{C(2k, k)}$	Maximum noise 2^k
1	2	1.414	2
2	6	2.449	4
3	20	4.472	8
4	70	8.367	16
5	252	15.875	32
6	924	30.397	64

PROBLEMS

10.4.1　Expand both sides of $(1 + t)^{a+b} = (1 + t)^a(1 + t)^b$ and equate like powers of t to get

$$C(a + b, r) = \sum_{s=0}^{a} C(a, s)C(b, r - s)$$

10.4.2　In Prob. 10.4.1 set $a = b = r$ to get

$$C(2r, r) = \sum_{s=0}^{r} C^2(r, s)$$

10.4.3　In Prob. 10.4.1, set $a = b = n$ and $r = n + 1$ to get

$$C(2n, n + 1) = \sum C(n, s)C(n, s - 1)$$

10.5 CORRELATION IN THE kTH DIFFERENCES

The consecutive values in a column of kth differences of a difference table use many of the same numbers, but an examination of how they enter shows that they have opposite signs in the consecutive differences—they are negatively correlated. We need to compute this correlation

$$\text{Av}\left\{\sum_{r=0}^{k} \frac{(-1)^r C(k, r) y_{k-r} \varepsilon_{k-r}}{\sqrt{C(2k, k)}} \cdot \sum_{r=1}^{k+1} \frac{(-1)^{r-1} C(k, r-1) y_{k-r} \varepsilon_{k-r}}{\sqrt{C(2k, k)}}\right\}$$

$$= -\frac{1}{C(2k, k)} \sum_{r=0}^{k} C(k, r) C(k, r-1) \sigma^2$$

$$= -\frac{C(2k, k+1)}{C(2k, k)} \sigma^2 = \frac{-k}{k+1} \sigma^2$$

where we have used the result from Prob. 10.4.3.

As we expected, the coefficient of correlation of the noise from successive entries in the kth column of a difference table is large and negative, $-k/(k+1)$. This is well known in the folklore of computing, but it is rarely estimated by theoretical considerations.

If the roundoff values are drawn from a normal Guassian distribution, then these correlation coefficients imply a probability of a sign change indicated in Table 10.5.

This alternation of sign will give us a way of estimating when the difference table column is noise rather than signal.

This is a theoretical result in the sense that we have averaged over the imagined ensemble of similar tables, and in practice all we have is the single table.   The

Table 10.5*

k	$-k/k+1$	Probability, percentage
0	$0 = 0$	.500
1	$-1/2 = -0.500$	.668
2	$-2/3 = -0.667$	.733
3	$-3/4 = -0.750$	.770
4	$-4/5 = -0.800$	.796
5	$-5/6 = -0.834$	.814
6	$-6/7 = -0.857$	.828
7	$-7/8 = -0.875$	.839
8	$-8/9 = -0.890$	.850
9	$-9/10 = -0.900$	.857
10	$-10/11 = -0.910$	.864
20	$-20/21 = -0.952$	.902

* From Roger Pinkham.

solution to this dilemma is the standard one in such situations; we appeal to the *ergodic hypothesis* which states that (for suitable situations) " the average over the ensemble is the same as the average over the function." A simple illustration of this is the very common habit of looking at, say, a table of 100,000 deaths and applying this to a single life. Or again, looking at the distribution of salaries paid (as of the current year) as a function of age and seeing the individual progress along the chart as he ages.

The ergodic hypothesis can be applied when the statistics are stationary (not changing in time or from place to place). In practice we rarely have perfectly stationary situations, and we must make judgments as to how much the results could be altered by the changes. In the two examples above, we sense that the health of the general population is not changing so rapidly that the table of 100,000 deaths is totally irrelevant, and the effect of the slow changes can be mentally added on. In the salary curve example, however, inflation and the changing status of jobs and job titles all make the curves dangerous to use as a basis for applying the ergodic hypothesis. Similarly, in tables and other computations we do not expect to find that all the results have the same level of roundoff noise. We can apply the hypothesis only to runs where the statistics do not seem to be changing too rapidly. In particular, the interval we use of the difference column should have numbers about the same size and about the same density as the sign changes.

10.6 ESTIMATION OF ROUNDOFF IN A TABLE

We have now examined all the parts needed in a theory of the estimation of roundoff in a table of results computed for equally spaced values of a variable or a parameter. The importance of being able to make an estimate of the random error in a table of results *from the results alone* should be recognized; it is not a perfect answer, but it is better than nothing.

First we have shown that the $(N + 1)$st differences of a polynomial are zero, and from experience we know that many functions are locally very like a polynomial. Thus we expect that the differences of many functions will approach zero as more and more differences are taken.

If the differences do not become small fairly rapidly as k becomes large, then it is likely that the spacing of the values is too large. It is easy to see that if the spacing is halved, then the first differences are about half as large as they were, the second differences one-quarter as large, the third, one-eighth, and in general the kth differences will be about $1/2^k$ times as large as they were. Thus it is generally true that, except near a singularity, for a smooth function the differences approach zero as k gets large.

On the other hand, we have just seen that random noise in a table rapidly approaches infinity and that the successive values in a particular column tend strongly to alternate in sign. Given a table of computed results, we expect a mixture of the two effects, the true differences approaching zero and the noise rapidly growing and alternating in sign.

In most reasonably planned computations the fourth or fifth differences begin to show the oscillation we identify with the noise. Very smooth computations, such as orbits of the planets in space, may have reasonably smooth differences up to at least the tenth order, and very rough computations may have the second or third difference showing alternating signs; but typically, as we said, it is at about the fourth or fifth difference that oscillation occurs. We will, for the sake of simplicity, suppose that we are interested in the fourth or fifth difference.

We need a specific test for a machine to make. Just what we use depends on how certain we wish to be, but as a general rule if we find two out of three or, safer, three out of four sign changes in consecutive values of a difference column, then we will call that the noise.

To estimate the noise we compute the average of squares (the variance) of this difference column, using all the values appearing to have somewhat the same oscillation rate and size. Using the ergodic hypothesis, we equate this number to the average over the imagined ensemble of tables. To get from the noise in the kth differences back to the noise in the table we divide by $C(2k, k) = (2k)!/(k!)^2$. This gives the square of the average noise in the original table.

Table 10.6.1

x	$\Gamma(x)$	Δ	Δ^2	Δ^3	Δ^4
1.0	1.000				
		-49			
1.1	0.951		$+16$		
		-33		-4	
1.2	0.918		$+12$		3
		-21		-1	
1.3	0.897		$+11$		-1
		-10		-2	
1.4	0.887		$+9$		$+2$
		-1		0	
1.5	0.886		$+9$		-2
		$+8$		-2	
1.6	0.894		$+7$		$+2$
		$+15$		0	
1.7	0.909		$+7$		$+2$
		$+22$		$+2$	
1.8	0.931		$+9$		
		$+31$			
1.9	0.962				

As an example consider Table 10.6.1 of the gamma function.   The average of the squares of the fourth differences is

$$\mathrm{Var}\{\Delta^4\} = \frac{9 + 1 + 4 + 4 + 4 + 4}{6} = \frac{26}{6} = \frac{13}{3}$$

We divide this by $C(8, 4) = 70$ to get $0.062 = \sigma^2$ which is an estimate of the noise in the table.

The table supposedly has only random roundoff in the last digit, and the corresponding model for the flat distribution is

$$\sigma_2{}^2 = \int_{-1/2}^{1/2} x^2 \, dx = \tfrac{1}{12} = 0.0833.$$

Thus we have the three values

$$\sigma_1 = 0.25$$
$$\sigma_2 = 0.29$$
$$\sigma_3 = 0.32$$

Table 10.6.2

x	$\zeta(x)$	$\Delta\zeta$	$\Delta^2\zeta$	$\Delta^3\zeta$	$\Delta^4\zeta$	$\Delta^5\zeta$
2	1.64493					
		-0.44287				
3	1.20206		0.32313			
		-0.11974		-0.24878		
4	1.08232		7435		0.20023	
		-0.04539		-4855		-0.16688
5	1.03693		2580		3335	
		-0.01959		-1520		-2403
6	1.01734		1060		932	
		-0.00899		-588		-596
7	1.00835		472		336	
		-0.00427		-252		-199
8	1.00408		220		137	
		-0.00207		-115		-75
9	1.00201		105		62	
		-0.00102		-53		-35
10	1.00099		52		27	
		-0.00050		-26		-16
11	1.00049		26		11	
		-0.00024		-15		-0
12	1.00025		11		11	
		-0.00013		-4		-11
13	1.00012		7		0	
		-0.00006		-4		$+3$
14	1.00006		3		3	
		-0.00003		-1		-4
15	1.00003		2		-1	
		-0.00001		-2		
16	1.00002		0			
		-0.00001				
17	1.00001					

where σ_3 is from a careful, exact computation of the roundoffs (knowing the correct values from a much more accurate table).   In view of the smallness of our sample the estimate is fairly good.

We observed that the theory need not apply near a singularity.   The zeta function

$$\zeta(x) = \sum_{k=1}^{\infty} \frac{1}{k^x}$$

clearly is infinite at $x = 1$ and has a singularity there.   Table 10.6.2 shows how the differences go.

We see signs of roundoff noise at the bottom of the fifth difference column, but at the top we see the effect of the singularity.

On the other hand, a table of a function like the error integral shows that the effects we have claimed usually happen (see Table 10.6.3)

Table 10.6.3

t	$f(t)$	Δ	Δ^2	Δ^3	Δ^4	Δ^5
0.00	0.0000					
		987				
0.25	0.0987		−59			
		928		−50		
0.50	0.1915		−109		+19	
		819		−31		+4
0.75	0.2734		−140		+23	
		679		−8		−10
1.00	0.3413		−148		+13	
		531		+5		+4
1.25	0.3944		−143		+17	
		388		+22		−11
1.50	0.4332		−121		+6	
		267		+28		−10
1.75	0.4599		−93		−4	
		174		+24		+4
2.00	0.4773		−69		0	
		105		+24		−7
2.25	0.4878		−45		−7	
		60		+17		+3
2.50	0.4938		−28		−4	
		32		+13		−4
2.75	0.4970		−15		−8	
		17		+5		+10
3.00	0.4987		−10		+2	
		7		+7		−9
3.25	0.4994		−3		−7	
		4		0		+9
3.50	0.4998		−3		+2	
		1		+2		
3.75	0.4999		−1			
		1				
4.00	0.5000					

PROBLEMS

10.6.1 Estimate the noise in Table 10.6.3.

10.6.2 Carry out the differences in Table 10.6.2 to order 7 and use the last few values to make a " desperate " estimate of the noise.

10.6.3 Discuss using the ergodic hypothesis to estimate the total high school population from the fraction of one's life that is spent in high school. Discuss the effects of nonstationarity in the population.

10.6.4 Discuss the roundoff in Table 10.5 of correlation coefficient probabilities.

10.7 ISOLATED ERRORS

It was very common in hand calculation to have an isolated error in a table; it is less common but by no means impossible in modern machine computation. The isolated error may well be due to the fact that for that particular value an indeterminate form arose and led to large roundoff errors.

It is very easy to locate the error by looking at the difference table. We know from Table 10.3 that this leads to a series of terms of the error times the binomial coefficients with alternating signs. For example, in reviewing Table 10.7, the author found an error by this simple device of looking at the difference table.

The pattern suggested placing the fourth differences opposite the value of 35°. Using the middle three values

$$41 = -4\varepsilon + C$$
$$-59 = 6\varepsilon + C$$
$$37 = -4\varepsilon + C$$

we solve the equations approximately. They suggest $\varepsilon = -10$. If we change the value at 35° to 6,421, we will have the corrections to the difference table

				—4
			8	
		—50		1
	—468		9	
6,421		—41		1
	—509		10	
		—31		—3
			7	
				0

and this looks much better (and the error represents a typical typographical error).

Occasionally it is necessary to supply a missing value in a table of equally spaced values.   For this we go back to the differences written out in terms of the original function values, Sec. 10.3.   If we pick an even-order difference, say 6, and suppose that this order of differences is zero, then we have, assuming for convenience that the error is at zero,

$$y_{-3} - 6y_{-2} + 15y_{-1} - 20y_0 + 15y_1 - 6y_2 + y_3 = 0$$

or

$$y_0 = \tfrac{1}{20}[(y_3 + y_{-3}) - 6(y_2 + y_{-2}) + 15(y_1 + y_{-1})]$$

Table 10.7

Degree	$C_n(z)$	Δ	Δ^2	Δ^3	Δ^4	Theoretical fourth difference due to error ε
0	8,346					
		−44				
5	8,302		−87			
		−131		4		
10	8,171		−83		3	
		−214		7		
15	7,957		−76		−1	
		−290		6		
20	7,667		−70		6	
		−360		12		
25	7,307		−58		−14	$+\varepsilon$
		−418		−2		
30	6,889		−60		41	-4ε
		−478		39		
35	6,411		−21		−59	$+6\varepsilon$
		−499		−20		
40	5,912		−41		37	-4ε
		−540		17		
45	5,372		−24		−10	$+\varepsilon$
		−564		7		
50	4,808		−17		−2	
		−581		5		
55	4,227		−12		0	
		−593		5		
60	3,634		−7		−2	
		−600		3		
65	3,034		−4		−2	
		−604		1		
70	2,430		−3		2	
		−607		3		
75	1,823		−0		−4	
		−607		−1		
80	1,216		−1		2	
		−608		1		
85	608		0			
		−608				
90	0					

PROBLEMS

10.7.1 Estimate the noise in the corrected table in this section.

10.7.2 Correct Table 10.7 by regarding the values of 35° as missing.

10.7.3 Discuss the dangers of trying to correct a bad value because the formula at that point was indeterminate.

10.8 SYSTEMATIC ERRORS

We have provided a method of estimating the random errors in a table. The method does nothing about systematic errors. One way to check an important table is to compute, by some *independent* method, several key values scattered across the range of values. This is easy to say (and hard to do in many cases) but without some redundant, independent calculations there can be no realistic checking of the results. The fact that the theoretician accepts the results is no measure of their correctness because, in a very real sense, a good theoretician can account for any new phenomenon he meets—it is his business to create new theories to fit new data !

The situation is the same as in experimental work. It is possible to estimate, from the measurements alone, the random component of the error, but the calculation of the systematic errors is another thing entirely and requires other methods.[1] The fact that it is hard is no reason for ignoring the problem. Any numerical analyst must, if he is a responsible scientist and not a hack technician, assume responsibility for the accuracy of his results and some of the burden for the correctness of the original formulation, as well as for the interpretation of the results. The job of a numerical analyst neither begins with the given equations nor ends with the output sheet. Realistic estimates of the total error in the answer are partly his responsibility, not exclusively that of the problem's proposer.

---

[1] W. J. Youden, Enduring Values, *Technometrics*, vol. 14, no. 1, February 1972, pp. 1–11.

11

*THE SUMMATION CALCULUS

11.1 INTRODUCTION

The difference and integral calculi are related to each other by two formulas

$$f(x) = \frac{d}{dx} \int_a^x f(x)\, dx$$

$$f(x) - f(a) = \int_a^x \frac{df}{dx}\, dx$$

Similarly, the difference and summation calculi are related by

$$f(n+1) = \Delta \sum_{k=0}^n f(k) = \sum_{k=0}^{n+1} f(k) - \sum_{k=0}^n f(k)$$

$$f(n+1) - f(0) = \sum_{k=0}^n \Delta f(k) = \sum_{k=0}^n [f(k+1) - f(k)]$$

where we have used the common summation notation

$$\sum_{k=a}^n f(k) = f(a) + f(a+1) + \cdots + f(n) \qquad \sum_{k=a}^a f(a) = f(a) \qquad \sum_{k=a}^{a-1} f(a) = 0$$

Because of the arguments $n + 1$ in the two formulas, Boole [4], Jordan [25], and others have used the notation for the sum

$$\sum_{k=a}^{n} f(k) = f(a) + f(a + 1) + \cdots + f(n - 1) \qquad \sum_{k=a}^{a} f(k) = 0$$

However, this differs from the common usage in mathematics, and we shall therefore avoid laying that trap for the unwary and instead accept the annoyance of the $n + 1$ that occurs in some of the arguments.   J. W. Tukey[1] has suggested the elegant notation

$$\sum_{k=a}^{<b} f(k) = f(a) + f(a + 1) + \cdots + f(b - 1)$$

as the solution to this dilemma, but we will not use it here.

In the integral calculus a table of derivatives of various functions is inverted to get the basic table of integrals.   For example, from

$$\frac{d}{dx} x^n = nx^{n-1}$$

we get

$$\int x^n \, dx = \frac{x^{n+1}}{n + 1} + C$$

In a similar way, from the table of first differences of various functions we get (using unit spacing in the differences)

$$\sum_{x=0}^{m} x^{(n)} = \frac{(m + 1)^{(n+1)}}{n + 1} \qquad n \neq -1$$

$$\sum_{x=0}^{m} a^x = \frac{a^{m+1} - 1}{a - 1} \qquad a \neq 1$$

$$\sum_{x=1}^{m} \ln x = \ln(m!)$$

$$\sum_{x=0}^{m} \sin(ax + b) = -\frac{\cos[a(m + \frac{1}{2}) + b] - \cos(-a/2 + b)}{2 \sin(a/2)}$$

$$= \frac{\sin[a(m + 1)/2]\sin(am/2 + b)}{\sin(a/2)}$$

$$\sum_{x=0}^{m} \cos(ax + b) = \frac{\sin[a(m + \frac{1}{2}) + b] - \sin(-a/2 + b)}{2 \sin(a/2)}$$

$$= \frac{\sin[a(m + 1)/2]\cos(am/2 + b)}{\sin(a/2)}$$

Many others are easily found, but we do not need them now.

[1] J. W. Tukey (1915–——).

PROBLEMS

11.1.1 Compute the sum $1 + 2 + 4 + \cdots + 2^{64}$.

11.1.2 Show that

$$\sum_{x=0}^{2N-1} \cos \frac{\pi x}{N} = 0 \qquad \sum_{x=0}^{2N-1} \sin \frac{\pi x}{N} = 0$$

11.1.3 Using Prob. 11.1.2, show that

$$\sum_{x=0}^{2N-1} \cos \frac{\pi k x}{N} \cos \frac{\pi m x}{N} = N\delta(m, k) \qquad 0 < m + k < 2N$$

$$\sum_{x=0}^{2N-1} \cos \frac{\pi k x}{N} \sin \frac{mx}{N} = 0$$

$$\sum_{x=0}^{2N-1} \sin \frac{\pi k x}{N} \sin \frac{\pi m x}{N} = N\delta(m, k) \qquad 0 < m + k < 2N$$

11.1.4 Do the same as in Prob. 11.1.3 except use $N - 1$ as the upper limit and 2π in place of π.

11.2 SUMMATION BY PARTS

An examination of the general processes used in the calculus of analytic integration reveals that there are two principal tools:

1 Change of variables
2 Integration by parts

Of these two methods, the first is not available in the summation calculus, which depends on equally spaced arguments, where usually $h = 1$. This makes computing a sum in an analytic, closed form much more difficult than computing an integral.

On the other hand the analog of integration by parts is summation by parts, and it plays the corresponding role in its field. Integration by parts in the calculus is based on the derivative of a product

$$d(uv) = u \, dv + v \, du$$

From the above equation the formula for integration by parts

$$\int u \, dv = uv - \int v \, du$$

is found by integration. Similarly, from the formula for the difference of a product

$$\Delta u(x)v(x) = u(x) \, \Delta v(x) + v(x + 1) \, \Delta u(x)$$

we obtain by summation

$$\sum_{x=0}^{m-1} u \, \Delta v = u(m)v(m) - u(0)v(0) - \sum_{x=0}^{m-1} v(x + 1) \, \Delta u(x)$$

We can choose $v(x)$ so that $v(0)$ equals zero (or any other value we please); just as in the calculus the constant of integration in finding the v can be any number we please (conventionally it is picked to be zero).

As an example of the use of summation by parts, consider

$$\sum_{x=0}^{m-1} xa^x = \frac{xa^x}{a-1}\bigg|_0^m - \sum_0^{m-1} \frac{a^{x+1}}{a-1}\cdot 1$$

$$= \frac{xa^x}{a-1} - \frac{a^{x+1}}{(a-1)^2}\bigg|_0^m = \frac{a}{(1-a)^2}[(m-1)a^m - ma^{m-1} + 1] \qquad a \neq 1$$

In general, the use of summation by parts closely resembles that of integration by parts. For example, summations such as

$$\sum \frac{1}{2^x}\sin\theta x$$

are accomplished by using summation by parts twice and then combining like terms. It can also be done by writing it as

$$\left\{\left(\frac{e^{i\theta}}{2}\right)^x - \left(\frac{e^{-i\theta}}{2}\right)^x\right\}\frac{1}{2i}$$

and summing directly.

PROBLEMS

11.2.1 Show that

$$\sum_{x=1}^{n} \frac{1}{2^x}\sin\theta x = \frac{2\sin\theta - 2^{-n}[2\sin\theta(n+1) - \sin\theta]}{1 + 8\sin^2(\theta/2)}$$

11.2.2 Sum $\displaystyle\sum_{x=1}^{n} x^2\cos x$.

11.2.3 Sum $\displaystyle\sum_{x=1}^{n} x\ln x$.

11.2.4 Sum $\displaystyle\sum_{x=0}^{n} x^2\cdot 2^x$.

11.2.5 Sum $\displaystyle\sum_{x=0}^{n} x^2\cdot x^{(2)}$.

11.3   SUMMATION OF POWERS OF x

The special case of summing the powers of the consecutive integers is of enough importance to justify a special examination. We could, of course, sum them by converting the power of x via the Stirling numbers of the second kind (Sec. 9.8) and summing the result by the obvious formula.

Instead, we shall approach the problem much as Jacob Bernoulli [3] did, and in the process we will introduce two valuable by-products: (1) the idea of a generating function, and (2) the Bernoulli numbers.   Both will have further use.

The summation of the first few cases of the powers of x is easily remembered if written in the form

$$\sum_{x=1}^{n} x = \sum_{1}^{n} x^{(1)} = \frac{(n+1)n}{2}$$

$$\sum_{x=1}^{n} x^2 = \sum_{1}^{n} (x^{(2)} + x^{(1)}) = \frac{(n+1)n(2n+1)}{6}$$

$$\sum_{x=1}^{n} x^3 = \sum_{1}^{n} (x^{(3)} + 3x^{(2)} + x^{(1)}) = \left[\frac{(n+1)n}{2}\right]^2$$

$$\sum_{x=1}^{n} x^4 = \frac{(n+1)n(2n+1)}{6} \frac{3n^2 + 3n - 1}{5}$$

$$\sum_{x=1}^{n} x^5 = \left[\frac{(n+1)n}{2}\right]^2 \frac{2n^2 + 2n - 1}{3}$$

$$\sum_{x=1}^{n} x^6 = \frac{(n+1)n(2n+1)}{6} \frac{3n^4 + 6n^3 - 3n + 1}{7}$$

$$\sum_{x=1}^{n} x^7 = \left[\frac{(n+1)n}{2}\right]^2 \frac{3n^4 + 6n^3 - n^2 - 4n + 2}{6}$$

The summation formula for $x^{(n)}$ is also valid for negative exponents ($n \neq -1$).   For example,

$$\sum_{x=1}^{m} \frac{1}{x(x+1)} = \sum_{x=1}^{m} (x-1)^{(-2)} = \frac{m^{(-1)} - 0^{(-1)}}{-1}$$

$$= -\frac{1}{m+1} + 1 = \frac{m}{m+1}$$

Similarly,

$$\sum_{1}^{m} \frac{1}{x(x+1)(x+2)} = \sum_{1}^{m} (x-1)^{(-3)} = \frac{m^{(-2)} - 0^{(-2)}}{-2}$$

$$= \frac{1}{2}\left[\frac{1}{1 \cdot 2} - \frac{1}{(m+1)(m+2)}\right]$$

PROBLEMS

11.3.1 Verify the formulas for x^3 and x^5, using the synthetic-division method.

11.3.2 Using $\Delta C(n, k) = C(n + 1, k) - C(n, k) = C(n, k - 1)$, show that

$$\sum_{x=k}^{m} C(x, k) = C(m + 1, k + 1)$$

11.3.3 Show that

$$\sum_{x=1}^{n} (2x - 1) = n^2$$

$$\sum_{x=1}^{n} (2x - 1)^2 - \frac{n(4n^2 - 1)}{3}$$

$$\sum_{y=1}^{n} (2x - 1)^3 = n^2(2n^2 - 1)$$

11.3.4 Compute

$$\sum_{x=1}^{n} \frac{1}{x(x + 1)(x + 2)(x + 3)}$$

11.3.5 Compute

$$\sum_{m=0}^{N-1} \sin^m \theta \qquad \theta \neq 0, \pm\pi, \ldots$$

Ans. $(1 - \sin^N \theta)/(1 - \sin \theta)$

11.4 GENERATING FUNCTIONS

The idea of a generating function is simple once it is well understood, but initially it seems to be baffling. We will therefore begin by a few simple examples.

The binomial coefficients of a given order n

$$C(n, k) \qquad k = 0, 1, 2, \ldots, n)$$

can be multiplied by the corresponding t^k and written as a sum. The binomial theorem shows that this sum is easily written as

$$(1 + t)^n = C(n, 0) + C(n, 1)t + C(n, 2)t^2 + \cdots + C(n, n)t^n$$

The expression $(1 + t)^n$ is said to be a *generating function* for the binomial coefficients.

The power of generating functions lies in the ease of manipulating them in certain processes. For example, again using the binomial coefficients, if we multiply a generating function of order m by a second of order n, we get the generating function of order $m + n$; that is,

$$(1 + t)^m (1 + t)^n = (1 + t)^{m+n}$$

Thus we have

$$\sum_{k=0}^{m} C(m, k)t^k \sum_{j=0}^{n} C(n, j)t^j = \sum_{i=0}^{m+n} C(m+n, i)t^i$$

Equating like powers of t, we get, as in Prob. 10.4.1, $m + n + 1$ identities.

$$\sum_{k=0}^{m} C(m, k)C(n, i-k) = C(m+n, i) \qquad i = 0, 1, \ldots, m+n$$

Consider next the Stirling numbers of the first kind

$$x^{(n)} = \sum_{k=0}^{n} s(n, k)x^k$$

as another example of a generating function for a set of numbers

$$s(n, k) \qquad k = 0, 1, \ldots, n$$

with x as the "dummy variable."

Instead of using only powers of t as the place holders we may use $t^k/k!$ if we please, or any other multiple of t^k. That is why we used the words *a generating function for*, though usually *the generating function* means just t^k. The use of $t^k/k!$ is often called "the exponential generating function."

PROBLEMS

11.4.1   Show that

$$\sum_{k=0}^{n}(-1)^k C(n, k) = 0$$

11.4.2   Show that

$$\sum_{k=0}^{n} C(n, k) = 2^n$$

11.4.3   In the generating function for the Stirling numbers of the first kind set the variable equal to 1 to obtain

$$\sum_{k=0}^{n} s(n, k) = 0 \qquad n \geq 1$$

11.5   SUMS OF POWERS OF n AGAIN

We propose to find the sums [where $S(n, p)$ is *not* a Stirling number]

$$S(n, p) = 1^p + 2^p + 3^p + \cdots + n^p = \sum_{k=1}^{n} k^p$$

using the method of generating functions.   In this method we multiply $S(n, p)$ by $t^p/p!$ and sum over all p to get

$$\sum_{p=0}^{\infty} \frac{S(n, p)t^p}{p!} = \sum_{p=0}^{\infty} \sum_{k=1}^{n} \frac{k^p t^p}{p!}$$

$$= \sum_{k=1}^{n} \sum_{p=0}^{\infty} \frac{(kt)^p}{p!} = \sum_{k=1}^{n} e^{kt}$$

$$= \frac{e^{(n+1)t} - e^t}{e^t - 1}$$

Thus we have the exponential generating function for the sums.

To find the coefficient of $t^p/p!$, we write

$$\frac{e^{(n+1)t} - e^t}{e^t - 1} = \frac{e^{nt} - 1}{t} \frac{te^t}{e^t - 1}$$

The first factor can be written in the form

$$\frac{e^{nt} - 1}{t} = \frac{\sum_{k=0}^{\infty} (nt)^k/k! - 1}{t} = \sum_{k=0}^{\infty} \frac{n^{k+1} t^k}{(k+1)!}$$

The second factor requires more attention.   We begin with the expansion (see Sec. 11.6 for details)

$$\frac{t}{e^t - 1} = \sum_{k=0}^{\infty} \frac{B_k t^k}{k!}$$

Solving for the first few coefficients, we find $B_0 = 1$ and $B_1 = -1/2$, so that

$$\frac{t}{e^t - 1} = 1 - t/2 + \sum_{k=2}^{\infty} \frac{B_k t^k}{k!}$$

Transposing the $t/2$ term, we have

$$t\left(\frac{1}{e^t - 1} + \frac{1}{2}\right) = 1 + \sum_{k=2}^{\infty} \frac{B_k t^k}{k!}$$

If we replace t by $-t$ on the left-hand side, we get

$$-t\left[\frac{1}{e^{-t} - 1} + \frac{1}{2}\right] = -\frac{t}{2} \frac{1 + e^{-t}}{e^{-t} - 1} = \frac{t}{2} \frac{e^t + 1}{e^t - 1}$$

$$= \frac{t}{e^t - 1} + \frac{t}{2}$$

which is the original expression.   Thus, except for B_1 which equals $-1/2$, all the other $B_{2k+1} = 0$ $(k = 1, 2, \ldots)$.

From this we see that the second factor

$$\frac{te^t}{e^t - 1} = t + \frac{t}{e^t - 1}$$

$$= t + 1 - \frac{t}{2} + \sum_{m=1}^{\infty} \frac{B_{2m} t^{2m}}{(2m)!}$$

$$= 1 + \frac{t}{2} + \sum_{m=1}^{\infty} \frac{B_{2m} t^{2m}}{(2m)!}$$

To find the expansion of the product of the two factors, we multiply the two series together and pick out the coefficient of $t^p/p!$

$$p! \left[\frac{n^{p+1}}{(p+1)!} + \frac{n^p}{p! \, 2} + \frac{n^{p-1} B_2}{(p-1)! \, 2!} + \frac{n^{p-3} B_4}{(p-3)! \, 4!} + \cdots \right]$$

$$= \frac{n^{p+1}}{p+1} + \frac{n^p}{2} + \frac{p}{2} B_2 n^{p-1} + \frac{p(p-1)(p-2)}{2 \cdot 3 \cdot 4} B_4 n^{p-3}$$

$$+ \frac{p(p-1)(p-2)(p-3)(p-4)}{2 \cdot 3 \cdot 4 \cdot 5 \cdot 6} B_6 n^{p-5} + \cdots$$

which is what Bernoulli obtained [3, p. 97].

PROBLEMS

11.5.1 We can write the generating function as

$$\frac{e^{(n+1)t} - e^{-t}}{e^t - 1} = \frac{e^{(n+1)t} - 1}{e^t - 1} - 1$$

$$= \frac{e^{(n+1)t} - 1}{t} \frac{t}{e^t - 1} - 1$$

and get an expansion for $S(n, p)$ in powers of $n + 1$.   Carry out the details.

11.5.2 Show that the formulas in Sec. 11.3 follow from

$$B_0 = 1 \qquad B_1 = -1/2 \qquad B_2 = 1/6 \qquad B_4 = -1/30 \qquad B_6 = 1/42.$$

11.6 THE BERNOULLI NUMBERS

The generating function for the Bernoulli numbers is

$$\frac{t}{e^t - 1} = \sum_{k=0}^{\infty} \frac{B_k t^k}{k!}$$

We have already found that

$$B_0 = 1 \qquad B_1 = -1/2 \qquad B_{2k+1} = 0 \qquad k = 1, 2, \ldots$$

To investigate these numbers further we write the first equation as

$$t = (e^t - 1) \sum_{k=0}^{\infty} \frac{B_k t^k}{k!}$$

$$= \sum_{m=1}^{\infty} \frac{t^m}{m!} \sum_{k=0}^{\infty} \frac{B_k t^k}{k!}$$

$$= \sum_{p=1}^{\infty} t^p \sum_{s=0}^{p-1} \frac{B_s}{s!(p-s)!}$$

For $p = 1$ we have, equating like powers of t,

$$t = t \cdot B_0 \rightarrow B_0 = 1$$

For $p > 1$

$$0 = \sum_{s=0}^{p-1} \frac{B_s}{s!(p-s)!}$$

Multiply by $p!$

$$0 = \sum_{s=0}^{p-1} \frac{p!}{s!(p-s)!} B_s = \sum_{s=0}^{p-1} C(p, s) B_s$$

This formula is easily remembered by the mnemonic of writing it as

$$(B + 1)^p - B^p = 0 \qquad p > 1$$

and then writing each B^k as B_k.

This type of symbolic manipulation is often called the "umbral calculus" and is widely used in formal manipulations in combinatorial problems.

To find more of the Bernoulli numbers we set $p = 2, 3, \ldots$:

$$p = 2 \qquad 0 = C(2, 0)B_0 + C(2, 1)B_1 \rightarrow B_1 = -1/2$$

$$p = 3 \qquad 0 = B_0 + 3B_1 + 3B_2 \rightarrow B_2 = 1/6$$

$$p = 4 \qquad 0 = B_0 + 4B_1 + 6B_2 + 4B_3 \rightarrow B_3 = 0$$

$$p = 5 \qquad 0 = B_0 + 5B_1 + 10B_2 + 10B_3 + 5B_4 \rightarrow B_4 = -1/30$$

Further calculations give

$$B_6 = 1/42 \qquad B_8 = -1/30 \qquad B_{10} = 5/66$$

A table of B_{2k} up to B_{60} can be found in Jordan [25].

We will see in the next chapter that the Bernoulli numbers occur in many places in analysis.

PROBLEMS

11.6.1 Compute B_6.

11.6.2 From the generating function for the reciprocal factorials

$$\sum_{k=0}^{\infty} \frac{t^k}{k!} = e^t$$

derive a relation for the binomial coefficients using

$$e^t \cdot e^t = e^{2t}$$

12

*INFINITE SERIES

12.1 INTRODUCTION

Most books on infinite series discuss convergence, divergence, and "summability" of series at great length, but almost completely neglect the actual computation (summation) of series. One reason for this is the paucity of methods for summing a series in closed form. Of course, if the indefinite summation can be done in closed form and if the series converges, then the infinite series can also be summed by letting the upper limit go to infinity. As an example, we had in Sec. 11.1

$$\sum_{x=1}^{n} \frac{1}{x(x+1)} = 1 - \frac{1}{n+1}$$

Hence

$$\sum_{x=1}^{\infty} \frac{1}{x(x+1)} = 1$$

In general,

$$\sum_{x=1}^{\infty} \frac{1}{x(x+1)\cdots(x+k-1)} = \frac{1}{k-1}\frac{1}{(k-1)!} \qquad k \geq 2 \qquad (12.1.1)$$

Frequently a problem in analysis can be reduced to that of evaluating an infinite series. Many times it is more work to evaluate the series than to do the original problem, but sometimes the series representation is an advantage. One of the reasons that the series is preferable is that the error committed, when only a finite number of terms are used, is more easily controlled than the errors made in the direct approach. Thus in the problem of evaluating

$$\int_0^x e^{-t^2}\, dt$$

for values of x, say less than 1, we can expand the exponential in an infinite series and integrate term by term to obtain

$$\int_0^x e^{-t^2}\, dt = \int_0^x \sum_{k=0}^{\infty} \frac{(-1)^k t^{2k}}{k!}\, dt = \sum_{k=0}^{\infty} \frac{(-1)^k x^{2k+1}}{(2k+1)k!}$$

If we are interested in values of x less than 1 and want eight figures correct, then taking 11 terms will suffice. This follows from these facts: the series is alternating in sign, the terms are monotone, the series is convergent, and the first neglected term relative to x has a denominator about 9.2×10^8. Were we to try to estimate the value of the integral by some approximate method of integration taken up later in the book, then the problem of estimating the error would be more difficult.

Occasionally infinite products occur. For example, consider the following problem: around a circle of radius 1 draw an equilateral triangle, around the triangle draw a circle, around the circle draw a square, around the square draw a circle, around the circle draw a regular pentagon, etc. See Fig. 12.1.1. What will happen to the radius of the circle as the process goes to infinity? For the

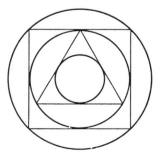

FIGURE 12.1.1

stage of drawing a circle around the N-sided polygon, let that radius be R_N. From the picture of one stage we easily see that

$$\frac{R_{N-1}}{R_N} = \cos \frac{180°}{N}$$

$$R_N = \frac{R_{N-1}}{\cos(180°/N)} = \frac{R_2}{\prod_{n=3}^{N} \cos(180°/n)}$$

$$= \frac{1}{\prod_{n=3}^{N} \cos(180°/n)}$$

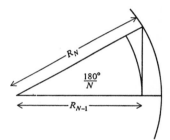

FIGURE 12.1.2

From Table 12.1 we easily see that the sequence of partial products converges to a value not far from 8.7.... In a sense we have experimentally determined the convergence and limiting value of the radius of the circle as the number of sides approaches infinity.

Table 12.1

N	R	N	R	N	R	N	R
						1,000	8.6572
		10	5.2467	100	8.2831	10,000	8.6959
		20	6.8367	200	8.4885	20,000	8.6981
3	2.0000	30	7.3997	300	8.5583	30,000	8.6993
4	2.8284	40	7.7017	400	8.5935	50,000	8.6996
5	3.4961	50	7.8900	500	8.6146	100,000	8.6996
6	4.0370	60	8.0184	600	8.6288	200,000	8.6996
7	4.4807	70	8.1118	700	8.6389		
8	4.8499	80	8.1826	800	8.6465		
9	5.2467	90	8.2383	900	8.6524		

PROBLEMS

12.1.1 Prove Eq. (12.1.1).

12.1.2 Write the series for

$$Si(x) = \int_0^x \frac{\sin t}{t}\, dt$$

How many terms do you need in order to compute $Si(2\pi)$ to eight decimal places?

12.2 KUMMER'S[1] COMPARISON METHOD

A method we shall repeatedly use is the following. Given a problem which we cannot solve easily, we look for a " nearby " problem which we can solve, and then we compute or estimate the difference between the two problems. The closer the nearby problem is to the given one, the smaller the difference we compute; hence with correspondingly less effort we can find an accurate answer to the given problem.

If a given series converges rapidly, then there is no serious problem in planning the computation of its sum. If the series converges slowly, then we look for a series with a known sum which converges at " about the same rate " as the given series. The difference between them will therefore converge more rapidly than the original series. Thus we mean by the words *about the same rate* that the difference approaches zero rapidly; and the more rapidly it reaches zero, the more it is at about the same rate. This is known as Kummer's method.

To fix notation, let the given series be

$$S = \sum_{x=r}^{\infty} a_x$$

and suppose we know the sum

$$S' = \sum_{x=r}^{\infty} C_x a_x$$

where $C_x \to C$ as $x \to \infty$. Then

$$S = \frac{S'}{C} + \sum_{x=r}^{\infty} \left(1 - \frac{C_x}{C}\right) a_x$$

The term in the parenthesis approaches 0 as $x \to \infty$, and so the series to be computed will converge more rapidly than the original series.

[1] Ernst Eduard Kummer (1810–1893).

For example, consider the series

$$S = \sum_{x=1}^{\infty} \frac{x}{(x^2 + 1)^2}$$

which converges like $1/x^3$.   This suggests using

$$S' = \sum_{x=2}^{\infty} \frac{1}{(x - 1)x(x + 1)} = \frac{1}{4}$$

as a comparison series.   Note that we had to start at $x = 2$ and omit the first term. We write

$$S = S' + (S - S')$$

$$= S' + \frac{1}{4} + \sum_{x=2}^{\infty} \left[\frac{x}{(x^2 + 1)^2} - \frac{1}{(x - 1)x(x + 1)} \right]$$

$$= \frac{1}{4} + \frac{1}{4} + \sum_{x=2}^{\infty} \frac{x^4 - x^2 - (x^4 + 2x^2 + 1)}{x(x^2 - 1)(x^2 + 1)^2}$$

$$= \frac{1}{2} - \sum_{x=2}^{\infty} \frac{3x^2 + 1}{x(x^2 - 1)(x^2 + 1)^2} = 0.39711677 \cdots$$

The new series converges like $1/x^5$ (and we could use a similar series for the $3x^2$ part of the numerator if we wished).

PROBLEMS

12.2.1   Use $\sum_{1}^{\infty} \frac{1}{x(x + 1)}$ to approximate $\sum_{1}^{\infty} 1/x^2$.

12.2.2   Approximate result of Prob. 12.2.1 by $\sum_{1}^{\infty} \frac{1}{x(x + 1)(x + 2)}$.   Continue k steps.

12.2.3   Approximate $\sum_{1}^{\infty} 1/x^3$ by $\sum_{2}^{\infty} \frac{1}{(x - 1)x(x + 1)}$

12.2.4   Approximate $\sum_{x=0}^{\infty} \frac{1}{1 + x^2}$ by $\sum_{x=1}^{\infty} \frac{1}{x(x + 1)}$

12.3   SOME STANDARD SERIES

Evidently Kummer's comparison method requires an arsenal of known series. There are numerous tables of series such as the classic table of Jolley [24] and the modern Russian translation of Gradshteyn and Ryshik [18].

Another source is the series one learns in the normal course of mathematics, such as the trigonometric and the hyperbolic series.   Many of these involve the

Bernoulli numbers introduced in Sec. 11.6.   There we defined Bernoulli numbers by the generating function

$$\frac{x}{e^x - 1} = \sum_{k=0}^{\infty} \frac{B_k x^k}{k!}$$

and noted that all B_k of odd order were 0 except $B_1 = -1/2$.   Thus

$$\frac{x}{e^x - 1} - \frac{x}{2} = \sum_{k=0}^{\infty} \frac{B_{2k} x^{2k}}{(2k)!}$$

But

$$\frac{x}{e^x - 1} + \frac{x}{2} = \frac{x}{2}\left(\frac{e^x + 1}{e^x - 1}\right) = \frac{x}{2}\left(\frac{e^{x/2} - e^{-x/2}}{e^{x/2} + e^{-x/2}}\right)$$

$$= \frac{x}{2} \coth \frac{x}{2} = \sum_{k=0}^{\infty} \frac{B_{2k} x^{2k}}{(2k)!}$$

or writing $2x$ in place of x,

$$\coth x = 2 \sum_{k=0}^{\infty} B_{2k} \frac{(2x)^{2k-1}}{(2k)!} = \frac{1}{x} + \frac{x}{3} - \frac{x^3}{45} + \frac{2x^5}{945} + \cdots$$

If we put $x = iz$ in this, we get

$$\cot z = \frac{1}{z} - \frac{z}{3} - \frac{z^3}{45} - \frac{2z^5}{945} + \cdots$$

$$= 2 \sum_{k=0}^{\infty} (-1)^k \left[\frac{(2z)^{2k-1} B_{2k}}{(2k)!}\right]$$

Now using the trigonometric identity

$$2 \cot 2t = \cot t - \tan t$$

in the form

$$\tan t = \cot t - 2 \cot 2t$$

we get

$$\tan t = \sum_{k=0}^{\infty} \frac{(-1)^{k-1} 2^{2k}(2^{2k} - 1) B_{2k}}{(2k)!} t^{2k-1}$$

Integrating this, we get

$$\ln \cos t = \sum_{k=0}^{\infty} \frac{(-1)^k 2^{2k}(2^{2k} - 1) B_{2k}}{2k(2k)!} t^{2k}$$

Again reverting to a trigonometric identity

$$\frac{1}{\sin t} = \cot t + \tan \frac{t}{2}$$

we get

$$\frac{t}{\sin t} = t \csc t = \sum_{k=0}^{\infty} \frac{(-1)^{k-1}(2^{2k}-2)B_{2k}}{(2k)!} t^{2k}$$

$$= 1 + \frac{t^2}{6} + \frac{7}{360} t^4 + \frac{31}{15,120} t^6 + \cdots$$

Evidently only ingenuity limits how far we can go in this matter.   A similar set of numbers, the Euler numbers, gives rise to a number of similar series, where

$$\sec t = \sum_{k=0}^{\infty} E_k \frac{t^k}{k!}$$

12.4   THE RIEMANN ZETA FUNCTION

The Riemann[1] zeta function is defined by

$$\zeta(z) = \sum_{k=1}^{\infty} \frac{1}{k^z}$$

and a short table of values (Table 10.6.2) was given in Sec. 10.6.   It is a very convenient series for comparison purposes.   The values of the zeta function for z an even integer are known in closed form

$$\zeta(2) = \frac{\pi^2}{6}$$

$$\zeta(4) = \frac{\pi^4}{90}$$

$$\zeta(6) = \frac{\pi^6}{945}$$

$$\zeta(8) = \frac{\pi^8}{9,450}$$

$$\zeta(2n) = \frac{(-1)^{n-1}B_{2n}(2\pi)^{2n}}{2(2n)!}$$

where B_{2n} are the Bernoulli numbers.

Note that from definition, $\zeta(n) \to 1$ as $n \to \infty$, and this indicates how B_{2n} grows as $n \to \infty$.

From the zeta function it is easy to derive the sums of other series.   Using the observation that

$$\frac{\zeta(n)}{2^n} = \frac{1}{2^n} + \frac{1}{4^n} + \frac{1}{6^n} + \cdots = \sum_{k=1}^{\infty} \frac{1}{(2k)^n}$$

[1] Bernhard Riemann (1826–1866).

we have

$$\left(1 - \frac{1}{2^n}\right)\zeta(n) = 1 + \frac{1}{3^n} + \frac{1}{5^n} + \cdots = \sum_{k=1}^{\infty} \frac{1}{(2k-1)^n}$$

and

$$\left(1 - \frac{1}{2^{n-1}}\right)\zeta(n) = 1 - \frac{1}{2^n} + \frac{1}{3^n} - \frac{1}{4^n} + \cdots = \sum_{k=1}^{\infty} \frac{(-1)^{k-1}}{k^n}$$

Again, for n an even number the sums can be expressed in terms of Bernoulli numbers.

PROBLEMS

12.4.1   For the series $\sum_{1}^{\infty} 1/(2k)^{2n}$

show that the series for $n = 1, 2, 3$ are

$$\frac{\pi^2}{24} \qquad \frac{\pi^4}{1{,}440} \qquad \frac{\pi^6}{64 \cdot 945}$$

12.4.2   For the series $\sum_{1}^{\infty} \frac{(-1)^{k-1}}{k^{2n}}$

show that the series for $n = 1, 2, 3$ are

$$\frac{\pi^2}{12} \qquad \frac{7\pi^4}{720} \qquad \frac{31\pi^6}{30{,}240}$$

12.5   ANOTHER INTEGRAL EQUATION

This problem, resembling the one in Sec. 9.10, arose in practice.   Given $g(T)$ defined by

$$g(T) = \int_0^{\infty} \frac{f(x)(x/T)^2 e^{x/T}}{(e^{x/T} - 1)^2} \, dx$$

find $f(x)$.

Again we begin by trying $f(x) = x^n$ and set $x = Ty$

$$g(T) = \int_0^{\infty} \frac{T^n y^n y^2 e^y}{(e^y - 1)^2} \, T \, dy$$

$$= T^{n+1} \int_0^{\infty} \frac{y^{n+2} e^{-y}}{(1 - e^{-y})^2} \, dy$$

$$= T^{n+1} \int_0^{\infty} y^{n+2} \sum_{k=1}^{\infty} k e^{-ky} \, dy$$

Setting $ky = z$,

$$g(T) = T^{n+1}\left(\sum_{k=1}^{\infty} \frac{1}{k^{n+2}}\right)\int_0^{\infty} e^{-z}z^{n+2}\,dz$$
$$= T^{n+1}(n+2)!\,\zeta(n+2)$$

Thus if

$$g(T) = T^{n+1}$$

then

$$f(x) = \frac{x^n}{(n+2)!\,\zeta(n+2)}$$

Now, *if* we can represent

$$g(T) = \sum_{n=0}^{N} a_n T^{n+1}$$

then the solution of the integral equation is

$$f(x) = \sum_{n=0}^{N} \frac{a_n x^n}{(n+2)!\,\zeta(n+2)}$$

Evidently the "convergence" of this finite series is rapid in comparison to the original series for $g(T)$. We postpone until Chaps. 14, 25, and 28 discussion of various ways we might find the a_n in the representation of $g(T)$.

Consider, now, how this resembles Kummer's method. We did not try to represent closely the given function $g(T)$ by a single function T^{n+1} whose solution we knew. Rather we tried a linear combination of functions T^{n+1}, each of whose solutions we knew, and then took the same linear combination of the known solutions. We forced all the approximation into the closeness of representation of the original function, and then we operated on the approximate function analytically to get the exact answer to the approximate problem.

In Kummer's method, and throughout much of numerical methods, we try to replace the given function which we cannot handle analytically by ones which we can, and as in the above case, we operate on these functions with the infinite operators of the problem. In Kummer's method we are concerned with the representation of the terms far out in the series by suitable approximate terms, and we then compute approximately the difference in our approximation, using mainly terms near the beginning of the series. However, it should be clear that instead of using a single comparison series we could use several simultaneously.

This method of simultaneous approximation in terms of several series differs somewhat from the earlier process of successive approximations, and provides an alternate approach, or if you prefer, an extension of the simple Kummer method.

For example, consider again the series

$$S = \sum_{x=1}^{\infty} \frac{x}{(x^2 + 1)^2}$$

We can use

$$S' = \sum_{2}^{\infty} \frac{1}{(x - 1)x(x + 1)}$$

$$S'' = \sum_{1}^{\infty} \frac{1}{x^3}$$

We therefore write

$$S = aS' + bS'' + S'''$$

where a, b, and S''' are to be determined. Ignoring for the moment the first term $x = 1$, we have for the general term

$$\frac{x}{(x^2 + 1)^2} = \frac{a}{x^3 - x} + \frac{b}{x^3} + \frac{F(x)}{(x^2 + 1)^2 x^3 (x^3 - x)}$$

$$x^7 - x^5 = a(x^7 - 2x^5 + x^3) + b(x^7 + x^5 - x^3 - x) + F(x)$$

Equating like powers of x

$$x^7: \qquad 1 = a + b \qquad a = -2$$
$$x^5: \qquad -1 = 2a + b \qquad b = 3$$

Thus we will have to compute

$$\sum_{2}^{\infty} \frac{5x^2 + 3}{(x^2 + 1)^2 x^3 (x^2 - 1)} \approx \frac{5}{x^7}$$

12.6 EULER'S METHOD

Another method of summing series numerically is Euler's method, which can be viewed in many ways. The author prefers the following approach.

Consider the finite series

$$\sum_{k=0}^{n-1} a_k t^k \qquad (12.6.1)$$

We apply summation by parts and recall that

$$\sum_{k=a}^{b-1} u(k)\, \Delta v(k) = [u(k)v(k)]_a^b - \sum_{k=a}^{b-1} v(k + 1)\, \Delta u(k)$$

We pick $u(k) = a_k$ and $\Delta v(k) = t^k$. Since we may use any additive constant, we can choose

$$v(k) = \sum_{x=0}^{k-1} t^x = \frac{1 - t^k}{1 - t}$$

Then

$$\sum_{k=0}^{n-1} a_k t^k = \left[a_k \frac{1 - t^k}{1 - t} \right]_0^n - \sum_{k=0}^{n-1} \frac{1 - t^{k+1}}{1 - t} \Delta a_k$$

$$= a_n \frac{1 - t^n}{1 - t} - \frac{1}{1 - t} \sum_{k=0}^{n-1} \Delta a_k + \frac{t}{1 - t} \sum_{k=0}^{n-1} t^k \Delta a_k \qquad (12.6.2)$$

but

$$\sum_{k=0}^{n-1} \Delta a_k = (a_1 - a_0) + (a_2 - a_1) + \cdots + (a_n - a_{n-1})$$

$$= a_n - a_0$$

and so Eq. (12.6.2) becomes

$$\sum_{k=0}^{n-1} a_k t^k = a_n \frac{1 - t^n}{1 - t} - \frac{a_n - a_0}{1 - t} + \frac{t}{1 - t} \sum_0^{n-1} t^k \Delta a_k$$

$$= \frac{a_0}{1 - t} - \frac{a_n t^n}{1 - t} + \frac{t}{1 - t} \sum_{k=0}^{n-1} t^k \Delta a_k \qquad (12.6.3)$$

We apply summation by parts to the third term, noting that it is of the same form as the original series, with a_k replaced by Δa_k in Eq. (12.6.3).

$$\frac{t}{1 - t} \sum_{k=0}^{n-1} t^k \Delta a_k = \frac{t}{1 - t} \left(\frac{\Delta a_0}{1 - t} - \frac{t^n \Delta a_n}{1 - t} + \frac{t}{1 - t} \sum_{k=0}^{n-1} t^k \Delta^2 a_k \right)$$

Thus, as a result of two applications of summation by parts, we have obtained

$$\sum_{k=0}^{n-1} a_k t^k = \frac{a_0}{1 - t} + \frac{t \Delta a_0}{(1 - t)^2} + \frac{t^2}{(1 - t)^2} \sum_{k=0}^{n-1} t^k \Delta^2 a_k$$

$$- \frac{a_n t^n}{1 - t} - \frac{1}{1 - t} \frac{\Delta a_n t^{n+1}}{1 - t}$$

After r applications, the expression (12.6.1) has the form

$$\sum_{k=0}^{n-1} a_k t^k = \frac{1}{1 - t} \sum_{i=0}^{r-1} \left(\frac{t}{1 - t} \right)^i \Delta^i a_0 + \left(\frac{t}{1 - t} \right)^r \sum_{k=0}^{n-1} t^k \Delta^r a_k$$

$$- \frac{t^n}{1 - t} \sum_{i=0}^{r-1} \left(\frac{t}{1 - t} \right)^i \Delta^i a_n$$

Since the original series converges, given an $\varepsilon > 0$, there exists an n_0 such that, for $n > n_0$, $|a_n| < \varepsilon/2^r$. Hence, by Sec. 10.3 $|\Delta^i a_n| \leq \varepsilon$. The last term in the above expression therefore goes to zero as $n \to \infty$, and we have

$$\sum_{k=0}^{\infty} a_k t^k = \frac{1}{1-t} \sum_{i=0}^{r-1} \left(\frac{t}{1-t}\right)^i \Delta^i a_0 + \left(\frac{t}{1-t}\right)^r \sum_{k=0}^{\infty} t^k \Delta^r a_k \qquad (12.6.4)$$

It can be shown that if the given series (12.6.1) converges, then the second term on the right also approaches zero as $r \to \infty$. We are then left with the series

$$\sum_{k=0}^{\infty} a_k t^k = \frac{1}{1-t} \sum_{i=0}^{\infty} \left(\frac{t}{1-t}\right)^i \Delta^i a_0 \qquad (12.6.5)$$

As an example of the above process, given a series $\sum_{k=0}^{\infty} u_k$, if u_{k+1}/u_k approaches t, we can write

$$\sum_{k=0}^{\infty} u_k = \sum_{k=0}^{\infty} a_k t^k$$

where a_{k+1}/a_k approaches 1, and apply Euler's method.

The most frequent case of application is when $t = -1$. Euler's method gives, from Eq. (12.6.5),

$$\sum_{k=0}^{\infty} (-1)^k a_k = \frac{1}{2} \sum_{i=0}^{\infty} \frac{(-1)^i}{2^i} \Delta^i a_0 \qquad (12.6.6)$$

Sometimes Euler's transformation makes the series converge faster, and sometimes it does not. Consider the following examples.

EXAMPLE 12.1

$$\sum_{k=0}^{\infty} \frac{(-1)^k}{2^k}$$

We have

$$a_n = \frac{1}{2^n}$$

$$\Delta a_n = \frac{1}{2^{n+1}} - \frac{1}{2^n} = \frac{-1}{2^{n+1}}$$

$$\Delta^i a_0 = \frac{(-1)^i}{2^i}$$

and so by Eq. (12.6.6)

$$\sum_{k=0}^{\infty} \frac{(-1)^k}{2^k} = \frac{1}{2} \sum_{i=0}^{\infty} \frac{(-1)^i}{2^i} \frac{(-1)^i}{2^i} = \frac{1}{2} \sum_{i=0}^{\infty} \frac{1}{4^i}$$

which converges more quickly.   ////

EXAMPLE 12.2

$$\sum_{k=0}^{\infty} \frac{(-1)^k}{3^k}$$

It is easy to see that

$$\Delta^i a_0 = \frac{(-2)^i}{3^i}$$

and so

$$\sum_{k=0}^{\infty} \frac{(-1)^k}{3^k} = \frac{1}{2} \sum_{i=0}^{\infty} \frac{(-1)^i}{2^i} \frac{(-2)^i}{3^i} = \frac{1}{2} \sum_{i=0}^{\infty} \frac{1}{3^i}$$

which converges somewhat more slowly.   ////

EXAMPLE 12.3 Similarly,

$$\sum_{k=0}^{\infty} \frac{(-1)^k}{4^k} = \frac{1}{2} \sum_{i=0}^{\infty} \left(\frac{3}{8}\right)^i$$

which converges more slowly.   ////

The " break point " for the application of Euler's method to such alternating series is seen to be at ratios of $|a_i|$ between 1/2 and 1/3.

In practice, what is usually done is to sum the first few (say 10) terms directly and then apply Euler's transformation to the rest.[1]

PROBLEMS

12.6.1 Apply Euler's method to show that $\log 2 = \sum_{k=1}^{\infty} \frac{(-1)^{k-1}}{k} = \sum_{k=1}^{\infty} \frac{1}{k 2^k}$

12.6.2 Show that

$$\arctan 1 = \sum_{k=0}^{\infty} \frac{(-1)^k}{2k+1} = \frac{1}{2} \sum_{k=0}^{\infty} \frac{k!}{1 \cdot 3 \cdot 5 \cdots (2k+1)}$$

12.6.3 Add the first eight terms of log 2 and apply Euler's method to the rest.
  Ans. $1 - 1/2 + 1/3 - \cdots - 1/8 = 0.63452381$; Euler part gives $= (1/2)(0.11724674)$

---

[1] See J. B. Rosser, Transformations to Speed and Convergence of Series, *J. Res. Natl. Bur. Std.*, vol. 46, 1951, and Bromwich [5].

12.7[1] IMPROVING THE CONVERGENCE OF SEQUENCES

Acceleration of the convergence of a series is only one example of a general technique of numerical analysis, the extrapolation of a slowly convergent or even divergent sequence to its "limit." Such sequences may be obtained as partial sums of series, as successive approximations to the zeros of a function by an iterative process, as estimates of an integral or derivative by finite difference methods using different values of the interval, and in a variety of other ways. In all these sequences, we may regard the terms of the sequence S_n as estimates of the limit S with an error R_n:

$$S_n = S + R_n \qquad (12.7.1)$$

If the form of dependence of R_n upon n is known, it is frequently possible to estimate the value of S quite accurately from a few values of S_n.

If, as frequently happens, R_n may be approximated by an expression of the form

$$R_n = \sum_{i=1}^{k} a_i (q_i)^n \qquad (12.7.2)$$

a family of nonlinear transformations discovered by Shanks[2] is particularly useful. The basic transformation applied to the partial sums S_n of a series

$$T(S_n) = \frac{S_{n+1} S_{n-1} - S_n^{\,2}}{S_{n+1} - 2S_n + S_{n-1}} \qquad (12.7.3)$$

is often very useful. If T were linear, we would have

$$T(CS_n) = CT(S_n)$$

and

$$T(S_n + U_n) = T(S_n) + T(U_n) \qquad (12.7.4)$$

Equation (12.7.4) is not true in general. However, we do have the weaker rule

$$T(S_n + C) = T(S_n) + C$$

If we write

$$S_n = \sum_{m=0}^{n} a_m$$

---

[1] This section was suggested by H. C. Thacher, Jr.
[2] D. Shanks, Nonlinear Transformations of Divergent and Slowly Convergent Sequences, *J. Math. Phys.*, vol. 34, pp. 1–42.

then $T(S_n)$ can be written

$$T(S_n) = \frac{(S_n + a_{n+1})(S_n - a_n) - S_n^2}{S_n + a_{n+1} - 2S_n + S_n - a_n}$$

$$= S_n + \frac{a_n a_{n+1}}{a_n - a_{n+1}} \qquad (12.7.5)$$

The repeated application of this transformation is most practically carried out by an algorithm due to Wynn.[1]   Let $\varepsilon_0(S_n) = S_n$, $\varepsilon_{-1}(S_n) = 0$, and for $r = 1$, 2, 3, ..., let

$$\varepsilon_{r+1}(S_n) = \varepsilon_{r-1}(S_n) + \frac{1}{\varepsilon_r(S_{n+1}) - \varepsilon_r(S_n)}$$

Then the $\varepsilon_{2k}(S_n)$ are equivalent to the results of applying the kth Shanks transformation to the sequence S_n.

As an example of the power of a single application, consider the Leibnitz series

$$\pi = 4 - 4/3 + 4/5 - 4/7 + \cdots$$

The convergence is so slow that it is practically valueless, but a single application gives the results shown in Table 12.7.

---

[1] P. Wynn, On a Device for Computing the $e_m(S_n)$ Transformation, *Math. Tables Aids Comp.*, vol. 10, pp. 91–96, 1956.

Table 12.7

n	S_n	$T(S_n)$
0	4.00000	
1	2.66667	3.16667
2	3.46667	3.13333
3	2.89524	3.14524
4	3.33968	3.13968
5	2.97605	3.14271
6	3.28374	3.14088
7	3.01707	3.14207
8	3.25237	3.14125
9	3.04184	

The ε_r transformations are members of a large family of lozenge, or rhombus, transformations. If the calculations are laid out in the form of a conventional difference table:

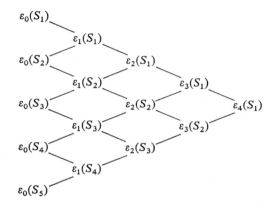

the values at the verticles of each rhombus are related by the algorithm. Hence, these transformations are collectively referred to as lozenge algorithms. Another important lozenge algorithm is the QD algorithm, which may be used for converting power series to equivalent continued fractions. A survey of these algorithms with extensive references was presented by Wynn.[1]

12.8 INTEGRALS AS APPROXIMATIONS TO SUMS

The definite integral is defined to be the limit of a sum. It should be evident, therefore, that integrals can be used as approximations to sums. One such actual formula for doing this appears in Sec. 18.9 and may be written as

$$f_0 + f_1 + \cdots + f_n = \int_0^n f(x)\, dx + 1/2(f_0 + f_n)$$

$$- \frac{1}{12}(\Delta f_0 - \Delta f_{n-1}) + \frac{1}{24}(\Delta^2 f_0 + \Delta^2 f_{n-2})$$

$$- \frac{19}{720}(\Delta^3 f_0 - \Delta^3 f_{n-3}) + \frac{3}{160}(\Delta^4 f_0 + \Delta^4 f_{n-4}) + \cdots$$

---

[1] P. Wynn, Acceleration Techniques in Numerical Analysis, with Particular Reference to Problems in One Independent Variable, *Proc. IFIPS*, Munich, 1962, pp. 149–156.

A closely related formula is the Euler-Maclaurin formula [63]

$$f_0 + f_1 + \cdots + f_n = \int_0^n f(x)\, dx + \frac{1}{2}(f_0 + f_n) - \frac{1}{12}(f_0' - f_n')$$

$$+ \frac{1}{720}(f_0''' - f_n''') - \frac{1}{30,240}(f_0^{iv} - f_n^{iv}) + \cdots$$

$$+ \frac{B_{2m}}{(2m)!}(f_0^{(2m-1)} - f_n^{(2m-1)}) + \cdots$$

where, again, the B_{2m} are the Bernoulli numbers.

In both formulas when dealing with infinite series, the upper limits are infinite and half the terms drop out.

12.9 THE DIGAMMA FUNCTION

For the final topic of this chapter, recall the formula

$$\int^x t^m\, dt = \frac{x^{m+1}}{m+1} + C$$

which does not apply when $m = -1$. Indeed, when $m = -1$, the integral can be said to define a new function, $\log x$.

Similarly, the summation formula

$$\sum_{t=1}^{x-1} t^{(m)} = \frac{x^{(m+1)}}{m+1} + C$$

does not apply when $m = -1$ and correspondingly can be said to define a new function, called the digamma function and denoted by $F(x)$:

$$F(x) = \sum_{r=1}^{x} \frac{1}{r} - \gamma$$

where γ = Euler's constant = $0.5772156649\ldots$ and hence $F(0) = -\gamma$. This formula applies when x is an integer. Another form,

$$F(x) = \sum_{r=1}^{\infty} \frac{x}{r(r+x)} - \gamma \qquad F(0) = -\gamma$$

provides a means of extending the definition of $F(x)$ to noninteger values. We need to show that this new definition is consistent with the old one. This amounts to showing that $F(x)$ satisfies the equation

$$\Delta F(x) = \frac{1}{x+1}$$

(This is analogous to the process of showing that a function satisfies an integration formula by showing that the derivative of the function is the integrand.) We have

$$\Delta F(x) = \sum_{r=1}^{\infty} \left[\frac{x+1}{r(r+x+1)} - \frac{x}{r(r+x)} \right]$$

$$= \sum_{r=1}^{\infty} \frac{1}{(r+x)(r+x+1)}$$

$$= \sum_{r=1}^{\infty} \left[\frac{1}{r+x} - \frac{1}{r+x+1} \right]$$

$$= \frac{1}{x+1}$$

The following relation exists between the digamma function and the natural logarithm:

$$F(x) = \lim_{n \to \infty} \left[\log(x+n+1) - \frac{1}{x+1} - \frac{1}{x+2} - \cdots - \frac{1}{x+n} \right]$$

Another relationship between known functions is

$$\frac{d}{dx} \log[\Gamma(1+x)] = F(x)$$

If we differentiate again, we get the so-called trigamma function

$$\frac{d^2}{dx^2} \log[\Gamma(1+x)] = \sum_{1}^{\infty} \frac{1}{(r+x)^2} = F'(x)$$

the tetragamma function

$$\frac{d^3}{dx^3} \log[\Gamma(1+x)] = -2 \sum_{1}^{\infty} \frac{1}{(r+x)^3} = F''(x)$$

the pentagamma function

$$\frac{d^4}{dx^4} \log[\Gamma(1+x)] = 6 \sum_{1}^{\infty} \frac{1}{(r+x)^4} = F'''(x)$$

etc.

These functions have been tabulated [47] and can be used to sum series whose general coefficient is a rational function. It should be recognized that much of systematic integration is based on the integration of rational functions and that this introduces the new functions ln x and arctan x. Similarly the summation of rational terms leads to the corresponding functions.

For example, supposing we have

$$S = \sum_{x=1}^{\infty} \frac{P_{n-2}(x)}{P_n(x)}$$

where $P_n(x)$ is a polynomial of degree n and $P_{n-2}(x)$ is at least two degrees less than $P_n(x)$. Using partial fractions, we can write [where the a_i are the zeros of $P_n(x)$]

$$S = \sum_{x=1}^{\infty} \left[\frac{A_1}{x-a_1} + \frac{A_2}{x-a_2} + \cdots + \frac{A_k}{x-a_k} + \frac{B_1}{(x-a_1)^2} + \frac{B_1}{(x-a_2)^2} \right.$$
$$\left. + \cdots + \frac{B_k}{(x-a_k)^2} + \cdots + \frac{M_1}{(x-a_1)^m} + \frac{M_2}{(x-a_2)^m} + \cdots + \frac{M_k}{(x-a_k)^m} \right]$$

We cannot rearrange the first group of terms since individually the series diverge. However, it is easy to see that $\sum_{i=1}^{k} A_i = 0$; hence we can write

$$\left[\left(\frac{A_1}{x-a_1} - \frac{A_1}{x} \right) + \left(\frac{A_2}{x-a_2} - \frac{A_2}{x} \right) + \cdots + \left(\frac{A_k}{x-a_k} - \frac{A_k}{x} \right) \right]$$

without changing the value of the term. We now have the series in a form that we can rearrange easily. As an example, consider [47]

$$S = \sum_{1}^{\infty} \frac{1}{(4x+2)(4x+1)(4x+3)^2}$$
$$= \sum_{1}^{\infty} \left[-\frac{1}{4x+2} + \frac{1}{4} \cdot \frac{1}{4x+1} + \frac{3}{4} \cdot \frac{1}{4x+3} + \frac{1}{2} \cdot \frac{1}{(4x+3)^2} \right]$$
$$= 1/4 F(1/2) - 1/16 F(1/4) - 3/16 F(3/4) + 1/32 F'(3/4)$$

13

DIFFERENCE EQUATIONS

13.1 INTRODUCTION

In the finite difference calculus the analogs of differential equations are difference equations. It might be supposed that corresponding to the differential equation

$$y' + 2y = x$$

there would be the difference equation

$$\Delta y + 2y = x$$

It is a curious fact that this is *not* the analogous form; instead we use

$$y(x + 1) + y(x) = x$$

which follows from writing

$$\Delta y(x) = y(x + 1) - y(x)$$

It is easy to pass from one form to the other, and they are equivalent in this sense. How we would go about solving the two forms, however, would be different.

One advantage of the form using different arguments, rather than the Δ operator, is that it more clearly reveals the nature of the problem.   The equation

$$\Delta^2 y + 2\Delta y + y = f(x)$$

looks as if it were of second order, but writing it in the function-value form

$$y(x + 2) - 2y(x + 1) + y(x) + 2[y(x + 1) - y(x)] + y(x) = f(x)$$

leads to

$$y(x + 2) = f(x)$$

which is trivial.   It is the maximum of the difference in the arguments of the unknown function that determines the order of the equation.

The close analogy between the difference and differential equations will be exploited to shorten the treatment.   We will also look at only the simpler problems in the field.

Among the various ways of approaching a field, one can give the general theory or instead give only a few well-chosen examples from which the reader can easily understand how the general theory goes.   There are many occasions when the second approach is more effective, and in this chapter we shall adopt it.   Thus the chapter is mainly a sequence of carefully chosen examples which illustrate how to solve typical difference equations of the forms we need.

13.2   FIRST-ORDER DIFFERENCE EQUATIONS WITH CONSTANT COEFFICIENTS

To solve the first-order differential equation

$$y' + y = x$$

we first consider the homogeneous equation

$$y' + y = 0$$

To solve this, we *guess* at a solution of the form

$$y = e^{mx}$$

and get

$$e^{mx}(m + 1) = 0$$
$$m = -1$$
$$y = e^{-x}$$

and the general solution of the homogeneous equation is

$$y(x) = Ce^{-x}$$

For the complete equation we again *guess*, this time at one particular solution. In this case it is reasonable to suspect that the right-hand side could only arise from a function like

$$y = ax + b$$

Trying this, we get

$$a + ax + b \equiv x$$

which determines $a = 1$ and $b = -1$. Adding this solution to the general solution of the homogeneous equation, we have the general solution of the original equation

$$y = Ce^{-x} + (x - 1)$$

All this should be familiar from the calculus course, and follows from the linearity of the operation of differentiation and the linearity of the equation. We expect, therefore, that the solution of the corresponding difference equation will be similar.

Given the difference equation

$$y(x + 1) + y(x) = x$$

we first solve the homogeneous equation

$$y(x + 1) + y(x) = 0$$

We guess at

$$y(x) = a^x$$

where we have used the conventional notation of writing $a = e^m$. The result of substituting this into the homogeneous equation is

$$a^x(a + 1) = 0$$
$$a = -1$$
$$y(x) = C(-1)^x$$

For the particular solution of the complete equation with the right-hand side x, we again *guess* at the form

$$y(x) = ax + b$$

from which we get

$$a(x + 1) + b + ax + b \equiv x$$
$$a = 1/2 \qquad b = -1/4$$

and the complete solution is

$$y = C(-1)^x + \frac{2x - 1}{4}$$

PROBLEMS

13.2.1   Solve the equation $y_{n+1} - y_n = n + 1$ $y(0) = 0$.

$$Ans. \quad y = \frac{n(n+1)}{2}$$

13.2.2   Solve $y_{n+1} + y_n = n^2$ $y(0) = 0$.

13.2.3   Explain what you must do to solve $y_{n+1} - ay_n = a^n$

$$Ans. \quad y = (n + C)a^{n-1}$$

13.3   THE GENERAL FIRST-ORDER LINEAR DIFFERENCE EQUATION

We next consider

$$y(x + 1) + A(x)y(x) = B(x)$$

where $A(x)$ and $B(x)$ are known functions.   The homogeneous equation is

$$y(x + 1) + A(x)y(x) = 0$$

thus

$$y(x + 1) = -A(x)y(x) = A(x)A(x - 1)y(x - 1) = \cdots$$
$$= (-1)^{x+1}A(x)A(x - 1) \cdots A(0)y(0)$$

where $y(0)$ is an arbitrary constant, say C.   Thus a solution is

$$y(x + 1) = \prod_{k=0}^{x} [-A(k)]$$

Let this solution be labeled $u(x)$.

As in the case of differential equations, we set

$$y(x) = u(x)v(x)$$

where $u(x)$ is the above solution.   Putting this into the complete equation, we get

$$u(x + 1)v(x + 1) + A(x)u(x)v(x) = B(x)$$

Substituting $v(x + 1) = v(x) + \Delta v(x)$, we get

$$v(x)[u(x + 1) + A(x)u(x)] + u(x + 1)\Delta v(x) = B(x)$$

The term within the square bracket is zero, since $u(x)$ is a solution of the homogeneous equation.   Thus

$$\Delta v(x) = \frac{B(x)}{u(x + 1)}$$

Summing this over the variable x from $x = 0$ to $x = x$, we get

$$v(x + 1) = \sum_{k=0}^{x} \frac{B(k)}{u(k + 1)} + C$$

where C is an arbitrary constant.

This is exactly analogous to the case of differential equations and is known as "the variation of parameters method."

PROBLEMS

Solve the following problems.

13.3.1 $ny(n + 1) - (n + 1)y(n) = 1$

13.3.2 $y(n + 1) - n^2y(n) = 0, \; y(1) = 1$ *Ans.* $y(n + 1) = (n!)^2$

13.3.3 $(n + 1)y(n + 1) - ny(n) = n^2$ *Ans.* $y(n + 1) = \dfrac{C}{n + 1} + \dfrac{2n^2 - 3n + 1}{6}$

13.3.4 $y(n + 1) - 3ny(n) = 3^n$

13.3.5 $y(n + 1) - ny(n) = n \quad (n > 0)$

13.3.6 $y(n + 1) - (n + 1)y(n) = 2^n(n - 1)$

13.4   THE FIBONACCI EQUATION

Leonardo of Pisa (1180–1250) was interested in the famous second-order difference equation with constant coefficients

$$y_{n+1} = y_n + y_{n-1} \qquad y_0 = 0 \qquad y_1 = 1$$

where we have now adopted the subscript notation.

We again try

$$y_n = a^n$$

and get

$$a^{n-1}(a^2 - a - 1) = 0$$

$$a^2 - a - 1 = 0$$

$$a = \frac{1 \pm \sqrt{5}}{2}$$

Hence the general solution is

$$y_n = C_1 \left(\frac{1 + \sqrt{5}}{2} \right)^n + C_2 \left(\frac{1 - \sqrt{5}}{2} \right)^n$$

To satisfy the two initial conditions, we have available the two constants C_1 and C_2. We get the two following equations for them:

$$y_0 = C_1 + C_2 = 0$$

$$y_1 = C_1\left(\frac{1 + \sqrt{5}}{2}\right) + C_2\left(\frac{1 - \sqrt{5}}{2}\right) = 1$$

This gives

$$C_1 = \frac{1}{\sqrt{5}} = -C_2$$

and the solution is

$$y_n = \frac{1}{\sqrt{5}}\left(\frac{1 + \sqrt{5}}{2}\right)^n - \frac{1}{\sqrt{5}}\left(\frac{1 - \sqrt{5}}{2}\right)^n$$

Note that the square root cancels out (as it must since from the original equation all the Fibonacci numbers are integers).

Mathematically speaking this is the solution, but practically it is generally easier to compute the numbers from the equation than from the solution. Thus we easily get

$$y_2 = y_1 + y_0 = 1 + 0 = 1$$
$$y_3 = y_2 + y_1 = 1 + 1 = 2$$
$$y_4 = y_3 + y_2 = 2 + 1 = 3$$
$$y_5 = y_4 + y_3 = 3 + 2 = 5$$
$$y_6 = y_5 + y_4 = 5 + 3 = 8$$

....

An alternate approach to the Fibonacci numbers uses generating functions. If we multiply the defining equation by t^n and sum $n = 1, 2, \ldots$, we get

$$\frac{1}{t}\left(\sum_{n=1}^{\infty} t^{n+1}y_{n+1} + ty_1\right) - y_1 = \sum_{n=1}^{\infty} t^n y_n + t\sum_{n=1}^{\infty} t^{n-1}y_{n-1}$$

Setting

$$\sum_{n=1}^{\infty} t^n y_n = Y(t)$$

we have the following equation for the generating function $Y(t)$

$$\frac{1}{t} Y(t) - y_1 = Y(t) + t Y(t)$$

$$Y(t)\left(\frac{1}{t} - 1 - t\right) = y_1 = 1$$

$$Y(t) = \frac{1}{1/t - 1 - t} = \frac{t}{1 - t - t^2}$$

If this rational function is broken up into two linear factors and these are suitably divided out, then the earlier expression for y_n is obtained.

PROBLEMS

13.4.1 Divide out the generating function to find the first five Fibonacci numbers.

13.4.2 Find a formula for the sum of the first N Fibonacci numbers.

$$\textit{Ans.} \quad \sum_{1}^{N} y_n = y_{N+1} + y_N - 1$$

13.4.3 Carry out the steps in the last paragraph in this section.

13.5 ANOTHER EXAMPLE OF A SECOND-ORDER LINEAR EQUATION

As a second example, consider the equation

$$x_{n+1} - 2x_n \cos \phi + x_{n-1} = 0$$

where ϕ is some constant.   We try

$$x_n = a^n$$

$$a^2 - 2a \cos \phi + 1 = 0$$

$$a = \cos \phi \pm \sqrt{\cos^2 \phi - 1}$$

$$= \cos \phi \pm i \sin \phi$$

$$= e^{i\phi}, e^{-i\phi}$$

Hence the solution is

$$x_n = C_1 e^{in\phi} + C_2 e^{-in\phi}$$

or, if you prefer,

$$x_n = \bar{C}_1 \cos n\phi + \bar{C}_2 \sin n\phi$$

This can, of course, be checked by direct substitution into the difference equation.

PROBLEMS

Solve

13.5.1 $y_{n+1} - 2y_n \cosh \phi + y_{n-1} = 0$
13.5.2 $y_{n+1} + 2ay_n - b^2 y_{n-1} = 0$
13.5.3 $y_{n+1} = a_1 y_n + a_2 y_{n-1}$ $y_0 = 0$ $y_1 = a$

13.6   AN EXAMPLE OF A SYSTEM OF EQUATIONS

A system of two first-order difference equations is equivalent to a single second-order equation.   Often all that is needed is the behavior of the solution, and we now give an example of how to find the behavior without actually solving the problem.   Consider the system of equations

$$\begin{aligned} p_{n+1} &= p_n + Kq_n \\ q_{n+1} &= p_n + q_n \end{aligned} \qquad (13.6.1)$$

We know that the solutions will be of the form

$$\begin{aligned} p_n &= Aa_1{}^n + Ba_2{}^n \\ q_n &= Ca_1{}^n + Da_2{}^n \end{aligned} \qquad (13.6.2)$$

with some constraints on the four coefficients $A, B, C,$ and D.   We assume $|a_1| > |a_2|$.   We can write

$$\frac{p_{n+1}}{q_{n+1}} = \frac{p_n + Kq_n}{p_n + q_n} = \frac{p_n/q_n + K}{p_n/q_n + 1} \qquad (13.6.3)$$

But

$$\frac{p_n}{q_n} = \frac{Aa_1{}^n + Ba_2{}^n}{Ca_1{}^n + Da_2{}^n} = \frac{A + B(a_2/a_1)^n}{C + D(a_2/a_1)^n} \to \frac{A}{C}$$

as $n \to \infty$.   Let $A/C = r$.   Then in the limit equation, Eq. (13.6.3) gives

$$r = \frac{r + K}{r + 1}$$

This is the same as

$$r^2 + r = r + K$$

$$r = \sqrt{K} = \lim \frac{p_n}{q_n}$$

Thus we have a way of computing square roots regardless of the initial choices p_0 and q_0 (not both zero).

PROBLEMS

13.6.1 In the above example, discuss the cases where

$$a_1 = a_2$$
$$a_1 = -a_2$$

13.6.2 Discuss the case

$$p_{n+1} = ap_n + bq_n$$
$$q_{n+1} = cp_n + dq_n$$

13.6.3 Discuss the possibilities of the starting values for p_0 and q_0 eliminating a_1, that is, $A = C = 0$, with due regard to roundoff.

13.7 A SYSTEM OF EQUATIONS WITH VARIABLE COEFFICIENTS

Consider finding the values of the two integrals

$$J(n) = \int_0^\infty x^n e^{-x} \sin x \, dx$$

$$n \geq 0$$

$$K(n) = \int_0^\infty x^n e^{-x} \cos x \, dx$$

We shall be particularly interested in showing that $J(4k + 3) = 0$, because using the transformation $x = t^{1/4}$, we get the integral

$$4J(4k + 3) = \int_0^\infty t^k (e^{-t^{1/4}} \sin t^{1/4}) \, dt = 0 \qquad k = 0, 1, \ldots$$

This shows that the kernel in the parentheses has *all* its moments equal to zero, but is not identically zero.   Thus the moments of a function do not uniquely determine the function.

The second purpose of this example is to show how a system of equations with variable coefficients can sometimes be reduced to a system with constant coefficients by a suitable choice of notation.

Using integration by parts on both integrals, we get the pair of simultaneous linear difference equations

$$J(n) = \frac{n}{2} \left[J(n - 1) + K(n - 1) \right]$$

$$K(n) = \frac{n}{2} \left[-J(n - 1) + K(n - 1) \right]$$

Standard integration tables give $J(0) = K(0) = 1/2$.

The variable coefficient $n/2$ suggests that $J(n)$ has a factor $n/2$, $J(n-1)$ has a factor $(n-1)/2$, etc., which suggests that $J(n)$ behaves like $n!/2^n$.

Thus we are led to the transformation

$$J(n) = \frac{n!}{2^n} j(n)$$

$$K(n) = \frac{n!}{2^n} k(n)$$

which leads to the system

$$j(n) = j(n-1) + k(n-1) \qquad j(0) = 1/2$$
$$k(n) = -j(n-1) + k(n-1) \qquad k(0) = 1/2$$

We reduce this to a single second-order equation by writing the top equation as

$$j(n+1) = j(n) + k(n)$$

and using the second equation of the pair, we eliminate $k(n)$

$$j(n+1) = j(n) + -j(n-1) + k(n-1)$$

Now using the first equation, we eliminate $k(n-1)$

$$j(n+1) - 2j(n) + 2j(n-1) = 0 \qquad j(0) = 1/2 \qquad j(1) = 1$$

The general solution is (where $i = \sqrt{-1}$)

$$j(n) = C_1(1+i)^n + C_2(1-i)^n$$

The initial conditions give

$$C_1 + C_2 = 1/2$$
$$(1+i)C_1 + (1-i)C_2 = 1$$

which leads to

$$C_1 = \frac{1-i}{4} \qquad C_2 = \frac{1+i}{4}$$

and finally

$$j(n) = \frac{1}{2}[(1+i)^{n-1} + (1-i)^{n-1}]$$

or

$$J(n) = \frac{n!}{2^{n+1}}[(1+i)^{n-1} + (1-i)^{n-1}]$$

(which is actually a real number).

To find $J(4k + 3)$, we observe that

$$(1 + i)^4 = 1 + 4i + 6i^2 + 4i^3 + i^4 = -4 = (1 - i)^4$$

hence

$$J(4k + 3) = \frac{(4k + 3)!}{2^{4k+4}} \{(-4)^k[(1 + i)^2 + (1 - i)^2]\} = 0$$

PROBLEMS

13.7.1 Show that $K(4k + 1) = 0$

13.7.2 If

$$I_k(\phi) = \int_0^\pi \frac{\cos k\theta - \cos k\phi}{\cos \theta - \cos \phi} \, d\theta$$

show that for $k =$ an integer, $I_k(\phi)$ satisfies

$$I_{k+2}(\phi) - 2 \cos \phi \, I_{k+1}(\phi) + I_k(\phi) = 0 \qquad I_1(\phi) = \pi$$

and hence

$$I_k(\phi) = \frac{\pi \sin k\phi}{\sin \phi}$$

13.7.3 Solve the system of equations in this section as a system.

13.8 SECOND-ORDER RECURRENCE RELATIONS

Many of the special functions that arise in practice satisfy second-order linear-recurrence relations. For example,

$$\sin n\theta: \quad \sin(n + 1)\theta - 2 \cos \theta \sin n\theta + \sin(n - 1)\theta = 0$$

$$\cos n\theta: \quad \cos(n + 1)\theta - 2 \cos \theta \cos n\theta + \cos(n - 1)\theta = 0$$

$$T_n(x): \quad T_{n+1}(x) - 2xT_n(x) + T_{n-1}(x) = 0$$

where $T_n(x)$ is the Chebyshev polynomial of order n (see Chaps. 28 and 29). Not all special functions satisfy such equations with constant coefficients. Most of the special functions have equations with variable coefficients; for example,

$$f_{n+1}(z) - \frac{2n}{z} f_n(z) + f_{n-1}(z) = 0$$

is the equation for the Bessel functions $J_n(z)$ and $Y_n(z)$ of the first and second kinds. The variable is, of course, n, and not z.

It is reasonable to expect that locally the solutions of an equation with variable coefficients will behave much like the solutions of an equation with constant coefficients, provided the constants are some reasonable local averages of the variable coefficients. Assuming that this is so, we recall that the solution was typically the sum of two solutions and of the form

$$y_n = C_1 a_1{}^n + C_2 a_2{}^n$$

Supposing that $|a_1| > |a_2|$, then even if we started with $C_1 = 0$, roundoff would bring in some of the first solution, and in time this would dominate so that the second solution would not be visible

$$y_n = C_1 a_1{}^n \left[1 + \frac{C_2}{C_1} \left(\frac{a_2}{a_1}\right)^n\right] \to C_1 a_1{}^n$$

Thus because of roundoff noise the calculation of the second solution is not practical this way.

This is what usually happens in practice—we want the solution that is growing less rapidly than the other. For example, the Bessel function $J_n(x)$ decreases as a function of n in the long run, while $Y_n(x)$ grows toward infinity. It is obvious that the way to counteract this effect is to compute from high values of n to low values so that the desired solution will grow and the unwanted one will decay. Thus starting out with a sufficiently high index n, we set the $(n + 1)$st solution to zero and the nth one to some value, say 1. Recurring down, we see the numbers grow (there may be local oscillation, but the amplitude will grow) until we get down to the first value. We will not have $J_0(x)$ but rather some constant multiple of it, and our problem is to find this constant. In the case of the Bessel function it is customary to use the known identity

$$J_0(z) + 2 \sum_{m=1}^{\infty} J_{2m}(z) = 1$$

to determine this constant. Thus the recurrence has given us not only $J_0(x)$ but also many of the values $J_n(x)$ (not all the way up to the starting value but quite a way up *provided* we started high enough in n).

This is an important fact to realize; it means that when we are presented with an expansion of a function in terms of a special function, we may many times find all the function values (for a given x) with a comparatively small amount of computation. Thus much of classical mathematics becomes practical when this device is used.

We have not discussed the way to pick the starting index n, as this is a complicated topic of some specialization, nor have we made the treatment rigorous.

When it is necessary to do some such recurrence, the reader can consult the literature.[1]

In the case of the trigonometric difference equations it has been observed that it is better to use the pair of first-order equations than the second-order equation. Furthermore, experience shows that the use of $\cos \theta = 1 - \phi(\theta)$ is preferable. Thus the equations to be used are

$$\sin(x + \Delta x) = \sin x - \phi(\Delta x)\sin x + \cos x \sin \Delta x$$

$$\cos(x + \Delta x) = \cos x - \phi(\Delta x)\cos x - \sin x \sin \Delta x$$

These are very useful in practice, especially when converting point by point from polar to rectangular coordinates. Experience shows that the accuracy is better than ± 3 in the sixth significant figure in an eight-digit computation *provided* the equations are reinitialized every 100 steps. If ± 3 in the fifth is acceptable, then reinitialize every 500 steps.

---

[1] See the classic papers by F. W. J. Olver, Numerical Solution of Second-Order Linear Difference Equations, *J. Res., Natl. Bur. Std.*, vol. 71, nos. 2 and 3, pp. 111–129, Apr.– Sept. 1967; and Bounds for the Solutions of Second-Order Linear Difference Equations, *J. Res., Natl. Bur. Std.*, vol. 71, no. 4, Oct.–Dec. 1967.

PART II

Polynomial Approximation-Classical Theory

14

POLYNOMIAL INTERPOLATION

14.1 ORIENTATION

Speaking in theatrical terms, we have just finished Act I in which various ideas and techniques are introduced and used to meet a number of short-range goals. In the second act, Part II of this book, the individual ideas and techniques will reappear in various ways, and a number of new ideas will also make their appearance. It will also include the further development of the basic themes with more of the classic "unity, emphasis, and coherence." Act III, in its turn, will give some new views of what happens in Act II, as well as introduce a few new themes. Regretfully for the analogy (and for the book), Parts IV and V are not the dramatic conclusion in which all the loose ends are gathered together, but rather they drift into a collection of miscellaneous topics needed to round out the whole. Thus they lack much of the central unity of Parts II and III, and only a few of the earlier themes are picked up and developed further.

In Part I we were almost exclusively concerned with finite, discrete problems and faced the infinite only in the very mild form of an infinite series. Thus the finite nature of the machine occurred mainly in the form of roundoff. In Part II we begin with interpolation, the simplest of the problems in which truncation error occurs, and we proceed to more difficult problems such as integration. We will also introduce (Chaps. 25 to 30) a pair of new ideas as to what is regarded as a solution to the problem of approximating an infinite operation with a finite amount of arithmetic on a finite computing machine.

We are more concerned with methods than with results, and therefore the results are often scattered in various chapters. For example, "Interpolation" is the title of this chapter, but it is also treated in Chaps. 15 and 17 through 20; the Index provides the cross references to subject material, while the chapters tend to be organized around the methods and ideas used.

14.2 INTERPOLATION

The first problem in interpolation that a person usually meets is that of "reading between the lines" of a published table, say a trigonometric or logarithmic table. In this process it is customary to use a straight line that passes through two adjacent table points—exactly the same as we did in the false-position method (Sec. 4.4). If we use the straight line to compute the values of the function in the interval $a \leq x \leq b$, then it is called "interpolation," while if the value to be computed from the line lies outside the interval, then it is called "extrapolation." The formulas in both cases are the same

$$f(x) = f(a) + \frac{x - a}{b - a} [f(b) - f(a)]$$

$$= \frac{(b - x)f(a) + (x - a)f(b)}{b - a}$$

it is only the value of x that differs.

In the case of Newton's method for finding the real zeros of a function (Sec. 4.6), we used *both* the value of the function and the value of the first derivative at the point to determine the approximating line and its zero. The formula was

$$x_{n+1} = x_n - \frac{f(x_n)}{f'(x_n)}$$

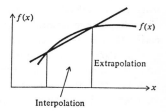

FIGURE 14.2.1

To generalize, we can use as our samples *both* function values and values of various orders of derivatives to determine a suitable-degree polynomial that will be used as the basis for interpolation. We have called this process "analytic substitution" (Sec. 4.6)—for the function we cannot handle we substitute an analytic approximation that we can.

The samples we use can arise directly from measurement, or they can come from computing values of a function that we cannot handle and are forced to sample. In the first case we generally have little or no control over the samples we are given (the information), and the samples are generally equally spaced in the independent variable. In the second case we can select what we will use, and this freedom is applied in later chapters, especially in Chap. 19 where we will use sampling places that in practice would be awkward to measure but are easy to compute from a function (on a computer, but not by hand).

Because of the wide variety of possible combinations of sampling places, and of values of functions and derivatives that are possible (and actually do occur in practice), it is hopeless to try to give a formula for each specific situation. Instead we give a general approach which finds the particular formula that takes advantage of whatever information is available. We also need a broad approach within which we can compare various formulas, so that we can both pick out the right one to use and compare alternate approaches to a situation.

The plan is, therefore, to examine the use of information in these forms:

Function values only—Chaps. 14 and 15
Values of the derivatives—Chap. 17
Differences of function values—Chap. 18
Arbitrarily placed samples—Chap. 19

Once we have decided on the information we are going to use (position as well as function and derivative values), then we face the problem of picking the class of approximation functions to be used. In Part II we will use polynomials exclusively, and this is the classical approach. In Part III we will use sines and cosines, which is the more modern approach. Exponential approximation occurs in Part IV.

Next we need to consider the criteria we will use to pick out the particular member of the class to use. We have so far used polynomials that exactly matched (agreed) with the information in the samples (within roundoff). In Chaps. 25 to 30, we will use two other criteria, least squares and the Chebyshev, or minimax, criteria.

Lastly, where shall we apply the test for matching? This question we will postpone for some time, though we faced it in finding zeros when we asked whether it was the closeness of the function to zero or the closeness of the computed root to the correct one that we wanted to achieve.   We will for now simply assume that it is the two functions, the given and approximating ones, that are to be close to each other.   However, in Part III we will see that there are very different answers to the question of where we should apply the criteria of matching.

To summarize, we have four questions which, it will turn out, determine the formula uniquely:

1   What samples (function and derivative values, etc.) shall we use?
2   What class of functions shall we use (polynomials here in Part II)?
3   What criterion shall we use (exact matching, least squares, Chebyshev, etc.)?
4   Where shall we apply the criterion?

14.3   INTERPOLATION USING ONLY FUNCTION VALUES

Our basic tool for finding formulas is the method of undetermined coefficients. The method assumes the form of the answer, and the conditions on the formula determine the arbitrary coefficients of the form.   For example, given $N + 1$ sample points (x_i, y_i) $(i = 1, 2, \ldots, N + 1)$, we can fit a polynomial of degree N

$$P_N(x) = a_0 + a_1 x + a_2 x^2 + \cdots + a_N x^N = \sum_{k=0}^{N} a_k x^k$$

to the data (since the polynomial form has exactly $N + 1$ coefficients to be determined by the $N + 1$ conditions).   The condition that the polynomial $P_N(x)$ pass through the sample point (x_i, y_i) is

$$y_i = P_N(x_i) = \sum_{k=0}^{N} a_k x_i^k \qquad i = 1, 2, \ldots, N + 1$$

These $N + 1$ equations determine the coefficients a_k.

The determinant of the coefficients of the unknows, a_k, is the Vandermonde[1] determinant

$$V_{N+1} = \begin{vmatrix} 1 & x_1 & x_1^2 & \cdots & x_1^N \\ 1 & x_2 & x_2^2 & \cdots & x_2^N \\ \cdots\cdots\cdots\cdots\cdots\cdots\cdots\cdots\cdots \\ 1 & x_{N+1} & x_{N+1}^2 & \cdots & x_{N+1}^N \end{vmatrix} = |x_i^k|$$

[1] Alexandre Theophile Vandermonde (1735–1796).

In the next section we will show that if $x_i \neq x_j$ for $i \neq j$, then the determinant cannot be zero; hence we can always solve for the a_k, and we have a solution to the interpolation problem for polynomials.

If we solve for the a_k by the use of determinants, substitute the results into the polynomial, and finally rearrange it suitably, we would have

$$
\begin{vmatrix}
y & 1 & x & x^2 & \cdots & x^N \\
y_1 & 1 & x_1 & x_1^2 & \cdots & x_1^N \\
y_2 & 1 & x_2 & x_2^2 & \cdots & x_2^N \\
\multicolumn{6}{c}{\cdots\cdots\cdots\cdots\cdots\cdots\cdots\cdots} \\
y_{N+1} & 1 & x_{N+1} & x_{N+1}^2 & \cdots & x_{N+1}^N
\end{vmatrix} = 0
$$

We can see directly that this is the solution. Expanding by the elements of the top row, it is clearly a polynomial of degree N. If x and y take on the values x_i and y_i, then two rows would be the same, and hence the determinant would be zero. This observation shows that the polynomial passes through the given points.

From this form it is immediately evident that if all the y_i were zero, then the polynomial would also be identically zero, $y(x) \equiv 0$. This result is important because it means that the *interpolating polynomial is unique (within roundoff)*. The interpolating polynomial can be written in many different forms and in many different notations, but if the same information is used in two different methods, then necessarily the two resulting polynomials are the same (within roundoff). It is also easy to see what the condition is for the coefficient of x_N to be zero. We will speak of the interpolating polynomial being of degree N even if, as sometimes happens, the actual polynomial is of lower degree.

As an example of finding an interpolating polynomial consider the problem where we are given the four points $(0, 2)$, $(1, 2)$, $(2, 0)$, $(3, 0)$ and wish to determine the cubic through them (see Fig. 14.3.1). We have the four equations

$$
\begin{aligned}
P(0) &= 2 = a_0 \\
P(1) &= 2 = a_0 + a_1 + a_2 + a_3 \\
P(2) &= 0 = a_0 + 2a_1 + 4a_2 + 8a_3 \\
P(3) &= 0 = a_0 + 3a_1 + 9a_2 + 27a_3
\end{aligned}
$$

It is easy to eliminate $a_0 = 2$. We next subtract twice the second equation from the third, and three times the second from the fourth to get

$$
\begin{aligned}
-2 &= 2a_2 + 6a_3 \\
-2 &= 6a_2 + 24a_3
\end{aligned}
$$

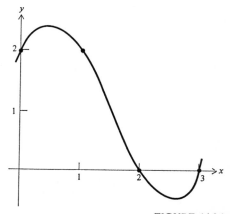

FIGURE 14.3.1

It follows that

$$4 = 6a_3 \qquad \begin{cases} a_3 = 2/3 \\ a_2 = -3 \\ a_1 = 7/3 \\ a_0 = 2 \end{cases}$$

and

$$P(x) = 2 + \tfrac{7}{3}x - 3x^2 + \tfrac{2}{3}x^3$$

As a second example, we find the interpolating polynomial for

$$y = \log x \qquad x = 1, 2, 3, 4$$

We have for the equations that determine the coefficients of the cubic

$$P_3 = a_0 + a_1 x + a_2 x^2 + a_3 x^3$$
$$\log 1 = 0.0000 = a_0 + a_1 + a_2 + a_3$$
$$\log 2 = 0.3010 = a_0 + 2a_1 + 4a_2 + 8a_3$$
$$\log 3 = 0.4771 = a_0 + 3a_1 + 9a_2 + 27a_3$$
$$\log 4 = 0.6021 = a_0 + 4a_1 + 16a_2 + 64a_3$$

Eliminating a_0 *pairwise* to preserve some symmetry, we get

$$0.3010 = a_1 + 3a_2 + 7a_3$$
$$0.1761 = a_1 + 5a_2 + 19a_3$$
$$0.1250 = a_1 + 7a_2 + 37a_3$$

Again eliminating pairwise, this time a_1, we get

$$-0.1249 = 2a_2 + 12a_3$$
$$-0.0511 = 2a_2 + 18a_3$$

Eliminating a_2, we have

$$0.0738 = 6a_3 \qquad \begin{cases} a_3 = 0.01230 \\ a_2 = -0.13625 \\ a_1 = 0.62365 \\ a_0 = -0.49970 \end{cases}$$

and

$$P_3(x) = 0.0123x^3 - 0.13625x^2 + 0.62365x - 0.4997$$

This, then, is the cubic approximating $y = \log x$.

PROBLEMS

14.3.1 Find a quadratic through $(0, 1)$, $(1, 2)$, and $(2, 3)$.

14.3.2 Find a quadratic through $(-1, a)$, $(0, b)$, and $(1, c)$.

$$\textit{Ans.} \quad y(x) = b + \frac{c-a}{2}x + \frac{c-2b+a}{2}x^2$$

14.3.3 Find a cubic through $(0, 1)$, $(1, 0)$, $(2, 1)$, and $(3, 0)$.

$$\textit{Ans.} \quad y(x) = \frac{-2x^3}{3} + 3x^2 - \frac{10x}{3} + 1$$

14.3.4 Find a quartic through $(-2, a)$, $(-1, b)$, $(0, c)$, $(1, b)$, and $(2, a)$. (Note that the symmetry removes the powers x and x^3.)

14.4   THE VANDERMONDE DETERMINANT

In Sec. 14.3 we met the important Vandermonde determinant, which plays a central role, along with its variants, in the theory of polynomial approximation. The determinant is defined by

$$V_{N+1}(x_1, x_2, \ldots, x_{N+1}) = |x_i^k| \qquad k = 0, 1, \ldots, N$$
$$i = 1, 2, \ldots, N+1$$

Consider it as a function of the variables x_i. It is clearly a polynomial in the x_i, and a count of the exponents shows that the degree is

$$0 + 1 + 2 + 3 + \cdots + N = \frac{N(N+1)}{2}$$

Now if $x_{N+1} = x_1$, then $V_{N+1} = 0$, and the determinant has the factor $x_{N+1} - x_1$. Similarly, if $x_{N+1} = x_2, x_3, \ldots, x_N$, then we have the corresponding factors. Thus we have, in total, the N factors

$$\prod_{j=1}^{N} (x_{N+1} - x_j)$$

Again, if $x_N = x_1, x_2, \ldots, x_{N-1}$, we have the $N - 1$ factors

$$\prod_{j=1}^{N-1} (x_N - x_j)$$

Repeating this argument, we see that we have all the factors

$$\prod_{k>j=1}^{N+1} (x_k - x_j)$$

This product is a polynomial of degree

$$N + (N - 1) + \cdots + 1 = \frac{N(N + 1)}{2}$$

and so we have found *all* the factors, and it remains to find the multiplicative constant by which the two representations might differ. To find this constant we compare the same term in both representations. The diagonal term in the determinant is

$$1 \cdot x_2 \cdot x_3{}^2 \cdots x_{N+1}^N$$

The term from the left-hand sides of the product is

$$x_{N+1}^N \cdot x_N^{N-1} \cdots x_2$$

These are the same, hence

$$V_{N+1}(x_1, x_2, \ldots, x_{N+1}) \equiv |x_i^k| \equiv \prod_{k>j=1}^{N+1} (x_k - x_j)$$

The Vandermonde determinant shows that the powers of x $(1, x, \ldots, x^N)$ are linearly independent over any set of $N + 1$ distinct points. Thus sampling a polynomial of degree N at $N + 1$ distinct points enables us to reconstruct (within roundoff) the polynomial from the samples alone. In this form it compares with the famous sampling theorem which we will examine in Chap. 34. The fundamental theorem of algebra shows that the powers of x are linearly independent in any interval; the Vandermonde shows that the first $N + 1$ powers are linearly independent over any $N + 1$ distinct points, and we know that there cannot be $N + 2$ functions linearly independent over $N + 1$ points.

PROBLEMS

14.4.1   By direct expansion of the determinant show that $V_3(x_1, x_2, x_3)$ factors properly.

14.4.2   Show directly that

$$V_{n+1} = V_N \cdot \prod_{i=1}^{N}(x_{N+1} - x_i)$$

and hence evaluate the Vandermonde determinant by recursion.

14.5   LAGRANGE[1] INTERPOLATION

In this section we give an alternate approach to the problem of finding the interpolating polynomial through $N + 1$ sample points.   The method, ideas, and notation are more important than the result.

Suppose we first find a polynomial of degree N through the $N + 1$ points whose y values are all zero except the ith which we will take as being 1.   We introduce a notation that we shall repeatedly use, namely,

$$\pi_i(x) = (x - x_1)(x - x_2) \cdots (x - x_{i-1})(x - x_{i+1}) \cdots (x - x_{N+1})$$

which is the product of all the factors *except* the ith one.   Using this notation, we easily see that the solution to the problem is

$$\frac{\pi_i(x)}{\pi_i(x_i)}$$

since for $i \neq j$

$$\pi_i(x_j) = 0 \qquad \pi_i(x_i) \neq 0 \qquad \text{and} \qquad \frac{\pi_i(x_i)}{\pi_i(x_i)} = 1$$

This *sampling polynomial* is of degree N and has the required values of being zero at all samples except the ith at which it is 1.

Now, just as we did in the first matrix inversion (Sec. 7.11), where we constructed the general solution out of the special unit solutions, so, too, we have by immediate inspection that the general solution is

$$P_N(x) = \sum_{i=1}^{N+1} y_i \left[\frac{\pi_i(x)}{\pi_i(x_i)} \right]$$

This technique of using the sampling polynomials is very useful and should be clearly understood.

---

[1] Joseph-Louis Lagrange (1736–1813).

The Lagrange interpolating polynomial is, of course, the same polynomial as before in Sec. 14.3, within roundoff. It is written in a different form which is more of theoretical than practical interest (since it has a great deal of roundoff in many cases).

14.6 ERROR OF POLYNOMIAL APPROXIMATIONS

Given a function $f(x)$, we took $N + 1$ sample points (x_i, y_i), $(i = 1, 2, \ldots, N + 1)$, and found a polynomial $P_N(x)$ through these points. We intend to use this polynomial in place of the original function, and it is therefore important to examine the question of how much the function and the polynomial differ at points other than the sample points (where they agree within roundoff error).

As an example, consider the function $y(x) = \log x$ (see Table 14.6). In Sec. 14.3 we found the polynomial approximating the function. As a measure of the difference $\log x - P(x)$, let us examine the values at the sample points and the midpoints between them where we would expect the error to be fairly large.

A theoretical expression for the difference between the original function $f(x)$ and the approximating polynomial $P(x)$ can be found if we observe that the difference is zero at all the sample points, and write

$$y(x) - P(x) = (x - x_1)(x - x_2) \cdots (x - x_{N+1})K(x)$$

where $K(x)$ is suitably chosen. We now choose an arbitrary x^*. We have

$$y(x^*) - P(x^*) - (x^* - x_1)(x^* - x_2) \cdots (x^* - x_{N+1})K(x^*) = 0$$

Now consider the function

$$\Phi(x) = y(x) - P(x) - (x - x_1)(x - x_2) \cdots (x - x_{N+1})K(x^*)$$

If $y(x)$ has an $(N + 1)$st derivative, we can differentiate $N + 1$ times, and since $P(x)$ is a polynomial of degree N and $K(x^*)$ is a constant, we get

$$\Phi^{(N+1)}(x) = y^{(N+1)}(x) - (N + 1)!K(x^*)$$

Table 14.6

x	$\log x$	$P(x)$	$\log x - P(x)$
1.0	0.0000	0.0000	0.0000
1.5	0.1761	0.1707	+0.0054
2.0	0.3010	0.3010	0.0000
2.5	0.3979	0.4000	−0.0021
3.0	0.4771	0.4771	0.0000
3.5	0.5441	0.5414	+0.0027
4.0	0.6021	0.6021	0.0000

But $\Phi(x)$ vanishes $N + 2$ times (at x^* and $x_1, x_2, \ldots, x_{N+1}$). Hence, by the mean value theorem, $\Phi'(x)$ vanishes at least $N + 1$ times in the interval containing all the x values (including x^*). Continuing to apply this theorem, we find that $\Phi^{(k)}(x)$ vanishes at least $N + 2 - k$ times and that $\Phi^{(N+1)}(x)$ vanishes at least once. Thus there is an $\bar{x}$ in the interval of the x values such that

$$y^{(N+1)}(\bar{x}) = (N + 1)!K(x^*)$$

We now have a value for the constant $K(x^*)$ and can put this back in the original expression to get

$$y(x^*) = P(x^*) + \frac{(x^* - x_1)(x^* - x_2) \cdots (x^* - x_{N+1})y^{(N+1)}(\bar{x})}{(N + 1)!}$$

But, since x^* was arbitrary, we may write x^* as x, and we have finally

$$y(x) = P(x) + \frac{(x - x_1)(x - x_2) \cdots (x - x_{N+1})y^{(N+1)}(\bar{x})}{(N + 1)!}$$

If we use this error term to estimate the error made in interpolating in the log table, we get

$$\frac{(x - 1)(x - 2)(x - 3)(x - 4)}{4!} y^{(4)}(\bar{x})$$

To estimate the error at $x = 3/2$, we obtain

$$\frac{(1/2)(-1/2)(-3/2)(-5/2)}{4!} \left(\frac{-6}{\bar{x}^4}\right) = \frac{15}{64} \frac{1}{\bar{x}_4}$$

All that we know is that $1 \leq x \leq 4$. At worst the error is $15/64 \simeq 0.23$, while at best it is 0.001—from Table 14.6 it is about 0.0054, which shows how little accuracy such an error term actually gives in this case.

It should be noted that the value $\bar{x} = \bar{x}(x)$ depends on x, and there may be several $\bar{x}$'s for some values of x. Also, $\bar{x}$ is not necessarily a continuous function of x. This latter effect is illustrated by applying the mean value theorem to the function $y(x) = x(1 - x)^2$ (Fig. 14.6.1) and choosing $a = 0$ in the usual form for the mean value theorem:

$$y'(\bar{x}) = \frac{y(x) - y(a)}{x - a} = \frac{y(x)}{x} = (1 - x)^2$$

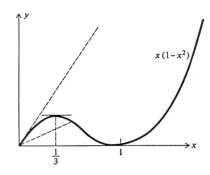

FIGURE 14.6.1

Thus $y'(\bar{x}) \geq 0$, and as

x goes to 0 to 1, $\bar{x}$ goes from 0 to 1/3
x goes from 1 to 2, $\bar{x}$ goes from 1/3 to 0
x goes beyond 2, $\bar{x}$ jumps over the interval where $y' < 0$, to $x > 1$

Hence $\bar{x} = \bar{x}(x)$ is not a continuous function of x.

14.7   DIFFICULTY OF POLYNOMIAL APPROXIMATION

It is customary, as in the previous section, to express the error of a polynomial approximation in terms of a suitable derivative of the function being approximated.   It is usually thought that for "most" reasonable functions such expressions for the error become small for sufficiently large n, but this is not so.

Suppose that we restrict our discussion to analytic functions, that is, functions having a convergent Taylor series

$$y(x) = \sum_{n=0}^{\infty} \frac{(x - x_0)^n}{n!} y^{(n)}(x_0)$$

at all points x_0 in our range of interest.   If, further, the function is an integral (entire) function, that is, converges everywhere in the finite part of the complex plane, as do $\sin x$, e^x, polynomials in x, etc., then it is indeed possible that all the higher derivatives are small.   But if the function has a singularity in the finite part of the complex plane, as do $\tan x$, $\log x$, rational functions in x, etc., then the Taylor series must have a finite radius of convergence R, and this in turn means that for an infinite number of values of n

$$\frac{(R + \varepsilon)^n}{n!} |y^{(n)}(x_0)| \geq 1 \qquad \varepsilon > 0$$

$$|y^{(n)}(x_0)| \geq \frac{n!}{(R + \varepsilon)^n}$$

In words, the upper bound on the nth derivative grows as $n!$. As an example, consider

$$y = \ln x$$

$$y' = \frac{1}{x}$$

$$y'' = \frac{-1}{x^2}$$

$$y''' = \frac{2!}{x^3}$$

$$\cdots\cdots\cdots\cdots\cdots\cdots\cdots$$

$$y^{(n)} = \frac{(-1)^{n-1}(n-1)!}{x^n}$$

Thus, even though the curve $y = \ln x$ looks smooth near some value x, nevertheless as n gets large, the derivatives at this point become very large in size and tend to behave as $n!$ or worse.

This is the general case; for "most" functions some of the higher-order derivatives tend to grow as $n!$. It is only for certain integral functions that all the derivatives can remain bounded.[1] Even for polynomials there is a tendency for the derivatives to grow in size until the Nth one which is $a_0 N!$, after which they suddenly all become zero. Of course, many of the higher derivatives of a function may be small while some of the others are large. For example, an even function, $f(x^2)$, has all its odd-order derivatives equal to zero at the origin, but if the function is not an integral function, then there are also an infinite number of the even-order derivatives which tend to behave as $n!$, or worse.

It would be helpful if we had to deal only with integral functions which can have nicely bounded derivatives, but the facts seem to be that if the function is an integral function, quite likely the whole problem can be solved analytically, whereas if it is necessary to use numerical methods, the function is likely to be rather poorly behaved.

The Weierstrass theorem which states, loosely speaking, that a continuous function in a closed interval can be uniformly approximated by a polynomial is often cited as justification for using a polynomial approximation [49, pp. 28–31].

---

[1] The converse is not true; the derivatives of an integral function do not need to be bounded. For example, if $y = xe^x$, then $y^{(n)}(0) = n$.

However, the method that we have been using of exact matching at the sample points is not the way in which the Weierstrass polynomial is defined; hence the theorem, while possibly suggestive, does not apply. Indeed, the simple function [56]

$$y(x) = \frac{1}{1 + x^2} \qquad -5 \le x \le 5$$

is well known as a bad example, since as a selected set of equally spaced points is increased in number, the approximating polynomial (in the sense of exact matching) diverges from the function between some of the sample points. Thus, even for equally spaced sample points, we cannot rely on a polynomial to be a good approximation if exact matching at the sample points is the criterion used to select the polynomial. The explanation of this phenomenon is, of course, that the derivatives grow too rapidly.

As an example of this effect, consider the Riemann zeta function given in Table 10.6.2 in Sec. 10.6. The top line of differences does not tend to become small very rapidly, which is caused by the obvious singularity at $x = 1$.

It should be noted, however, that if a table is given at one-half the spacing of a second table of the same function, then the first differences are one-half as large, the second one-fourth as large, etc., as those of the second table. This suggests an empirical rule: If the differences of a table approach zero rapidly, then probably too small a spacing was used and the table was overcomputed, whereas if the differences do not become small, then a finer spacing should be considered.

A final, and perhaps much more basic, objection to polynomial approximation is that it rarely has any physical implications that will lead to useful insights. Remember: The purpose of computing is insight, not numbers.

On the other hand, the theory is simple and well developed, requires a minimum of computation, and is useful. Experience shows that polynomials often do a good job, although the error term may be either unobtainable or pessimistic. The values of the difference table can be highly suggestive of the values of the derivative, and they indicate the smallness of the contribution due to taking more terms in the Newton interpolation polynomial. However, the use of the differences as a guide to the values of the derivatives can be dangerous; thus, for integer x the values of $\sin \pi x$ are all zero, so that the difference table is also all zeros, suggesting zero error in the approximation

$$\sin \pi x = 0$$

which is hardly true. The fallacy is obvious here, but in some situations it might be overlooked.

PROBLEMS

14.7.1 Compute the nth derivative of

$$\frac{1}{1+x^2} \equiv \frac{1}{2i}\left(\frac{1}{x-i} - \frac{1}{x+i}\right) \qquad i = \sqrt{-1}$$

and evaluate at $x = 0$.

14.8 ON SELECTING SAMPLE POINTS

The problem of which sample points to select occurs when we try to interpolate. If we had complete freedom in choosing, we would select the value x for which we were going to interpolate, thus making the problem trivial. But it often happens that we have an extensive set of values (x_i, y_i) which we might use, none of which coincides with the desired value x. In the absence of knowledge of the size of the derivative occurring in the error term we can only pick our samples to minimize the factor

$$(x - x_1)(x - x_2) \cdots (x - x_{n+1})$$

in front of the derivative term.

Common sense and the minimization of this factor coincide, that is, use information close to the place at which we want to interpolate.

14.9 SUBTABULATION

Subtabulation means interpolating values into a table at equally spaced points. This is often far cheaper than computing the values at the points. Of course it requires the use of some standard subtabulating routine, which is additional programming, but it can often save machine time. For example, Table 14.9 for cubic interpolation gives the values to use to multiply the original table values to get the subtabulated values at one-tenth the spacing. Other formulas are easily found.

Since the results of a computation are often to be used by humans, it is necessary to have the final results at a fine enough spacing to support linear interpolation (most humans will not go into quadratic or cubic interpolation themselves). When the computation of the individual values is very expensive, that is the time to consider subtabulation, a highly developed field due to past hand-calculation efforts of table makers. Before making any table, consult Fox [12].

Table 14.9

$$f(x) = A_{-1}(x)f(-1) + A_0(x)f(0) + A_1(x)f(1) + A_2(x)f(2)$$

x	A_{-1}	A_0	A_1	A_2	
0.0	0	1.0	0.0	0.0	1.0
0.1	−0.0285	0.9405	0.1045	−0.0165	0.9
0.2	−0.0480	0.8640	0.2160	−0.0320	0.8
0.3	−0.0595	0.7735	0.3315	−0.0455	0.7
0.4	−0.0640	0.6720	0.4480	−0.0560	0.6
0.5	−0.0625	0.5625	0.5625	−0.0625	0.5
	A_2	A_1	A_0	A_{-1}	x

PROBLEMS

14.9.1   Using these Lagrange interpolating coefficients, compute the error integral at 1.40 from Table 10.6.3.

14.9.2   Using Table 10.6.3 of the error integral (Sec. 10.6), compute the entries in the table for the midpoints.

15

FORMULAS USING FUNCTION VALUES

15.1 INTRODUCTION

The purpose of this chapter is to present a uniform method for finding a wide class of formulas based on sample values of the function. Chapters 17, 18, and 19 will elaborate the basic method for formulas that use other kinds of information besides the function values. Chapter 16 is devoted to a uniform method for finding the corresponding error terms of the formulas.

The emphasis is on a *uniform* approach to the problem of finding a formula to meet a given situation. We are not interested in giving an exhaustive list of formulas because, among other reasons, it is hopeless. The situation is something like the problem of trigonometric identities. There are so many possible identities that after giving a short list of basic ones and assuming a modest amount of experience in deriving others, we drop the subject. One does not seriously propose to make a large table of trigonometric identities—it is generally easier to derive the one you need from a few basic ones than to look it up in some large table. Similarly, after giving a few standard formulas, we shall be content with a description of *how* to find any reasonable formula you might want from the given class of possible formulas.

The problem may be viewed as one of information retrieval; we wish to find a particular piece of information, not knowing for sure that it has ever been recorded, and instead of trying to provide a catalog of known results (which would necessarily be finite and to a great extent represent past interests), we prefer a method of creating it as needed. *Instead of information retrieval we prefer information regeneration.* Of course this attitude will not work in all areas of knowledge, but in a mature science the principles should show how to recreate the information on demand. We will end these few chapters with a flow diagram indicating how to derive formulas. Currently on computers we recreate the elementary, special function values when needed rather than use a table, and so the approach is not novel.

Our basic approach is through these four questions (given in the preceding chapter):

1   What samples?
2   What class of functions?
3   What criterion of goodness of fit?
4   Where is the test to be applied?

Here in Part II we use polynomials as our approximating functions. This, as we said before, is the classical approach. In Parts III and IV we will use other classes.

In the next few chapters we will use the criterion of the formula being true for the successive powers of x, as high as we can go. Later we will try other criteria. We will also apply the test of accuracy to the functions themselves, and only much later will we consider other ways of applying the criteria of accuracy.

Returning to the first question, What samples?, we will confine ourselves in this chapter to function values, in Chap. 17 to function values plus various orders of derivatives of the functions, in Chap. 18 to function values and their differences, and in Chap. 19 we shall take the location of the samples as parameters which gives further freedom in the choice of a formula to fit a given situation.

Thus, the four questions provide the structure of the chapters and the basis for our uniform method which is used throughout much of the book.

15.2   FORMULAS USING INTERPOLATION

The classical approach to finding a formula for some operation, typically integration, is to first find the interpolating polynomial and then apply the operation, typically integration, to the polynomial. This is the method of *analytic substitution*.

As an example, consider finding the formula for integrating a function between the two given endpoints a and b. We have the two points $(a, y(a))$ and $(b, y(b))$, and the interpolating line through them is

$$y(x) = \frac{(b - x)y(a) + (x - a)y(b)}{b - a}$$

Integrating this curve, we get

$$\int_a^b y(x)\,dx = \int_a^b \frac{(b - x)y(a) + (x - a)y(b)}{b - a}\,dx$$

$$= \frac{-(b - x)^2 y(a) + (x - a)^2 y(b)}{2(b - a)}\Big|_a^b$$

$$= \frac{(b - a)^2 y(b) + (b - a)^2 y(a)}{2(b - a)}$$

$$= \left[\frac{y(b) + y(a)}{2}\right](b - a)$$

which is the well-known trapezoid rule.

As a second example, consider finding a quadratic through three equally spaced points and integrating to find the area. We take the points $(-1, y(-1))$, $(0, y(0))$, and $(1, y(1))$. Using undetermined coefficients, we assume the form

$$y = A_0 + A_1 x + A_2 x^2$$
$$y(-1) = A_0 - A_1 + A_2$$
$$y(0) = A_0$$
$$y(1) = A_0 + A_1 + A_2$$

It follows easily that

$$A_0 = y(0)$$
$$A_1 = \tfrac{1}{2}[y(1) - y(-1)]$$
$$A_2 = \tfrac{1}{2}[y(1) - 2y(0) + y(-1)]$$

The integral of this quadratic is

$$I = \int_{-1}^1 (A_0 + A_1 x + A_2 x^2)\,dx = 2A_0 + \frac{2A_2}{3}$$

$$= 2y_0 + \tfrac{1}{3}[y(1) - 2y(0) + y(-1)]$$
$$= \tfrac{1}{3}[y(-1) + 4y(0) + y(1)]$$

which is the well-known Simpson's formula.

Suppose we apply this formula to

$$\int_0^1 e^x \, dx$$

We need to shift either the formula or the integral so that the ranges match; we move the formula

$$\int_0^1 y \, dx = \tfrac{1}{6}[y(0) + 4y(\tfrac{1}{2}) + y(1)]$$

$$
\begin{array}{ll}
e^0 = 1.0000 & 1 \cdot e^0 = 1.0000 \\
e^{1/2} = 1.6487 & 4 \cdot e^{1/2} = 6.5948 \\
e^1 = 2.7183 & 1 \cdot e^1 = \underline{2.7183} \\
 & 10.3131
\end{array}
$$

$$I_{\text{calc}} = \tfrac{1}{6}(10.3131) = 1.7188$$

$$I_{\text{anal}} = \int_0^1 e^x \, dx = e^1 - 1 = 1.7183 \qquad \text{error} = 0.0005$$

PROBLEMS

15.2.1   Compute $\int_0^1 e^{-x} \, dx$ using Simpson's formula.

15.2.2   Compute $\int_0^\pi \sin x \, dx$ using Simpson's formula.

15.2.3   Find the midpoint formula $\int_0^1 f(x) \, dx = f(1/2)$

15.3   THE TAYLOR-SERIES METHOD OF FINDING FORMULAS

There is a second method of finding formulas that is occasionally used.   In this method the function, typically an integrand, is expanded into a Taylor series about the midpoint (usually), and then the unknown coefficients are determined so that as many terms of the series exactly cancel as possible; the rest is regarded as the error, the *truncation error*.   Thus the truncation error is the rest of the infinite series that did not exactly cancel out.

As an example of the Taylor-series approach, consider again finding Simpson's formula.

$$\int_{-1}^1 f(x) \, dx = af(-1) + bf(0) + cf(1)$$

We write $f(x)$ as a Taylor series about $x = 0$

$$f(x) = f(0) + xf'(0) + \frac{x^2}{2!}f''(0) + \frac{x^3}{3!}f'''(0) + \frac{x^4}{4!}f^{\mathrm{iv}}(0) + \cdots$$

We get on the two sides of the formula

$$2f(0) + \frac{2}{3!}f''(0) + \frac{2}{5!}f^{\mathrm{iv}}(0) + \cdots$$

$$= a\left[f(0) - f'(0) + \frac{f''(0)}{2!} - \frac{f'''(0)}{3!} + \frac{f^{\mathrm{iv}}(0)}{4!} - \cdots\right]$$

$$+ b[f(0)]$$

$$+ c\left[f(0) + f'(0) + \frac{f''(0)}{2!} + \frac{f'''(0)}{3!} + \frac{f^{\mathrm{iv}}(0)}{4!} + \cdots\right]$$

Since we want this formula to be true for a wide class of functions, we must regard the derivative values at $x = 0$ as being independent variables, and we therefore set their coefficients equal to zero. This gives the equations [using $k!f^{(k)}(0)$]

$$2 = a + b + c$$
$$0 = -a \quad + c$$
$$\tfrac{2}{3} = a \quad\quad + c$$
$$0 = -a \quad + c$$
$$\tfrac{2}{3} = a \quad\quad + c + \text{error}$$

The solution of these equations is $a = 1/3 = c$ and $b = 4/3$, as before. We see that Simpson's formula is exactly true for cubics even though it was originally

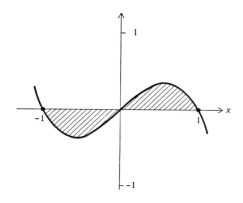

FIGURE 15.3.1

derived only for quadratics.   This is true because due to symmetry the cubic part of the function *exactly* cancels out.   Thus the cubic

$$y(x) = x(1 - x^2)$$

which vanishes at the sample points has its area equal to zero, and any multiple of this can be subtracted from a function without changing the answer (see Fig. 15.3.1).

The value of the error E_4 is, from the last equation,

$$E_4 = \tfrac{2}{5} - \tfrac{2}{3} = -\tfrac{4}{15}$$

15.4   THE DIRECT METHOD OF FINDING FORMULAS

Instead of finding formulas from the interpolating polynomial, which is the classic method, we shall adopt another method, the direct approach to the formula. This method is not only, as its name implies, more direct, it is also more powerful in that it can find formulas that the interpolation method cannot—by going directly to the formula, we can avoid those cases in which there is no interpolating polynomial but for which there is a formula of the required form.   But far more important is the fact that the direct method provides a natural basis for generalization to other classes of functions.   It will be our basic approach to finding formulas.   For polynomials, the direct method resembles the Taylor-series method, but has a different approach to the error.

To illustrate the direct method, suppose we again find Simpson's formula. We want a formula for the integral that uses the function values $f(-1), f(0)$, and $f(1)$

$$\int_{-1}^{1} f(x)\, dx = af(-1) + bf(0) + cf(1)$$

where a, b, and c, are the undetermined coefficients.   We make the formula exactly correct for $f(x)$ equals to 1, x, and x^2.

$$f(x) = 1:\qquad 2 = a + b + c \qquad (15.4.1)$$

$$f(x) = x:\qquad 0 = -a \; + c \qquad (15.4.2)$$

$$f(x) = x^2:\qquad \tfrac{2}{3} = a \; + c \qquad (15.4.3)$$

These equations are exactly those of the Taylor-series approach, and so the answer is the same.

We will, for the moment, rate formulas by the error found from the first higher power of x that does not exactly cancel; in Chap. 16 we will in part justify this approach. Putting x^3 in the formula, we get

$$0 = -a + c$$

which is true from Eq. (15.4.2).
Using x^4, we get

$$\tfrac{2}{3} = a + c + E_4$$

$$E_4 = \tfrac{2}{3} - \tfrac{1}{3} - \tfrac{1}{3} = -\tfrac{4}{15}$$

as we should.

As a second example, consider finding a formula of the form

$$\int_{-1}^{1} f(x)\sin\frac{\pi}{2}x\,dx = w_{-1}f(-1) + w_0 f(0) + w_1 f(1)$$

With three parameters w_{-1}, w_0, and w_1 we can make the formula exact for $f(x)$ equal to 1, x, and x^2. We get

$$0 = w_{-1} + w_0 + w_1$$

$$\frac{8}{\pi^2} = -w_{-1} \qquad + w_1$$

$$0 = w_{-1} \qquad + w_1$$

where we have used

$$\int_{-1}^{1} \sin\frac{\pi}{2}x\,dx = 0 \qquad \int_{-1}^{1} x\sin\frac{\pi}{2}x\,dx = \frac{8}{\pi^2} \qquad \int_{-1}^{1} x^2\sin\frac{\pi}{2}x\,dx = 0$$

The solution of the three equations is

$$w_1 = \frac{4}{\pi^2} = -w_{-1} \qquad w_0 = 0$$

and the formula is

$$\int_{-1}^{1} f(x)\sin\frac{\pi}{2}x\,dx = \frac{4}{\pi^2}\,[f(1) - f(-1)]$$

If we try $f(x) = x^3$, we get (on setting $x = \dfrac{2}{\pi}y$)

$$\int_{-1}^{1} x^3 \sin\frac{\pi}{2}x\,dx = \left(\frac{2}{\pi}\right)^4 \int_{-\pi/2}^{\pi/2} y^3 \sin y\,dy$$

$$= \left(\frac{2}{\pi}\right)^4 [(3y^2 - 6)\sin y - (y^3 - 6y)\cos y]\Big|_{-\pi/2}^{\pi/2}$$

$$= \left(\frac{2}{\pi}\right)^4 \left[3\left(\frac{\pi}{2}\right)^2 - 6\right]2 = \frac{24}{\pi^4}(\pi^2 - 8) = E_3 + \frac{8}{\pi^2}$$

Hence E_3 measures the error.

As a final example of the direct method, consider a formula of the form

$$\int_{-1}^{1} f(x)\,dx = w_{-2/3}f(-2/3) + w_0 f(0) + w_{2/3}f(2/3)$$

We try successive powers of x

$f(x) = 1$:	$2 = w_{-2/3}$	$+ w_0 + w_{2/3}$	(15.4.4)
$f(x) = x$:	$0 = -\frac{2}{3}w_{-2/3}$	$+ (\frac{2}{3})w_{2/3}$	(15.4.5)
$f(x) = x^2$:	$\frac{2}{3} = (-\frac{2}{3})^2 w_{-2/3}$	$+ (\frac{2}{3})^2 w_{2/3}$	(15.4.6)
$f(x) = x^3$:	$0 = (-\frac{2}{3})^3 w_{-2/3}$	$+ (\frac{2}{3})^3 w_{2/3}$	(15.4.7)
$f(x) = x^4$:	$\frac{2}{3} = (\frac{2}{3})^4 w_{-2/3}$	$+ (\frac{2}{3})^4 w_{2/3} + E_4$	(15.4.8)

From Eq. (15.4.5)

$$w_{2/3} = w_{-2/3}$$

Then from Eqs. (15.4.4) and (15.4.6)

$$2 = 2w_{2/3} + w_0$$
$$1 = 2/3 \cdot 2w_{2/3}$$

we get

$$w_{2/3} = 3/4 = w_{-2/3}$$
$$w_0 = 2/4$$

Equation (15.4.7) follows from Eq. (15.4.5), and Eq. (15.4.8) gives

$$E_4 = \frac{14}{135}$$

These three examples should make clear the direct method; there may, as in the second case, be some difficult integrations to get some numbers, but the method is simple.

PROBLEMS

15.4.1 Find the formula for using

$$\int_{-\pi}^{\pi} f(x)\sin x\,dx$$

using $f(-\pi), f(-\pi/2), f(0), f(\pi/2),$ and $f(\pi)$.

Ans. $\int_{-\pi}^{\pi} f(x)\sin x\,dx = \left(1 - \dfrac{8}{\pi^2}\right)[f(\pi) - f(-\pi)] + \dfrac{16}{\pi^2}[f(\pi/2) - f(-\pi/2)]$

15.4.2 Find a formula for

$$-\int_0^1 f(x)\ln x \, dx$$

using $f(0)$, $f(1/2)$, and $f(1)$.

Ans. $-\int_0^1 f(x)\ln x \, dx = \dfrac{1}{36}\,[17f(0) + 20f(1/2) - f(1)]$ $E_3 = 1/48$

15.4.3 Find $\displaystyle\int_0^1 f(x)\, dx = w_0\, f(1/3) + w_1 f(2/3)$

15.4.4 Find $\displaystyle\int_0^\infty e^{-x}f(x)\, dx = w_0\, f(0) + w_2\, f(2)$

15.4.5 Find $\displaystyle-\int_0^1 f(x)\ln x \, dx = w_0\, f(0) + w_1 f(1/3) + w_2\, f(2/3) + w_3\, f(1)$

15.5 THE INVERSE VANDERMONDE

The method illustrated in the last section consists of the following steps:

1 Write down the proposed formula with undetermined coefficients.
2 For $f(x)$ substitute 1, x, x^2, x^3, ... until there are as many equations as there are coefficients—later we may find that we need more equations.
3 Solve the resulting linear equations.

The equations can always be solved since the determinant of the unknowns is the Vandermonde determinant, which we proved (Sec. 14.4) cannot be zero if the sample points are distinct.

The equations define the formula and will be called the *defining equations*. If we write the matrix of unknowns as

$$X = \begin{pmatrix} 1 & 1 & 1 & \cdots & 1 \\ x_1 & x_2 & x_3 & \cdots & x_N \\ x_1^2 & x_2^2 & x_3^2 & \cdots & x_N^2 \\ \hdotsfor{5} \\ x_1^{N-1} & x_2^{N-1} & x_3^{N-1} & \cdots & x_N^{N-1} \end{pmatrix}$$

and write the *weights*, which we have generally been labeling w_i, as a column vector w, then the left-hand sides are the *moments* of the operator for which we are finding a formula and are the column vector m in the equation

$$m = Xw$$

From this we see that there is a separation of the sample points used and the operator. The sample points occur only in the matrix, while the moments of the operator, which are all that we ever use of the operator in the method, are the resultant vector.   If we could find in some direct fashion the inverse matrix X^{-1}, which depends *only* on the sample points, then questions of the loss of accuracy in the solution process would be easy to examine.   We therefore set about to invert the matrix corresponding to the Vandermonde determinant.

The basis for the inversion process is the *sample polynomials* introduced in the Lagrange interpolation process (Sec. 14.5), namely,

$$\frac{\pi_i(x)}{\pi_i(x_i)}$$

Notice that

$$\frac{\pi_i(x_j)}{\pi_i(x_i)} = \begin{cases} 0 & i \neq j \\ 1 & i = j \end{cases}$$

We will work with the $\pi_i(x)$ which have the property

$$\pi_i(x_j) = \begin{cases} 0 & i \neq j \\ \pi_i(x_i) \neq 0 & i = j \end{cases}$$

Expanding out the $\pi_i(x)$, we have the polynomials

$$\pi_i(x) = c_{i, N-1}x^{N-1} + c_{i, N-2}x^{N-2} + \cdots + c_{i, 0} \qquad c_{i, N-1} = 1$$

where the $c_{i, j}$ are the elementary symmetric functions of the sample points x_j *except* for the ith sample point.   The matrix of coefficients times the Vandermonde matrix gives

$$\begin{pmatrix} c_{1,0} & c_{1,1} & c_{1,2} & \cdots & c_{1,N-1} \\ c_{2,0} & c_{2,1} & c_{2,2} & \cdots & c_{2,N-1} \\ \cdots\cdots\cdots\cdots\cdots\cdots\cdots\cdots\cdots \\ c_{N,0} & c_{N,1} & c_{N,2} & \cdots & c_{N,N-1} \end{pmatrix} \begin{pmatrix} 1 & 1 & 1 & \cdots & 1 \\ x_1 & x_2 & x_3 & \cdots & x_N \\ x_1^2 & x_2^2 & x_3^2 & \cdots & x_N^2 \\ x_1^{N-1} & x_2^{N-1} & x_3^{N-1} & \cdots & x_N^{N-1} \end{pmatrix}$$
$$= (\pi_i(x_j))$$

which for our initial purposes is close enough to the inverse.   To get the inverse, evidently we only need to divide the $c_{i, j}$ by $\pi_i(x_i)$, or else we can divide the answer in the ith position by $\pi_i(x_i)$ to get w_i.   Let C be the inverse matrix

$$C = X^{-1}$$

then

$$w = Cm$$

is the solution of the system of defining equations.

It is important to notice that the inverse is found by forming the elementary symmetric functions of the sample points, excluding one sample point each time, and using these as the coefficients of the inverse, *provided* they are normalized by the constant $\pi_i(x_i)$. There are no awkward questions of the solution of the system of equations and ill-conditioning of the matrix; the method is fairly straightforward and should be clearly understood since it will be developed further in later chapters.

15.6 UNIVERSAL MATRICES

The special case of equally spaced samples arises frequently in practice, and it is worth tabulating these *universal matrices*. We have labeled them S_n with the sample points used in the following parentheses; thus $S_4(-3/2, -1/2, 1/2, 3/2)$ means that four sample points at $-3/2$, $-1/2$, $1/2$, and $3/2$ are used. When this universal matrix is multiplied by the vector of moments of the operator we are approximating, we get the weights of the formula.

Universal Matrices

$$S_2 = S_2(-\tfrac{1}{2}, \tfrac{1}{2}) = \begin{pmatrix} \tfrac{1}{2} & -1 \\ \tfrac{1}{2} & 1 \end{pmatrix} = \tfrac{1}{2}\begin{pmatrix} 1 & -2 \\ 1 & 2 \end{pmatrix}$$

$$S_3 = S_3(-1, 0, 1) = \begin{pmatrix} 0 & -\tfrac{1}{2} & \tfrac{1}{2} \\ 1 & 0 & -1 \\ 0 & \tfrac{1}{2} & \tfrac{1}{2} \end{pmatrix} = \tfrac{1}{2}\begin{pmatrix} 0 & -1 & 1 \\ 2 & 0 & -2 \\ 0 & 1 & 1 \end{pmatrix}$$

$$S_4 = S_4(-\tfrac{3}{2}, -\tfrac{1}{2}, \tfrac{1}{2}, \tfrac{3}{2}) = \tfrac{1}{48}\begin{pmatrix} -3 & 2 & 12 & -8 \\ 27 & -54 & -12 & 24 \\ 27 & 54 & -12 & -24 \\ -3 & -2 & 12 & 8 \end{pmatrix}$$

$$S_5 = S_5(-2, -1, 0, 1, 2) = \tfrac{1}{24}\begin{pmatrix} 0 & 2 & -1 & -2 & 1 \\ 0 & -16 & 16 & 4 & -4 \\ 24 & 0 & -30 & 0 & 6 \\ 0 & 16 & 16 & -4 & -4 \\ 0 & -2 & -1 & 2 & 1 \end{pmatrix}$$

$$S_6 = S_6(-\tfrac{5}{2}, -\tfrac{3}{2}, \ldots, \tfrac{5}{2})$$

$$= \tfrac{1}{3{,}840}\begin{pmatrix} 45 & -18 & -200 & 80 & 80 & -32 \\ -375 & 250 & 1{,}560 & -1{,}040 & -240 & 160 \\ 2{,}250 & -4{,}500 & -1{,}360 & 2{,}720 & 160 & -320 \\ 2{,}250 & 4{,}500 & -1{,}360 & -2{,}720 & 160 & 320 \\ -375 & -250 & 1{,}560 & 1{,}040 & -240 & -160 \\ 45 & 18 & -200 & -80 & 80 & 32 \end{pmatrix}$$

$$S_7 = \tfrac{1}{720}\begin{pmatrix} 0 & -12 & 4 & 15 & -5 & -3 & 1 \\ 0 & 108 & -54 & -120 & 60 & 12 & -6 \\ 0 & -540 & 540 & 195 & -195 & -15 & 15 \\ 720 & 0 & -980 & 0 & 280 & 0 & -20 \\ 0 & 540 & 540 & -195 & -195 & 15 & 15 \\ 0 & -108 & -54 & 120 & 60 & -12 & -6 \\ 0 & 12 & 4 & -15 & -5 & 3 & 1 \end{pmatrix}$$

To make sure we understand how these universal matrices work and their power, let us apply them to a number of specific formulas.   Again let us derive Simpson's formula as a check on the method.   We evidently want to use $S_3(-1, 0, 1)$.   The moments of the operator are

$$m_0 = \int_{-1}^{1} 1 \cdot dx = 2$$

$$m_1 = \int_{-1}^{1} x \, dx = 0$$

$$m_2 = \int_{-1}^{1} x^2 \, dx = 2/3$$

Therefore we have

$$S_3 m = \begin{pmatrix} 0 & -\frac{1}{2} & \frac{1}{2} \\ 1 & 0 & -1 \\ 0 & \frac{1}{2} & \frac{1}{2} \end{pmatrix} \begin{pmatrix} 2 \\ 0 \\ \frac{2}{3} \end{pmatrix} = \begin{pmatrix} \frac{1}{3} \\ \frac{4}{3} \\ \frac{1}{3} \end{pmatrix} = w$$

which is, of course, the correct answer.

Next consider finding the Simpson's "half-formula"

$$\int_{-1}^{0} f(x) \, dx = w_{-1} f(-1) + w_0 f(0) + f(1)$$

The moments are

$$m_0 = \int_{-1}^{0} 1 \cdot dx = 1$$

$$m_1 = \int_{-1}^{0} x \, dx = -1/2$$

$$m_2 = \int_{-1}^{0} x^2 \, dx = 1/3$$

Hence

$$\begin{pmatrix} 0 & -\frac{1}{2} & \frac{1}{2} \\ 1 & 0 & -1 \\ 0 & \frac{1}{2} & \frac{1}{2} \end{pmatrix} \begin{pmatrix} 1 \\ -\frac{1}{2} \\ \frac{1}{3} \end{pmatrix} = \frac{1}{12} \begin{pmatrix} 5 \\ 8 \\ -1 \end{pmatrix} = w$$

Again, suppose we had tried to find a formula of the form

$$\frac{df}{dx}\bigg|_{x=0} = w_{-1} f(-1) + w_0 f(0) + w_1 f(1)$$

The moment vector is

$$\begin{pmatrix} 0 \\ 1 \\ 0 \end{pmatrix}$$

and we have

$$\begin{pmatrix} 0 & -\frac{1}{2} & \frac{1}{2} \\ 1 & 0 & -1 \\ 0 & \frac{1}{2} & \frac{1}{2} \end{pmatrix} \begin{pmatrix} 0 \\ 1 \\ 0 \end{pmatrix} = \begin{pmatrix} -\frac{1}{2} \\ 0 \\ \frac{1}{2} \end{pmatrix}$$

In the case of

$$\int_{-1}^{1} f(x) \sin \frac{\pi}{2} x \, dx$$

the moments were 0, $8/\pi^2$, and 0 so that

$$\begin{pmatrix} 0 & -\frac{1}{2} & \frac{1}{2} \\ 1 & 0 & -1 \\ 0 & \frac{1}{2} & \frac{1}{2} \end{pmatrix} \begin{pmatrix} 0 \\ \frac{8}{\pi^2} \\ 0 \end{pmatrix} = \begin{pmatrix} \dfrac{-4}{\pi^2} \\ 0 \\ \dfrac{4}{\pi^2} \end{pmatrix}$$

Last, suppose we wanted to interpolate at a point x using the same three sample points. This time the moment vector depends on x. We have, after a little thought,

$$\begin{pmatrix} 0 & -\frac{1}{2} & \frac{1}{2} \\ 1 & 0 & -1 \\ 0 & \frac{1}{2} & \frac{1}{2} \end{pmatrix} \begin{pmatrix} 1 \\ x \\ x^2 \end{pmatrix} = \begin{pmatrix} \dfrac{x^2 - x}{2} \\ 1 - x^2 \\ \dfrac{x^2 + x}{2} \end{pmatrix}$$

or more usually

$$f(x) = \frac{x^2 - x}{2} f(-1) + (1 - x^2)f(0) + \frac{x^2 + x}{2} f(1)$$

as the interpolating quadratic, which is clearly the correct answer.

PROBLEMS

Using the universal matrices, do
15.6.1   Prob. 15.4.2.
15.6.2   Prob. 15.4.5.   Use S_4 and first move the formula to proper sample points.
15.6.3   Prob. 15.4.4.

15.6.4 Prob. 15.4.3.

15.6.5 Prob.15.4.1.

15.6.6 Find a quartic interpolation formula through $(-2, -1, 0, 1, 2)$.

15.7 SUMMARY OF THE DIRECT METHOD

The direct method consists of first deciding what *information* (samples) to use and then writing a linear combination of this information with undetermined coefficients. We have chosen in this chapter to confine the discussion to function values at the chosen sample points.

We now impose the conditions that the formula be true for $f(x)$ equal to 1, x, x^2, ... as far as we can go. This produces the *defining equations*. On one side are the moments of the operator; typically this operator is integration, but we did one problem estimating the derivatives at the origin and another problem estimating an interpolated value at the point x. On the other side of the equation is the Vandermonde matrix of sample points. The inverse of the Vandermonde matrix can be found, if desired, as a combination of sums and products of the sample points in the form of the elementary symmetric functions, each time excluding one sample.

In the special case of equally spaced values the inverse *universal matrices* were tabulated. They show clearly how the method separates the information used (the samples) from the particular formula sought.

The fact that the Vandermonde determinant cannot be zero shows that any formula we want can be found *provided* we can find the moments of the operator we are approximating.

Conventionally we impose the conditions that the formula be true for 1, x, x^2, x^3, ..., but a little thought will show that we could take any sequence of polynomials for which the kth polynomial was exactly of degree k (and no lower). The reason is that if the formula is true for all powers lower than k, then putting in either x^k or a polynomial of degree k is the same since automatically all the lower powers of the polynomial will cancel out. Thus there are times when for the convenience of the algebra we will use a sequence of polynomials of degree k. Note that in the case of finding the error we must use a polynomial whose leading coefficient is 1 so that we get the right scale factor.

Further thought shows that *any* set of N linearly independent polynomials could be used in place of 1, x, x^2, ..., x^{N-1}, and occasionally this freedom is worthwhile.

We have not given long lists of integrals to evaluate; rather we have concentrated on finding formulas. The formulas are meant to be used, but the process of substituting function values into a formula is so straightforward that it should not require further drill at this stage.   But it is important to realize that the formulas are to be used and not merely admired.

15.8   APPENDIX

A number of integrals arise in the course of finding formulas of the form

$$\int_b^a K(x)f(x)\,dx$$

Among them are:

$$\int_{-0}^{\infty} e^{-x}x^n\,dx = n! \qquad n \text{ an integer}$$

$$\int_{-\infty}^{\infty} e^{-x^2}x^{2n}\,dx = \frac{1\cdot3\cdot5\cdots(2n-1)}{2^n}\sqrt{\pi} \qquad n \geq 1$$

$$\int_{-\infty}^{\infty} e^{-x^2}\,dx = \sqrt{\pi} \qquad n = 0$$

$$-\int_0^1 (\ln x)x^n\,dx = \frac{1}{(n+1)^2}$$

$$\int_0^1 \frac{x^n}{\sqrt{1-x^2}}\,dx = \int_0^{\pi/2} \cos^n x\,dx = \int_0^{\pi/2} \sin^n x\,dx$$

$$= \begin{cases} \dfrac{1\cdot3\cdot5\cdots(n-1)}{2\cdot4\cdots(n)}\dfrac{\pi}{2} & n = 2, 4, \ldots \\[3mm] \dfrac{2\cdot4\cdots(n-1)}{1\cdot3\cdots n} & n = 1, 3, 5, \ldots \end{cases}$$

$$\int_0^{\pi/2} \sin^n x \cos^m x\,dx = \frac{\Gamma[(n+1)/2]\Gamma[(m+1)/2]}{2\Gamma[(m+n+2)/2]}$$

where $\Gamma(x) = (x-1)\Gamma(x-1)$

$\Gamma(1/2) = \sqrt{\pi}$

16

ERROR TERMS

16.1 THE NEED OF AN ERROR ESTIMATE

We have adopted an algorithmic approach to finding formulas; we intend to give a rather detailed description of how any of a wide variety of formulas can be found. With this approach we also need a uniform method for finding the error term of the formula, because without an estimate of the error it is difficult to rely on an answer produced by a machine. The error term is also called the *truncation error* or *remainder term*, from the Taylor-series form of derivation (Sec. 15.3). The error is expressed in terms of a high-order derivative, just as in the case of interpolation formulas, and all the faults of higher derivatives (Sec 14.7) apply; in practice they are both hard to find and likely to be large. The derivatives occur because a polynomial approximation is used in the classical approach.

In this chapter we are concerned with finding the error term, not only for the kinds of formulas we have already examined, but also for those we will take up in the next few chapters. Therefore, we treat the general case, and, as so often happens, the general case is more difficult to carry out than are many trick methods which work in special cases. It is worth accepting this cost in these days of almost infinite knowledge. We also ignore, for the moment, the question of roundoff in the *use* of the formulas; in the derivations themselves we will watch out for roundoff effects.

Our approach is to first find a formula in terms of an *influence function* which gives the exact expression for the error. But let us be clear about this: if we had the exact error in a useful form, then we would know the exact right answer! When there is an exact expression for an error, it is generally useless, and the more we give up in precision of stating the error, the more useful we can hope it will be. From the influence-function representation of the error we will pass in many cases to a form resembling the mean value theorem with its unknown place of evaluation, θ—the form of the error will be some known constant times a high-order derivative evaluated at some θ in a known interval. From this it is easy to pass to bounds in terms of bounds on the derivative.

When we want to compare formulas, one with another, and both have same-order derivatives in their error terms, it is natural to suppose that the one with the smaller coefficient has the smaller error; but it does not follow. Indeed, we realize that a relatively poor formula will for certain functions give more accurate answer than that of a better formula. Evidently we mean that one formula is better than another formula when measured over some class of functions. Because of the unknown point θ in the error terms we are in fact unable to find the variance over any reasonable ensemble of functions [if $f(x)$ and $-f(x)$ are both in the ensemble of functions with equal weight or probability, then the average would be zero]. Thus we are forced to compare formulas having same-order derivatives by comparing the coefficients of the derivatives, and we will at times make the wrong choice—we hope to be right most of the time.

16.2 THREE BACKGROUND IDEAS

The method of finding the error term for a formula rests on three important results, all of which we have seen in the past, but it is worth reexamining them to be sure we understand what is going on. This is especially necessary because the method, while simple once it is understood, seems to baffle the beginner.

The first result was used in Sec. 4.7. The general expansion of a function in a Taylor series with an integral remainder has the form

$$f(x) = f(a) + (x - a)f'(a) + \frac{(x - a)^2}{2!} f''(a) + \cdots$$

$$+ \frac{(x - a)^{m-1}}{(m - 1)!} f^{(m-1)}(a) + \frac{1}{(m - 1)!} \int_a^x f^{(m)}(s)(x - s)^{m-1} \, ds$$

If the last term is integrated by parts, we get

$$u = (x - s)^{m-1} \qquad du = -(m - 1)(x - s)^{m-2}\, ds$$

$$dv = f^{(m)}(s)\, ds \qquad v = f^{(m-1)}(s)$$

$$\frac{1}{(m - 1)!}\left[(x - s)^{m-1}f^{(m-1)}(s)\Big|_a^x + (m - 1)\int_a^x f^{(m-1)}(s)(x - s)^{m-2}\, ds\right]$$

$$= \frac{1}{(m - 1)!}\left[-(x - a)^{m-1}f^{(m-1)}(a)\right] + \frac{1}{(m - 2)!}\int_a^x f^{(m-1)}(s)(x - s)^{m-2}\, ds$$

The integrated term cancels the preceding term of the expansion, while the integral is the same as before *except* that the index m is reduced by 1.   Thus, a Taylor series with an integral remainder is simply the repeated application of integration by parts (applied in the reverse way, of course).

The second result we need is the weighted mean value theorem for integrals (proved in Sec. 4.7), which states that if $f(x)$ and $g(x)$ are continuous and if $f(x) \geq 0$ in $a \leq x \leq b$, then for some θ in $a < \theta < b$

$$\int_a^b f(x)g(x)\, dx = g(\theta)\int_a^b f(x)\, dx$$

The third item is the index m for which x^m first *fails* to satisfy the formula. We began our derivations by making the formula exact for $f(x)$ equal to 1, x, x^2, ..., etc., and continued until we had enough equations to determine the coefficients in the formula.   It sometimes happens that an additional (in principle any number but in practice generally only one) power of x will fit the formula.   For example, in Simpson's formula we used 1, x, and x^2 to determine the three coefficients, and we later found that x^3 would also fit the formula exactly.   In the Simpson example, $m = 4$.

We need a test to see if by chance the next power of x exactly fits.   In the cases we have examined we are given function values at the sample points x_i (which are assumed known), and then all the arithmetic is rational in terms of the moments m_k (Sec. 15.5) and the sample points x_i.   In the case of the next chapter, where we are also given derivatives of the function as well as function values, the situation is the same.   In Chap. 19 where the x_i are not given in advance, we will have to examine this point further.   Thus we assume we can find the value of m which is the lowest power of x that does *not* fit the formula.   If we let E_k be the error in the formula when x^k is substituted into it, we have

$$E_0 = E_1 = \cdots = E_{m-1} = 0 \qquad E_m \neq 0$$

PROBLEMS

Find *m* for:

16.2.1 Prob. 15.4.1
16.2.2 Prob. 15.4.3
16.2.3 Prob. 15.4.4

16.3 THE BASIC METHOD APPROACH

The class of formulas we are going to examine has two properties:

1 The LHS (left-hand side) is a linear operator.
2 The estimation of this operator is made by a linear combination of function values and derivatives of order $\leq m - 2$ where the formula is made exact for $f(x)$ equal to $1, x, x^2, \ldots, x^{m-1}$.

It is easiest to think of the operator as integration, but we are not confined to it. We have the operator equation

$$\text{LHS} = \text{RHS} + \text{remainder} = \text{RHS} + R$$

or

$$R = \text{LHS} - \text{RHS}$$

Supplying the function $f(x)$, we get

$$R[f(x)] = (\text{LHS} - \text{RHS})[f(x)]$$

We next represent $f(x)$ in the Taylor series with an integral remainder

$$f(x) = f(a) + (x - a)f'(a) + \frac{(x - a)^2}{2}f''(a) + \cdots$$

$$+ \frac{(x - a)^{m-1}}{(m-1)!}f^{(m-1)}(a) + \frac{1}{(m-1)!}\int_a^x f^{(m)}(s)(x - s)^{m-1}\, ds$$

Now let A be the smallest of *all* the values of x that occur any place in the formula, and let B be the largest. In the Taylor expansion we see that (on setting $a = A$)

$$f(x) = P_{m-1}(x) + \frac{1}{(m-1)!}\int_A^x f^{(m)}(s)(x - s)^{m-1}\, ds$$

where $P_{m-1}(x)$ is a polynomial, in x, of degree $m - 1$.

We made the formula (determined the coefficients and later in Chap. 19 possibly the sample points x_i) so that

$$E_0 = E_1 = \cdots = E_{m-1} = 0 \qquad E_m \neq 0$$

hence

$$(\text{LHS} - \text{RHS})[P_{m-1}(x)] \equiv 0$$

We have, therefore, only to work with the remainder term of the Taylor series

$$R[f(x)] = (\text{LHS} - \text{RHS})\left[\frac{1}{(m-1)!}\int_A^x f^{(m)}(s)(x-s)^{m-1}\,ds\right]$$

16.4 THE INFLUENCE FUNCTION

We now introduce the notation

$$(x-s)_+^j = \begin{cases} 0 & \text{if } s \geq x \\ (x-s)^j & \text{if } s \leq x \end{cases} \quad j \geq 1$$

For $j = 0$ (see Fig. 16.4.1a):

$$(x-s)_+^0 = \begin{cases} 0 & \text{if } s \geq x \\ 1 & \text{if } s < x \end{cases}$$

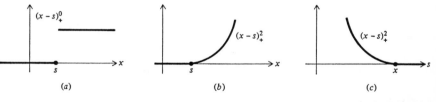

(a)   (b)   (c)

FIGURE 16.4.1

Note that (see Fig. 16.4.1b and c):

$$\frac{d}{dx}(x-s)_+^{m-1} = (m-1)(x-s)_+^{m-2}$$

$$\frac{d^2}{dx^2}(x-s)_+^{m-1} = (m-1)(m-2)(x-s)_+^{m-3}$$

etc., up to derivatives less than[1] than $m - 1$.   Note also (see Fig. 16.4.2) that

$$\int_a^b (x-s)_+^{m-1}\,dx = \frac{(b-s)_+^m - (a-s)_+^m}{m} \quad m > 0$$

---

[1] We can use derivatives of order $m - 1$ if we are not frightened of differentiating $(x-s)_+^1$ to get $(x-s)_+^0$.

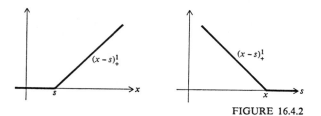

FIGURE 16.4.2

Using this new notation, we may increase the upper limit of the integral to B (B was defined in Sec. 16.3); that is, we have

$$R(f) = (\text{LHS} - \text{RHS})\left[\frac{1}{(m-1)!}\int_A^B f^{(m)}(s)(x-s)_+^{m-1}\,ds\right]$$

since we have added nothing in the integrand.

Because everything is linear and the operations in LHS and RHS are on the variable x, not s, we have

$$R(f) = \frac{1}{(m-1)!}\int_A^B f^{(m)}(s)(\text{LHS} - \text{RHS})[(x-s)_+^{m-1}]\,ds$$

We set

$$(\text{LHS} - \text{RHS})\left[\frac{(x-s)_+^{m-1}}{(m-1)!}\right] \equiv G(s)$$

$G(x)$ is called the *influence function*.

Hence

$$R[f(x)] = \int_A^B f^{(m)}(s)G(s)\,ds$$

Notice that in this expression we have separated the dependence on the formula $G(s)$ from the dependence on the function $f(x)$ we are using.

For the particular function

$$f(x) = x^m$$

we get

$$R[f(x)] = \int_A^B m!\,G(s)\,ds \equiv E_m$$

where E_m is, of course, the result of substituting x^m in the formula. Thus we know that

$$\int_A^B G(s)\,ds = \frac{E_m}{m!} \neq 0$$

An examination of the structure of $G(s)$ shows that RHS $\lfloor f(x) \rfloor$ is piecewise a polynomial inside (A,B) and that it is continuous. For many operators the LHS $\{f(x)\}$ is also piecewise continuous.

16.5 WHEN $G(s)$ HAS A CONSTANT SIGN

In most of the important cases it turns out that the influence function $G(s)$ has a constant sign. When this is so, we can use the mean value theorem for integrals and write the remainder, or error term, in the form

$$ R = R(f) = \int_A^B f^{(m)}(s) G(s) \, ds = f^{(m)}(\theta) \int_A^B G(s) \, ds = \frac{E_m}{m!} f^{(m)}(\theta) $$

where m is the first power of x that does not satisfy the formula.

As a trivial example of this, consider the Taylor series itself. We have

$$ G(s) = \frac{(x - s)^{m-1}}{(m - 1)!} $$

We then apply the weighted mean value theorem and get

$$ f(x) - f(a) - (x - a)f'(a) - \frac{(x - a)^2}{2!}f''(a) - \dots $$

$$ - \frac{(x - a)^{m-1}}{(m - 1)!} f^{(m-1)}(a) = \frac{1}{(m - 1)!} \int_a^x f^{(m)}(s)(x - s)^{m-1} \, ds $$

$$ = \frac{f^{(m)}(\theta)}{(m - 1)!} \int_a^x (x - s)^{m-1} \, ds $$

$$ = \frac{f^{(m)}(\theta)}{(m - 1)!} \frac{(x - a)^m}{m} = \frac{(x - a)^m f^{(m)}(\theta)}{m!} $$

This form of the Taylor series is widely used in applications.

The central problem is, therefore, how we can determine when $G(s)$ has a constant sign in the interval (A,B). As a first approach, we examine a number of special cases. Consider first the trapezoid rule

$$ R(f) = \int_0^1 f(x) \, dx - \frac{f(1) + f(0)}{2} $$

which is exact for $f(x)$ equal to 1 and x.   Thus $m = 2$, and

$$G(s) = \frac{\text{LHS} - \text{RHS}}{1!} [(x - s)_+^1]$$

$$1!\,G(s) = \int_0^1 (x - s)_+ \, dx - \frac{(1 - s)_+ + (0 - s)_+}{2}$$

$$= \frac{(1 - s)_+^2 - (0 - s)_+^2}{2} - \frac{(1 - s)_+ + (0 - s)_+}{2}$$

For $s \geq 1$ (see Fig. 16.5.1)

$$1!\,G(s) \equiv 0$$

since all the terms vanish.   For $1 \geq s \geq 0$

$$1!\,G(s) = \frac{1}{2}[(1 - s)^2 - (1 - s)] = \frac{-s(1 - s)}{2} \leq 0$$

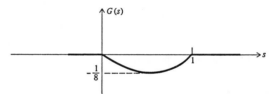

FIGURE 16.5.1

For $0 \geq s$

$$1!\,G(s) = \frac{1}{2}(1 - 2s + s^2 - s^2 - 1 + s + s) \equiv 0$$

Therefore

$$R(f) = \frac{E_2}{2!}\, f''(\theta) = -\frac{f''(\theta)}{12}$$

As a second example, consider Simpson's formula, where $m = 4$,

$$3!\,G(s) = (\text{LHS} - \text{RHS})\{(x - s)_+^3\}$$

$$= \frac{(1 - s)_+^4 - (-1 - s)_+^4}{4} - \frac{1}{3}[(-1 - s)_+^3 + 4(0 - s)_+^3 + (1 - s)_+^3]$$

We have four regions to consider (see Fig. 16.5.2): $s > 1$; $1 \geq s \geq 0$; $0 \geq s \geq -1$; and $-1 \geq s$.

$s \geq 1$: $\quad 3! \, G(s) \equiv 0$

$1 \geq s \geq 0$: $\quad 3! \, G(s) = \dfrac{(1-s)^4}{4} - \dfrac{(1-s)^3}{3} = (1-s)^3 \left(\dfrac{1-s}{4} - \dfrac{1}{3} \right)$

$\qquad\qquad\qquad = (1-s)^3 \left(\dfrac{-3s - 1}{12} \right) < 0$

$0 \geq s \geq -1$: $\quad 3! \, G(s) = \dfrac{(1-s)^4}{4} - \dfrac{(1-s)^3}{3} - \dfrac{4(-s)^3}{3}$

$\qquad\qquad\qquad = \dfrac{-1 + 6s^2 + 8s^3 + 3s^4}{12}$

$\qquad\qquad\qquad = (1+s)^3 \left(\dfrac{3s - 1}{12} \right) < 0$

$-1 \geq s$: $\quad 3! \, G(s) \equiv 0 \qquad$ by tedious algebra

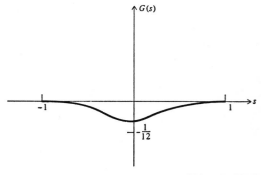

FIGURE 16.5.2

Thus using $E_4 = -\frac{4}{15}$ (from Sec. 15.3), we have

$$R(f) = -\frac{4}{15} \frac{f^{(4)}(\theta)}{4!} = -\frac{f^{(4)}(\theta)}{90}$$

as the error of Simpson's formula.

As a third example, consider the formula

$$\int_0^\infty e^{-x} f(x) \, dx = \frac{f(0) + f(2)}{2}$$

To find m, we try

$$f(x) = 1: \qquad \int_0^\infty e^{-x}\, dx = 1 = \frac{1+1}{2} = 1$$

$$f(x) = x: \qquad \int_0^\infty xe^{-x}\, dx = 1 = \frac{0+2}{2} = 1$$

$$f(x) = x^2: \qquad \int_0^\infty x^2 e^{-x}\, dx = 2 = \frac{0+2^2}{2} = 2$$

$$f(x) = x^3: \qquad \int_0^\infty x^3 e^{-x}\, dx = 3! \neq \frac{0+2^3}{2} = 4$$

Hence $m = 3$, and $E_3 = 3! - 4 = 2$. For $s \geq 0$ (see Fig. 16.5.3)

$$2!\, G(s) = (\text{LHS} - \text{RHS})[(x-s)_+^2]$$

$$= \int_0^\infty e^{-x}(x-s)_+^2\, dx - \frac{(2-s)_+^2 + (0-s)_+^2}{2}$$

$$= \int_s^\infty e^{-x}(x-s)_+^2\, dx - \frac{(2-s)^2}{2}$$

$$= \left[\frac{(x-s)_+^2\, e^{-x}}{-1} + \frac{2(x-s)_+\, e^{-x}}{-1} + \frac{2e^{-x}}{-1} \right]\Bigg|_s^\infty - \frac{(2-s)_+^2}{2}$$

$$= 2e^{-s} - \frac{(2-s)_+^2}{2}$$

For $s \geq 2$

$$2G(s) = 2e^{-s} > 0$$

For $2 \geq s \geq 0$

$$2G(s) = 2e^{-s} - \frac{(2-s)^2}{2} = \frac{s^2}{2} - \frac{s^3}{3} + \cdots$$

requires some examination.

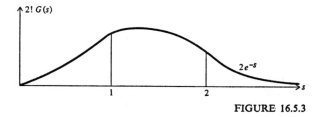

FIGURE 16.5.3

From the series expansion we see that at $x = 0$ the curve is tangent to the s axis and that the second derivative is positive so that the curve is concave upward. A little more effort convinces one that $G(s) \geq 0$ for all s. We had $E_3 = 2$, so that

$$R(f) = \frac{2}{3!} f'''(\theta) = \frac{f'''(\theta)}{3}$$

16.6 THE PRACTICAL EVALUATION OF $G(s)$

The three previous examples show that the method is straightforward, but can be tedious. We need, therefore, to examine the process to see how it can be simplified.

In the first place, it should be clearly recognized that $G(s)$ is piecewise continuous and needs to be evaluated *only once* for a formula. Thus the simple, crude method of using a computer to plot the points with small spacing and looking at the resulting picture can be used. As we did in the third example, a little bit of the calculus can also be used to aid in plotting $G(s)$.

There are further items to be noted. First, outside the range of all the function values $G(s) = 0$. This follows easily from noting that if s is greater than every sample point and value used in the operator, then all the values will be zero. At the other end of the range of s, all the terms will be there, and since the formula is true for a polynomial in x of degree $m - 1$ and in that range we have $(x - s)^{m-1}$ which is a polynomial of degree $m - 1$, we know that $G(s)$ will vanish identically.

We notice that as we progress from right to left, taking more and more terms, the algebra becomes more difficult. This can be easily remedied by using a second notation

$$(x - s)^j_- = \begin{cases} 0 & s \leq x \\ (x - s)^j & s > x \end{cases} \qquad j \geq 0$$

whose meaning is obvious. Note that

$$(x - s)^j_- + (x - s)^j_+ \equiv (x - s)^j$$

Now, since

$$(\text{LHS} - \text{RHS})[(x - s)^j] \equiv 0 \qquad j < m$$

$$(\text{LHS} - \text{RHS})\left[\frac{(x - s)^{m-1}_+}{(m-1)!}\right] = -(\text{LHS} - \text{RHS})\left[\frac{(x - s)^{m-1}_-}{(m-1)!}\right]$$

we can start from the other end of the interval and use fewer terms for a while.

When the formula is symmetric, as Simpson's formula was, then $G(s)$ is also symmetric [notice how replacing s by $-s$ transforms one expression for $G(s)$ into the other] so that only half the range need be examined. The formula may be symmetric about a point $x \neq 0$, and $G(s)$ will also be symmetric about that point.

We need, in the method, the value of E_m. We can find this directly from Chap. 15 from knowing the sample points, without finding the weights. All we need to do is compute, using $\pi(x) = (x - x_1)(x - x_2) \cdots (x - x_n)$, (or $x\pi(x)$),

$$R[\pi(x)] = (\text{LHS} - \text{RHS})[\pi(x)] = \text{LHS}[\pi(x)] = E_m$$

since $\pi(x)$ is zero at each sample point on the right-hand side. Thus *before* we get involved in the details of the formula and in finding $G(s)$, we can find the error [if $G(s)$ turns out to be of constant sign].

PROBLEMS

16.6.1 Show that the three-eighths rule is valid.

$$\int_0^3 f(x)\, dx = \frac{3}{8}\, [f(0) + 3f(1) + 3f(2) + f(3)] - \frac{3}{80}\, f^{\text{iv}}(\theta)$$

16.6.2 Show that the midpoint and trapezoid rules have errors which are in the ratio of $-1/2$ to each other.

16.6.3 Show that

$$\int_{-1}^1 f(x)\, dx = \frac{1}{4}\left[3f\left(-\frac{2}{3}\right) + 2f(0) + 3f\left(\frac{2}{3}\right) \right] + \frac{7}{1,620}\, f^{\text{iv}}(\theta)$$

16.6.4 Find the error term of Prob. 15.4.3.
16.6.5 Find the error term of Prob. 15.4.4.
16.6.6 Find the error term of Prob. 15.4.5.

16.7 WHEN $G(s)$ IS NOT OF CONSTANT SIGN

When $G(s)$ is not of constant sign, we *cannot* express the error as we did in the previous section. Indeed, for some functions the error cannot be so expressed. To show this, suppose that $G(s)$ is positive except for some small interval and that the integral of $G(s)$ is positive (see Fig. 16.7.1). Consider now a function $f(x)$ whose mth derivative is positive and continuous but which is small outside and large inside the interval in which $G(s)$ is negative. Thus the integral

$$\int_A^B f^{(m)}(s) G(s)\, ds$$

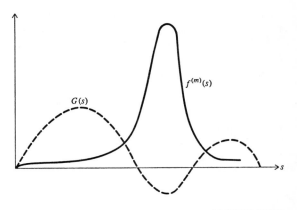

FIGURE 16.7.1

will be negative, but the product

$$f^{(m)}(s) \int_A^B G(s) \, ds$$

will be positive. Hence there can be no θ in such a case, and we have proved

Theorem 16.7.1 If $G(s)$ is not of constant sign, then there are functions which cannot have their error terms written in the form

$$\frac{E_m}{m!} f^{(m)}(\theta) = f^{(m)}(\theta) \int_A^B G(s) \, ds$$

The second point to observe is that when $G(s)$ changes sign, we can still get a bound on the error. The bound follows readily by taking the absolute value of $G(s)$:

$$\int_A^B f^{(m)}(s) |G(s)| \, ds = f^{(m)}(\theta) \int_A^B |G(s)| \, ds$$

Hence

$$\left| \int_A^B f^{(m)}(s) G(s) \, ds \right| \le \max\{|f^{(m)}(x)|\} \int_A^B |G(s)| \, ds$$

where the max is over all x ($A \le x \le B$). The computation of the integral

$$\int_A^B |G(s)| \, ds$$

can be done analytically or estimated numerically since, again, it need be done only once for any particular formula.

The third observation is that when the sign of $G(s)$ changes, then we can still find an error term.   As an illustration, consider the function $G(s)$ as shown in Fig. 16.7.2.   The point $x = C$ has been chosen so that the two shaded areas are equal in size; that is,

$$\int_C^B G(s)\, ds = 0$$

We have

$$R = R(f) = \int_A^B f^{(m)}(s)G(s)\, ds$$

$$= \int_A^C + \int_C^B$$

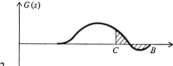

FIGURE 16.7.2

We integrate the second term by parts and set

$$\int_C^s G(s)\, ds = H(s) \qquad Note: \quad H(B) = H(C) = 0.$$

Then

$$R = \int_A^C f^{(m)}(s)G(s)\, ds + f^{(m)}(s)H(s)\Big|_C^B - \int_C^B f^{(m+1)}(s)H(s)\, ds$$

$$= \int_A^C f^{(m)}(s)G(s)\, ds - \int_C^B f^{(m+1)}(s)H(s)\, ds$$

We can now apply the mean value theorem to each piece.

$$R = f^{(m)}(\theta_1)\int_A^C G(s)\, ds - f^{(m+1)}(\theta_2)\int_C^B H(s)\, ds$$

If $G(s)$ is as shown in Fig. 16.7.3, we could simply set

$$H(s) = \int_A^s G(s)\, ds$$

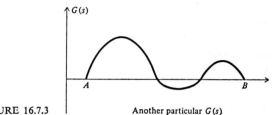

FIGURE 16.7.3   Another particular $G(s)$

and integrate this by parts

$$R(f) = f^{(m)}(s)H(s)\Big|_{A}^{B} - \int_{A}^{B} f^{(m+1)}(s)H(s)\,ds$$

$$= f^{(m)}(B)H(B) - f^{(m+1)}(\theta) \int_{A}^{B} H(s)\,ds$$

If we put $f(x) = x^{m+1}$ in this formula, we get

$$R(f) \equiv E_{m+1} = (m+1)!\, BH(B) - (m+1)! \int_{A}^{B} H(s)\,ds$$

so that

$$\int_{A}^{B} H(s)\,ds = \frac{E_{m+1}}{(m+1)!} - B\frac{E_m}{m!}$$

Therefore we have in general

$$R(f) = \frac{f^{(m)}(B)E_m}{m!} - \frac{f^{(m+1)}(\theta)}{(m+1)!}[E_{m+1} - (m+1)BE_m]$$

We shall not analyze the general case, since the results are not used in the text, but shall merely observe that such forms can be useful at times.

16.8   THE FLAW IN THE TAYLOR-SERIES APPROACH

Since there exists an error term from the Taylor-series of the form[1]

$$f(a+x) = f(a) + (x-a)f'(a) + \frac{(x-a)^2}{2!}f''(a) + \cdots$$

$$+ \frac{(x-a)^{m-1}}{(m-1)!}f^{(m-1)}(a) + \frac{(x-a)^m f^{(m)}(a+\theta x)}{m!}$$

[1] See Sec. 16.5 for this Lagrange form.

it is natural to ask why we cannot use the Taylor-series approach (second method, Sec. 15.3) to obtain the error of the formula.   Let us suppose that we do use the Taylor-series approach.   The coefficients of the formula have been determined to make the powers of x cancel on both sides up to the mth power, which is assumed not to cancel.   Let us fix the x_i values in our mind.   We therefore get a series of terms in $f^{(m)}(a + \theta_i x_i)$ for various θ_i

$$\alpha_1 f^{(m)}(a + \theta_1 x_1) + \alpha_2 f^{(m)}(a + \theta_2 x_2) + \cdots + \alpha_k f^{(m)}(a + \theta_k x_k)$$

where, of course, the θ_i depend on x_i, $\theta_i = \theta_i(x_i)$.   We now want to replace this expression by

$$(\alpha_1 + \alpha_2 + \cdots + \alpha_k) f^{(m)}(a + \theta x)$$

for some θ (and we expect some α_i to be positive and some to be negative and hence some cancellation).   Under what circumstances does there exist such a θ?

We know from the influence-function approach that if $G(s)$ is of constant sign, then there is such a θ and that if $G(s)$ is not of constant sign, then there are some functions $f(x)$ for which there is no such θ.   In this indirect manner we can answer the question.   There seems to be no direct way of finding the answer, but if one were found, then it might provide an alternative, and perhaps better, approach to the theory of the error term.

16.9   A CASE STUDY

The idea that $G(s)$ does not change sign except in very special circumstances is apt to emerge from the study of classical formulas where it is always of constant sign. The following example is therefore illuminating.

Consider the family of integration formulas

$$\int_{-1}^{1} f(x)\, dx = w_{-1} f(-t) + w_0 f(0) + w_1 f(t)$$

where t is a parameter (see Fig. 16.9.1).   This family includes a number of interesting cases.   We are sampling the integrand at the points $-t$, 0, and t and are weighting the samples by w_{-1}, w_0, and w_1.   We will only consider $0 \le t \le 1$ and ignore $t > 1$.

By symmetry it is easy to see that $w_{-1} = w_1$.   The defining equations are, therefore,

$$
\begin{aligned}
f \equiv 1: &\quad 2 = 2w_1 + w_0 \\
f \equiv x^2: &\quad 2/3 = 2w_1 t^2 \\
f \equiv x^4: &\quad 2/5 = 2w_1 t^4 + E_4
\end{aligned}
$$

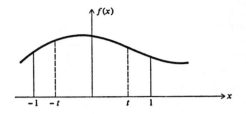

FIGURE 16.9.1

From the second equation we have

$$w_1 = \frac{1}{3t^2}$$

and hence from the first

$$w_0 = 2\left(1 - \frac{1}{3t^2}\right)$$

and from the third we get

$$E_4 = 2\left(\frac{1}{5} - \frac{t^2}{3}\right)$$

We need to examine the influence function $G(s)$ (see Fig. 16.9.2). Again by symmetry we need only examine the two intervals $1 \geq s \geq t$ and $t \geq s \geq 0$. We have, since $m = 4$,

$$3! \, G(s) = \frac{(1-s)^4_+ - (-1-s)^4_+}{4} - \frac{1}{3t^2}[(t-s)^3_+ - (-t-s)^3_+]$$
$$- 2\left(1 - \frac{1}{3t^2}\right)(-s)^3_+$$

For $1 \geq s \geq t$

$$3! \, G(s) = \frac{(1-s)^4}{4} \geq 0$$

For $t \geq s \geq 0$

$$3! \, G(s) = \frac{(1-s)^4}{4} - \frac{(t-s)^3}{3t^2}$$

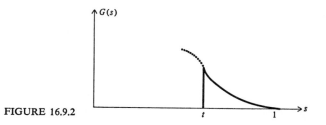

FIGURE 16.9.2

Can this change sign?   To find out, we evaluate it at $s = 0$

$$3! \, G(0) = 1/4 - \frac{t}{3}$$

Hence for $t = 3/4$, $G(0) = 0$; for $t > 3/4$, $G(0) < 0$, and the influence function $G(s)$ changes sign(see Fig. 16.9.3).   Thus

for $3/4 < t < 1$ $\qquad\qquad$ $G(s)$ changes sign

for $0 \leq t \leq 3/4$ and $t = 1$ $\qquad$ $G(s)$ has constant sign

The case $t = 1$ requires special investigation and is Simpson's formula

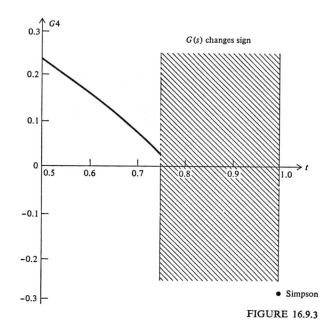

FIGURE 16.9.3

If $G(s)$ does not change sign, then the error is

$$\frac{E_4 f^{(4)}(\theta)}{4!} = \frac{1}{12}\left(\frac{1}{5} - \frac{t^2}{3}\right) f^{(4)}(\theta)$$

The case $E_4 = 0$, which is

$$t = \sqrt{3/5} = 0.7745967\ldots$$

immediately attracts attention. It falls in the region $3/4 < t < 1$ of sign change. And this is typical; in the region of sign change there is apt to be a place where the total area under $G(s)$ is exactly zero, meaning that $E_m = 0$, and the formula will have extra accuracy [and a different $G(s)$ will have to be investigated using the appropriate m, in this particular case, two higher than before].

Table 16.9 shows a number of special cases of interest in the family of formulas. A study of this table shows how the error changes as we change where we sample the integrand. The beginner should study families of formulas to get a feel for the effects of where samples are taken and how the weights change. From this can come a feeling of what can be expected from some new formula before it is investigated.

Table 16.9

t	w_0	w_1	E_4		Remarks
$1/2$	$-2/3$	$4/3$	$7/30$ =	0.233	Bad roundoff properties
$1/\sqrt{3}$	0	1	$8/45$ =	0.1778	2 point Gauss (Chap. 19)
$2/3$	$2/4$	$3/4$	$14/135$ =	0.1037	Better than twice as accurate as Simpson's formula
$1/\sqrt{2}$	$2/3$	$2/3$	$1/15$ =	0.0667	Chebyshev (Chap. 19)
$3/4$	$22/27$	$16/27$	$1/40$ =	0.0250	Better than 10 times as accurate as Simpson's formula
$\sqrt{3/5}$	$8/9$	$5/9$	0 =	0.0000	3 point Gauss (Chap. 19)
1	$4/3$	$1/3$	$-4/15$ =	-0.2667	Simpson's forumla

17

FORMULAS USING DERIVATIVES

17.1 INTRODUCTION

In the past three chapters we have studied, respectively, interpolation, finding formulas, and finding their error terms, using only function values. In this chapter we will repeat the work, this time using values of the function and of one or more derivatives.

Why should we care about using derivatives? Experimentally they are hard to come by, but when the function is given in a mathematical form, then they are easy to get. This remark contradicts one's initial impression from having taken a calculus course where differentiation tends to spread out the formula over more and more paper as the derivatives get higher and higher. This spreading out is indeed true, but it is not really relevant to modern computing. When an examination is made of how machine time is spent, it turns out that the evaluations of square roots, exponentials, logarithms, arctangents, and sines and cosines are done by subroutines which usually take more time than 50 arithmetic operations. But differentiation *does not* introduce any new radicals, exponentials, logarithms, or arctangents, and at worst a sine goes into a cosine and a cosine goes into a sine of the same angle. Thus, while the formula for the derivative may appear much longer when written out on paper, it will usually take much less time to evaluate *provided* the costly parts of the function evaluation have been saved to be used again.   As we shall see, the value of a derivative at a point is worth almost as much as another function value at some new point.

We can expect, therefore, that when pressed for machine time, it will be worth considering formulas which use one or more derivatives as an alternative to using more function values.   In this chapter, we show how to derive such formulas.

17.2   HERMITE[1] INTERPOLATION

Hermite interpolation uses both the function value and the value of the first derivative at each point.   Thus the interpolating polynomial is tangent to the function at the sample points, and Hermite interpolation is sometimes known as "osculating interpolation."   The first approach to Hermite interpolation will be similar to our approach to Lagrange interpolation (Sec. 14.5).   In place of the sampling polynomials $\pi_i(x)$ of the Lagrange formula we need to find two such families of functions, the $\sigma_i(x)$, which are zero at all the data except the ith function value, and the $\tau_i(x)$, which are zero at all the data except the ith first-derivative value.   In mathematical symbols

$$\sigma_i(x_j) = \begin{cases} 0 & i \neq j \\ \neq 0 & i = j \end{cases} \qquad \tau_i(x_j) = 0 \qquad \text{for all } j$$

$$\sigma_i'(x_j) = 0 \qquad \text{for all } j \qquad \tau_i'(x_j) \begin{cases} = 0 & i \neq j \\ \neq 0 & i = j \end{cases}$$

We begin by constructing the $\tau_i(x)$.   Evidently, for all the sample points, except for x_i, we have double zeros, and at x_i we have a single zero.   Therefore, our sampling polynomial is

$$\tau_i(x) = \pi_i{}^2(x)(x - x_i)$$

where as before

$$\pi_i(x) = (x - x_1)(x - x_2) \cdots (x - x_{i-1})(x - x_{i+1}) \cdots (x - x_{N+1})$$

The degree of $\tau_i(x)$ in x is $2N + 1$, as it should be.

The $\sigma_i(x)$ are less obvious, so we assume the reasonable form

$$\sigma_i(x) = \pi_i{}^2(x)(x + b)$$

where $x_i + b \neq 0$.   Differentiating, we get

$$\sigma_i'(x) = 2\pi_i(x)\pi_i'(x)(x + b) + \pi_i{}^2(x)$$

[1] Charles Hermite (1822–1901).

But

$$\sigma_i'(x_i) = 0$$

so we have

$$\frac{\sigma_i'(x_i)}{\sigma_i(x_i)} = \frac{2\pi_i(x_i)\pi_i'(x_i)(x_i + b) + \pi_i^{\,2}(x_i)}{\pi_i^{\,2}(x_i)(x_i + b)} = 0$$

$$= 2\frac{\pi_i'(x_i)}{\pi_i(x_i)} + \frac{1}{x_i + b} = 0$$

$$= 2\sum_{j \neq i} \frac{1}{x_i - x_j} + \frac{1}{x_i + b} = 0$$

which determines b.

Again, the important part is *how* we found the $\sigma_i(x)$ and $\tau_i(x)$, not the result.

Using these functions makes it easy to see that the Hermite interpolating polynomial is

$$y(x) = \sum_{i=1}^{N} \left[\frac{\sigma_i(x)}{\sigma_i(x_i)} y_i + \frac{\tau_i(x)}{\tau_i'(x_i)} y_i' \right]$$

To find the error term, we proceed *exactly* as we did in Sec. 14.6. We know that the difference between the function $y(x)$ and the approximating polynomial $P_{2N+1}(x)$ has double zeros at each x_i, and so we write the difference as

$$y(x) = P_{2N+1}(x) + \pi^2(x)K(x)$$

where, as usual,

$$\pi(x) = (x - x_1)(x - x_2) \cdots (x - x_{N+1})$$

We now pick an arbitrary $x^* \neq x_j$ (all j) such that

$$y(x^*) = P_{2N+1}(x^*) + \pi^2(x^*)K(x^*)$$

and then consider the expression

$$\phi(x) = y(x) - P_{2N+1}(x) - \pi^2(x)K(x^*)$$

This has double zeros at each x_i and a simple zero at x^*. Using a simple extension of the mean value theorem, we have that the first derivative has at least $2N + 2$ zeros, the second derivative $2N + 1$ zeros, ..., the $(2N + 2)$nd derivative at least 1 zero, say $\bar{x}$. Hence

$$y^{(2N+2)}(\bar{x}) = (2N + 2)! \, K(x^*)$$

and putting this in the original expression for $\phi(x)$, we get

$$y(x^*) = P_{2N+1}(x^*) + \frac{\pi^2(x^*)}{(2N + 2)!} y^{(2N+2)}(\bar{x})$$

But, since x^* was arbitrary, we may write x^* as x, and we have the formula

$$y(x) = P_{2N+1}(x) + \frac{\pi^2(x)}{(2N+2)!} \, y^{(2N+2)}(\bar{x})$$

for the error of Hermite interpolation.

PROBLEMS

17.2.1   Construct the Hermite interpolating polynomial on $x = 0$ and $x = 1$.

17.2.2   Apply Hermite interpolation to

$$y = \sin x \quad \text{using} \quad \begin{array}{ll} \sin 0° = 0.0000 & \cos 0° = 1.0000 \\ \sin 30° = 0.5000 & \cos 30° = 0.8660 \end{array}$$

(beware of radians and degrees!) and compute $\sin 15°$.

17.2.3   Using $0°$, $30°$, and $60°$, by the Hermite interpolation find $\sin 15°$.

17.2.4   Carry out the theory for y, y', and y''.

17.3   THE DIRECT METHOD

In the direct method for finding the Hermite interpolating polynomial of degree $2N + 1$ we simply write down the form

$$P_{2N+1}(x) = a_0 + a_1 x + \cdots + a_{2N+1} x^{2N+1}$$

and impose the conditions on the polynomial to get the defining equations (where we have put the derivative values after the function values)

$$
\begin{aligned}
P_{2N+1}(x_1) &= a_0 + a_1 x_1 + a_2 x_1{}^2 + \cdots + a_{2N+1} x_1^{2N+1} \\
P_{2N+1}(x_2) &= a_0 + a_1 x_2 + a_2 x_2{}^2 + \cdots + a_{2N+1} x_2^{2N+1} \\
& \cdots \cdots \cdots \cdots \cdots \cdots \cdots \cdots \\
P_{2N+1}(x_{N+1}) &= a_0 + a_1 x_{N+1} + a_2 x_{N+1}^2 + \cdots + a_{2N+1} x_{N+1}^{2N+1} \\
P'_{2N+1}(x_1) &= \quad\;\; a_1 + 2a_2 x_1 + \cdots + (2N+1)a_{2N+1} x_1^{2N} \\
& \cdots \cdots \cdots \cdots \cdots \cdots \cdots \cdots \\
P'_{2N+1}(x_{N+1}) &= \quad\;\; a_1 + 2a_2 x_{N+1} + \cdots + (2N+1)a_{2N+1} x_{N+1}^{2N}
\end{aligned}
$$

We will later show that the determinant of these equations is not zero, and hence we can solve directly for the interpolating polynomial.

We recognize that if we were to integrate this polynomial, we would get a linear combination of the data

$$\int_a^b P_{2N+1}(x) \, dx = \sum_{k=1}^{N+1} [w_k f(x_k) + w'_k f'(x_k)]$$

Knowing the form, let us apply the direct method for finding the formula for integration that uses both the function values and the values of the first derivative at each point. Using $f(x) = x^j$, we get the defining equations

$$x^j: \qquad m_j = \sum_{k=1}^{N+1} [w_k x_k{}^j + w_k' j x_k^{j-1}] \qquad j = 0, 1, \ldots, 2N+1$$

The matrix of these equations is exactly the *transpose* of the earlier set of equations used to find the Hermite interpolating polynomial.

We shall solve these equations by what amounts to finding the inverse determinant. If we write the $\sigma_i(x)$ as polynomials, we get

$$\sigma_i(x) = s_{i,0} + s_{i,1}x + \cdots + s_{i,2N+1}x^{2N+1}$$

Now if for each j we multiply the jth row by $s_{i,j}$ and add all the resulting equations, we get, after some careful thought about what the $\sigma_i(x)$ actually are [note that $\sigma_i(x_j) = 0$ for $i \neq j$ and $\sigma_i'(x_j) = 0$ for all j],

$$\sum_{j=0}^{2N+1} s_{i,j} m_j = \sum_{k=1}^{N+1} [w_k \sigma_i(x_k) + w_k' \sigma_i'(x_k)] = w_i \sigma_i(x_i)$$

Similarly, expanding the

$$\tau_i(x) = t_{i,0} + t_{i,1}x + \cdots + t_{i,2N+1}x^{2N+1}$$

we are led to

$$\sum_{j=0}^{2N+1} t_{i,j} m_j = w_i' \tau_i'(x_i)$$

Thus we have, for all practical purposes, the inverse of the corresponding determinant. Since we have found the inverse, the determinant could not have been zero.

The method seems a bit magical so we will illustrate it in the next section by applying it to two particular cases. Again, it is the method more than the result that we care about.

17.4 THE HERMITE UNIVERSAL MATRICES

The process of using the sampling polynomials is so useful in the practical work of finding formulas, and the inverse matrices are so useful, that it is worth developing the cases of two and three equally spaced data points.

In the case of two points, we use $x = -1/2$ and $x = 1/2$ as the sample points. For any linear operator $L[f(x)]$ we have the defining equations.

$$
\begin{aligned}
L(1) &= m_0 = w_1 && + w_2 \\
L(x) &= m_1 = w_1(-1/2) + w_2(1/2) + w_1' && + w_2' \\
L(x^2) &= m_2 = w_1(-1/2)^2 + w_2(1/2)^2 + 2[w_1'(-1/2) + w_2'(1/2)] \\
L(x^3) &= m_3 = w_1(-1/2)^3 + w_2(1/2)^3 + 3[w_1'(-1/2)^2 + w_2'(1/2)^2]
\end{aligned}
$$

From symmetry it is obvious that the two $\sigma_i(x)$ are related, $\sigma_{1/2}(x) = \sigma_{-1/2}(-x)$

$$
\begin{aligned}
\sigma_{-1/2} &= (x - 1/2)^2(x + b) \\
\sigma_{-1/2}' &= 2(x - 1/2)(x + b) + (x - 1/2)^2
\end{aligned}
$$

We know $\sigma_{-1/2}'(-1/2) = 0$, and so we have

$$-2(-1/2 + b) + 1 = 0$$
$$b = 1$$

and therefore

$$\sigma_{-1/2}(-1/2) = 1/2$$

Hence

$$\frac{\sigma_{-1/2}(x)}{\sigma_{-1/2}(-1/2)} = 2x^3 - \frac{3}{2}x + \frac{1}{2}$$

gives the elements of the inverse matrix (namely, $1/2$, $-3/2$, 0, and 2). Correspondingly, we have for $\tau_{1/2}(x) = -\tau_{-1/2}(-x)$

$$\tau_{-1/2} = (x - \tfrac{1}{2})^2(x + \tfrac{1}{2}) = x^3 - \frac{1}{2}x^2 - \frac{1}{4}x + \frac{1}{8}$$

$$\tau_{-1/2}' = 2(x - \tfrac{1}{2})(x + \tfrac{1}{2}) + (x - \tfrac{1}{2})^2$$

$$\tau_{-1/2}'(-1/2) = 1$$

$$\frac{\tau_{-1/2}(x)}{\tau_{-1/2}(-1/2)} = x^3 - \frac{1}{2}x^2 - \frac{1}{4}x + \frac{1}{8}$$

Using these coefficients, we have the inverse matrix on the points $x = -1/2$ and $x = 1/2$

	1	x	x^2	x^3
$\dfrac{\sigma_{-1/2}(x)}{\sigma_{-1/2}(-1/2)}$	1/2	-3/2	0	2
$\dfrac{\sigma_{1/2}(x)}{\sigma_{1/2}(1/2)}$	1/2	3/2	0	-2
$\dfrac{\tau_{-1/2}(x)}{\tau_{-1/2}(-1/2)}$	1/8	-1/4	-1/2	1
$\dfrac{\tau_{1/2}(x)}{\tau_{1/2}(1/2)}$	-1/8	-1/4	1/2	1

For three points x is equal to $-1, 0$, and 1 and we have for the inverse

$$\frac{1}{4}\begin{pmatrix} 0 & 0 & 4 & -5 & -2 & 3 \\ 4 & 0 & -8 & 0 & 4 & 0 \\ 0 & 0 & 4 & 5 & -2 & -3 \\ 0 & 0 & 1 & -1 & -1 & 1 \\ 0 & 4 & 0 & -8 & 0 & 4 \\ 0 & 0 & -1 & -1 & 1 & 1 \end{pmatrix}$$

We derive only one row, say the fifth, since the others are similar.

$$\tau_0(x) = (x-1)^2(x+1)^2 x = 0 + x + 0x^2 - 2x^3 + 0x^4 + x^5$$
$$\tau_0'(x) = 1 - 6x^2 + 5x^4$$
$$\tau_0'(0) = 1$$

Hence the coefficients are $0, 1, 0, -2, 0$, and 1, and the $1/4$ in front adjusts the entries correctly.

Higher-order cases of the inverse matrix can be done but are not likely to be useful.

PROBLEMS

17.4.1 Find $\displaystyle\int_0^{2h} f(x)\,dx = \frac{h}{15}[7f(0) + 16f(h) + 7f(2h)]$

$$+ \frac{h^2}{15}[f'(0) - f'(2h)] + \frac{h^7 f^{(6)}\theta}{4{,}725}$$

17.4.2 Using both matrices, find $-\displaystyle\int_0^1 f(x)\ln x\,dx$.

17.4.3 Find $\displaystyle\int_0^1 f(x)\,dx$ for 2 points.

17.4.4 Find $\displaystyle\int_0^1 f(x)\,dx$ using $x = 1/4$ and $x = 3/4$ as the sample points for $f(x)$ and $f'(x)$.

17.5 SOME EXAMPLES

We propose to find some typical formulas as illustrations of the method of using derivatives as well as function values.

EXAMPLE 17.1   Find

$$\int_0^1 f(x)\, dx = w_0\, f(1/2) + w_0'\, f'(1/2)$$

if $f(x) = 1$ then $1 = w_0$

if $f(x) = x$ then $1/2 = w_0(1/2) + w_0',\ w_0' = 0$

To get E_2 we use $f(x) = x^2$

$$1/3 = w_0(1/2)^2 + 2w_0'(1/2) + E_2$$
$$E_2 = 1/3 - 1/4 = 1/12$$

The $G(s)$ is $(m = 2)$

$$1!\, G(s) = \frac{(1 - s)_+^2 - (0 - s)_+^2}{2} - (1/2 - s)_+$$

For $1 \geq s \geq 1/2$

$$G(s) = \frac{(1 - s)^2}{2} \geq 0$$

By symmetry, or further algebra, we conclude that $G(s) \geq 0$ $(0 \leq s \leq 1)$, and we have the midpoint formula (again)

$$\int_0^1 f(x)\, dx = f(1/2) + \frac{1}{24}\, f''(\theta)$$ ////

EXAMPLE 17.2   Find

$$\int_0^1 f(x)\, dx = w_0 f(0) + w_1 f(1) + w_0' f'(0) + w_1' f'(1)$$

We can use the inverse in Sec. 17.4 as a check.   The defining equations are

$$1 = w_0 + w_1$$
$$1/2 = \quad + w_1 + w_0' + \ w_1'$$
$$1/3 = \quad + w_1 \quad\quad + 2w_1'$$
$$1/4 = \quad + w_1 \quad\quad + 3w_1'$$
$$1/5 = \quad + w_1 \quad\quad + 4w_1' + E_4$$

The fourth equation minus the third gives

$$1/4 - 1/3 = w_1' = -1/12$$

and the rest is easy, $w_1 = 1/2$, $w_0 = 1/2$, $w_0' = 1/12$, $E_4 = 1/30$

$$\int_0^1 f(x)\, dx = \frac{1}{12}\, [6f(0) + 6f(1) + f'(0) - f'(1)] + \frac{f^{(4)}(\theta)}{720}$$

To check $G(s)$, we have

$$3! \, G(s) = (\text{LHS} - \text{RHS})[(x - s)_+^3]$$

and for $1 \geq s \geq 0$

$$3! \, G(s) = \frac{(1 - s)_+^4}{4} - \frac{1}{12} [6(1 - s)_+^3 - 3(1 - s)_+^2]$$

$$= \frac{(1 - s)^2}{4} [1 - 2s + s^2 - 2 + 2s + 1] = \left[\frac{s(1 - s)}{2} \right]^2 \geq 0 \qquad ////$$

EXAMPLE 17.3 Find

$$\int_{-1}^{1} \frac{f(x)}{\sqrt{1 - x^2}} \, dx = a_{-1} f(-1) + a_0 f(0) + a_1 f(1)$$

$$+ \, b_{-1} f'(-1) + b_0 f'(0) + b_1 f'(1)$$

By now we can begin to recognize symmetry and realize that if in this problem we set

$$a_{-1} = a_1$$

$$b_{-1} = -b_1 \qquad b_0 = 0$$

then the formula will be true for any odd power of x. Thus we only need to use $f(x)$ equal to 1, x^2, and x^4:

$$f(x) = 1 : \qquad \pi = 2a_1 + a_0$$

$$f(x) = x^2 : \qquad \frac{\pi}{2} = 2a_1 \qquad + 2(2b_1)$$

$$f(x) = x^4 : \qquad \frac{3\pi}{8} = 2a_1 \qquad + 4(2b_1)$$

from which it is easy to find

$$b_1 = \frac{-\pi}{32} \qquad a_1 = \frac{5\pi}{16} \qquad a_0 = \frac{6\pi}{16}$$

Using x^6

$$\frac{5\pi}{16} = 2a_1 + 6(2b_1) + E_6$$

$$E_6 = \frac{\pi}{16}$$

Thus we have

$$\int_{-1}^{1} \frac{f(x)}{\sqrt{1-x^2}}\, dx = \frac{\pi}{32}\left[10f(-1) + 12f(0) + 10f(1)\right.$$

$$\left. + f'(-1) - f'(1)\right] + \frac{\pi}{16 \cdot 6!} f^{(6)}(\theta)$$

We can check the algebra by using the universal matrix

$$\frac{1}{4}\begin{pmatrix} 0 & 0 & 4 & -5 & -2 & 3 \\ 4 & 0 & -8 & 0 & 4 & 0 \\ 0 & 0 & 4 & 5 & -2 & -3 \\ 0 & 0 & 1 & -1 & -1 & 1 \\ 0 & 4 & 0 & -8 & 0 & 4 \\ 0 & 0 & -1 & -1 & 1 & 1 \end{pmatrix}\begin{pmatrix} \pi \\ 0 \\ \pi/2 \\ 0 \\ 3\pi/8 \\ 0 \end{pmatrix} = \frac{\pi}{32}\begin{pmatrix} 10 \\ 12 \\ 10 \\ 1 \\ 0 \\ -1 \end{pmatrix}$$

We need, still, to check $G(s)$ and due to symmetry we have only to examine

$$1 \geq s \geq 0$$

We leave the details to the reader.   ////

PROBLEMS

17.5.1   Derive

$$\int_{0}^{\infty} e^{-x}f(x)\, dx = w_0\, f(0) + w_2\, f(2) + w_0'\, f'(0) + w_2'\, f'(2)$$

Ans. $w_0 = 1,\ w_2 = 0,\ w_0' = w_2' = 1/2$

17.5.2   Derive

$$\int_{0}^{\infty} e^{-x}f(x)\, dx = w_0 f(0) + w_2 f(2) + w_4 f'(4) + w_0' f'(0) + w_2' f'(2) + w_4' f'(4)$$

17.5.3   Derive

$$-\int_{0}^{1} (\ln x)f(x)\, dx$$

using x equal to 0, 2/3, and 1, and both f and f'.

17.5.4   Derive

$$\int_{0}^{a} \frac{f(x)}{\sqrt{2ax - a^2}}\, dx$$

using x equal to 0, and a, and both f and f'.   Find the error term.

17.5.5 Derive and check $G(s)$

$$\int_0^{3h} f(x)\, dx = \frac{3h}{80}\, [13f(0) + 27f(h) + 27f(2h) + 13f(3h)]$$

$$+ \frac{3h^2}{40}\, [f'(0) - f'(3h)] + \frac{9h^7}{11{,}200}\, y^{(7)}(\theta)$$

17.5.6 Derive and check $G(s)$

$$\int_0^{4h} f(x)\, dx = \frac{2h}{945}\, \{305[f(0) + f(4)] + 604[f(1) + f(3)]\}$$

$$- 2h\{27[f'(4) - f'(0)] + 96[f'(3) - f'(1)]\} + \frac{52h^9}{99{,}225}\, y^{(9)}(\theta)$$

17.6 BIRKHOFF INTERPOLATION AND FORMULAS

A natural generalization of the previous methods of interpolation is to allow the function and a variable number of derivatives to be used as data at each point; we can use $f(x_i), f'(x_i), \ldots, f^{(k_i - 1)}(x_i)$ which are k_i consecutive derivatives $(0, 1, \ldots, k_i - 1)$ at the point x_i. This we shall call the Birkhoff[1] interpolation.

The methods used before show that we can find the formula by the method of undetermined coefficients. The error term will turn out to be

$$[(x - x_1)^{k_1}(x - x_2)^{k_2} \cdots (x - x_{n+1})^{k_{n+1}}]\, \frac{f^{(m)}(\theta)}{m!}$$

where $m = k_1 + k_2 + \cdots + k_{n+1}$.

When we try to find formulas using these same data we are led, in the direct method, to a system of equations whose matrix is related to the Vandermonde, and this can again be inverted by the same technique of finding the sampling polynomials. The existence of the inverse shows that the determinant of the system of equations is not zero; therefore we can indeed find the interpolating polynomials and the formulas of the desired form (provided we can find the moments m_k of the operator).

As an example of the method and the kinds of formulas that can exist, consider

$$\int_{-1}^1 f(x)\, dx = w_{-1} f(-1) + w_0 f(0) + w_1 f(1) + w_0' f'(0) + w_1' f'(1) + w_1'' f''(1)$$

The coefficients of the defining equations can be arranged as in Table 17.6.

[1] George David Birkhoff (1884–1944).

Producing all the sampling polynomials is awkward, but we can use those that are easy to find (a typical compromise we shall often make).

$$\text{I:} \qquad x^2(x-1)^3 = x^5 - 3x^4 + 3x^3 - x^2 + 0x + 0$$
$$\text{II:} \qquad (x+1)x(x-1)^3 = x^5 - 2x^4 + 0x^3 + 2x^2 - x + 0$$
$$\text{III:} \qquad (x+1)(x^2)(x-1)^2 = x^5 - x^4 - x^3 + x^2 + 0x + 0$$

The coefficients are in the *respective* columns headed I, II, and III.   Multiplying the equations by these numbers and adding, we get the three equations

$$\text{I:} \qquad -2/3 - 6/5 = -8w_{-1} \qquad \text{thus } w_{-1} = \quad 7/30$$
$$\text{II:} \qquad 4/3 - 4/5 = -w_0' \qquad \text{thus } w_0' = -16/30$$
$$\text{III:} \qquad 2/3 - 2/5 = \quad 4w_1'' \qquad \text{thus } w_1'' = \quad 2/30$$

Table 17.6

$f(x)$	m_k	w_{-1}	w_0	w_1	w_0'	w_1'	w_1''	I	II	III
1	2	1	1	1	0	0	0	0	0	0
x	0	-1	0	1	1	1	0	0	-1	0
x^2	2/3	1	0	1	0	2	2	-1	2	0
x^3	0	-1	0	1	0	3	6	3	0	-1
x^4	2/5	1	0	1	0	4	12	-3	-2	-1
x^5	0	-1	0	1	0	5	20	1	1	1

From these three, the other constants are easily found, and we have the formula

$$\int_{-1}^{1} f(x)\,dx = \frac{1}{30}\left[7f(-1) + 16f(0) + 37f(1) - 16f'(0) - 14f'(1) + 2f''(1)\right]$$

It should now be apparent that we can find formulas of a wide class—all we need to know are the moments m_k of the operator we are using, and the rest is tedious algebra.   The sample points need not be in the range of integration.

PROBLEMS

17.6.1 Derive

$$\int_0^1 f(x)\,dx = \frac{f(1) + f(0)}{2} - \frac{f'(1) - f'(0)}{10} + \frac{f''(1) + f''(0)}{120} \qquad E_7 = -1/20$$

and check $G(s)$.

17.6.2 Derive

$$\int_0^1 f(x)\,dx = \frac{f(1) + f(0)}{2} - \frac{3}{28}\left[f'(1) - f'(0)\right] + \frac{1}{84}\left[f''(1) + f''(0)\right]$$

$$+ \frac{1}{1,680}\left[f'''(1) - f'''(0)\right] \qquad E_9 = 1/70$$

and check $G(s)$.

17.6.3 Find

$$\int_0^\infty e^{-x}f(x)\,dx = a_0 f(0) + a_1 f(4) + b_0 f'(0) + b_1 f'(4) + c_0 f''(0) + c_1 f''(4)$$

17.6.4 Derive

$$-\int_0^1 (\ln x)f(x)\,dx = \frac{15}{16}f(0) + \frac{1}{16}f(1) + \frac{3}{16}f'(0) + \frac{7}{288}f''(0)$$

17.6.5 Derive

$$\int_0^\infty e^{-x}f(x)\,dx = \frac{1}{4}f(0) + \frac{3}{4}f(2) - \frac{1}{2}f'(0) - \frac{1}{2}f''(0)$$

17.7 AN EXAMPLE OF A NONINTERPOLATORY FORMULA

From the Birkhoff interpolation one is apt to draw the conclusion that almost any reasonable set of information can be used as a basis for interpolation. This is false. The simple example of three equally spaced points with both the function and second derivatives (position and acceleration, if you wish) does *not* determine a quintic (fifth-degree) polynomial. To prove this, let the points be $-1, 0,$ and 1. The function and defining equations are:

$$
\begin{aligned}
f(x) \;&=\; A_0 + A_1 x + A_2 x^2 + A_3 x^3 + A_4 x^4 + A_5 x^5 \\
f(-1) \;&=\; A_0 - A_1 + A_2 \quad\; - A_3 \quad\;\; + A_4 \quad\; - A_5 \\
f(0) \;&=\; A_0 \\
f(1) \;&=\; A_0 + A_1 + A_2 \quad\; + A_3 \quad\;\; + A_4 \quad\; + A_5 \\
f''(-1) \;&=\; \qquad\qquad\quad 2A_2 - 6A_3 + 12A_4 - 20A_5 \\
f''(0) \;&=\; \qquad\qquad\quad 2A_2 \\
f''(1) \;&=\; \qquad\qquad\quad 2A_2 - 6A_3 + 12A_4 - 20A_5
\end{aligned}
$$

The determinant of the A_i is

$$D = \begin{vmatrix} 1 & -1 & 1 & -1 & 1 & -1 \\ 1 & 0 & 0 & 0 & 0 & 0 \\ 1 & 1 & 1 & 1 & 1 & 1 \\ 0 & 0 & 2 & -6 & 12 & -20 \\ 0 & 0 & 2 & 0 & 0 & 0 \\ 0 & 0 & 2 & 6 & 12 & 20 \end{vmatrix} = 0$$

To prove that this is zero, we expand by the elements of the second row and then the elements of the fifth row

$$D = (-1)(-2) \begin{vmatrix} -1 & -1 & 1 & -1 \\ 1 & 1 & 1 & 1 \\ 0 & -6 & 12 & -20 \\ 0 & 6 & 12 & 20 \end{vmatrix} = 0$$

Now add the second row to the first and the fourth to the third to get

$$D = 2 \begin{vmatrix} 0 & 0 & 2 & 0 \\ 1 & 1 & 1 & 1 \\ 0 & 0 & 24 & 0 \\ 0 & 6 & 12 & 20 \end{vmatrix} = 0$$

The first and third rows are proportional to each other, and so the determinant is 0.

Thus there is no interpolating polynomial. In a sense the data put four conditions on the even part and two on the odd part of the polynomial.

However, there are infinitely many formulas of the form

$$\int_{-1}^{1} f(x)\,dx = a_{-1}f(-1) + a_0 f(0) + a_1 f(1)$$

$$+ b_{-1}f''(-1) + b_0 f''(0) + b_1 f''(1)$$

which use the same data, and the formulas are exact for quintics. To show this, we write the defining equations

$$
\begin{array}{lll}
1: & 2 = & a_{-1} + a_0 + a_1 \\
x: & 0 = -a_1 & + a_1 \\
x^2: & \tfrac{2}{3} = & a_{-1} + a_1 + 2(b_{-1} + b_0 + b_1) \\
x^3: & 0 = -a_{-1} & + a_1 + 6(-b_{-1} + b_1) \\
x^4: & \tfrac{2}{5} = & a_{-1} + a_1 + 12(b_{-1} + b_1) \\
x^5: & 0 = -a_{-1} & + a_1 + 20(-b_{-1} + b_1)
\end{array}
$$

Using a_1 as a parameter (since we know from immediately above that the determinant has rank 5), we get

$$\int_{-1}^{1} f(x)\, dx = 2f(0) + \frac{1}{60} \left[f''(-1) + 18f''(0) + f''(1) \right]$$

$$+ a_1 \{ [f(-1) - 2f(0) + f(1)]$$

$$- \frac{1}{12} [f''(-1) + 10f''(0) + f''(1)] \}$$

If we impose the condition that the formula be exact for x^6 as well, then we get

$$\frac{2}{7} = \frac{1}{60}(30 + 30) + a_1 \left[(1 + 1) - \frac{1}{12}(30 + 30) \right]$$

$$\frac{2}{7} = 1 + a_1(2 - 5)$$

or

$$a_1 = \frac{5}{21}$$

Thus we have

$$\int_{-1}^{1} f(x)\, dx = \frac{1}{21} [5f(-1) + 32f(0) + 5f(1)]$$

$$- \frac{1}{315} [f''(-1) - 32f''(0) + f''(1)]$$

which is exact for sixth degree polynomials.

The complete explanation of this apparent paradox is nontrivial,[1] but it should be no surprise that occasionally the two-step process [finding the interpolating polynomial and then applying the operator (integration from -1 to $+1$ in this case)] fails even though the direct one-step process succeeds. What is surprising is that it happens in this "reasonable" case.

PROBLEMS

17.7.1 Discuss the importance of

$$Q(x) = 3x^5 - 10x^3 + 7x$$

[which has $Q(-1) = Q(0) = Q(1)$ and $Q''(-1) = Q''(0) = Q''(1)$] to the example of noninterpolatory polynomials.

17.7.2 Give another example of a noninterpolatory polynomial.

17.7.3 If the integration was from 0 to 1 instead of -1 to 1, show that there is no such integration formula.

[1] M. P. Epstein and R. W. Hamming, "Non-Interpolatory Quadrature Formulas," SIAM.

17.8   AN EXPERIMENT IN COMPARING THE VALUE OF DERIVATIVES

The existence of the inverse determinant in the Birkhoff interpolation case shows that the use of *consecutive* derivatives at a point (omitting none avoids the example in Sec. 17.7) gives about as much information as only function values at points; the linear independence of the powers of x is preserved.   Or, as in information theory, a polynomial of degree N can be reconstructed from the $N + 1$ Birkhoff samples.

Can we be more precise in this comparison of more sample points versus more derivatives at the old sample points?   One approach is to examine the size of the error terms for various families of formulas.   Assuming that the $G(s)$ function is of constant sign, then the E_m would provide one measure of the relative merit—though let us remember we are assuming that over the unspecified ensemble of functions we are considering, the value of the θ does not favor one formula over another.

We shall use the integral

$$\int_{-1}^{1} f(x)\, dx$$

as the basis for an experiment.

The E_m can be found by integrating *any* polynomial of degree m with leading coefficient 1.   By picking a polynomial with zeros at the sample points (meaning a multiplicity at x_i great enough to make the highest derivative used at that point zero at that point) we do not have to consider the right-hand side of the formula since all the values will then be zero.   When the degree of the polynomial is odd, then the formula is clearly accurate for a power one higher, and we must add another factor to the polynomial we are using.   It is simplest to pick another factor x to do this.   An example of this effect is Simpson's formula which is accurate for one higher power of x than it was designed for.

The first family we shall examine is the one-point family

$$\int_{-1}^{1} f(x)\, dx = w_0 f(0) + w_1 f'(0) + w_2 f''(0) + \cdots + w_n f^{(n)}(0)$$

For N an even number $E_{N+1} = 0$ from symmetry; hence we need only examine the odd cases for N, and m is even.   We have

$$E_{2p} = \int_{-1}^{1} x^{2p}\, dx = \frac{2}{2p + 1}$$

In particular

$$E_2 = 2/3 \qquad E_4 = 2/5 \qquad E_6 = 2/7 \qquad E_8 = 2/9 \qquad E_{10} = 2/11$$

A natural second case is the family of two-point formulas

$$\int_{-1}^{1} f(x)\, dx = \sum_{k=0}^{N} [w_{-1}^{(k)} f^{(k)}(-1) + w_{1}^{(k)} f^{(k)}(1)]$$

We have

$$E_{2p} = \int_{-1}^{1} (x^2 - 1)^p\, dx = (-1)^p 2 \int_{0}^{1} (1 - x^2)^p\, dx$$

Setting $x = \sin t$,

$$E_{2p} = (-1)^p 2 \int_{0}^{\pi/2} \cos^{2p+1} t\, dt = (-1)^p 2 \frac{(2p)(2p-2) \cdots 2}{(2p+1)(2p-1) \cdots 3}$$

$$= \frac{(-1)^p 2^{2p+1} (p!)^2}{(2p+1)!} \approx \frac{(-1)^p \sqrt{\pi p}}{p + 1/2}$$

In particular

$$E_2 = -4/3$$
$$E_4 = 16/15$$
$$E_6 = -25/35$$
$$E_8 = 256/315$$
$$E_{10} = -512/693$$

The third family is

$$\int_{-1}^{1} f(x)\, dx = \sum_{k=0}^{N} [w_{-1}^{(k)} f^{(k)}(-1) + w_0^{(k)} f^{(k)}(0) + w_1^{(k)} f^{(k)}(1)]$$

with

$$E_{3p} = \begin{cases} \int_{-1}^{1} x^p (x^2 - 1)^p\, dx & p \text{ even} \\ \int_{-1}^{1} x^{p+1}(x^2 - 1)^p\, dx & p \text{ odd} \end{cases}$$

Setting $t = x^2$, we get

$$E_{3p} = \begin{cases} (-1)^p \int_{0}^{1} t^{(p-1)/2}(1 - t)^p\, dt & \text{even} \\ (-1)^p \int_{0}^{1} t^{p/2}(1 - t)^p\, dt & p \text{ odd} \end{cases}$$

These are beta functions whose values are

$$E_{3p} = \begin{cases} \dfrac{\Gamma[(p+1)/2]\Gamma(p+1)}{\Gamma[3(p+1)/2]} & p \text{ even} \\ \dfrac{\Gamma[(p+2)/2]\Gamma(p+1)}{\Gamma[(3p+4)/2]} & p \text{ odd} \end{cases}$$

Table 17.8*

Family	E_2	E_4	E_6	E_8	E_{10}	E_{12}	E_{14}	E_{16}
1-point	0.666667	0.400000	0.285714	0.222222	0.181818	0.153846	0.133333	0.117647
2-point	−1.333333	1.066667	−0.914286	0.812698	0.738817	0.681985	−0.636519	0.589077
3-point		0.266667	0.152381		−0.027706	0.017050		−0.003431
4-point		−0.118519		−0.030100		−0.003929		0.000796
5-point			−0.047619		0.007215			0.000468
6-point			−0.026819			0.002017		
7-point				−0.013169			0.000460	
8-point				−0.008075				0.0002000
9-point					−0.004380			
10-point					−0.002806			
11-point					−0.001616			
12-point						−0.001633		

* Computed by Mrs. Anne Duke.

Table 17.8 gives more results. The columns contain the values of E_m, and hence the error is of the form

$$\frac{E_m f^{(m)}(\theta)}{m!}$$

provided $G(s)$ is of constant sign.

We see that as we go up a column, the E_m tend to get larger, meaning that using more derivatives is not as good as using more points insofar as the size of the error coefficient is concerned; it still may be worth the machine time saved.

It is important to realize that Table 17.8 refers to one type of formula and represents an isolated experiment, not a definitive conclusion.

PROBLEMS

17.8.1   Check $G(s)$ for the three-point formulas:
   (a) function value only, E_4
   (b) function value plus first derivative, E_6.
17.8.2   Discuss other reasonable experiments that might be tried.
17.8.3   Derive the four-point integration formulas for E_4 and E_8.
17.8.4   (Harder)   Write an essay on the effect on the weight used of the function value at that point when more derivatives are given there.

18

FORMULAS USING DIFFERENCES

18.1 USE OF DIFFERENCES

Before the age of large-scale digital computers most hand calculations were carried out in terms of function values and their differences. There were at least three reasons for this. First, differences generally use fewer digits than function values do—and so with hand calculations the arithmetic is that much easier. Second, the differences give immediate clues to the values of the derivatives, which are generally unobtainable. In particular we have the obvious approximations

$$\frac{\Delta y_n}{\Delta x} \approx \frac{dy}{dx}\bigg|_{x=x_{n+1/2}}$$

$$\frac{\Delta^2 y_n}{\Delta x^2} \approx \frac{d^2 y}{dx^2}\bigg|_{x=n+1}$$

Third, the difference table also gives a clue to the roundoff in the function values.

When computing machines came into widespread use, it became fairly clear that the first reason, fewer digits leading to faster arithmetic, was not very relevant, and the time to find the differences in a machine greatly exceeds any possible machine savings. Under the second reason, we still use differences as estimates of derivatives, though to some extent this has fallen into disuse because the differences are often not found. The third reason of roundoff estimation was ignored chiefly because it soon became evident that the hand-computing experts had only an intuitive idea of the connection between roundoff and the differences. We have had to gradually work out a more explicit theory as given in Chap. 10. In particular, we had to quantify their impression of sign changes in the higher differences by first computing the correlation coefficient (Sec. 10.5) together with the corresponding probabilities of observing sign changes. We also had to work out the relation of the noise in the higher differences to the noise in the original function values (Sec. 10.4). This theory was intuitively understood, but the hand-computing experts could not say, with scientific detail, what they did so that it could be put on a computer. Indeed, they constantly maintained that it was an art and that it could not be quantified. Needless to say, in the theory presented in Chap. 10 some of their art was indeed lost, but some of it has been converted from intuition to measurable effects.

Today we find that much of current computing still ignores the approach through differences—the feeling persists that the differences take too much time and space in the computer and are not worth the trouble. The author believes, however, that going back to using them is a good idea and that many of the currently popular algorithms are poor substitutes for the direct use of the differences. For example, Romberg integration[1] effectively builds a table of differences of the answers at different spacings and tries to use this table to estimate the error. But the error comes from two distinct sources, the truncation error and the roundoff error. Taking more points in an integrand reduces the truncation error but builds up the roundoff, and in the behavior of the Romberg process it is hard to separate the two effects. With the direct use of the differences of the integrand values, the corresponding difference table enables us to separate the two effects and to make an intelligent guess about which is currently dominating and which strategy to adopt to improve the result—more function values may only increase the roundoff error!

18.2 NEWTON'S INTERPOLATION FORMULA

The first two methods for finding an interpolating polynomial through $N + 1$ points (the Vandermonde and Lagrange approaches) tacitly assumed that the number of points to be used was known. Often what is known is the accuracy desired, and the number of points to be used is determined as information about

---

[1] The purpose of Romberg integration is to eliminate the low-order-error terms of the error expansion in powers of Δx. The basic technique is usually attributed to Richardson.

the function is computed. Newton's interpolation formula, which we will now develop, is simply another way of writing the interpolating polynomial. It is useful because the number of points being used can easily be increased or decreased without repeating all the computation. The Newton formula of Sec. 9.6 is really the special case of equally spaced sample points.

As before, let the polynomial passing through the $N + 1$ points

$$(x_i, y_i) \qquad (i = 1, 2, \ldots, N + 1)$$

be labeled $P_N = P_N(x)$. We can write

$$P_N(x) = y_1 + (x - x_1)P_{N-1}(x)$$

where $P_{N-1}(x)$ is some polynomial of degree $N - 1$. It is clear that $P_N(x_1) = y_1$, and we are reduced to taking care of only N points ($i = 2, 3, \ldots, N + 1$). The above equation can be written as

$$P_{N-1}(x) = \frac{P_N(x) - y_1}{x - x_1}$$

and hence we want

$$P_{N-1}(x_i) = \frac{P_N(x_i) - P_N(x_1)}{x_i - x_1} = \frac{y_i - y_1}{x_i - x_1} \qquad i = 2, 3, \ldots, N + 1$$

That is to say, we want $P_{N-1}(x)$ to pass through the points

$$\left(x_i, \frac{y_i - y_1}{x_i - x_1} \right) \qquad i = 2, 3, \ldots, N + 1$$

The quantities

$$\frac{y_i - y_1}{x_i - x_1} \equiv [x_i, x_1] \equiv [x_1, x_i]$$

are called "divided differences" and are customarily written with brackets.[1]

The next step is, of course, to write

$$P_{N-1}(x) = [x_1, x_2] + (x - x_2)P_{N-2}(x)$$

and require $P_{N-2}(x)$ to take on the values of the divided differences of the divided differences,

$$[[x_i, x_1], [x_2, x_1]] = \frac{[x_i, x_1] - [x_2, x_1]}{x_i - x_2} = [x_i, x_2, x_1]$$

---

[1] Many other notations are used, for example $f[x_i, x_1]$ and $\rho[x_i, x_1]$.

It was easy to see that the first-order divided differences were independent of the order of the arguments in the brackets. We now show that the same is true of the second-order divided differences. If we start with three points

$$(x_1, y_1) \qquad (x_2, y_2) \qquad (x_3, y_3)$$

we get a unique polynomial (quadratic) through the three points. This may be written

$$y = y_1 + (x - x_1)\{[x_2, x_1] + (x - x_2)[x_3, x_2, x_1]\}$$

If we now choose the points in the order x_a, x_b, and x_c, we have

$$y = y_a + (x - x_a)\{[x_b, x_a] + (x - x_b)[x_c, x_b, x_a]\}$$

Since these are both the same quadratic, the coefficients of the x^2 term must be the same; that is, the two symbols

$$[x_3, x_2, x_1] = [x_c, x_b, x_a]$$

are in fact the same thing in alternative forms.

In general, we define

$$[x_1, x_2, x_3, \ldots, x_n] \equiv \frac{[x_1, x_2, \ldots, x_{n-1}] - [x_1, x_2, \ldots, x_{n-2}, x_n]}{x_{n-1} - x_n}$$

and in exactly the same manner show that it is independent of the order of the x_i. Note that the denominator is the difference of the nonrepeated x's in each $[\cdots]$ taken in the same order.

One way to make a table of the needed values is as follows:

x_1	y^*			
		$[x_2, x_1]^*$		
x_2	y_2		$[x_3, x_2, x_1]^*$	
		$[x_3, x_1]$		$[x_4, x_3, x_2, x_1]^*$
x_3	y_3		$[x_4, x_2, x_1]$	.
		$[x_4, x_1]$	.	
x_4	y_4	.		.
.	.		.	
.	.	.		.
.	.		.	
.	.			

The asterisks indicate entries used as pivots in calculating succeeding values.

From this table we can write

$$y(x) = y_1 + (x - x_1)([x_2, x_1] + (x - x_2)\{[x_3, x_2, x_1] + (x - x_3)\{\cdots\}\})$$

As a specific example, consider again the log table

x	$\log x$	[,]	[,,]	[,,,]
1	0.0000*			
		0.30100*		
2	0.3010		−0.06245*	
		0.23855		+0.01230*
3	0.4771		−0.05015	
		0.20070		
4	0.6021			

The asterisks indicate the pivotal values.

Hence we get

$$y(x) = 0 + (x - 1)\{0.3010 + (x - 2)[(-0.06245) + (x - 3)(0.01230)]\}$$

In particular,

$$y(2.5) = 3/2\{0.3010 + 1/2[(-0.06245) + (-1/2)(0.01230)]\}$$
$$= 0.40001$$

The correct value for log 2.5 is 0.3979

PROBLEMS

18.2.1 Given a divided difference table, show how to add one more data point at the botton of the table. Also show how to add one at the top. In the first case how does one change the Newton polynomial?

18.2.2 Make the divided difference table for $\Gamma(x)$ in Sec. 10.6.

18.2.3 Make the divided difference table for $\zeta(x)$ in Sec. 10.6.

18.2.4 Make the divided difference table for the error integral in Sec. 10.6.

18.3 AN ALTERNATIVE FORM FOR THE DIVIDED DIFFERENCE TABLE

The divided difference table, which lies at the heart of Newton's interpolation formula, may be written in an alternative and sometimes more useful form. This is based on the observation that

$$[x_1, x_2, \ldots, x_n] = \frac{[x_1, x_2, \ldots, x_{n-1}] - [x_1, x_2, \ldots, x_{n-2}, x_n]}{x_{n-1} - x_n}$$
$$= \frac{[x_1, x_2, \ldots, x_{n-1}] - [x_2, x_3, \ldots, x_n]}{x_1 - x_n}$$

In particular,

$$[x_1, x_2, x_3] = \frac{[x_1, x_2] - [x_1, x_3]}{x_2 - x_3}$$

$$= \frac{[x_1, x_2] - [x_2, x_3]}{x_1 - x_3}$$

Thus we can write

x	y	[,]	[,,]	[,,,]
x_1	y_1			
		$[x_2, x_1]$		
x_2	y_2		$[x_3, x_2, x_1]$	
		$[x_3, x_2]$		$[x_4, x_3, x_2, x_1]$
x_3	y_3		$[x_4, x_3, x_2]$	
		$[x_4, x_3]$		
x_4	y_4			

While the entries in this table differ from those in the previous section, the entries in the top row, which are the ones used in Newton's formula, are the same.

Using the sample example of log x, we get the following table:

x	$\log x$	[,]	[,,]	[,,,]
1	0.0000			
		0.3010		
2	0.3010		-0.06245	
		0.1761		$+0.01230$
3	0.4771		-0.02555	
		0.1250		
4	0.6021			

In this form, we may easily add a line at either end of the table and still expect the entries in the body of the table to vary smoothly. While there is no theoretical need to choose the points in numerical order, the smoothness in the table is lost if they are not. The first divided differences are secant lines and hence approximations to the first derivative in the interval (x_i, x_{i+1}). Similarly, the second divided differences are local approximations to the second derivatives, etc.

PROBLEMS

18.3.1 Using the table

x	2	0	3	1
	8	0	27	1

compute the Newton interpolating polynomial. Show that $y = x^3$.

18.3.2   Compute the difference table both ways (Secs. 18.2 and 18.3) for $y = \sin x$ at a spacing of 30°:

x	$\sin x$
0°	0.0000
30°	0.5000
60°	0.8660
90°	1.0000

Find the approximating polynomial.

18.3.3   If you have only the top row of the divided difference table in storage in a computer, how can you add one point at the bottom of·the table and compute the next divided difference?   How do you add one point at the top and find the new line?

18.4   NEWTON'S FORMULA AT EQUAL SPACES

Very often our information about a function (in the form of samples) is given at a set of equally spaced values of x.   This simplifies much of the notation and computation as well as the ideas involved.

For equally spaced data it is customary to use the notation of Part 1:

$$\Delta y_n = y_{n+1} - y_n$$

This is the familiar notation of the difference calculus except that we have fixed $\Delta x = h$ at all times:

$$\Delta x_n = x_{n+1} - x_n = h$$

We also have

$$\Delta^2 y_n = y_{n+2} - 2y_{n+1} + y_n$$

etc.

These differences are related to the divided differences as follows:

$$[x_2, x_1] = \frac{y_2 - y_1}{x_2 - x_1} = \frac{\Delta y_1}{h}$$

$$[x_3, x_2, x_1] = \frac{[x_3, x_2] - [x_2, x_1]}{x_3 - x_1} = \frac{\Delta y_2/h - \Delta y_1/h}{2h}$$

$$= \frac{\Delta^2 y_1}{2! h^2}$$

and, in general,

$$[x_1, x_2, \ldots, x_n] = \frac{\Delta^{n-1} y_1}{(n-1)! h^{n-1}}$$

The differences are approximations to derivatives at the middle of the range of the samples used:

$$\Delta y_1 \approx h \frac{dy(x + h/2)}{dx}$$

$$\Delta^2 y_1 \approx h^2 \frac{d^2 y(x + h)}{dx^2}$$

$$\cdots\cdots\cdots\cdots\cdots$$

$$\Delta^n y_1 \approx h^n \frac{d^n y(x + nh/2)}{dx^n}$$

It is these relationships that enable the finite difference calculus to approximate expressions in the differential calculus.

Newton's formula in this new notation for equally spaced sample points is (Secs. 9.6 and 18.2)

$$y = y_0 + (x - x_0)\frac{\Delta y_0}{h} + (x - x_0)(x - x_0 - h)\frac{\Delta^2 y_0}{2h^2}$$

$$+ (x - x_0)(x - x_0 - h)(x - x_0 - 2h)\frac{\Delta^3 y_0}{3!\,h^3} + \cdots$$

If we suppose that $x_0 = 0$, we get

$$y = y_0 + x\frac{\Delta y_0}{h} + x(x - h)\frac{\Delta^2 y_0}{2h^2} + x(x - h)(x - 2h)\frac{\Delta^3 y_0}{3!\,h^3} + \cdots$$

and if we further assume $h = 1$, we have

$$y = y_0 + x\,\Delta y_0 + x(x - 1)\frac{\Delta^2 y_0}{2!} + x(x - 1)(x - 2)\frac{\Delta^3 y_0}{3!} + \cdots$$

PROBLEMS

18.4.1 Write the error term for Newton's formula when the data are given at $-n$, $-(n - 1), \ldots, 0, 1, 2, \ldots, n$ and a $(2n)$th difference is used in the formula.

18.5 INTERPOLATION IN TABLES

One of the major uses of equally spaced interpolation formulas is for interpolation in tables of equally spaced data. A typical example is the error integral whose difference table is given in Sec. 10.6. The differences are well behaved in the sense that they first tend toward zero, being fairly smooth in the third-difference column, but beginning with the fourth difference, and clearly in the fifth difference where the sign is almost exactly alternating, we find that roundoff noise is dominating. For this table we would not expect, then, to go beyond fourth differences in any interpolation formula.

In principle, Newton's formula for interpolation gives the answer, and except for roundoff effects the answer is exactly the same as could be found by any other interpolating polynomial using the same sample points. Nevertheless, in practice, many other formulas are used, and in the interest of general education, so that the reader can understand the methods given in the prefaces of most tables, we shall discuss some of those which are most popular.

18.6 THE LOZENGE DIAGRAM

The lozenge diagram is a device for showing that a large number of formulas which appear to be different are really all the same. We have used the notation for the binomial coefficients

$$C(u + k, n) = \frac{(u + k)(u + k - 1)(u + k - 2) \cdots (u + k - n + 1)}{n!}$$

There are n factors in the numerator and n in the denominator. Viewed as a function of u, $C(u + k, n)$ is a polynomial of degree n.

Figure 18.6.1 shows the lozenge diagram. A line starting at a point on the left edge and following some path across the page defines an interpolation formula if the following rules are used.

1a For a *left*-to-*right* step, *add*.

1b For a *right*-to-*left* step, *subtract*.

2a If the *slope* of the step is *positive*, use the product of the difference crossed times the factor immediately *below*.

2b If the *slope* of the step is *negative*, use the product of the difference crossed times the factor immediately *above*.

3a If the step is *horizontal* and passes through a *difference*, use the product of the difference times the *average* of the factors *above and below*.

3b If the step is *horizontal* and passes through a *factor*, use the product of the factor times the *average* of the differences *above and below*.

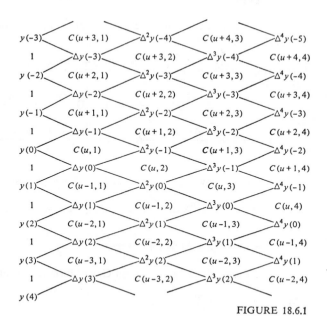

FIGURE 18.6.1

As an example of rules 1*a* and 2*b*, consider starting at $y(0)$ and going down to the right.   We get, term by term,

$$y(u) = y(0) + C(u, 1)\, \Delta y(0) + C(u, 2)\, \Delta^2 y(0) + C(u, 3)\, \Delta^3 y(0) + \cdots$$

$$= y(0) + u\, \Delta y(0) + \frac{u(u - 1)}{2}\, \Delta^2 y(0) + \frac{u(u - 1)(u - 2)}{3!}\, \Delta^3 y(0) + \cdots$$

which is Newton's formula.

Had we gone up and to the right, we would have used rules 1*a* and 2*a* to get Newton's backward formula:

$$y(u) = y(0) + C(u, 1)\, \Delta y(-1) + C(u + 1, 2)\, \Delta^2 y(-2)$$
$$+ C(u + 2, 3)\, \Delta^3 y(-3) + \cdots$$

$$= y(0) + u\, \Delta y(-1) + \frac{(u + 1)u}{2}\, \Delta^2 y(-2) + \frac{(u + 2)(u + 1)u}{3!}\, \Delta^3 y(-3) + \cdots$$

$$(18.6.1)$$

To get Stirling's formula, we start at $y(0)$ and go horizontally to the right, using rules $3a$ and $3b$:

$$y(u) = y(0) + u \frac{\Delta y_0 + \Delta y_{-1}}{2} + \frac{C(u+1, 2) + C(u, 2)}{2} \Delta^2 y_{-1}$$

$$+ C(u+1, 3) \frac{\Delta^3 y_{-2} + \Delta^3 y_{-1}}{2} + \cdots$$

$$= y_0 + u \frac{\Delta y_0 + \Delta y_{-1}}{2} + \frac{u^2}{2} \Delta^2 y_{-1} + \frac{u(u^2 - 1)}{3!} \frac{\Delta^3 y_{-2} + \Delta^3 y_{-1}}{2} + \cdots$$

$$(18.6.2)$$

If we start midway between $y(0)$ and $y(1)$, we get Bessel's formula:

$$y(u) = 1 \frac{y_0 + y_1}{2} + \frac{C(u, 1) + C(u-1, 1)}{2} \Delta y_0 + C(u, 2) \frac{\Delta^2 y_{-1} + \Delta^2 y_0}{2} + \cdots$$

$$= \frac{y_0 + y_1}{2} + (u - \tfrac{1}{2}) \Delta y_0 + \frac{u(u-1)}{2} \frac{\Delta^2 y_{-1} + \Delta^2 y_0}{2} + \cdots \qquad (18.6.3)$$

If we zigzag properly, we can get Gauss' formula for interpolation:

$$y(u) = y_0 + u \Delta y_0 + \frac{u(u-1)}{2} \Delta^2 y(-1) + \frac{u(u^2-1)}{3!} \Delta^3 y(-1) + \cdots \qquad (18.6.4)$$

All sorts of paths can be chosen, and each will give some formula. What we need to show is that these are all valid interpolation formulas. Such a proof requires showing the following:

1 At least one valid formula results, and since we have found Newton's interpolation formula from the diagram, this step is completed.

2 The contribution around any closed path is zero; hence we may deform a path into any other path that we wish.

3 If two formulas end at the same place, they are the same. It is necessary to prove this fact since the entry points into the lozenge diagram need not be the same for different formulas.

To prove step 2, we take a single complete lozenge (Fig. 18.6.2).

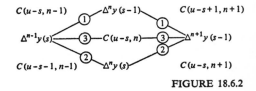

FIGURE 18.6.2

Path 1: $\quad C(u-s, n)\,\Delta^n y(s-1) + C(u-s+1, n+1)\,\Delta^{n+1} y(s-1)$

Path 2: $\quad C(u-s, n)\,\Delta^n y(s) + C(u-s, n+1)\,\Delta^{n+1} y(s-1)$

Path 3: $\quad C(u-s, n)\,\dfrac{\Delta^n y(s-1) + \Delta^n y(s)}{2}$

$$+ \frac{C(u-s+1, n+1) + C(u-s, n+1)}{2}\,\Delta^{n+1} y(s-1)$$

Recalling that going from right to left produces negative terms (rule 1*b*), we have to show only that the three paths are the same.

Path 1 − path 2

$$= C(u-s, n)[\Delta^n y(s-1) - \Delta^n y(s)]$$
$$+ [C(u-s+1, n+1) - C(u-s, n-1)]\,\Delta^{n+1} y(s-1)$$
$$= C(u-s, n)[-\Delta^{n+1} y(s-1)]$$
$$+ C(u-s, n)\,\frac{(u-s+1) - (u-s-n)}{n+1}\,\Delta^{n+1} y(s-1)$$
$$= C(u-s, n)[-\Delta^{n+1} y(s-1) + \Delta^{n+1} y(s-1)] = 0$$

Also, path 3 equals the average of paths 1 and 2. Hence step 2 of the proof is complete.

To examine step 3 of the proof, we use Fig. 18.6.3.

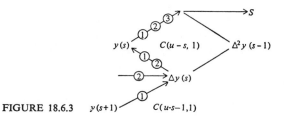

FIGURE 18.6.3

Path 1 $= y(s+1) + C(u-s-1, 1)\,\Delta y(s) - C(u-s, 1)\,\Delta y(s) + S$

Path 2 $= \dfrac{y(s+1) + y(s)}{2} + \dfrac{C(u-s-1, 1) + C(u-s, 1)}{2}\,\Delta y(s)$

$\qquad - C(u-s, 1)\,\Delta y(s) + S$

Path 3 $= y(s) + S$

where S is the rest of the formula.   Now using
$$y(s + 1) - y(s) = \Delta y(s)$$
we obtain
Path 1 $= y(s) + \Delta y(s) + (u - s)\,\Delta y(s) - \Delta y(s) - (u - s)\,\Delta y(s) + S$
$= y(s) + S = \text{path } 3$

Similarly, path 2 may be reduced to path 3.   Thus we conclude that the interpolation formula depends on the ending values, not on the path used to reach them.

18.7   REMARKS ON THESE FORMULAS

How do these different formulas compare with each other and with the Lagrange formula found earlier?   The value obtained in an interpolation depends on the polynomial used, and the polynomial depends on the sample points actually used (some formulas might appear to use values not needed).   The error term has the form (see Sec. 14.6)
$$\frac{(x - x_1)(x - x_2) \cdots (x - x_{n+1}) y^{(n+1)}(\bar{x})}{(n + 1)!}$$
The coefficient of the derivative is minimized when x is in the middle of the range of samples.   Thus there is a tendency to use an even number of samples when the interpolation point is in the middle of an interval and an odd number when it is near a sample point.

The methods derived by the lozenge diagram explicitly exhibit the differences, and these differences give some idea of the accuracy but require more arithmetic than does the Lagrange method of interpolating.   The Lagrange method, however, does not directly use the differences, and so no indication of the accuracy is available.   This is one of the fundamental problems in numerical computation—the use of differences gives some clues to the accuracy of the method being used, while the Lagrange approach minimizes the amount of arithmetic.   As a result, the difference methods tend to be used in exploratory work, and the Lagrange methods in well-understood, routine work.

18.8   MISCELLANEOUS INTERPOLATION FORMULAS

A number of other formulas which are not directly obtained from the lozenge diagram are occasionally useful.   They usually rest on the fact that any particular difference may be eliminated by using
$$\Delta^n y(s + 1) - \Delta^n y(s) = \Delta^{n+1} y(s)$$

The cost of this elimination is the inclusion of an extra difference of some other order.

For example, in Bessel's interpolation formula we may eliminate all odd-order differences; this gives

$$y(u) = (1 - u)y(0) + uy(1) + \frac{(2 - u)(1 - u)(-u)}{3!} \Delta^2 y(-1)$$
$$+ \frac{(u + 1)u(u - 1)}{3!} \Delta^2 y(0) + \cdots$$

which is known as Everett's formula and is quite popular since the table maker need only publish the function and even-order differences.

Similarly we may start with almost any formula and eliminate differences of any order that we please (at the cost of several differences of the orders that we leave in the formula). If we wish, we could eliminate, say, both the second- and third-order differences, thus using the function, the first differences, and the fourth and higher differences. Carried to the extreme, this leads to the Lagrange form which uses no differences but many function values. The purpose of eliminating the differences is mainly to save type and space in printing the table, and this is done at the cost of extra work on the part of the user. The proper balance depends on circumstances and cannot be given once and for all.

A second device often used is called "throwback." This idea, due mainly to Comrie, uses the fact that the coefficients of the successive differences tend to be proportional to each other in various interpolation formulas, such as Everett's; consequently, if suitable amounts of the higher differences are combined with those of the lower when the table is printed, then much of the effect of the higher-order differences in the interpolation is automatically achieved by using the lower-order formula.

Thus, in Bessel's formula, the ratio of the coefficients of the Δ^4 to the coefficient of Δ^2 is

$$\frac{B^{IV}}{B^{II}} = \frac{(u + 1)(u - 2)}{12} \qquad 0 \le u \le 1$$

which varies from $-1/6$ to $-3/16$. Hence if we put c units of Δ^4 with Δ^2 to form the modified second difference column, we make an error of

$$(B^{IV} - cB^{II}) \Delta^4$$

The number c is often taken as -0.184 as an average compromise.

We again remind the reader that table making is a highly skilled art and refer the reader to Fox [12] as well as to the standard text of Ralston [49].

18.9 THE HAMMING-PINKHAM INTEGRATION FORMULA

We now turn to a family of integration formulas whose *correction terms* (truncation errors) are expressed as differences of the function values. The effect of taking more differences is to increase the accuracy of the formula as judged by the degree of the polynomial for which it is exactly true (within roundoff). Thus we can *sequentially* try successive members of the family to find the most appropriate one instead of more or less committing ourselves to the accuracy *before* starting (as we must with other types of formulas). Because we have the difference table of the integrand values we can make an estimate of the roundoff errors independently of the estimate of the truncation error.

These formulas are closely related to the composite integration formulas discussed in Chap. 20, and so it is worth a few moments to explain the idea of a composite formula. When we find that the formula such as Simpson's is not accurate enough, we usually decide to use more points; with more points we can make the formula more accurate in the sense of being exact for a higher order polynomial. Instead of using this method we can apply the same basic formula to a sequence of intervals and add the results of the individual integrals. This second approach leads to the idea of a *composite formula*. The two best-known examples are the composite trapezoid rule

$$\int_0^{nh} f(x)\,dx = h\left\{\frac{1}{2}f(0) + f(h) + f(2h) + \cdots + f[(n-1)h] + \frac{1}{2}f(nh)\right\}$$

and the composite Simpson's formula

$$\int_0^{2nh} f(x)\,dx = \frac{h}{3}\{f(0) + 4f(h) + 2f(2h) + 4f(3h) + \cdots + 4f[(2n-1)h] + f(2nh)\}$$

where in both cases we have used the basic integration formula repeatedly.

The classic formula which led to the investigation of this family of formulas is the Gregory[1] formula (1670)

$$\int_0^n f(x)\,dx = \frac{1}{2}f(0) + f(1) + f(2) + \cdots + f(n-1) + \frac{1}{2}f(n)$$

$$-\frac{1}{12}[\Delta f(n-1) - \Delta f(0)] - \frac{1}{24}[\Delta^2 f(n-2) + \Delta^2 f(0)]$$

$$-\frac{19}{720}[\Delta^3 f(n-3) - \Delta^3 f(0)] - \frac{9}{480}[\Delta^4 f(n-4) + \Delta^4 f(0)]$$

$$-\frac{863}{60,480}[\Delta^5 f(n-5) - \Delta^5 f(0)] + \cdots$$

[1] James Gregory (1638–1675).

This formula is clearly the composite trapezoid rule *with end corrections* that make the formula as accurate as the differences included indicate.   The alternating sign between the differences has the effect of causing the accuracy to jump by 2 or 0 as one more term of differences is included; thus we tend to stop on an even-order difference and get the next odd-order one "free."

A similar formula based on Simpson's composite formula is

$$\int_0^{2n} f(x)\, dx = \frac{1}{3} f(0) + \frac{4}{3} f(1) + \frac{2}{3} f(2) + \cdots + \frac{4}{3} f(2n - 1) + \frac{1}{3} f(2n)$$

$$- \frac{4}{720} [\Delta^3 f(2n - 3) - \Delta^3 f(0)] - \frac{4}{480} [\Delta^4 f(2n - 4) + \Delta^4 f(0)]$$

$$- \frac{548}{60,480} [\Delta^5 f(2n - 5) - \Delta^5 f(0)] + \cdots$$

where the coefficients of the differences are *all* smaller in size than those of the Gregory formula (see Table 18.10 for more coefficients).

A still more interesting formula is the Hamming-Pinkham formula

$$\int_0^{2n} f(x)\, dx = 2[f(1) + f(3) + \cdots + f(2n - 1)]$$

$$+ \frac{1}{6} [\Delta f(2n - 1) - \Delta f(0)] + \frac{1}{12} [\Delta^2 f(2n - 2) + \Delta^2 f(0)]$$

$$+ \frac{13}{360} [\Delta^3 f(2n - 3) - \Delta^3 f(0)] + \frac{1}{80} [\Delta^4 f(2n - 4) + \Delta^4 f(0)]$$

$$+ \frac{41}{30,240} [\Delta^5 f(2n - 5) - \Delta^5 f(0)] + \cdots$$

where only *half* of the integrand values in the basic summation and comparatively few more at the ends are needed to form the differences.   Thus the coarse sampling rate is used in the main part of the integrand, and double the rate is used at the ends to estimate the differences.   This means that about half the computing is saved for the same order of accuracy.

The first two cases, the Gregory and Simpson composite formulas, can be used in an adaptive mode of computation by halving the spacing if the truncation error is too large; the third formula requires taking one-third of the spacing if the old integrand values are to be used in the next step.

In order to compare these three formulas (and some others) we shall, as usual, consider a family of formulas which includes them as special cases. This family is

$$\int_0^{2n} f(x)\,dx = af(0) + bf(1) + 2af(2) + \cdots + af(2n)$$

$$+ \sum_{s=0}^{\infty} q_s\{\Delta^s f(2n-s) + (-1)^s\,\Delta^s f(0)\}$$

where we want both the coefficients a and b and the q_s to be independent of n, which is the number of double intervals used.

In the polynomial approach to finding formulas we require the formula to be exact for successive powers of x; that is, we set down the defining equations

$$\int_0^{2n} x^p\,dx = a(0)^p + b(1)^p + 2a(2)^p + b(3)^p + \cdots + a(2n)^p$$

$$+ \sum_{s=0}^{\infty} q_s\{\Delta^s(2n-s)^p + (-1)^s\,\Delta^s(0)^p\} \qquad p = 0, 1, 2, \ldots$$

For $p = 0$ we have

$$2n \equiv nb + [2 + 2(n-1)]a = 2na + nb$$

or

$$b = 2(1-a)$$

which means we have a one parameter family of formulas depending on a.

18.10   THE DERIVATION OF THE FORMULAS

The derivation of the Hamming-Pinkham formula is based on the method of generating functions (Sec. 11.4). We multiply the pth defining equation by $t^p/p!$ and sum to get (using the power-series expansion of the exponential)

$$\int_0^{2n} e^{tx}\,dx = ae^0 + be^t + 2ae^{2t} + \cdots + ae^{2nt}$$

$$+ \sum_{s=0}^{\infty} q_s(\Delta^s[e^{2n-s}]^t + (-1)^s\,\Delta^s[e^0]^t)$$

But noting that this is a composite formula with weights

a	b	a					
	a	b	a				
		a	b	a			
			a	b	a		
					a	$\cdots$	

$$a \quad b \quad 2a \quad b \quad 2a \quad b \quad 2a \quad \cdots$$

we see that the first sum on the right-hand side is

$$(ae^{-t} + b + ae^t)(e^t + e^{3t} + \cdots + e^{(2n-1)t})$$

$$= (ae^{-t} + b + ae^t)\frac{e^{2nt} - 1}{e^t - e^{-t}}$$

The differences that occur in the formula can be written as

$$\Delta^s(e^{(2n-s)t}) = \Delta^{s-1}(e^{(2n-s+1)t} - e^{(2n-s)t}) = \Delta^{s-1}(e^{(2n-s+1)t})(1 - e^{-t})$$

$$\vdots$$

$$= e^{2nt}(1 - e^{-t})^s$$

$$(-1)^s \Delta^s(e^{0t}) = (-1)^s \Delta^{s-1}(e^t - e^0) = (-1)^s \Delta^{s-1}(e^t - 1) = (-1)^{s-1} \Delta^{s-1}(1 - e^t)$$

$$\vdots$$

$$= (1 - e^t)^s$$

Performing the integration on the left-hand side of the formula and assembling the results, we have

$$\frac{e^{2nt} - 1}{t} = (ae^t + b + ae^{-t})\frac{e^{2nt} - 1}{e^t - e^{-t}} + \sum_{s=0}^{\infty} q_s[e^{2nt}(1 - e^{-t})^s + (1 - e^t)^s]$$

If we now define $Q(t)$ as

$$Q(t) = \frac{1}{t} - \frac{ae^t + b + ae^{-t}}{e^t - e^{-t}} - \sum_{s=0}^{\infty} q_s(1 - e^{-t})^s$$

then we see that we can write the equation as

$$e^{2nt}Q(t) + Q(-t) = 0$$

This equation is true, if $Q(t) \equiv 0$. If $Q(t) = 0$, then

$$\sum_{s=0}^{\infty} q_s(1 - e^{-t})^s = \frac{1}{t} - \frac{ae^t + b + ae^{-t}}{e^t - e^{-t}}$$

and defines the q_s.

It is natural to set

$$1 - e^{-t} \equiv v \qquad \text{or} \qquad e^{-t} = 1 - v \qquad t = -\ln(1 - v)$$

so that we have a power series in v with coefficients q_s. Using $b = 2(1 - a)$, we have

$$\sum_{s=0}^{\infty} q_s v^s = \frac{-1}{\ln(1 - v)} - \frac{a/(1 - v) + 2 - 2a + a(1 - v)}{1/(1 - v) - (1 - v)}$$

$$= \frac{-1}{\ln(1 - v)} - \frac{2 - 2v + av^2}{2v - v^2}$$

$$= \frac{-1}{\ln(1 - v)} + a - \frac{1}{v} - \frac{2a - 1}{2 - v}$$

The choice of $a = 1/2$ eliminates one term and gives the Gregory formula (as a check on the work so far). For $a = 1/2$ we have

$$\sum_{s=0}^{\infty} q_s v^s = \frac{-1}{\ln(1-v)} - \frac{1}{v} + \frac{1}{2}$$

to determine the q_s. Using

$$-\ln(1-v) = v + \frac{v^2}{2} + \frac{v^3}{3} + \frac{v^4}{4} + \cdots$$

and dividing out $1/\ln(1-v)$, we get

$$\sum q_s v^s = -\frac{v}{12} - \frac{v}{24} - \frac{19}{720} v^3 - \cdots$$

which checks with Gregory's formula.

If we label the coefficients of Gregory's formula g_s, we have for the general case

$$\sum_{s=0}^{\infty} q_s v^s = \frac{-1}{\ln(1-v)} + a - \frac{1}{v} - \frac{2a-1}{2-v}$$

$$= \sum_{s=0}^{\infty} g_s v^s + (a - \tfrac{1}{2}) - \frac{2a-1}{2-v}$$

$$= \sum_{s=0}^{\infty} g_s v^s - (a - \tfrac{1}{2}) \frac{v}{2-v}$$

Dividing out $v/(2-v)$, we get

$$\sum_{s=0}^{\infty} q_s v^s = \sum_{s=0}^{\infty} g_s v^s - (a - \tfrac{1}{2}) \sum_{s=1}^{\infty} \left(\frac{v}{2}\right)^s$$

Equating coefficients of v^s, we have $q_0 = g_0$ and for $s = 0$

$$q_s = g_s - (a - \tfrac{1}{2}) 2^{-s}$$

From this, Table 18.10 is easily compiled and is displayed in Fig. 18.10.1.

A remarkable formula results by setting $a = 0$ ($b = 2$) since now half the integrand values have weight zero and so do not enter into the summation.

$$\int_0^{2n} f(x)\,dx = 2 \sum_{r=0}^{n-1} f(2r+1) + \frac{1}{6} [\Delta f(2n-1) - \Delta f(0)]$$

$$+ \frac{1}{12} [\Delta^2 f(2n-2) + \Delta^2 f(0)] + \frac{13}{360} [\Delta^3 f(2n-3) - \Delta^3 f(0)]$$

$$+ \frac{1}{80} [\Delta^4 f(2n-4) + \Delta^4 f(0)] + \frac{41}{30,240} [\Delta^5 f(2n-5) - \Delta^5(0)] + \cdots$$

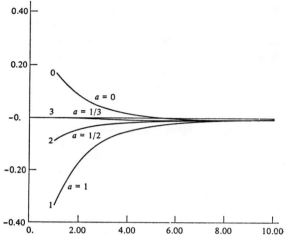

FIGURE 18.10.1

Table 18.10

s	Gregory $a = 1/2$	Simpson $a = 1/3$	$a = 0$	$a = 1$
1	$-\dfrac{1}{12}$	0	$\dfrac{2}{12}$	$-\dfrac{4}{12}$
2	$-\dfrac{1}{24}$	0	$\dfrac{2}{24}$	$-\dfrac{4}{24}$
3	$-\dfrac{19}{720}$	$-\dfrac{4}{720}$	$\dfrac{26}{720}$	$-\dfrac{64}{720}$
4	$-\dfrac{9}{480}$	$-\dfrac{4}{480}$	$\dfrac{6}{480}$	$-\dfrac{24}{480}$
5	$-\dfrac{863}{60480}$	$-\dfrac{548}{60480}$	$\dfrac{82}{60480}$	$-\dfrac{1808}{60480}$
6	$-\dfrac{275}{24192}$	$-\dfrac{212}{24192}$	$-\dfrac{86}{24192}$	$-\dfrac{464}{24192}$
7	$-\dfrac{33953}{3628800}$	$-\dfrac{29228}{3628800}$	$-\dfrac{19778}{3628800}$	$-\dfrac{48128}{3628800}$
8	$-\dfrac{8183}{1036800}$	$-\dfrac{7508}{1026800}$	$-\dfrac{6158}{1036800}$	$-\dfrac{10208}{1036800}$
9	$-\dfrac{3250433}{479001600}$	$-\dfrac{3094508}{479001600}$	$-\dfrac{2782658}{479001600}$	$-\dfrac{3718208}{479001600}$
10	$-\dfrac{14013}{2365440}$	$-\dfrac{13628}{2365440}$	$-\dfrac{12858}{2365440}$	$-\dfrac{15168}{2365440}$

Experimental results show that this formula compares favorably with other members of the family; indeed the results are somewhat better on the average. We can, using this formula, estimate the truncation error made by dropping the differences past some order k, and we can estimate the roundoff from the difference table.

PROBLEMS

18.10.1 Derive Gregory's formula when n is an odd number.

18.10.2 Consider using five points (four intervals) in the basic formula and derive a two-parameter family of formulas.

18.10.3 Derive the midpoint formula (Bickley's formula). See [8].

$$\int_0^n f(x)\,dx = \sum_{k=0}^{k-1} f\left(k + \frac{1}{2}\right) + \frac{1}{24}[\Delta f_{n-3/2} - \Delta f_{1/2}] + \frac{1}{24}[\Delta^2 f_{n-5/2} + \Delta^2 f_{1/2}]$$
$$+ \frac{223}{5,760}[\Delta^3 f_{n-7/2} - \Delta^3 f_{1/2}] + \frac{103}{2,880}[\Delta^4 f_{n-9/2} + \Delta^4 f_{1/2}]$$
$$+ \frac{32,119}{967,680}[\Delta^5 f_{n-11/2} - \Delta^5 f_{1/2}] + \cdots$$

18.10.4 Derive

$$\int_0^1 f(x)\,dx = \sum_{k=0}^{\infty} A_k[f^{2k}(0) + f^{2k}(1)]$$

where

$$A_{k-1} = \frac{2(2^{2k} - 1)}{(2k)!}\,B_{2k}$$

19

*FORMULAS USING THE SAMPLE POINTS AS PARAMETERS

19.1 INTRODUCTION

Up to now we have been using the weights in the formulas as the parameters. Since the weights occurred linearly in the defining equations, they were easy to find (the determinant of the system was shown to be nonzero, and we could use the sampling polynomials to find the w_i directly if we wanted to). We now consider using the position x_i of the samples as *additional* parameters to make the formulas exact for more powers of x. The new parameters do *not* occur linearly, and so the defining equations will be harder to solve. But again it should be pointed out that the defining equations are solved only once when the formula is being developed, not each time the formula is used. The gain in making the formula exact for more powers of x while using *the same number* of sample points is obvious.

   The resulting formulas generally have awkward values for the sample positions x_i and the weights w_i, and they were therefore neglected in hand-calculating days. This is no reason to neglect them when computing machines are readily available. The principal remaining objection is that the peculiar values of the x_i and w_i must be provided to the routine by keypunching and storing them from published tables[1] or from other routines which found them and saved the results.

The mathematical theory is both elegant and extensive, and we will give only a minimum of material to indicate what is possible.

19.2   SOME EXAMPLES

Before plunging into the theory associated with the general case, we shall examine a number of special cases to orient ourselves.   First consider estimating the following integral from two samples:

$$\int_0^\infty e^{-x} f(x)\, dx = w_1 f(x_1) + w_2 f(x_2)$$

where the parameters are w_1, w_2, x_1, and x_2.   The four defining equations are, using the first four powers of x,

1:	$m_0 = 1 = w_1 + w_2$	c_0	
x:	$m_1 = 1 = w_1 x_1 + w_2 x_2$	c_1	c_0
x^2:	$m_2 = 2 = w_1 x_1^2 + w_2 x_2^2$	1	c_1
x^3:	$m_3 = 6 = w_1 x_1^3 + w_2 x_2^3$		1

We now use the polynomial

$$\pi(x) = (x - x_1)(x - x_2) \equiv x^2 + c_1 x + c_0$$

and multiply the top equation by c_0, the second equation by c_1, and the third equation by 1, and then add the three equations.   We get

$$2 + c_1 + c_0 = w_1(x_1^2 + c_1 x_1 + c_0) + w_2(x_2^2 + c_1 x_2 + c_0)$$
$$= w_1 \pi(x_1) + w_2 \pi(x_2) = 0$$

since $\pi(x_i) = 0$.   Shifting the three multipiers down one equation and repeating the process, we get

$$6 + 2c_1 + c_0 = w_1 x_1 \pi(x_1) + w_2 x_2 \pi(x_2) = 0$$

Thus we have the pair of equations

$$c_1 + c_0 = -2$$
$$2c_1 + c_0 = -6$$

[1] For $\int_{-1}^{1} f(x)\, dx$, $\int_0^\infty e^{-x} f(x)\, dx$, $\int_{-\infty}^{\infty} e^{-x^2} f(x)\, dx$ and many others, see National Bureau of Standards, "Handbook of Mathematical Functions," and Dover Publications, Inc., New York, 1964.   Also see Mineur [41] and Stroud and Secrest [57].

whose solution is

$$c_1 = -4$$
$$c_0 = 2$$

Since these are the coefficients of $\pi(x)$, we have

$$\pi(x) = x^2 - 4x + 2 = 0$$

and the sample points are

$$x_1 = 2 - \sqrt{2}$$
$$x_2 = 2 + \sqrt{2}$$

We are now reduced to the case of Chap. 15 with known sample points. The first two defining equations are:

$$1 = w_1 + w_2$$
$$1 = (2 - \sqrt{2})w_1 + (2 + \sqrt{2})w_2$$

whose solutions are:

$$w_1 = \frac{\sqrt{2} + 1}{2\sqrt{2}}$$

$$w_2 = \frac{\sqrt{2} - 1}{2\sqrt{2}}$$

Our formula is, therefore,

$$\int_0^\infty e^{-x} f(x)\, dx = \frac{\sqrt{2} + 1}{2\sqrt{2}} f(2 - \sqrt{2}) + \frac{\sqrt{2} - 1}{2\sqrt{2}} f(2 + \sqrt{2})$$

which is exact for cubics using only two samples of the integrand.

How did we know that the weights we found from the first two defining equations would satisfy the last two equations? The answer is simple; we chose the x_i so that the last two equations were linear combinations (with coefficients c_0, c_1, and 1) of the first two equations; hence the w_i automatically satisfied the last two equations.

As a second example, consider

$$\int_{-1}^1 f(x)\, dx = w_1 f(x_1) + w_2 f(x_2) + w_3 f(x_3)$$

With six parameters we can write six defining equations:

$$
\begin{array}{lllll}
2 = w_1 & + w_2 & + w_3 & c_0 \\
0 = w_1 x_1 & + w_2 x_2 & + w_3 x_3 & c_1 & c_0 \\
2/3 = w_1 x_1{}^2 & + w_2 x_2{}^2 & + w_3 x_3{}^2 & c_2 & c_1 & c_0 \\
0 = w_1 x_1{}^3 & + w_2 x_2{}^3 & + w_3 x_3{}^3 & 1 & c_2 & c_1 \\
2/5 = w_1 x_1{}^4 & + w_2 x_2{}^4 & + w_3 x_3{}^4 & & 1 & c_2 \\
0 = w_1 x_1{}^5 & + w_2 x_2{}^5 & + w_3 x_3{}^5 & & & 1
\end{array}
$$

We again use the appropriate $\pi(x)$,

$$\pi(x) \equiv (x - x_1)(x - x_2(x - x_3) \equiv x^3 + c_2 x^2 + c_1 x + c_0$$

By multiplying by the c_i as shown on the right of the defining equations and adding the results, we get three equations for c_0, c_1, and c_2:

$$
\begin{aligned}
2c_0 + 0c_1 + \tfrac{2}{3}c_2 + 0 \cdot 1 &= 0 \\
0c_0 + \tfrac{2}{3}c_1 + 0c_2 + \tfrac{2}{5} \cdot 1 &= 0 \\
\tfrac{2}{3}c_0 + 0c_1 + \tfrac{2}{5}c_2 + 0 \cdot 1 &= 0
\end{aligned}
$$

Clearly $c_0 = c_2 = 0$ and $c_1 = -3/5$. Hence

$$\pi(x) = x^3 + 0x^2 - \tfrac{3}{5}x + 0 = 0$$

$$x_1 = -\sqrt{3/5}$$
$$x_2 = 0$$
$$x_3 = \sqrt{3/5}$$

Knowing the sample points x_i, we have the defining equations

$$
\begin{aligned}
2 &= \quad w_1 + w_2 + \quad w_3 \\
0 &= -\sqrt{3/5}\, w_1 \quad + \sqrt{3/5}\, w_3 \\
2/3 &= \quad \tfrac{2}{3}w_1 \quad + \tfrac{2}{3}w_3
\end{aligned}
$$

It follows from the second that $w_1 = w_3$, from the third equation that

$$w_1 = w_3 = 5/9$$

and finally from the first equation $w_2 = 8/9$.   The formula is, therefore,

$$\int_{-1}^{1} f(x)\, dx = \frac{5}{9} f(-\sqrt{3/5}) + \frac{8}{9} f(0) + \frac{5}{9} f(\sqrt{3/5})$$

and is exact for fifth-degree polynomials (see Sec. 16.9).

If we had used the obvious symmetry, we would have realized that $w_1 = w_3$, $x_1 = -x_3$, and $x_2 = 0$ would make the formula true for *all* odd powers of x. Hence we would have had the defining equations

$$
\begin{array}{rl}
1: & 2 = 2w_1 \quad + w_2 \\
x^2: & 2/3 = 2w_1 x_1{}^2 \\
x^4: & 2/5 = 2w_1 x_1{}^4
\end{array}
$$

The ratio of the last two equations gives

$$
\begin{aligned}
x_1{}^2 &= 2/5 \div 2/3 = 3/5 \\
x_1 &= -\sqrt{3/5} \\
x_3 &= \sqrt{3/5}
\end{aligned}
$$

From the second equation

$$ w_1 = \frac{1}{3x_1{}^2} = \frac{5}{9} $$

and from the first $w_2 = 8/9$.   The use of symmetry clearly makes things easier!

PROBLEMS

19.2.1   Derive $\displaystyle\int_{-1}^{1} f(x)\,dx = f(-1/\sqrt{3}) + f(1/\sqrt{3})$ $E_4 = 8/45$
(See Sec. 16.5.)

19.2.2   Derive $\displaystyle -\int_{0}^{1} \ln x\, f(x)\,dx = w_1 f(x_1) + w_2 f(x_2)$

19.2.3   Derive $\displaystyle\int_{0}^{1} x\, f(x)\,dx = w_1 f(x_1) + w_2 f(x_2)$ $x_i = \dfrac{5 \pm \sqrt{40/7}}{9}$

19.2.4   Derive $\displaystyle\int_{-\pi/2}^{\pi/2} \sin x\, f(x)\,dx = w_1 f(x_1) + w_2 f(x_2)$

19.2.5   Derive $\displaystyle\int_{-1}^{1} f(x)\,dx = w_1 f(x_1) + w_2 f(x_2) + w_3 f(x_3) + w_4 f(x_4)$

Use symmetry to show that $w_1 = w_4$, $x_1 = -x_4$, $x_2 = -x_3$, and $w_2 = w_3$, so that $\pi(x)$ is a quadratic in x^2.

19.2.6   Derive $\displaystyle\int_{-\infty}^{\infty} e^{-x^2} f(x)\,dx = w_1 f(x_2) + w_2 f(x_2) + w_3 f(x_3)$

Use symmetry to show that $w_1 = w_{-1}$, $x_1 = -x_3$, and $x_2 = 0$.

19.2.7   Derive $\displaystyle\int_{-\pi}^{\pi} \sin x\, f(x)\,dx = \sum_{i=1}^{4} w_i f(x_i)$

Use odd symmetry to show that $w_1 = -w_4$ and $w_2 = -w_3$.

19.2.8   Derive $\int_{-1}^{1} x f(x)\,dx = \sqrt{5/27}[f(-\sqrt{3/5}) + f(\sqrt{3/5})]$

19.2.9   Derive $\int_{-1}^{1} \sqrt{1 - x^2} f(x)\,dx = \dfrac{\pi}{4}\,[f(-1/2) + f(1/2)]$

19.3   GAUSS' QUADRATURE (INTEGRATION)—FORMAL

The best-known case of using the sample positions as parameters in an integration formula is Gauss' quadrature on N points:

$$\int_{-1}^{1} f(x)\,dx = \sum_{k=1}^{N} w_k f(x_k)$$

where both the weights and the sample points are regarded as parameters.   Thus the examples in the previous section were Gaussian quadratures.   With $2N$ parameters we can write down the $2N$ defining equations:

$$1:\quad \int_{-1}^{1} 1 \cdot dx \quad = 2 = m_0 \quad = w_1 \quad\quad + w_2 \quad\quad + \cdots + w_N$$

$$x:\quad \int_{-1}^{1} x\,dx \quad = 0 = m_1 \quad = w_1 x_1 \quad + w_2 x_2 \quad + \cdots + w_N x_N$$

$$x^2:\quad \int_{-1}^{1} x^2\,dx \quad = \frac{2}{3} = m_2 \quad = w_1 x_1{}^2 \quad + w_2 \quad x_{22} \quad + \cdots + w_N x_N{}^2$$

$$\cdots \cdots \cdots \cdots \cdots \cdots \cdots \cdots \cdots \cdots \cdots \cdots \cdots \cdots$$

$$x^{2N-1}:\quad \int_{-1}^{1} x^{2N-1}\,dx = 0 = m_{2N-1} = w_1 x_1{}^{2N-1} + w_2 x_2{}^{2N-1} + \cdots w_N x_N{}^{2N-1}$$

We again use the polynomial

$$\pi(x) = (x - x_1)(x - x_2) \cdots (x - x_N)$$
$$= x^N + c_{N-1} x^{N-1} + \cdots + c_0$$

We now multiply the jth equation by c_j and sum to get

$$\sum_{j=0}^{N-1} m_j c_j + m_N = \sum_{k=1}^{N} w_k \pi(x_k) = 0$$

If we shift the multipliers c_j down one line and repeat, we get

$$\sum_{j=0}^{N-1} m_{j+1} c_j + m_{N+1} = \sum_{k=1}^{N} w_k x_k \pi(x_k) = 0$$

Repeating this shifting process, we are finally led to a total of N equations

$$\sum_{j=0}^{N-1} m_{j+k} c_j + m_{N+k} = 0 \qquad k = 0, 1, \ldots, N - 1$$

In the next section we will show that under very reasonable assumptions the persymmetric determinant [44] of these equations is not zero, and hence the system can be solved for the c_j. We then put these c_j into the equation for the sampling places

$$\pi(x) = x^N + c_{N-1}x^{N-1} + \cdots + c_0 = 0$$

This will, as we show in the next section, yield N real, distinct zeros. With these zeros we are now reduced to the case in Chap. 15 with known sample points, and we can solve the first N of the defining equations for the weights. Thus we have our formula.

An examination of the process shows that we first solve N simultaneous linear equations involving the moments m_k to obtain the c_j. These c_j are then used to form a polynomial, and we have to find the N zeros of it. Using these N zeros, we now solve another set of N simultaneous linear equations for the N weights. If all goes well, we will have our formula. We next turn to showing that all will go well in certain important cases.

19.4 GAUSS' QUADRATURE—ANALYSIS

The purpose of this section is to show that for a wide class of formulas the process of finding a Gaussian integration formula will go through without a hitch, in the sense that mathematically everything will be as we said it would be. Consider the class of integration formulas

$$\int_a^b K(x)f(x)\,dx = \sum_{k=1}^N w_k f(x_k)$$

where we assume that

$$K(x) \geq 0 \qquad a \leq x \leq b$$

The moments are

$$m_j = \int_a^b K(x)x^j\,dx$$

The defining equations are:

$$m_j = \sum_{k=1}^N w_k x_k^j \qquad j = 0, 1, \ldots, 2N-1$$

If we multiply the jth equation by the c_j defined by

$$\pi(x) = (x - x_1)(x - x_2) \cdots (x - x_N)$$
$$= x^N + c_{N-1}x^{N-1} + \cdots + c_0 \qquad c_N = 1$$

and sum, we get

$$\sum_{j=0}^{N-1} m_j c_j + m_N = \sum_{k=0}^N w_k \pi(x_k) = 0$$

Shifting and repeating, we get, in all, the system of equations as before

$$\sum_{j=0}^{N-1} m_{j+k} c_j + m_{N+k} = 0 \qquad k = 0, 1, \ldots, N-1$$

To prove that the determinant of these equations is not zero, we show that the only solution the *homogeneous* equations can have is $c_j = 0$ for all j. Using the set of N *homogeneous* equations, we multiply the kth equation by c_k and add, to get

$$\sum_{k=0}^{N-1}\sum_{j=0}^{N-1} m_{j+k} c_j c_k = 0$$

Recalling the definition of m_k, we have

$$\int_a^b K(x)\left[\sum_{k=0}^{N-1} c_k x^k \sum_{j=0}^{N-1} c_j x^j\right] dx = 0$$

or

$$\int_a^b K(x)\left[\sum_{k=0}^{N-1} c_k x^k\right]^2 dx = 0$$

Since $K(x) \geq 0$, the above equation can be zero only if $\sum c_k x^k \equiv 0$, that is, when all the $c_j = 0$ (as determined by the homogeneous equations). This in turn means that the determinant cannot be zero, and hence that we can solve the original set of nonhomogeneous linear equations for the c_j.

We now have the polynomial $\pi(x)$ whose zeros are the sample points. We need to show that when $K(x) \geq 0$, this particular polynomial has exactly N real, distinct zeros in the interval $a \leq x \leq b$. We assume, to the contrary, that it does not, that possibly there are complex zeros, real zeros outside the interval, or multiple zeros. Take all the odd-order zeros in the interval and for each write a simple factor in the product

$$p(x) = (x - x_1)(x - x_2) \cdots (x - x_k)$$

By our assumption (that, we will prove, leads to a contradiction) the degree k of $p(x)$ is less than N. Consider now

$$\int_a^b K(x)\pi(x)p(x)\, dx = \sum_{j=1}^{N} w_j \pi(x_j)p(x_j) = 0$$

The right-hand side of the formula is zero because $\pi(x_j) = 0$ at all the sample points. Furthermore we made the formula exact for all powers of x up to $2N$ and hence for all polynomials of degree less than $2N$. But $\pi(x)p(x)$ is of degree less than $2N$, and so the integral must be zero.   Finally, the integrand is always non-negative, and we have a contradiction since

$$\int_a^b K(x)\pi(x)p(x)\,dx > 0$$

To avoid the contradiction, we must have $k = N$, and thus there are N real, distinct zeros in the interval $a \le x \le b$.

When we solve the first N defining equations for the weights w_j (the Vandermonde determinant is not zero), we can reasonably wonder if they will also be appropriate for the last N equations.   The equations that defined the c_i show how the higher moments are linearly related to the lower moments, and we can apply these relations to show that the last N equations are consistent with the first N since each (after the first N) is a linear combination of the earlier ones.

It is an important fact that when $K(x) \ge 0$, the weights w_i are all positive numbers.   This follows from picking $f(x) = \pi_i^2(x)$ which is a polynomial of degree $2(N - 1)$ for which the error is zero.   We get

$$\int_a^b K(x)\pi_i^2(x)\,dx = \pi_i^2(x_i)w_i$$

and hence $w_i > 0$.   Thus roundoff propagation is not severe in these formulas (see Sec. 19.8).

19.5   THE ERROR TERM

It would clearly be awkward to prove directly that the associated $G(s)$ is of constant sign, and so we proceed in an indirect fashion.   Consider the Hermite interpolation formula (Sec. 17.2) using the N sample points.   We have

$$f(x) = P_{2N-1}(x) + \pi^2(x)\frac{f^{(2N)}(\theta)}{(2N)!} \qquad \theta = \theta(x)$$

We know that the integration formula is exact for all polynomials of degree less than $2N$, and so the polynomial part of the Hermite interpolation formula will exactly cancel out on both sides of the formula.   That leaves the error term to be considered.   On the right-hand side it cancels out due to the factor $\pi(x)$.   On the left we have

$$\int_a^b K(x)\pi^2(x)\frac{f^{(2N)}(\theta)}{(2N)!}\,dx \qquad K(x) \ge 0$$

where $\theta = \theta(x)$.   Note that the factor $K(x)\pi^2(x)$ is of constant sign, and so we may apply the mean value theorem for integrals (Sec. 4.7) and get

$$\frac{f^{(2N)}(\theta^*)}{(2N)!} \int_a^b K(x)\pi^2(x)\,dx$$

But the integral is the same as

$$\int_a^b K(x)\pi^2(x)\,dx = E_{2N} \neq 0$$

since all the lower powers of x do not contribute to the error.   Thus we have for the error of Gauss' quadrature, when $K(x) \geq 0$,

$$\frac{E_{2N}}{(2N)!} f^{(2N)}(\theta)$$

We earlier proved (Sec. 16.7) that if the form of the error was as indicated, then the $G(s)$ must be of constant sign.

All this reasoning has been based on the assumption that in the interval, $K(x)$ does not change sign.   If it does change sign, then all kinds of pathological cases can arise.   The extreme case is perhaps the kernel

$$K(x) = e^{-x^{1/4}} \sin x^{1/4}$$

For the integral

$$\int_0^\infty e^{-x^{1/4}} \sin x^{1/4} f(x)\,dx$$

all the moments

$$m_k = \int_0^\infty e^{-x^{1/4}} \sin(x^{1/4}) x^k\,dx = 0$$

by Sec. 13.7.   This shows that the defining equations are satisfied if all the weights are zero, and it gives rise to the formula

$$\int_0^\infty e^{-x^{1/4}} \sin x^{1/4} f(x)\,dx = 0$$

which is true for all polynomials (of finite degree).   But for the function

$$f(x) = \sin x^{1/4}$$

the integral clearly has a positive value.   Fortunately in practice the kernel $K(x)$ of the integral is usually of constant sign, so that the above analysis is relevant and the particular example is not.

PROBLEMS

19.5.1 Discuss the kernel

$$K(x) = e^{-x^{1/4}}[1 + \sin x^{1/4}] \qquad 0 \le x \le \infty$$

which does not change sign. See [27].

19.6 THREE SPECIAL CASES

There are three important special cases for which the sample points x_i and the corresponding weights w_i are known in a closed, analytical form. For these cases there is no necessity for reading in these numbers from some other source—they can be programmed in. These cases are:

$$\int_{-1}^{1} \frac{f(x)}{\sqrt{1-x^2}}\,dx = \frac{\pi}{N}\sum_{k=1}^{N} f\left(\cos\frac{2k-1}{2N}\pi\right) + \frac{\pi}{2^{2N-1}}\frac{f^{(2N)}(\theta)}{(2N)!}$$

$$\int_{-1}^{1} \sqrt{1-x^2}f(x)\,dx = \frac{\pi}{N+1}\sum_{k=1}^{N}\sin^2\frac{\pi k}{N+1} f\left(\cos\frac{k\pi}{N+1}\right) + \frac{\pi}{2^{2N}}\frac{f^{(2N)}(\theta)}{(2N)!}$$

$$\int_{-1}^{1} \sqrt{\frac{1-x}{1+x}}f(x)\,dx = \frac{4\pi}{2N+1}\sum_{k=1}^{N}\sin^2\frac{k\pi}{2N+1} f\left(\cos\frac{2k\pi}{2N+1}\right) + \frac{\pi}{2^{2N}}\frac{f^{(2N)}(0)}{(2N)!}$$

We shall derive the first case [30]. The defining equations are:

$$\int_{-1}^{1} \frac{x^m}{\sqrt{1-x^2}}\,dx = \sum_{j=1}^{N} w_j x_j^m \qquad m = 0, 1, \ldots, 2N-1$$

Setting

$$x = \cos\theta \qquad x_i = \cos\theta_i$$

then

$$\int_{0}^{\pi} \cos^m\theta\,d\theta = \sum_{j=1}^{N} w_j \cos^m\theta_j$$

We know

$$\cos N\theta = \text{polynomial of degree } N \text{ in } \cos\theta$$

$$= \text{a linear combination of } (\cos\theta)^m$$

Taking exactly that same linear combination of the first $N+1$ defining equations, we get

$$\int_{0}^{\pi} \cos N\theta\,d\theta = \sum_{i=1}^{N} w_i \cos N\theta_i$$

With small modifications of this trick we can get the $2N$ equations

$$\int_0^\pi \cos^m \theta \, d\theta = \sum_{j=1}^N w_j \cos^m \theta_j$$

$$\int_0^\pi \cos N\theta \cos^m \theta \, d\theta = \sum_{j=1}^N w_j \cos N\theta_j \cos^m \theta_j \qquad m = 0, 1, \ldots, N-1$$

which are equivalent to the original $2N$ equations.

Since $\cos m\theta$ equals the polynomial in $\cos \theta$, we can similarly convert these equations to the equivalent set (see Sec. 26.2):

$$\int_0^\pi \cos m\theta \, d\theta = 0 = \sum_{j=1}^N w_j \cos m\theta_j$$

$$\int_0^\pi \cos N\theta \cos m\theta \, d\theta = 0 = \sum_{j=1}^N w_j \cos N\theta_j \cos m\theta_j$$

The second N equations will be satisfied if

$$\cos N\theta_j = 0$$

or

$$\theta_j = \frac{2j-1}{2N} \pi \qquad j = 1, 2, \ldots, N$$

From the orthogonality of the cosines over the equally spaced θ_j (Prob. 11.1.3), we see that the first N equations will be satisfied if $w_j \equiv w$ and from the first equation that

$$w = \frac{\pi}{N}$$

Thus we have our formula.

Similar, more elaborate trigonometry will derive the other two formulas, although there are other, perhaps preferable, ways to find them.

PROBLEMS

19.6.1   Derive the second formula.
19.6.2   Derive the third formula.

19.7   GIVEN SOME SAMPLE POINTS

It often happens that the values of the function are known at one or both ends of the interval of integration.   Formulas which take advantage of this are known as Radau and Lobatto integrations, respectively.

The effect of fixing a sample point is to decrease the number of defining equations by 1. Let us examine the case of one endpoint, say at $x = a$. For the formula

$$\int_a^b K(x)f(x)\,dx = w_a f(a) + \sum_{j=1}^{N-1} w_j f(x_j)$$

we have the defining equations

$$m_k = w_a a^k + \sum_{j=1}^{N-1} w_j x_j^k \qquad k = 0, 1, \ldots, 2N - 2$$

If we multiply the kth equation ($k = 0, 1, \ldots, 2N - 3$) by a and subtract from the $(k + 1)$st equation, we get

$$m_{k+1} - a m_k = \sum_{j=1}^{N-1} w_j(a - x_j)x_j^k \qquad k = 1, 0, \ldots, 2N - 3$$

Set

$$m_{k+1} - a m_k = M_k$$
$$w_j(a - x_j) \quad = W_j$$

and we have, in effect, eliminated the known point and are reduced to the pure Gaussian case of all unknown points. Evidently we can iterate the process and eliminate another given sample point each time until all we have left are the unknown sample points. We then find the unknown x_i. With these we are then back to the earlier case of all sample points being known.

Again we may ask, How does it go in practice? Will the sample points be real, distinct, and in the interval? For the two cases, a sample point at one end of the interval (Radau) or a sample point at each end (Lobatto) and for $K(x) \geq 0$ ($a \leq x \leq b$), we can carry out all the steps. Corresponding to the analysis in Sec. 19.4 where we used $\pi^2(x)$, we now use

$$(x - a)\pi^2(x)$$

[where $\pi(x)$ is the product over the unknown sample points] in the case of one endpoint, and

$$(x - a)\pi^2(x)(b - x)$$

in the case of two endpoints. Thus if we can find the c_i, then the zeros of the polynomial are real, distinct, and in the interval $a \leq x \leq b$. The analysis that we can solve for the c_i is a bit messy, but follows readily from Sec. 19.4.

PROBLEMS

Find (where only endpoints are given):

19.7.1 $\int_{-1}^{1} f(x)\,dx = \frac{1}{6}[f(-1) + f(1) + 5f(-\sqrt{1/5}) + 5f(\sqrt{1/5})]$ $E_6 = -2/3^3 \cdot 5^3 \cdot 7$

19.7.2 $\int_{0}^{\infty} e^{-x} f(x)\,dx = \frac{1}{2}[f(0) + f(2)]$

19.7.3 $\int_{0}^{\infty} e^{-x} f(x)\,dx = w_0 f(0) + w_1 f(x_1) + w_2 f(x_2)$

19.7.4 $-\int_{0}^{1} f(x) \ln x\,dx = w_0 f(0) + w_1 f(1) + w_2 f(x_2)$

19.7.5 $\int_{0}^{\infty} e^{-x} f(x)\,dx = w_0 f(0) + w_1 f(x_1) + w_2 f(4) = \frac{1}{12}[3f(0) + 8f(1) + f(4)]$

19.7.6   Carry out the details of the analysis in the case of Lobatto integration.

19.8   CHEBYSHEV INTEGRATION

It sometimes happens that we are either given or would like to put linear restraints on the weights of a formula.   Such restraints are likely to involve the weights associated with the known sample points, since they are not easily applied to positions that have not even been specified in advance.   The main exception is the important case of the same restraint applied to all the weights of the sample.   The reason that this restraint arises is that this is the one that minimizes the effect of the noise in the samples; all other sets have some increase in the variance of the weights since the sum of the weights is given by the first moment m_0.   The proof goes as follows.   To minimize

$$\sum_{i=1}^{N} w_i^2$$

subject to the restraint (the first defining equation) that

$$\sum_{i=1}^{N} w_i = m_0$$

we use Lagrange multipliers and minimize the function of the weights

$$F(w_i) \equiv \sum_{i=1}^{N} w_i^2 - \lambda \sum_{i=1}^{N} w_i$$

We get

$$\frac{\partial F}{\partial w_j} = 2w_i - \lambda = 0$$

$$w_i = \frac{2}{\lambda} = \frac{m_0}{N} \qquad \text{for all } i$$

When all the weights are equal, the formula is called "Chebyshev."

The best-known Chebyshev case is $K(x) = 1$ with the range $-1 \leq x \leq 1$. The $w_i = w = 2/N$.   The defining equations are:

$$m_k = w \sum_{i=1}^{N} x_i^{\,k} \qquad k = 0, 1, \ldots, N$$

or

$$\sum_{i=1}^{N} x_i^{\,k} = \frac{m_k}{w}$$

The sums

$$s_k = \sum_{i=1}^{N} x_i^{\,k} = \frac{m_k}{w}$$

are symmetric functions in the x_i and hence by the fundamental theorem [60] of symmetric functions theory are expressible in terms of the elementary symmetric functions

$$p_1 = -\sum x_1$$
$$p_2 = \sum x_1 x_2$$
$$\cdots\cdots\cdots\cdots\cdots\cdots\cdots$$
$$p_N = (-1)^N x_1 x_2 \cdots x_N$$

Indeed, the Newton identities

$$s_1 + p_1 = 0$$
$$s_2 + p_1 s_1 + 2p_2 = 0$$
$$s_3 + p_1 s_2 + p_2 s_1 + 3p_3 = 0$$
$$\cdots\cdots\cdots\cdots\cdots\cdots\quad\cdots\cdots$$
$$s_N + p_1 s_{N-1} + \cdots + N p_N = 0$$

express exactly this relationship.

Thus, since the p_i are our c_{N-i}, we can convert these defining equations to equations in terms of the c_i.   It is known that these equations can be solved for real x_i only for the cases[1] of $N = 1, 2, \ldots, 7, 9$; all other cases give complex x_i.

But let us be reasonable about this; the complex zeros will occur, in this and other cases, only as conjugate pairs (we are considering integrating only real functions), and the weights will be such that the sum is real.   Therefore, for the labor of finding one complex function value we get the accuracy equivalent, in some sense, to that of two real points.   And why, given a mathematically expressed

---

[1] For tables to 10 decimal places see National Bureau of Standards, "Handbook of Mathematical Functions," *op. cit.*

integrand, should we refuse to use complex sample points? An experiment integrating a function with a known analytic answer for the cases $N = 1, 2, \ldots, 12$ showed that the errors fell on a smooth curve, as expected.

The Chebyshev case

$$K(x) = \frac{1}{\sqrt{1 - x^2}}$$

is given in Sec. 19.6 and can be solved for all N with real samples—indeed it is both a Gaussian and a Chebyshev formula at the same time!

19.9   RALSTON INTEGRATION

A. Ralston has published[1] an interesting example of a mixed type of integration in which two sample points and one linear restraint are assigned. Specifically, the formula

$$\int_{-1}^{1} f(x) \, dx = \sum_{i=1}^{n} w_i f(x_i)$$

has the conditions

$$x_1 = -1$$
$$x_n = 1$$
$$w_1 = -w_n$$

We have, therefore, $2N - 3$ parameters and can make the formula exact for $1, x, \ldots, x^{2N-4}$. The defining equations are

$$2 = \sum_{i=2}^{n-1} w_i$$
$$0 = -2w_1 + \sum w_i x_i$$
$$2/3 = \sum w_i x_i^2$$
$$0 = -2w_1 + \sum w_i x_i^3$$
$$\cdots\cdots\cdots\cdots\cdots\cdots\cdots\cdots\cdots\cdots\cdots\cdots$$
$$\frac{2}{2N - 3} = \sum w_i x_i^{2N-4}$$

In order to follow the subsequent algebra more easily and at the same time not oversimplify it, we take the case of $N = 5$ which gives $2N - 3 = 7$ equations.

[1] A. Ralston, *J. Assoc. Computing Machinery*, vol. 6, pp. 384–394, July, 1959.

The sample polynomial is

$$\pi(x) = (x - x_2)(x - x_3)(x - x_4) = x^3 + c_2 x^2 + c_1 x + c_0$$

The usual elimination process leads to $N - 1$ (in this case, four) equations:

$$c_0 \;\; + \tfrac{1}{3}c_2 = \;\; -w_1 c_1 \qquad -w_1$$
$$\tfrac{1}{3}c_1 + \tfrac{1}{5} \;\; = -w_1 c_0 \qquad -w_1 c_2$$
$$\tfrac{1}{3}c_0 \;\; + \tfrac{1}{5}c_2 = \;\; -w_1 c_1 \qquad -w_1$$
$$\tfrac{1}{5}c_1 + \tfrac{1}{7} \;\; = -w_1 c_0 \qquad -w_1 c_2$$

It is easy to see that we can form $N - 3\,(= 2)$ equations which do not involve the quadratic terms $w_i c_i (i = 0, 1, 2)$,

$$\tfrac{2}{3}c_0 + \tfrac{2}{15}c_2 = 0$$
$$\tfrac{2}{15}c_1 + \tfrac{2}{35} = 0$$

We have, therefore, $N - 3$ linear, nonhomogeneous equations in the unknowns $c_0, c_1, \ldots, c_{N-3}$, or one more unknown than equations. Solving the equations in terms of, say, c_0, we have

$$c_1 = -3/7$$
$$c_2 = -5c_0$$

These are put in the first two quadratic equations of the set to give

$$-\tfrac{1}{3}c_0 = -\tfrac{2}{7}w_1$$
$$\tfrac{1}{35} = \;\; 2c_0 w_1$$

Thus in the general case, as well as in the specific case of $N = 5$, when we eliminate c_0, we come to a quadratic in w_1. In the case of $N = 5$,

$$w_1 = \pm \sqrt{1/60}$$

The plus or minus sign has the effect of reversing the formula from left to right or right to left.

Knowing w_1, we can easily find the c_i and the sample polynomial

$$\pi(x) = x^3 - 5(\tfrac{3}{245})x^2 - \tfrac{3}{7}x + \tfrac{3}{245}$$

Reasoning similar to that of Sec. 19.4 shows that the zeros are real, distinct, and in the interval of integration.

PROBLEMS

19.9.1 Show that the Ralston case of $N = 6$ has the sample polynomial

$$\pi(x) = x^4 + \frac{4x^3}{3\sqrt{6}} - \frac{2}{3}x^2 - \frac{4x}{7\sqrt{6}} + \frac{1}{21}$$

19.10   GAUSSIAN INTEGRATION USING DERIVATIVES

It is natural to turn to Gaussian integration using both function and derivative values.   If only the function and first-derivative values are used, then the sample points often (not always) turn out to be complex; but if the function and the first and second derivatives are used, then much of the analytical part of Sec. 19.4 can be carried through using $\pi^4(x)$ in place of $\pi^2(x)$.

Rather than examine the messy general case, let us examine the particular case

$$\int_{-1}^{1} f(x)\, dx = w_1 f(-x_1) + w_0 f(0) + w_1 f(x_1)$$
$$- w_1' f'(-x_1) + w_1' f'(x_1)$$
$$+ w_1'' f''(-x_1) + w_0'' f''(0) + w_1'' f''(1)$$

where we have used symmetry to reduce the complexity of the algebra and have made it exact for *all* odd powers of x.

The defining equations are, on dividing by 2,

1:	$1 = w_1$	$+ 1/2 w_0$		0
x^2:	$1/3 = w_1 x_1^2$	$+ 2w_1' x_1 + 2w_1''$	$+ w_0''$	0
x^4:	$1/5 = w_1 x_1^4$	$+ 4w_1' x_1^3 + 12 w_1'' x_1^2$		$+ x_1^6$
x^6:	$1/7 = w_1 x_1^6$	$+ 6w_1' x_1^5 + 30 w_1'' x_1^4$		$- 3x_1^4$
x^8:	$1/9 = w_1 x_1^8$	$+ 8w_1' x_1^7 + 56 w_1'' x_1^6$		$+ 3x_1^2$
x^{10}:	$1/11 = w_1 w_1^{10}$	$+ 10 w_1' x_1^9 + 90 w_1'' x_1^8$		-1

Multiplying these equations by the constants on the right and adding, we get (dropping the subscripts on the x)

$$\frac{x^6}{5} - \frac{3}{7} x^4 + \frac{3}{9} x^2 - \frac{1}{11} = 0$$

Regarding this as a function of x,

$$f(x) \equiv \frac{x^6}{5} - \frac{3}{7} x^4 + \frac{3}{9} x^2 - \frac{1}{11} = \text{a cubic in } x^2$$

we see that $f(0) = -1/11$ and $f(1) = 8(1/33 - 1/35) > 0$, and hence there is a real x that is the solution to our problem of finding a sample point.   The rest now follows easily.

PROBLEMS

19.10.1   Carry out the details of the analysis mentioned in the text.
19.10.2   Complete the above special case.
19.10.3   Discuss the case where some points use $f(x)$, some $f(x)$, $f'(x)$, $f''(x)$, and some f, f', f'', f''', f^{iv}.
19.10.4   Show that

$$\int_{-1}^{1} f(x)\,dx = w_1 f(x_1) + w_2 f(x_2) + w_1' f'(x_1) + w_2' f'(x_2)$$

leads to complex sample points x_i.
19.10.5   Derive the two-point case using f, f', and f'' for $\int_{-1}^{1} f(x)\,dx$.

19.11   AN ALGORITHMIC APPROACH TO FINDING FORMULAS

Computer science emphasizes the algorithmic approach to various fields, and we shall adopt it for the problem of finding formulas.   In order not to obscure matters too much we shall omit a lot of the trivial details that are easily supplied (especially exits from the flow diagram when various steps fail to work).

The formulas we are considering are of the form of a linear operator which is being estimated by a linear combination of samples of the function, and possibly one or more derivatives, using weights w_i, w_i', w_i''. . . .   We have examined the problem mainly in terms of the operation of integration over a definite interval, but much of the material applies to *any* linear operator.

To fix notation, let

n_1 = the number of weights w_i, w_i', w_i'', . . . to be determined
n_2 = the number of sample points x_i that are given
n_3 = the number of sample points x_i to be found
n_4 = the number of linear restraints on the weights
$N = n_2 + n_3$

We are finding formulas in the classical sense of making them correct for successive powers of x, that is, putting in 1, x, x^2, . . . as far as we can to form the defining equations, Sec. 15.5.   How far can we go?   We have $n_1 + n_3 - n_4$ free parameters and can therefore expect that the same number of defining equations will *usually* determine all the parameters.   We will, therefore, need the same number of moments m_i of the linear operator, and later we will need more m_i, both for the possibility of higher accuracy (as, for example, in Simpson's formula, Sec. 15.3) and for the error term, Chap. 16.

Before writing down the defining equation, box 1 in Fig. 19.11.1, it is well to look for symmetry about some point $x = a$, either even or odd symmetry (Sec. 17.5, Example 17.3, and Sec. 19.2). If there is symmetry, then using powers of $x - a$, instead of powers of x, will simplify the problem (though it should be realized that anything that can be done along these lines can be done later by manipulating the equations properly). When we use the symmetry we have:

Even symmetry	Odd symmetry
$m_{2k+1} = 0$ all k	$m_{2k} = 0$ all k
$x_i - a = -(x_{N-i} - a)$	$x_i - a = -(x_{N-i} - a)$
$w_i = w_{N-i}$	$w_i = -w_{N-i}$
$w_i' = -w_{N-i}'$	$w_i' = w_{N-i}'$
$w_i'' = w_{N-i}''$	$w_i'' = -w_{N-i}''$
. . . .	. . . .

These choices make the corresponding defining equations exact for *all* values of the index k, so that we need only write the defining equations for

$$x^{2k} \quad \text{or} \quad x^{2k+1}$$

Thus there will be about half as many defining equations to be solved and about one-eighth as much algebraic manipulation to solve them.

To the defining equations we append the n_4 equations giving the linear restraints on the weights. The combined set of $n_1 + n_3$ equations are to be solved for both the x_i that are not known and for the weights w_i, w_i', w'', ...

If $n_3 = 0$ (all sample points are known), then we solve these equations (box 2 in Fig. 19.11.1). There are three approaches possible:

1 Direct elimination—the hard way.

2 Solve for some of the weights using the sampling polynomials π_i, σ_i, τ_i, etc., to obtain them directly, Secs. 15.5, 17.4, and 17.6.

3 In the Birkhoff cases we can construct the inverse matrix as combinations of sums and products of the x_i (Secs. 14.5, 14.6, 17.4, 17.6).

These equations can be solved when the system has Birkhoff data (consecutive derivatives) (Sec. 17.6). In other cases we are not certain that they can be solved; indeed we know from the example of Sec. 17.7 that they can be inconsistent or that they can admit of a one-parameter (or more) family of solutions. In the later case we can add further defining equations and hope for consistency when the rank rises to full order (Sec. 17.7).

When we have the formula, we next turn to finding the error term (box 3 of Fig. 19.11.1) which requires additional moments m_i. We first find the value of m such that $E_m \neq 0$. Note that the value of E_m can be found *directly*, without finding the weights, as a linear combination of the moments m_i. The coefficients of this linear combination are polynomials in the x_i only (Sec. 17.8). We then construct the $G(s)$, and if it is of constant sign, we have the structure of the error

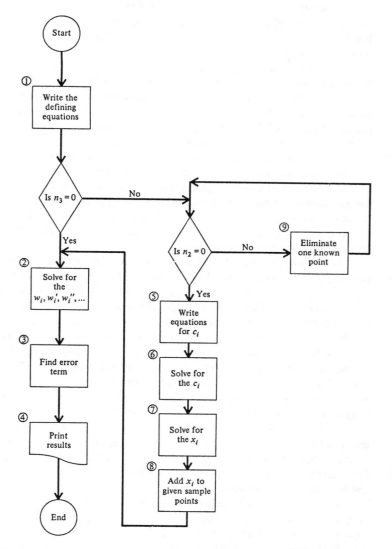

FIGURE 19.11.1

term (Sec. 16.5), while if it is not of constant sign, we turn to Sec. 16.7 for guidance. For some Gaussian cases we can use the interpolation formula's error term instead (Sec. 19.5).

Finally it is necessary to organize our scattered results and print out the formula, or why we did not get it (box 4 in Fig. 19.11.1).

If $n_3 \neq 0$ but $n_2 = 0$, we have the pure Gaussian case of all unknown sample points.   For the case of only function values we first construct the equations for the coefficients c_i of the polynomial $\pi(x)$ whose zeros are the unknown sample points.   This we can always do (box 5 in Fig. 19.11.1).

To solve these equations for the c_i is another matter (box 6 in Fig. 19.11.1). We need to consider the rank of the equations and what to do if this is less than full rank.   Adding more defining equations is one approach, provided inconsistency has not shown that such a formula cannot exist.   For the case of $K(x) \geq 0$, integration with no constraints on the weights, and at most only endpoints as assigned samples, we know that the equations are of full rank (Sec. 19.4).   The general case using derivatives is obscure.

In box 7 of Fig. 19.11.1, we solve this polynomial for the unknown sample points x_i.   Again we are assured that in the case of integration with $K(x) \geq 0$, and at most assigned sample points at the ends of the interval, we will find real, distinct zeros in the interval of integration.   This can happen in other cases too, and it can fail to happen.   But the conventional view that complex zeros or multiple zeros are always fatal should be resisted, and the question judged on its own merits.   It is easy to show that in the case of complex zeros only the corresponding weights can be complex.   The complex sample points give rise to a real quadratic factor in the $\pi(x)$ and most of the $\pi_i(x)$; and almost all the usual proofs go through as before, with, at most, minor modification.   If multiple roots occur, then thought should be given to using a one-higher derivative at that place.

In box 8 of Fig. 19.11.1, we add these new sample points to the old ones (if any) and look for duplicates.   If there are none, then using a suitable number of defining equations, we carry out box 2 (Fig. 19.11.1) and know that the other defining equations will automatically be satisfied.   Duplicate points are unlikely and have not been discussed.

If $n_3 \neq 0$, then $n_2 \neq 0$, and we go to box 9 (Fig. 19.11.1) and eliminate a given sample point (Sec. 19.7).   Going around the loop enough times, we end up with all unknown sample points x_i and go to box 5.

We have ignored many details, and much research can be done to increase the number of circumstances under which we can be sure that the method will go through without any problems in going from the defining equations to the formula.   With modern symbol-manipulating routines available, this process could be put on a machine.

COMPOSITE FORMULAS

20.1 INTRODUCTION

The previous Chaps. 14 to 19 examined the process of approximating a linear operator as a linear combination of samples of the function, and possibly some its derivatives, and used integration as a typical linear operator. These chapters concentrated on *methods*, as the title of the book suggests, and did not attempt to apply the resulting formulas to specific functions. The present chapter looks a little closer at the actual application of the methods to the practical problem of integration, although we shall also look once more at the problem of interpolation. In both cases we will look at how formulas can be combined (composite formulas) to achieve the desired result.

Books on numerical methods often give many examples of formulas applied to specific functions. We do not intend to do this for a number of reasons. First, it supposes that the reader cannot evaluate a function, and possibly some of its derivatives, at a given set of sample points and then compute a weighted sum of these results—presumably drill in such matters is not necessary at this late stage in the reader's mathematical development. Second, for a given formula for integration it is easy to find a function for which it will give a very good result, while at the same time a "better" formula will give a worse result. Formulas are *not* to be judged on how well they do on a specific function, but rather how well they do over some class of functions. We cannot use the *average* error over any reasonable ensemble of functions as a measure because, among other reasons, the ensemble is very likely to have both $f(x)$ and $-f(x)$ in it with equal probability, so that the average error will automatically be zero—we probably want to use the variance over the ensemble if only we could agree on one or more suitable ensembles of functions! In the absence of such tests, we are forced to use the size of the coefficient E_m in the error term as a guide for comparing formulas having the same-order error term (the same m). As noted earlier, we are not certain about the effect of the mean value theorem in placing the θ value more favorably in one formula than in another, and so our choice is not certain, only plausible. Third, with large-scale computation it is more and more true that the user does not get to look at the results—they are buried deep in some inner loop and at best he sees the results of a few early test cases. But worse than this, he is often forced to pick the method *before* any real data are available and later can modify his choice only slightly. Commitments in large projects must be made in advance of definite knowledge, and at much later times cannot be broken easily without jeopardizing the whole project. Thus it is necessary to train numerical analysts to know in advance what will probably happen. Of course tests can and should often be made of specific cases. Of course somewhat similar problems can be tried before making the final choice, but many times there is no real hope of knowing more than a general description of what the data will be like when they become available.

For all these reasons we will stick to the somewhat abstract approach of trying to reason what will happen in practice rather than regularly appealing to specific examples—this approach is necessary in order to get training in this difficult art of being right even when you are not sure what the problem is. This puts a burden on testing whatever seems, at the moment, to be relevant material whenever you can. Unfortunately the design of experiments lies outside the course, and the art of guessing is very difficult to teach.

20.2 POLYNOMIAL APPROXIMATION AGAIN

By now it should be very clear that what we are doing by our formulas is equivalent (with a very few exceptions) to first finding an interpolating polynomial and then applying a linear operator to this interpolating polynomial—we called it analytic substitution. We need, therefore, to think a bit about what kinds of functions can be approximated well by low-order polynomials, as well as the chief characteristics of high-order polynomials.

First, a polynomial is finite, along with all orders of its derivatives, at all finite points. This means that integrals like

$$\int_0^h \sqrt{x} e^{-x} \, dx \qquad \int_0^h \frac{\arctan x}{\sqrt{x}} \, dx \qquad -\int_0^h \ln x \arctan x \, dx$$

usually are not going to be very accurately done, except by chance or by h being small. One frequently sees proof in the literature that if a sufficiently small spacing is used, then some particular composite formula will give accurate results (in the limit), but it seems foolish to try to imitate the fundamental definition of integration by adding enough small intervals to get the sum accurately.

Second, polynomials go to infinity as the argument goes to both $+\infty$ and $-\infty$. Thus integrals of the form

$$\int_0^\infty \frac{dx}{1 + x^2} \qquad \int_e^\infty \frac{dx}{x^2 \ln x} \qquad \int_0^\infty e^{-x} f(x) \, dx$$

are apt to be troublesome, unless the kernel smothers the growth at infinity, as is likely to be true in the third example.

Third, polynomials love to wiggle, especially high-order ones. But for integration with a constant kernel $K(x) = 1$, the wiggles are apt to be more awful-looking than they are in practice. There is a tendency for the successive loops to cancel out more than might seem reasonable at first sight. For example, in Simpson's formula applied to a cubic the area between the interpolating quadratic and the cubic had a net area of zero (Sec. 15.3).

The problem, then, is to discover the nature of any nonpolynomial behavior and to incorporate this into the kernel $K(x)$; it is foolish to do otherwise. A very high percentage of the problems that arise in practice will have some peculiar behavior at one or both ends of the interval, but only occasionally will there be trouble inside the range (being bounded at infinity is real trouble for a polynomial). It is because such troubles arise so often that we have given all the material for developing formulas so that an appropriate one can be found for almost any kind of difficulty you face. Only the most common cases, or perhaps the easiest ones to discuss, have been reported in the literature, and the others are awaiting you when you start practicing the difficult art of practical computing. Thus you will occasionally have to develop your own formulas to fit the peculiar needs of your current work.

20.3   THE NEWTON-COTES FORMULAS

The most obvious family of formulas, and the most commonly discussed, but not used in practice, is the Newton-Cotes formulas for integrating a function, given a set of equally spaced sample points (including the endpoints).   The trapezoid rule (Sec. 16.5)

$$\int_0^h f(x)\, dx = \frac{h}{2}\,[f(0) + f(h)] - \frac{1}{12}\,h^3 f''(\theta)$$

Simpson's formula (Sec.16.5),

$$\int_0^{2h} f(x)\, dx = \frac{h}{3}\,[f(0) + 4f(h) + f(2h)] - \frac{1}{90}\,h^5 f^{iv}(\theta)$$

and the three-eighths rule (Prob. 15.4.4 in disguise)

$$\int_0^{3h} f(x)\, dx = \frac{3h}{8}\,[f(0) + 3f(h) + 3f(2h) + f(3h)] - \frac{3}{80}\,h^5 f^{iv}(\theta)$$

are the first three members of the family.   Table 20.3 gives the coefficients for the general form

$$\int_0^{nh} f(x)\, dx = Ah[B_0 f(0) + B_1 f(h) + B_2 f(2h) + \cdots + B_n f(nh)] + R_n$$

where n is the number of intervals, *not* the number of sample points as we have been doing.   Evidently there are $n + 1$ sample points.   Since the coefficients are symmetric, we need not give all the table.

Some of the coefficients become negative for $n = 8$.   For $n = 9$ they are all positive, but for $n \geq 10$ there are negative coefficients.   This tends to produce poor roundoff properties, and as a result the higher-order Newton-Cotes formulas are seldom used.   The order of the error terms jumps by 2 in going from an odd number to the next even number, which tends to favor the even-order formulas.

Notice the appearance of the powers of h in the error term.   When we use other than unit spacing, the powers of h automatically appear in the defining equations.   We may either compute with the h values present or else pretend we have unit spacing and supply the proper power of h later (for equally spaced samples); the latter is usually preferable.

If these formulas are to be compared for the same interval, then it must be remembered that h = range of integration/n, and the corresponding adjustments must be made in the error terms

The Newton-Cotes formulas can be derived in many ways.   Perhaps the simplest way to find the actual coefficients is to observe that the Gregory formula, when written to include all the differences that can be computed from the sample

Table 20.3

n	A	B_0	B_1	B_2	B_3	B_4	B_5	R_n
1	$\frac{1}{2}$	1	1					$-(1/12)h^3f''(\theta)$
2	$\frac{1}{3}$	1	4	1				$-(1/90)h^5f^{IV}(\theta)$
3	$\frac{3}{8}$	1	3	3	1			$-(3/80)h^5f^{IV}(\theta)$
4	$\frac{2}{45}$	7	32	12	32	7		$-(8/945)h^7f^{VI}(\theta)$
5	$\frac{5}{288}$	19	75	50	50	75	19	$-(275/12,096)h^7f^{VI}(\theta)$
6	$\frac{1}{140}$	41	216	27	272	27	216	$-(9/1,400)h^9f^{VIII}(\theta)$
7	$7/17,280$	751	3,577	1,323	2,989	2,989	1,323	$-(8,183/518,400)\,h^9f^{VIII}(\theta)$
8	$4/14,175$	989	5,888	−928	10,496	−4,540	10,496	
9	$9/89,600$	2,857	15,741	1,080	19,344	5,778	5,778	
10	$5/299,376$	16,067	106,300	−48,525	272,400	−260,550	427,368	

points, is exact for polynomials of the maximum degree, and hence when written in the Lagrange form, must be the same as if it were derived directly. Thus for the case of $n = 4$ we have

$$\int_0^4 f(x)\, dx = \tfrac{1}{2}f_0 + f_1 + f_2 + f_3 + \tfrac{1}{2}f_4 + \tfrac{1}{12}(\Delta f_0 - \Delta f_3)$$

$$\tfrac{1}{24}(\Delta^2 f_0 + \Delta^2 f_2) + \tfrac{19}{720}(\Delta^3 f_0 - \Delta^3 f_1)$$

$$-\tfrac{3}{160}(\Delta^4 f_0 + \Delta^4 f_0) = \tfrac{14}{45}f_0 + \tfrac{64}{45}f_1 + \tfrac{24}{45}f_2 + \tfrac{64}{45}f_3 + \tfrac{14}{45}f_4$$

We have not derived the error terms [by showing that the influence function $G(s)$ is of constant sign] because as we said earlier, the formulas beyond Simpson's are seldom used.

PROBLEMS

20.3.1   Apply the $n = 2, 4, 6, 8$ Newton-Cotes cases to $\int_0^1 e^{-x}\, dx$.   Compare with the correct answer.

20.3.2   Discuss the noise amplification of the Newton-Cotes formulas.

20.3.3   Derive the Newton-Cotes formula from the Gregory formula for $n = 6$.

20.4   REMARKS ON SOME FORMULAS

Why are the Newton-Cotes formulas seldom used for high orders? Partly because the roundoff appears to be so bad. If we assume that the roundoff error ε_i is independent at successive points x_i where the function is evaluated and that it has a noise level of σ^2, then the variance of the sum will be

$$\mathrm{Av}\{\textstyle\sum w_i \varepsilon_i^2\} = \sum_i \sum_j w_i w_j \,\mathrm{Av}\{\varepsilon_i \varepsilon_j\}$$

$$= (\textstyle\sum w_i^2)\sigma^2$$

since

$$\mathrm{Av}\{\varepsilon_i \varepsilon_j\} = 0 \qquad i \neq j$$

But this effect should not be exaggerated, though it is serious when the weights have different signs since their sum is fixed in advance at m_0.

The Gaussian formulas require peculiar weights and sample points, and this needlessly discourages people from using them; however the weights being all positive for $K(x) \geq 0$ is encouraging. The Chebyshev formulas do not seem to be worth the trouble of the awkward sample points *except* when they are given by analytical formulas. Thus we are forced to look more closely at the composite formulas that were briefly mentioned in Sec. 18.9.

However, the Gaussian formulas are often very useful. Among other places they are often used is in solving integral equations. For example, given the integral equation

$$y(x) = f(x) + \int_{-1}^{1} K(x,s)y(s)\ ds$$

we apply Gaussian quadrature and are led to a system of simultaneous linear equations:

$$y(x_i) = f(x_i) + \sum_{j=1}^{N} w_j K(x_i, s_j)y(s_j) \qquad i = 1, 2, \ldots, N$$

Since the cost of solving linear equations is proportional to N^3, the use of Gaussian quadrature, which reduces the number of equations by a factor of 2 for the same order of truncation error, is clearly worth the awkwardness of the peculiar sample points and weights.

20.5 COMPOSITE FORMULAS

Since high-order formulas are not often used, composite formulas must be the major formulas in practice. In the case of the Hamming-Pinkham formulas the error term is given by the sum of *all* the neglected differences and in practice is usually not far from the first neglected difference.

In the case of composite formulas based directly on some formula, say Simpson's, we find that the error term is the sum of derivatives, and we need the following theorem.

Theorem 20.5.1 If $g(x)$ is continuous and the $c_i \geq 0$, then for some value θ in the interval of all the arguments

$$g(\theta) \sum c_i = \sum_{i=1}^{N} c_i g(\theta_i)$$

PROOF The proof follows by induction from the simple case of two Consider

$$\phi(\theta) \equiv c_1 g(\theta_1) + c_2 g(\theta_2) - (c_1 + c_2)g(\theta)$$

In the interval of the θ_i, $g(\theta)$ takes on both a maximum and a minimum. If $g(\theta)$ is not constant, then $\phi(\theta)$ has opposite signs at these two points: hence since it is continuous, it takes on the value 0. The induction follows easily.   ////

This theorem is the discrete analog of the theorem

$$\int_a^b f(x)g(x)\,dx = g(\theta)\int_a^b f(x)\,dx \qquad f(x) \geq 0$$

(Secs. 7.4 and 16.2).   Thus we get for the composite Simpson's formula

$$\int_0^{2nh} f(x)\,dx = \frac{h}{3}\,[f(0) + 4f(h) + 2f(2h) + \cdots + f(2nh)] - \frac{nh^5\,f^{(4)}(\theta)}{90}$$

The h^5 appears in the error term because we have shifted from unit spacing in t to the spacing h in x by the transformation

$$ht = x$$

$$dt = \frac{1}{h}\,dx$$

and

$$\frac{d^n}{dt^n} \equiv \frac{h^n d^n}{dx^n}$$

The error is often written as

$$-\frac{(b-a)h^4 f^{(4)}(\theta)}{45}$$

where $b - a = $ range of integration $= 2nh$.

In formulas like Ralston's, the opposite signs of the function values at the ends clearly cause them to cancel out, except at the ends of the range, and this is their chief merit.

In formulas using derivatives there is a strong tendency to have the odd-order derivatives cancel out.   Problem 17.5.2 derives the formula

$$\int_0^1 f(x)\,dx = \frac{1}{12}\,[6f(0) + 6f(1) + f'(0) - f'(1)] + \frac{f^{(4)}(\theta)}{720}$$

or

$$\int_0^h f(x)\,dx = \frac{h}{12}\,[6f(0) + 6f(h) + f'(0) - f'(h)] + \frac{h^5 f^{(4)}(\theta)}{720}$$

The composite formula is, therefore,

$$\int_0^{nh} f(x)\,dx = \frac{h}{2}\,\{f(0) + 2f(h) + \cdots + 2f[(n-1)h] + f(nh)\}$$
$$+ \frac{h}{12}\,[f'(0) - f'(nh)] + nh\,\frac{h^4 f^{(4)}(\theta)}{720}$$

Thus only the end derivatives are used (this formula can also be found from the Euler-Maclaurin formula in Sec. 12.8).

PROBLEMS

Find the corresponding composite formula from the following problems:

20.5.1 *18.10.4*
20.5.2 *17.4.1*
20.5.3 *17.5.3*
20.5.4 *17.5.4.*

20.6 COMPOSITE OR HIGH-ACCURACY FORMULA?

How shall we choose between the two alternate choices of integration, a low-order composite formula or a single high-accuracy formula? The answer is both very simple and very complex. In principle it is a question of the rate of growth of the higher derivatives of the integrand [ignoring the kernel $K(x)$]. Section 14.7 discussed the growth of the higher derivatives for functions that have singularities in the finite part of the complex plane [recall that $1/(1 + x^2)$ has singularities at $x = \pm i$] and showed that some of the high-order derivatives must behave like $n!$ or worse; otherwise the series would converge everywhere.

Since, except for a few elementary cases, the high-order derivatives are not available in closed form, it is usual to estimate them from a difference table. Of course this has its dangers, especially near a singularity. While we try to incorporate all the singularity we can into the kernel, there is usually some singularity left, not necessarily in the interval of integration, to cause trouble.

There is no magic number, but experience seems to indicate that for moderate engineering and scientific accuracies of, say, five decimal places, fourth- and fifth-order derivatives are about as far as it is wise to go in the absence of knowledge that the higher derivatives are not excessively large. On the other hand, experience with Gaussian integration indicates that good answers can be obtained from high-accuracy formulas. Apparently it is the equal spacing that is not favorable to the use of high-order derivatives in the Newton-Cotes, and similar, formulas.

20.7 GREGORY-TYPE FORMULAS

The trouble in estimating the value of the derivative occurring in the error term leads to sequential methods of estimating the accuracy of a formula. The use of several different spacings and the comparison of the resultant numbers provide a popular way of approaching the problem. One of the more popular methods is to start with a low-order formula with a known error-term structure, usually expressed in the Taylor-series form with many terms being kept, and then using tighter and tighter spacings, build up a table of values of estimates of the integral. The next step is to eliminate pairwise the lowest-order derivative that appears in the error term; thus from n estimates we get $n - 1$ estimates with one of the derivatives eliminated. Then from these we eliminate the next derivative, etc. This is not the same as directly finding the most accurate formula you can, but it tends to avoid oscillation in the weights (the effective weights, once all the elimination is done) that occurs in the Newton-Cotes formulas.

It would seem, however, that the direct use of the Gregory, or Hamming-Pinkham, formula (which was invented for this purpose) would be better, and a number of experiments back up the theory. The approach through the differences allows almost independent estimates of *both* truncation and roundoff error, so that a comparison of the effects of using either a higher-order formula (using more differences) or more points can be made accurately.

The Gregory and Hamming-Pinkham formulas have an additional feature worth noting. When the integrand becomes flat at the ends of the interval, for example,

$$\int_{-\infty}^{\infty} e^{-x^2} \, dx$$

then the differences all approach zero and can be ignored.

Hartree[1] gives Table 20.7 for results obtained when computing

$$I = \int_0^\infty e^{-x^2} \, dx = h\left(1/2 + \sum_{k=1}^{\infty} e^{-h^2 k^2}\right)$$

The remarkable accuracy for $h = 1/2$ is unfortunately accompanied by a lack of knowledge of when it occurs or when h is too large; the error analysis of Gregory's formula is a difficult topic.

Table 20.7   COMPUTATION OF $\int_0^\infty e^{-x^2} \, dx$*

	I			
0.5	0.88622	69254	5	Correct to 11 decimals
0.6		69254	8	
0.7		69285		
0.8	0.88622	72808		
0.9	23	598		
1.0	32	0		
1.1	0.88674			

[1] See Hartree [19].   See also R. A. Fisher, *Phil. Trans. Roy. Soc. London, Ser. A*, 1922, for general h and error theory.

PROBLEMS

20.7.1 Make a flow diagram for the Gregory formula which involves the machine choice of when to stop using more differences. Discuss your reasoning for terminating as you do.

20.8 COMPOSITE INTERPOLATION

For interpolation the problem of whether to use one-high-order polynomial for the whole interval or to use a sequence of lower-order polynomials for various intervals is exactly the same problem as when to use a single high-order integration formula or a different composite formula. The chief difference is that integration is a "smoothing" operation and is less vulnerable to the inevitable wiggles of a polynomial, and so there is more of a tendency to go to composite formulas when interpolating. However, among other difficulties in the interpolation process, when we shift from one interval to the next, we have the situation that there are sharp corners.

Spline interpolation was designed to cope with this problem, and for most practical purposes spline interpolation means cubic spline interpolation. In cubic spline interpolation the function and the first and second derivatives are required to agree at the ends of the intervals where one cubic meets the next. Thus we have a continuous second derivative throughout the whole range of interpolation.

Spline interpolation is useful in many situations which require a smooth interpolating function through given data. Examples include interpolating data to drive automatic milling machines, many draughting applications where "smoothness" is highly desirable, and in computer science, graphical output of smooth curves from given data.

Spline interpolation is a large, well developed field, and we can, per usual, give only a brief introduction [2].

20.9 THE CUBIC SPLINE EQUATIONS[1]

Let the ordinates y_i be given at x_i ($i = 1, \ldots, n$), respectively. Let $h_i = x_{i+1} - x_i$ denote the mesh-point spacing. Let $y(x)$ be an interpolation curve through these points and define y_i' and y_i'' as the first and second derivatives, respectively, of $y(x)$ at $x = x_i$. Let $y(x)$ be expressed in piecewise fashion as

$$y(x) = f_i(x) \begin{cases} \text{for } x_i \leq x < x_{i+1} & i = 1, \ldots, n - 2 \\ \text{for } x_{n-1} \leq x \leq x_n & i = n - 1 \end{cases} \qquad (20.9.1)$$

[1] Much of this material is taken from a Bell Telephone Laboratories memorandum by D. G. Schweikert.

with the following conditions on function value and first and second derivatives:

$$f_i(x_i) = y_i \qquad i = 1, \ldots, n-1$$
$$f_{i-1}(x_i) = y_i \qquad i = 2, \ldots, n \tag{20.9.2}$$

$$f'_{i-1}(x_i) = f'_i(x_i) \qquad i = 2, \ldots n-1 \tag{20.9.3}$$
$$f''_{i-1}(x_i) = f''_i(x_i) \qquad i = 2, \ldots, n-1 \tag{20.9.4}$$

The individual cubic polynomial $f_i(x)$ can be expressed using the interval values y_i, y_{i+1} and either y'_i, y'_{i+1}, or y''_i, y''_{i+1} to represent the cubic coefficients. Both forms yield the same $y(x)$, of course, but since using the mesh-point second-derivative values as unknowns has some minor advantages in most practical computation, the very similar derivation of the slope form is omitted here.

Assuming $y'''(x) = $ constant in each interval means that $y''(x)$ is linear, we get

$$f''_i(x) = y''_i\left(\frac{x_{i+1} - x}{h_i}\right) + y''_{i+1}\left(\frac{x - x_i}{h_i}\right) \tag{20.9.5}$$

Integrating twice more and selecting the constant of integration such that Eqs. (20.9.2) are satisfied yields

$$f_i(x) = y_i\left(\frac{x_{i+1} - x}{h_i}\right) + y_{i+1}\left(\frac{x - x_i}{h_i}\right)$$
$$- \frac{h_i^2}{6} y''_i\left[\frac{x_{i+1} - x}{h_i} - \left(\frac{x_{i+1} - x}{h_i}\right)^3\right]$$
$$- \frac{h_i^2}{6} y''_{i+1}\left[\frac{x - x_i}{h_i} - \left(\frac{x - x_i}{h_i}\right)^3\right] \tag{20.9.6}$$

which identically satisfies the continuity condition [Eq. (20.9.4)] on the second derivative. This form does not, in general, satisfy the continuity conditions on slope [Eq. (20.9.3)]. Expanding the LHS and RHS of Eq. (20.9.3) by differentiating Eq. (20.9.6) and evaluating yields, respectively,

$$f'_i(x_i) = \frac{y_{i+1} - y_i}{h_i} - \frac{h_i}{6}(2y''_i + y''_{i+1}) \tag{20.9.7}$$

$$f'_{i-1}(x_i) = \frac{y_i - y_{i-1}}{h_{i-1}} + \frac{h_{i-1}}{6}(y''_{i-1} + 2y''_i) \tag{20.9.8}$$

which, when equated, produce the condition

$$h_{i-1}y''_{i-1} + 2(h_{i-1} + h_i)y''_i + h_i y''_{i+1} = 6\left(\frac{y_{i+1} - y_i}{h_i} - \frac{y_i - y_{i-1}}{h_{i-1}}\right)$$
$$i = 2, \ldots, n-1 \tag{20.9.9}$$

which must be satisfied at $n - 2$ points by the n unknown quantities y_i''. Two more conditions are required on the y_i'', and these are obtained by specifying one end condition at each end.

Specifying the end second derivatives y_1'' and y_n'' leaves $n - 2$ unknowns, which can be obtained by solving the $n - 2$ equations of Eq. (20.9.9). The coefficient matrix is tridiagonal, and the diagonal term is dominant, which assures us that the matrix is never singular. An example of Eq. (20.9.9) in matrix form is shown for six equally spaced points, where $d_i = (y_{i+1} - 2y_i + y_{i-1})/(2h^2)$:

$$
\begin{pmatrix}
4 & 1 & 0 & 0 \\
1 & 4 & 1 & 0 \\
0 & 1 & 4 & 1 \\
0 & 0 & 1 & 4
\end{pmatrix}
\begin{pmatrix}
y_2'' \\
y_3'' \\
y_4'' \\
y_5''
\end{pmatrix}
=
\begin{pmatrix}
12d_2 - \dfrac{y_1''}{h} \\
12d_3 \\
12d_4 \\
12d_5 - \dfrac{y_6''}{h}
\end{pmatrix}
\tag{20.9.10}
$$

Instead of specifying y_1'', we could specify the end slope y_1'—or a linear combination of the two. If y_1' is specified, then y_1'' is an unknown, and we must find an equation involving both. Evaluating Eq. (20.9.7) for $i = 1$ yields the required equation,

$$
2h_1 y_1'' + h_1 y_2'' = 6\left(\frac{y_2 - y_1}{h_1} - y_1'\right) \tag{20.9.11}
$$

which, when included in the above example, alters Eq. (20.9.10) slightly to the form

$$
\begin{pmatrix}
2 & 1 & 0 & 0 & 0 \\
1 & 4 & 1 & 0 & 0 \\
0 & 1 & 4 & 1 & 0 \\
0 & 0 & 1 & 4 & 1 \\
0 & 0 & 0 & 1 & 4
\end{pmatrix}
\begin{pmatrix}
y_1'' \\
y_2'' \\
y_3'' \\
y_4'' \\
y_5''
\end{pmatrix}
=
\begin{pmatrix}
12d_1 \\
12d_2 \\
12d_3 \\
12d_4 \\
12d_5 - \dfrac{y_6''}{h}
\end{pmatrix}
$$

where $d_1 \equiv (y_2 - y_1 - y_1'h)/(2h^2)$. If y_n' is specified instead of y_n'', an equation similar to Eq. (20.9.11) appears at the bottom of the matrix.

If the appropriate end conditions are not actually known, the simplest choice is $y_1'' = 0$. Another, and smoother, choice is $y_1'' = 0.5y_2''$.

Although the spline is a "global" rather than a "local" curve, i.e., altering a single y_i or end condition affects the spline throughout $[x_1, x_n]$, the dominant diagonal term causes such effects to become small rapidly as the distance from the altered point increases, virtually eliminating roundoff problems. In addition, since the solution of the tridiagonal equations involves βn arithmetic operations, there are no computational difficulties or disadvantages in fitting hundreds of points at a time.

The storage of the spline coefficients y_i, y_i'' $(i = 1, \ldots, n)$ involves $2n$ locations compared with only n locations for the coefficients of an n-point interpolation polynomial.

20.10 COMPARISON WITH POLYNOMIAL INTERPOLATION

The results of the widely used polynomial interpolation are frequently perfectly acceptable, as illustrated in Fig. 20.10.1. Generally speaking, the spline does not have any particular advantages when used for (1) the approximation of a well-behaved, easily calculated mathematical function (unless the derivatives are also important) or (2) the curve-fitting of experimental data which are "dense." The term *dense* will be loosely defined here as meaning that the number of points, in any subregion, is more than an order of magnitude larger than the number of inflection points expected in the fitted curve for that subregion, and that there are no "abrupt" changes in the expected second derivative.

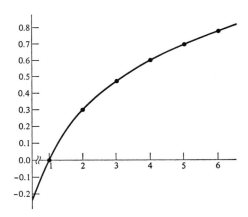

FIGURE 20.10.1
An example of smooth polynomial interpolation.

The spline function's smoothness properties give it significant advantages over polynomials when dealing with "sparse" data; i.e., the number of points is less than an order of magnitude larger than the number of expected inflection points, or there is a large change in the expected second derivative over a small number of points.

One of the polynomial characteristics which causes great difficulty is the manner in which an oscillation caused by one bad point or "difficult" region is reflected, usually undiminished, throughout $[x_1, x_n]$.   Figure 20.10.2 shows such an example.

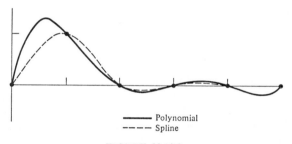

————— Polynomial
– – – – Spline

FIGURE 20.10.2
Comparison of oscillation characteristics.

All but one of the six points are in a straight line, but the solid line corresponding to the interpolating polynomial continues oscillating with an amplitude which would not diminish, even if there were 50 more data points in a line to the right.   The amplitudes of the oscillation for the interpolating cubic spline function (with $y_1'' = y_n'' = 0$), indicated by the dashed line, are reduced by about 1/3 in each successive interval.

Frequently in engineering problems the first or second derivative of the fitted curve is of more interest than the function value.   Figure 20.10.3 is an example of sparse data for which the second derivative was of interest.   Probably largely due to the point at x_5, the interpolating polynomial (Fig. 20.10.3) is unacceptable, giving large excursions of the second derivative and, as we shall discuss later, an unnecessary number of them.   It should be emphasized that the fifth point was not a piece of bad data but, in fact, indicated a legitimate localized oscillation of considerable interest.   Of course it is easy to dismiss this by insisting that more points should have been measured in that region.   However, one cannot dismiss the feeling that something smoother can be fitted to these points.

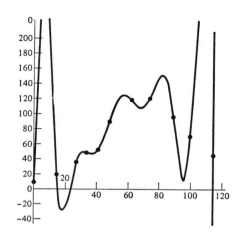

FIGURE 20.10.3
Polynomial interpolation of data.

The interpolation spline for these data (with $y_1'' = y_n'' = 0$), shown in Fig. 20.10.4, demonstrates that the high-degree continuity associated with polynomials should not be confused with the smoothness required in the typical physical problem. We are, in fact, sacrificing higher-derivative continuity to obtain second-derivative smoothness. Where third derivative smoothness is important, we note that the quintic spline function (Ref. 2, pp. 143–148) minimizes $\int (y''')^2 \, dx$, and similarly for higher odd-degree splines. The resulting degradation of second-derivative smoothness is unacceptable in the typical physical problem, but the higher odd-integer spline functions are of interest in function approximation since all nonzero derivatives of a spline converge to the corresponding function derivative (Ref. 2, pp. 135–143).

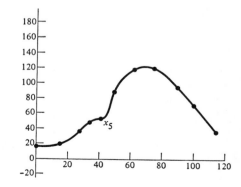

FIGURE 20.10.4
Spline interpolation of data in
Fig. 20.10.3.

The possible failure of polynomial interpolation can be anticipated by looking at a divided difference table (Sec. 18.2). Differences of the jth order can be associated with the jth derivative of an interpolating polynomial through $j + 1$ adjacent points. Scanning a column in the difference table gives an indication of how stable the corresponding derivative should be. If the higher-order differences do not become smooth, polynomial interpolation will likely be unacceptable. The difference table for the points in Fig. 20.10.3 is shown in Table 20.10. The strong plus-minus oscillation in the fourth-difference column is a sure sign of trouble.

While the term *smoothness* is mathematically vague, there is one necessary condition which can be formally defined and is useful here in comparing the spline and polynomial.

The points in Fig. 20.10.1 have negative second differences. An obvious requirement for a smooth interpolating function is to have inflection points only where the second differences change sign.[1] In Table 20.10 the second differences have four changes of sign, while the interpolating polynomial (Fig. 20.10.3) has eight inflection points—four *extraneous* inflection points. The spline (Fig. 20.10.4) has four inflection points, satisfying this necessary smoothness requirement. While in practice the cubic spline almost always satisfies this requirement, there exist data for which it will fail. The more general spline-in-tension function always succeeds and is useful in these exceptional cases.

[1] D. G. Schweikert, An Interpolation Curve Using a Spline in Tension, *J. Math. Phys.*, 45, pp. 312–317, 1966.

Table 20.10 DIVIDED DIFFERENCE TABLE FOR DATA IN FIG. 20.10.3

x	y	1st	2nd $\times 10^1$	3rd $\times 10^2$	4th $\times 10^3$	5th $\times 10^5$	6th $\times 10^6$	7th $\times 10^7$	8th $\times 10^9$	9th $\times 10^{11}$	10th $\times 10^{12}$
0	16										
		0.2									
14	19		0.4								
		1.3		−0.0							
27	36		0.4		−0.12						
		2.0		−0.5		1.7					
33	48		−1.0		0.72		−0.86				
		0.6		2.0		−3.6		0.27			
41	53		3.1		−1.01		1.12		−0.56		
		5.3		−1.6		3.1		−0.23		0.98	
48	90		−1.5		0.45		−0.64		0.40		−0.13
		2.1		0.2		−0.9		0.11		−0.55	
62	119		−0.8		−0.04		0.14		−0.15		
		0.1		0.0		0.1		−0.02			
74	120		−0.6		0.00		−0.00				
		−1.6		0.1		0.0					
89	96		−0.4		0.00						
		−2.5		0.1							
99	71		0.1								
		−2.3									
114	36										

356

INDEFINITE INTEGRALS—FEEDBACK

21.1 INTRODUCTION

There is a significant difference between computing a definite integral

$$\int_a^b f(x)\, dx$$

and computing an indefinite integral, which may be written in the form

$$\int_a^x f(x)\, dx$$

The result of the first computation is a single number, while the result of the second is a table of numbers.

The main purpose of this chapter is to introduce, after a brief survey, two new concepts. The first concept is that of *instability*, which can occur when function values that have already been computed are used to compute the next value. This repeated use of the function values may be compared to "feedback" and produces instability in the same manner. Feedback is the use of part of the output at a particular time as input at a slightly later time. Sometimes this time delay can be neglected, and sometimes it cannot. Perhaps the simplest physical example is the feedback amplifier which is widely used in control circuits.   (See Fig. 21.1.1.)   The output voltage is the sum of the two inputs on the left times the gain of the amplifier, which is taken as -10^9, and one-tenth of the output is fed into the input.

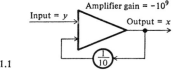

FIGURE 21.1.1

We have

$$(y + \tfrac{1}{10}x)(-10^9) = x$$

or

$$x = -\frac{10^{10}y}{10^9 + 10} = -\frac{10y}{1 + 10^{-8}} \approx -10y$$

Thus the output is approximately -10 times the input, quite independent of small changes in the characteristics of the amplifier; that is, it is not sensitive to the exact value of the gain.

Under some circumstances, if there is enough delay in returning the signal from the output to the input, the system oscillates—*sings* is the word often used. Such situations arise frequently in public auditoriums which have a microphone (y) and an amplifier system output (x).   The output from the loudspeaker is delayed by the time that it takes the sound to get from the speaker to the microphone, and the result is heard by the audience unless care is taken to control it. The same unstable situation sometimes occurs in adjusting a shower to the desired temperature.

The theory of feedback systems is well developed and cannot be treated in detail here.   We stress the point that when we use an output result as an input to the same process at a later stage, we run the risk of producing an unwanted oscillation, whose period and behavior depend on the particular process.

The second new concept is an extension of the technique that we have been using to find formulas.   In the past we have used all the parameters that we had in order to make the formula exact for as many powers of x as we could; or, in other words, we have concentrated on making the truncation error of as high an order as possible.   In order to cope with instability and other effects, such as roundoff propagation, it is necessary to leave some of the coefficients of the general formula undetermined at the time we are finding the error term, and later select values for them which will produce suitable formulas. The problem of calculating an indefinite integral

$$y(x) = \int_a^x f(x)\, dx$$

may be recast in the form of a simple differential equation

$$y'(x) = f(x) \qquad y(a) = 0$$

In this form the integrand values are labeled y', while the answer is labeled y. We shall adopt the differential equation approach because it blends in with later notation when the more general differential equation

$$y'(x) = f(x, y) \qquad y(a) = b$$

is examined. (See Chap. 23.) This approach somewhat limits the class of formulas that we find.

We shall also make a slight change in our notation. Up to this time we have been using $y(x)$ and y_x to mean the same thing. We shall now use the subscript notation to count; that is, we shall number the points at which we compute the answer $0, 1, 2, \ldots$. But for reasons of clarity it is desirable to keep the spacing of the problem at h and to let the origin fall where it may. Thus $y(a + nh)$ and y_n are the same quantity. No confusion should arise once the convention is understood, and the notation becomes a great deal simpler.

21.2 SOME SIMPLE FORMULAS FOR INDEFINITE INTEGRALS

A simple formula for approximating one step of an indefinite integral is

$$y_{n+1} = y_n + hy'[a + (n + \tfrac{1}{2})h] = y_n + hy'_{n+1/2}$$

where

$$y' = f(x)$$

or

$$y = \int_a^x y'\, dx$$

This is clearly the midpoint formula, and it has one-half the error of the usual trapezoid rule

$$y_{n+1} = y_n + \frac{h}{2}(y'_{n+1} + y'_n)$$

The midpoint formula also has better roundoff properties.

A more accurate formula, which may be found by using the universal matrices of Sec. 15.6, is

$$y_{n+1} = y_n + \frac{h}{24}(-y'_{n-1} + 13y'_n + 13y'_{n+1} - y'_{n+2})$$

where the error term is

$$\tfrac{11}{720}h^5 y^{(5)}(\theta) \approx 0.0153h^5 y^{(5)}(\theta)$$

If this formula is used[1] to compute an integral, then two extra points which lie outside the range of integration have to be computed (one at each end). If the values of the integrand cannot be found at such points because of singularities in the formula, then the assumption that the integrand can be approximated accurately by a polynomial is probably false. An alternative formula can be found for the ends of the interval, namely,

$$y_1 = y_0 + \frac{h}{24}(9y'_0 + 19y'_1 - 5y'_2 + y'_3)$$

Simpson's formula is widely used for computing indefinite integrals. In our present notation this takes the form

$$y_{n+1} = y_{n-1} + \frac{h}{3}(y'_{n-1} + 4y'_n + y'_{n+1})$$

with an error term

$$-\frac{h^5 y^{(5)}(\theta)}{90} \approx -0.0111h^5 y^{(5)}(\theta)$$

which is slightly less than the above formula.

There is one often overlooked trouble that occurs when Simpson's formula is used in the conventional manner, and although the trouble is not serious, it can be annoying. Simpson's formula carries the value of the integral two steps forward, and the special Simpson's half-formula (Sec. 15.6) is used to start the chain for the odd-numbered sample points. The result of this jumping two steps ahead is that the accumulated errors at the odd- and even-numbered points are rather independent of each other, especially as the weights attached to the computed integrand values are different. All this tends to produce an oscillation. While this oscillation is due to errors being committed, and hence gives some

---

[1] We of course actually compute $(24/h)y_n$ and multiply by $h/24$ when we print out values of y_n. In this way, while we do the same amount of computing, the roundoff effects do not build up so rapidly, since the final multiplication by $h/24$ does not propagate further.

measure of the accuracy of the results, it can be very annoying at times. The oscillation can be avoided if only the even-numbered-point chain is computed and Simpson's half-formula is used to produce each odd-numbered point.

PROBLEMS

21.2.1 Derive the four-point integration formulas given in this section with their error terms.

21.3 A GENERAL APPROACH

Many special formulas could be investigated one at a time, but we shall turn to a general approach and examine a whole class at once. In estimating the next value y_{n+1} of the integral, we can use old values of the integral $y_n, y_{n-1}, \ldots$, as well as both new values of the integrand $y'_{n+1}, \ldots$ and old values $y'_n, y'_{n-1}, \ldots$. We shall arbitrarily limit the present study to the form

$$y_{n+1} = a_0 y_n + a_1 y_{n-1} + a_2 y_{n-2} + h(b_{-1} y'_{n+1} + b_0 y'_n + b_1 y'_{n-1} + b_2 y'_{n-2})$$
$$+ E_5 \frac{h^5 y^{(5)}(\theta)}{5!}$$

which uses three old values of the integral together with one new and three old values of the integrand. The question immediately arises, Why not try the more reasonable approach of using integrand values symmetrically placed about the value being computed? Such an approach would probably give more accurate formulas, but we shall not do so for reasons that will become apparent when we investigate the general differential equation.

We have written the equations so that the defining equations will be homogeneous in h; *we can therefore act as if $h = 1$*, most of the time, and remember to supply the h factors at the proper moment. This involves a slight change in the meaning of E_5, but no serious confusion will occur in practice.

With seven parameters in the general form, we shall impose only five conditions—namely, that it be exact for $y = 1, x, \ldots, x^4$, or, what is the same thing, exact for $y = 1, (x - n), \ldots, (x - n)^4$—and we shall save two parameters for other purposes. Earlier we made the formula exact when the *integrand* was $1, x, x^2, \ldots$; now we make the *answer* exact for $1, x, x^2, \ldots$. Since the integrals of powers of x are powers one higher in x, we need account only for why we have added the condition of exactness for $y = 1$. The consequence of not doing so is that $y' = 0$ will not integrate into $y = $ constant.   Using these five conditions and taking a_1 and a_2 as parameters, we get from the defining equations

$$a_0 = 1 - a_1 - a_2 \qquad b_0 = \tfrac{1}{24}(19 + 13a_1 + 8a_2)$$
$$a_1 = a_1 \qquad\qquad b_1 = \tfrac{1}{24}(-5 + 13a_1 + 32a_2)$$
$$a_2 = a_2 \qquad\qquad b_2 = \tfrac{1}{24}(1 - a_1 + 8a_2)$$
$$b_{-1} = \tfrac{1}{24}(9 - a_1) \qquad E_5 = \tfrac{1}{6}(-19 + 11a_1 - 8a_2)$$

Note that the equations are linear in a_1 and a_2.   Each point in the a_1, a_2 plane corresponds to a formula—a common mathematical device for studying families of formulas.

PROBLEMS

21.3.1   Derive the above formulas for the coefficients.

21.4   TRUNCATION ERROR

We have written the general form of our formula as if the influence function $G(s)$ (see Chap. 16) had a constant sign (Sec. 16.5).   It is necessary to examine this assumption and find which values of a_1, a_2 make this true.

Using Chap. 16 as our guide, we first notice that, except for special values of a_1, a_2 which make $E_5 = 0$, we have $m = 5$.   The influence function of Sec. 16.4 has an $A = -2h$, so that the Taylor expansion is (Sec. 16.2)

$$y(x) = y(-2h) + (x + 2h)y'(-2h) + \frac{(x + 2h)^2}{2!} y''(-2h)$$
$$+ \frac{(x + 2h)^3}{3!} y'''(-2h) + \frac{(x + 2h)^4}{4!} y^{(4)}(-2h)$$
$$+ \frac{1}{4!} \int_{-2h}^{x} y^{(5)}(s)(x - s)^4 \, ds$$

If this is substituted in the general form of Sec. 21.3, we pick $B = h$ and get

$$R(y) = \int_{-2h}^{h} y^{(5)}(s)G(s) \, ds$$

where $G(s)$ is the influence function.   (Note that no integration is needed in constructing it.)

$$G(s) = \frac{1}{4!} \{(h - s)_+^4 - a_0(-s)_+^4 - a_1(-h - s)_+^4 - a_2(-2h - s)_+^4$$
$$- 4h[b_{-1}(h - s)_+^3 + b_0(-s)_+^3 + b_1(-h - s)_+^3 + b_2(-2h - s)_+^3]\}$$

Now, if $G(s)$ is of constant sign (Sec. 16.5), then we have

$$R(x^5) = 5! \int_{-2h}^{h} G(s) \, ds = E_5 h^5$$

where E_5 is the result of substituting $y(x) = x^5$ for unit spacing.

The problem is to find those values of a_1, a_2 such that $G(s)$ has a constant sign. Thus we need to find the zeros of $G(s)$ for each pair (a_1, a_2). Since this is difficult to do, we resort to a common mathematical trick and invert the problem:[1] we set $G(s) = 0$ and examine the resulting values of a_1, a_2. $G(s)$ is a *linear function* in a_1, a_2

$$G(s) = G_0(s) + a_1 G_1(s) + a_2 G_2(s)$$

where

$$G_0(s) = \frac{1}{4!} \left[(h - s)_+^4 - (-s)_+^4 - \frac{3}{2} h(h - s)_+^3 - \frac{19}{6} h(-s)_+^3 \right.$$
$$\left. + \frac{5}{6} h(-h - s)_+^3 + \frac{h}{6} (-2h - s)_+^3 \right]$$

$$G_1(s) = \frac{1}{4!} \left[(-s)_+^4 - (-h - s)_+^4 + \frac{h}{6} (h - s)_+^3 - \frac{13}{6} h(-s)_+^3 \right.$$
$$\left. - \frac{13}{6} h(-h - s)_+^3 + \frac{h}{6} (-2h - s)_+^3 \right]$$

$$G_2(s) = \frac{1}{4!} \left[(-s)_+^4 - (-2h - s)_+^4 - \frac{4}{3} h(-s)_+^3 - \frac{16}{3} (-h - s)_+^3 \right.$$
$$\left. - \frac{4}{3} h(-2h - s)_+^3 \right]$$

There are three intervals of s in which we must examine $G(s)$: $h \geq s \geq 0$, $0 \geq s \geq -h$, and $-h \geq s \geq -2h$. In the first interval

$$(h - s)^3 \left(h - s - \frac{3}{2} h + a_1 \frac{h}{6} \right) = 0$$

leads to

$$a_1 = 3 + \frac{6s}{h}$$

which describes a family of vertical lines in the a_1, a_2 plane. The line moves from $a_1 = 9$ to $a_1 = 3$ as s goes from h to 0 (Fig. 21.4.1).

---

[1] Carl Gustav Jacob Jacobi (1804–1851) said "Always invert."

In the second interval we have

$$\left[(h-s)^4 - s^4 - \frac{3}{2} h(h-s)^3 + \frac{19}{6} hs^3 \right]$$

$$+ a_1 \left[s^4 + \frac{h}{6}(h-s)^3 + \frac{13}{6} hs^3 \right] + a_2 \left(s^4 + \frac{4h}{3} s^3 \right) = 0$$

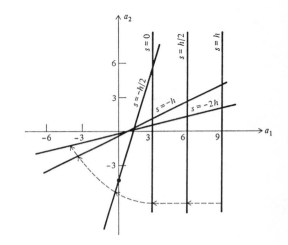

FIGURE 21.4.1
Region of constant sign for $G(s)$. (The lines are those along with $G(s) = 0$ for the value s marked on the line.)

This describes a line which continues to move to the left but gradually tilts until, at $s = -h$,

$$a_2 = \frac{a_1 - 1}{2}$$

This line passes through the point $(1,0)$ with a slope of $1 : 2$.

In the third interval we have the line rotating about the point $(1,0)$ and ending up as

$$a_2 = \frac{a_1 - 1}{8}$$

Above and to the left of these lines (see Fig. 21.4.1), we have a $G(s)$ which has no zeros[1] and hence is of constant sign. Thus in this region we have the error term of the form

$$\frac{E_5 h^5}{5!} y^{(5)}(\theta)$$

Outside this region we do not have such an error term, and in view of future work, we shall not investigate it further.

In the region the error is measured by E_5 (Sec. 21.3). The "equi-error" lines are shown as slant lines in Fig. 21.4.2. These lines show that, other things

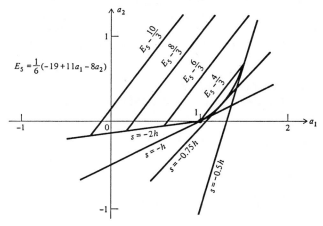

$$E_5 = \frac{1}{6}(-19 + 11a_1 - 8a_2)$$

FIGURE 21.4.2
Equitruncation curves. (The value of E along the line gives the truncation error on that line.)

being the same, we should try to stay down toward the lower right where the error term is small. The slope of the lines, 11 : 8, indicates the "exchange ratio" between the two parameters; if we increase a_2 by 11 units and a_1 by 8, then the error is the same.

---

[1] There is also a region in the lower right for which $G(s)$ does not change sign, but both the size of the error term and the roundoff effects keep us from using such formulas, and we shall ignore it in the future. There is also a line for which $E_5 = 0$ and where we need a different $G(s)$ and we can ignore it for stability reasons.

21.5   STABILITY

We first examine the formulas by considering the trivial case

$$y' = 0$$

If this case does not behave favorably, then the more general case of

$$y' = f(x)$$

will also give trouble.

If we choose a point such as $a_1 = -1$ and $a_2 = 0$ in the a_1, a_2 plane, then the formula becomes

$$y_{n+1} = 2y_n - y_{n-1}$$

Now consider what happens to the solution $y = 0$ when a small error ε (which we can imagine coming from roundoff) is put in the solution at y_0. It is easy to compute Table 21.5.1. Thus the effect of the small error grows steadily.

As another example, choose $a_1 = 1$, and $a_2 = 1$. The equation is

$$y_{n+1} = -y_n + y_{n-1} + y_{n-2}$$

and we get Table 21.5.2.

This time the error oscillates as well as grows.

Table 21.5.1

$\vdots$
$y_{-2} = 0$
$y_{-1} = 0$
$y_0 = \varepsilon$
$y_1 = 2\varepsilon$
$y_2 = 3\varepsilon$
$y_3 = 4\varepsilon$
$\vdots$

Table 21.5.2

$\vdots$
$y_{-2} = 0$
$y_{-1} = 0$
$y_0 = \varepsilon$
$y_1 = -\varepsilon$
$y_2 = 2\varepsilon$
$y_3 = -2\varepsilon$
$y_4 = 3\varepsilon$
$y_5 = -3\varepsilon$
$\vdots$

These two examples indicate that we need to examine the phenomena further; clearly some points (a_1, a_2) are very poor choices.

The general form of the difference equation that we are investigating is a linear difference equation with constant coefficients, and we may apply the methods of Chap. 13.   Since the values of y' are the integrand values which we compute independently of the y values, we consider the difference equation as being in the form

$$y_{n+1} - a_0 y_n - a_1 y_{n-1} - a_2 y_{n-2} = F_n$$

where F_n is a function of the derivative values y'_K which are the computed integrand samples.

We first solve the homogeneous equation

$$y_{n+1} - (1 - a_1 - a_2)y_n - a_1 y_{n-1} - a_2 y_{n-2} = 0$$

To solve this, we set $y_n = \rho^n$ to get the characteristic equation

$$\rho^3 - (1 - a_1 - a_2)\rho^2 - a_1 \rho - a_2 = 0$$

or

$$(\rho - 1)[\rho^2 + (a_1 + a_2)\rho + a_2] = 0 \qquad a_2 \neq 0$$

Let the zeros of the quadratic factor be ρ_1 and ρ_2.   The solution of the homogeneous equation is

$$y_n = C_1(\rho_1)^n + C_2(\rho_2)^n + C_3$$

provided that no two roots are equal.   If two roots are equal, as they happen to be in the two examples above, then the solution needs to be changed.   If either of the roots is greater in absolute value than 1, then the solution will grow similarly to a geometric progression as n increases (provided that the corresponding coefficient is not zero).

We therefore wish to explore the region of the a_1, a_2 plane for which $|\rho_i| < 1$.   Again we invert the problem and determine the boundaries of the region.   We first examine the boundary along which a root becomes equal to 1. To do this, we set $\rho = 1$ in the quadratic factor.   This defines the line

$$1 + a_1 + 2a_2 = 0$$

$$a_2 = -\frac{1 + a_1}{2}$$

Along this line $\rho = 1, 1, a_2$ (the characteristic roots), and the solution has the form

$$y_n = C_1 + nC_2 + C_3(a_2)^n \qquad a_2 \neq 1$$

Next we examine where $\rho = -1$.   To do this, we set $\rho = -1$ in the quadratic factor.   This defines the line

$$a_1 = 1$$

Along this line $\rho = 1, -1, -a_2$.

Next we examine where the roots become complex.   This curve is defined by

$$(a_1 + a_2)^2 - 4a_2 = 0$$

or

$$a_1 = \pm 2\sqrt{a_2} - a_2$$

which is shown in Fig. 21.5.1.   Inside this parabola the roots are complex conjugates of each other and both have the modulus

$$|\rho| = \sqrt{a_2}$$

Thus the region in which $|\rho_i| < 1$ is bounded by the line

$$a_2 = 1$$

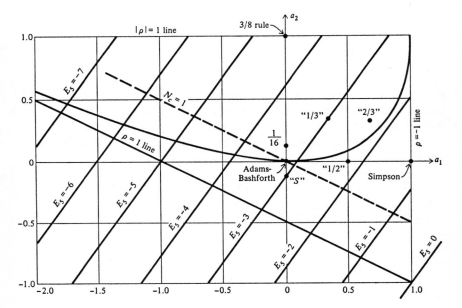

FIGURE 21.5.1
Stability region in a_1, a_2 plane.

which just cuts off the parabola where it is tangent to the other lines. These three straight lines in Fig. 21.5.1, then, define the triangle within which the characteristic roots must lie on order for the method of integration to have an error that does not grow as a geometric progression. Outside this region some $|\rho_i| > 1$. It is easy to see that the two illustrative examples in this section correspond to points which lie on the edge of the region and give double roots in the characteristic equation; the multiple roots, in turn, give rise to the linear growth. When the $|\rho_i| < 1$, we say that the method is *stable*; when $|\rho_i| = 1$, we say that it is *conditionally stable* (for simple roots only). The solution of the homogeneous equation is to be added to the solution of the complete equation. In solving the difference equation, any isolated roundoff will start the various solutions of the homogeneous equation, and unless the $|\rho_i| \leq 1$, there will be generated a solution that tends to infinity regardless of what F_n does.

The unwanted solutions of the difference equation occur because we have replaced a first-order differential equation by a third-order difference equation. The root $\rho = 1$ corresponds to the desired solution, and the other two ρ_i are the unwanted solutions. By requiring the $|\rho_i| \leq 1$ we are protected against their giving rise to large effects in the final answer. For some values of ρ such that $|\rho| = 1$, bounded solutions are possible.

PROBLEMS

21.5.1   Discuss the example where $a_1 = -1$ and $a_2 = 0$ and show that the computed table agrees with the theory when the proper equation is used.

21.5.2   Do the same as in Prob. 21.5.1 for $a_1 = 1$ and $a_2 = 1$.

21.5.3   Discuss the case of $a_1 = 2$ and $a_2 = 0$. Show that the theory and computation agree in giving a geometric growth in the error.

21.5.4   Discuss the behavior of the error as you cross each of the three bounding lines of stability region.

21.6   CORRELATED ROUNDOFF NOISE

The y_n values that we compute will, in general, have a roundoff noise level σ. At first thought the amplification factor of the roundoff noise would be taken as

$$N_a = (a_0{}^2 + a_1{}^2 + a_2{}^2)^{1/2}$$

However, because the y_{n+1} value is computed from the y_n, y_{n-1}, and y_{n-2} values, the noise in v_{n+1} is related to (correlated with) the noise in y_n, y_{n-1}, and y_{n-2}, and we need to give the matter more thought, although we still do not want N_a to be too big.

Suppose that we have a solution of

$$y' = 0$$

which is identically zero, and a small roundoff error occurs.   The solution of the difference equation is

$$y_n = C_1(\rho_1)^n + C_2(\rho_2)^n + C_3$$

Assuming that the method is stable, $|p_i| < 1$, then

$$y_n \rightarrow C_3 \qquad \text{as } n \rightarrow \infty$$

Let us compute C_3 from the initial data

$$y_{-2} = 0$$
$$y_{-1} = 0$$
$$y_0 = \varepsilon$$

We have

$$\frac{C_1}{\rho_1{}^2} + \frac{C_2}{\rho_2{}^2} + C_3 = 0$$

$$\frac{C_1}{\rho_1} + \frac{C_2}{\rho_2} + C_3 = 0$$

$$C_1 + C_2 + C_3 = \varepsilon$$

Solving for C_3, we get

$$C_3 = \frac{\varepsilon}{1 - (\rho_1 + \rho_2) + \rho_1\rho_2}$$

Using the quadratic factor in the characteristic equation, we have

$$\rho_1 + \rho_2 = -(a_1 + a_2)$$
$$\rho_1\rho_2 = a_2$$

Hence

$$C_3 = \frac{\varepsilon}{1 + a_1 + 2a_2} = N_c \varepsilon$$

where

$$N_c = \frac{1}{1 + a_1 + 2a_2}$$

Thus, if the roundoff error ε is not to end up greater than the starting value ε, then

$$|N_c| = |1 + a_1 + 2a_2|^{-1} \le 1$$

and the smaller the better. Since we are mainly concerned with positive

$$1 + a_1 + 2a_2$$

we find that the condition for an isolated error ε not to grow is that (a_1, a_2) lie above the line

$$a_2 = -\frac{a_1}{2}$$

This line is shown in Fig. 21.5.1 as a dashed line.

PROBLEMS

21.6.1 The theory of roundoff leading to the result was developed for a single, isolated error. Each step will produce an error. How, if at all, does this change the argument?

21.7 SUMMARY

Table 21.7 lists a number of well-known methods which fall in the general class that we are investigating.

The choice $a_1 = 1$, $a_2 = 0$ is Simpson's method and lies on the edge of the stability region.

The choice $a_1 = 0$, $a_2 = 1$ is the three-eighths rule and also falls on the edge of the stability region.

The choice $a_1 = 0$, $a_2 = 0$ is labeled "Adams-Bashforth" because it is used in the Adams-Bashforth method of integrating differential equations (see Chap. 23).

All other methods of the general form are linear combinations of these three.

Figure 21.5.1 presents the relevant information. Each point in the a_1, a_2 plane corresponds to a method of integration. First, we want to choose the point inside the stability triangle, although we may venture onto the edge for some cases, such as Simpson's and the three-eighths-rule methods. Second, we want to stay toward the lower right to have a small truncation error (Fig. 21.4.2). Third, we wish to stay above the dashed line to keep the amplification of isolated error small.

Table 21.7 SOME METHODS FOR INTEGRATION

	Adams-Bashforth	Simpson	Three-eighths rule	1/3	1/2	2/3	−1/8*	1/16
a_0	1	0	0	1/3	1/2	0	9/8	15/16
a_1	0	1	0	1/3	1/2	2/3	0	0
a_2	0	0	1	1/3	0	1/3	−1/8	1/16
b_{-1}	9/24	1/3	3/8	13/36	17/48	25/72	3/8	18/48
b_0	19/24	4/3	9/8	39/36	51/48	91/72	6/8	39/48
b_1	−5/24	1/3	9/8	15/36	3/48	−43/72	−3/8	−6/48
b_2	1/24	0	3/8	5/36	1/48	9/72	0	3/48
E_5	−19/6	−4/3	−9/2	−3	−9/4	−43/18	−3	−39/12
Noise amplification N_c	1	1, 0, 1, 0, …	1, 0, 0, …	1/2	2/3	3/7	1.13	8/9
N_a	1	1	1	$1/\sqrt{3}$	$1/\sqrt{2}$	$\sqrt{5}/3$	$\sqrt{82}/8$	0.94

* Taken from R. Hamming, Stable Predictor-Corrector Methods for Ordinary Differential Equations, *J. Assoc. Computing Machinery*, vol. 6, no. 1, pp. 37–47, January, 1959.

Thus, except for the annoying tendency toward oscillation, Simpson's formula is one of the best of the general form examined (which does *not* include the first example of Sec. 21.2). The user should select his own method to suit the particular balance of factors in each problem, and no single best formula can be offered for all circumstances. The methods in Table 21.7 labeled 1/3, 1/2, and 2/3 are mixtures of the basic methods and are of interest in some applications.

PROBLEMS

21.7.1 Develop the theory of Secs. 21.3 to 21.7 (using only one parameter a_1) for the general form

$$y_{n+1} = a_0 y_n + a_1 y_{n-1} + h(b_{-1} y'_{n+1} + b_0 y'_n + b_1 y'_{n-1}) + E_4 \frac{h^4 y^{(4)}(\theta)}{4!}$$

and compare with known results.

Ans.

	$a_1 = 0*$	Simpson	$a_1 = \frac{1}{5}$	
$a_0 = 1 - a_1$	1	0	$\frac{4}{5}$	$\rho_1 = -a_1$
$a_1 = a_1$	0	1	$\frac{1}{5}$	hence
$b_{-1} = (5 - a_1)/12$	$\frac{5}{12}$	$\frac{1}{3}$	$\frac{1}{15}$	stability
$b_0 = (8 + 8a_1)/12$	$\frac{8}{12}$	$\frac{4}{3}$	$\frac{8}{10}$	range
$b_1 = (-1 + 5a_1)/12$	$-\frac{1}{12}$	$\frac{1}{3}$	0	$-1 < a_1 < 1$
$E_4 = -1 + a_1$	-1	0	$-\frac{4}{5}$	E_4 not valid
Noise amplification				for
$N_c = 1/(1 + a_1)$	1	1, 0, 1, 0, ..., 1	$\frac{5}{6}$	$\frac{1}{5} < a_1 < 5$
N_a	1	1	$\sqrt{17}/5$	But $a_1 = 1$ higher order

* Southard and Yowell, *Math. Tables and Aids to Comp.*, vol. 6, pp. 253–254, 1952. [Clearly it is Simpson's half-formula see (15.6)]

21.7.2 Discuss the case of $a_1 = -1$.

21.8 SOME GENERAL REMARKS

The reader should be careful to note that only one particular class of formulas was investigated; other formulas are possible. For example, we could investigate formulas for which we used both integrand values (y') and derivatives of integrand values (y''). We would expect much better formulas in the sense of efficient computation, since frequently the derivative is readily calculated from various pieces of the computation of the integrand. However, this would involve more coding (and more mistakes) and for the occasional problem would probably be a poor strategy. For a recurring problem such formulas could save considerable machine time, and they should be investigated. See Prob. 21.8.1.

The powerful Gauss-type formulas have not been adapted to indefinite integral calculation because of the difficulty of fitting the peculiar spacing to consecutive steps.

The choice of E_5 as the error term has found wide favor in practice and represents a general compromise between different factors. Special situations, of course, require other choices.

The main purpose of this chapter was to introduce the idea of stability and the method of saving parameters to cope with it. We have not, therefore, gone into the involved subject of how to start an integration until enough back values are available to use the general formulas. The formulas of Sec. 21.2 can sometimes be used for this purpose.

PROBLEMS

21.8.1 Investigate the class of formulas

$$y_{n+1} = y_n + h(b_{-1}y'_{n+1} + b_0y'_n) + h^2(c_{-1}y''_{n+1} + c_0y''_n) + E_4\frac{h^4y^{(4)}(\theta)}{4!}$$

using b_{-1} as a parameter.   Note the ease of starting.

Ans.

	General solution		A Milne method*		
b_{-1}	b_{-1}	1	$\frac{1}{2}$	0	$\frac{1}{3}$	$\frac{2}{3}$
b_0	$1 - b_{-1}$	0	$\frac{1}{2}$	1	$\frac{2}{3}$	$\frac{1}{3}$
c_{-1}	$(1 - 3b_{-1})/6$	$-\frac{2}{6}$	$-\frac{1}{12}$	$\frac{1}{6}$	0	$-\frac{1}{6}$
c_0	$(2 - 3b_{-1})/6$	$-\frac{1}{6}$	$\frac{1}{12}$	$\frac{2}{6}$	$\frac{1}{6}$	0
E_4	$-1 + 2b_{-1}$	1	0	-1	$-\frac{1}{3}$	$+\frac{1}{3}$
E_5	··········	···	$\frac{1}{6}$			

*W. E. Milne, "Numerical Solution of Differential Equations," p. 76, John Wiley & Sons, Inc., New York, 1953, Dover, 1970.

The $G_4(s)$ changes sign for $1/3 < b_{-1} < 2/3$, but for the case $b_{-1} = 1/2$ the error is

$$\frac{E_5}{5!}h^5y^{(5)}(\theta) = \frac{h^5}{720}y^{(5)}(\theta)$$

since

$$G(s) = \frac{(h - s)^2 s^2}{4!} \neq 0 \qquad 0 < s < h$$

and is particularly favorable.

21.8.2 Develop a method for starting an integration of the general form with due care to questions of accuracy.

21.9   EXPERIMENTAL VERIFICATION OF STABILITY

The reader should not become dismayed at this elaborate theoretical development. If a broad class of formulas is to be examined, some such approach is necessary; but if a *particular* formula is under consideration, then a simple experimental approach is often satisfactory.

Suppose that we are considering the formula

$$y_{n+1} = -y_n + \frac{3}{2} y_{n-1} + \frac{y_{n-2}}{2} + \frac{h}{48} (15y'_{n+1} + 85y'_n + 61y'_{n-1} + 7y'_{n-2})$$

It is easy to substitute $y = 1, x, x^2, x^3, x^4$ to see if they fit exactly.

With regard to stability, we simply try

$$y_{-2} = 0$$
$$y_{-1} = 0$$
$$y_0 = \varepsilon$$

and compute

$$y_1 = -\varepsilon$$
$$y_2 = \frac{5\varepsilon}{2}$$
$$y_3 = -\frac{7\varepsilon}{2}$$

$$\cdots\cdots\cdots\cdots$$

In this case it is clear that the method is unstable.   A slight amount of care must be exercised to be sure that the experiments tried actually start all the possible terms (make each of the $C_i \neq 0$ at some time on some trial).   Usually a single trial starts a linear combination of *all* the terms.

The direct experimental check for an unstable method is convincing and readily applied.   The conclusion from negative evidence that a method is stable clearly requires some care, but often a particular formula is easy to handle.

The approach is illustrated in the next section.

*21.10   AN EXAMPLE OF A CONVOLUTION INTEGRAL WHICH ILLUSTRATES THE CONCEPT OF STABILITY

Oscillation with a geometric growth in amplitude is typical of stability troubles, and the following example shows how the various ideas can be applied in other situations.

The problem[1] was to compute $\Phi(t)$ when the data $\psi(t)$ were given at values of $\log t = -14.4(0.2)(0.8)$, and the equation connecting $\Phi(t)$ and $\psi(t)$ was

$$\int_0^t \Phi(t - \tau)\psi(\tau) \, d\tau = t \quad (21.10.1)$$

or the equivalent

$$\int_0^t \Phi(\tau)\psi(t - \tau) \, d\tau = t \quad (21.10.2)$$

The data $\psi(t)$ were monotone increasing.

A reasonable computing form was set up and tried. The results showed an oscillation with growing amplitude. The period of the oscillation was fixed, and the growth was more or less geometric, and so the conjecture was made that it was due to instability. An examination of the computing process was made with data which were uniformly $\psi(t) = 1$ but with a single perturbation; these data gave rise to oscillations (Sec. 21.9).

A new computing plan was sought. After some study, the following was adopted. We set

$$f(t) = \int_0^t \psi(\tau) \, d\tau$$

Thus $f(0) = 0$ and $f'(\tau) = \psi(\tau)$. Next we introduce notation t_i, $(i = 0, 1, \ldots, n)$ with $t_0 = 0$, $t_1 = 10^{-14.4}, \ldots, t_i = 10^{(-14.6 + 0.2i)}$ and *assumed* that $\psi(t)$ was constant between t_0 and t_1.

In order to calculate $f(t)$, we used the trapezoid rule for integration,

$$f(t_{n+1}) = f(t_n) + \tfrac{1}{2}[\psi(t_{n+1}) + \psi(t_n)](t_{n+1} - t_n)$$

since the data did not justify any more elaborate method.

We now had, from Eq. (21.10.2),

$$t_{n+1} = \int_0^{t_{n+1}} \Phi(\tau)\psi(t_{n+1} - \tau) \, d\tau$$

$$= \sum_{i=0}^n \int_{t_i}^{t_{i+1}} \Phi(\tau)\psi(t_{n+1} - \tau) \, d\tau$$

In each of the integrals in the sum we made the approximation

$$\int_{t_i}^{t_{i+1}} \Phi(\tau)\psi(t_{n+1} - \tau) \, d\tau = \Phi(t_{i+1/2}) \int_{t_i}^{t_{i+1}} f'(t_{n+1} - \tau) \, d\tau$$

$$= -\Phi(t_{i+1/2})[f(t_{n+1} - t_{i+1}) - f(t_{n+1} - t_i)]$$

---

[1] I. L. Hopkins and R. W. Hamming, On Creep and Relaxation, *J. Appl. Phys.*, vol. 28, pp. 906–909, August, 1957.

where $t_{i+1/2}$ is the midvalue $\frac{1}{2}(t_{i+1} + t_i)$. Thus

$$t_{n+1} = -\sum_{i=0}^{n-1} \Phi(t_{i+1/2})[f(t_{n+1} - t_{i+1}) - f(t_{n+1} - t_i)] + \Phi(t_{n+1/2})f(t_{n+1} - t_n)$$

Solving for $\Phi(t_{n+1/2})$, we got

$$\Phi(t_{n+1/2}) = \frac{t_{n+1} - \sum_{i=1}^{n-1} \Phi(t_{i+1/2})[f(t_{n+1} - t_i) - f(t_{n+1} - t_{i+1})]}{f(t_{n+1} - t_n)} \qquad (21.10.3)$$

This provided a method for calculating each value $\Phi(t_{n+1/2})$ in turn, where the first value was given by

$$\Phi_{1/2} = \frac{t_1}{f(t_1)} \qquad (21.10.4)$$

The reason that Eqs. (21.10.3) and (21.10.4) are successful, while a number of other ways of approximating the integral are not, is that an error in calculating a $\Phi(t)$ (due to faulty data or even just numerical roundoff of a product) does not grow in size in subsequent stages of the computation.   To see this intuitively, we calculate the coefficient of $\Phi(t_{n-1/2})$ in the expression for $\Phi(t_{n+1/2})$. This coefficient is

$$-\frac{f(t_{n+1} - t_{n-1}) - f(t_{n+1} - t_n)}{f(t_{n+1} - t_n) - f(t_{n+1} - t_{n+1})}$$

[since $f(0) = 0$].   We apply the mean value theorem to both the numerator and the denominator separately.

$$\frac{\psi(t_{n+1} - \theta_1)(t_n - t_{n-1})}{\psi(t_{n+1} - \theta_2)(t_{n+1} - t_n)} = -\left[\frac{\psi(t_{n+1} - \theta_1)}{\psi(t_{n+1} - \theta_2)}\right]\left[\frac{t_n - t_{n-1}}{t_{n+1} - t_n}\right] \qquad (21.10.5)$$

where $t_n > \theta_1 > t_{n-1}$ and $t_{n+1} > \theta_2 > t_n$.   The terms in each bracket are less than 1 in magnitude, and so the error in $\Phi(t_{n-1/2})$ is multiplied by a number much less than 1 in size.   Thus the effect of the error oscillates and dies out rapidly.

Since the answer was in the same form as the data and bore the same relationship [see Eqs. (21.10.1) and (21.10.2)], the answers were used as input data, and the original data reproduced well within experimental error.   [We risked the fact that $\Phi(t)$ was *decreasing*, and hence the first factor in Eq. (21.10.5) was greater than 1; the second factor more than compensated for this for our data.]

In this example we see that the same phenomenon of instability occurs even though we have no fixed characteristic equation [compare Eq. (21.10.3)].   The general idea of feedback still applies.   The above examination covered the feedback from only a single term; in practice it is usually necessary to make some examination to see that the additional feedback from other terms does not cause instability.

21.11 INSTABILITY IN ALGORITHMS

Iterative algorithms often show instability in much the same manner as illustrated in the last few sections.

For example, an iteration scheme may give a sequence of y_n values that oscillate and grow in size like a geometric progression.   Thus we assume that

$$y_n = a + b\rho^n$$

where a is the " true " value.   We write this for three consecutive values

$$y_{n-1} = a + b\rho^{n-1}$$
$$y_n = a + b\rho^n$$
$$y_{n+1} = a + b\rho^{n+1}$$

Transpose each a and divide the first equation by the second and the second by the third to get

$$\frac{y_{n-1} - a}{y_n - a} = \frac{1}{\rho} = \frac{y_n - a}{y_{n+1} - a}$$

Solving for a we have

$$a = \frac{y_{n+1}y_{n-1} - y_n^2}{y_{n+1} - 2y_n + y_{n-1}}$$

which reminds us of Shank's method [Eq. (12.7.3)].

Note that the process will improve convergence as well as cope with geometric divergence.   After applying the method we iterate two more steps and reapply, etc., until the error is sufficiently small.   If we have a function $f(t)$ being found by iteration we can apply the method at each point t_k.

22

INTRODUCTION TO DIFFERENTIAL EQUATIONS

22.1 THE SOURCE AND MEANING OF DIFFERENTIAL EQUATIONS

Our present views [10] of the fundamental laws of nature are often stated in the form of differential equations.   For example, Newton's law

$$f = ma = m\frac{d^2x}{dt^2}$$

is a second-order differential equation whenever the force depends on position. As another example, consider a colony of bacteria growing in a medium which furnishes an adequate food supply.   When the colony is studied in the large, it is natural to suppose that the rate of growth is proportional to the number $y(t)$ present at time t; that is,

$$y'(t) = ky(t)$$

Thus differential equations often provide our most primitive mathematical description of a natural phenomenon.

When possible, the differential equation is solved in closed form.   Textbooks on differential equations often give the impression that most differential equations can be solved in closed form, but experience does not bear this out. Various analytical tools are available for obtaining approximate solutions, and these also should be examined before resorting to a numerical solution.

Given the differential equation

$$y' = x^2 - y^2$$

what do we mean by a solution? Setting aside the formal mathematical definition, we intuitively mean a curve $y = y(x)$ at each point (x, y) of which the slope $y'(x)$ of the curve is given by the differential equation. This is a *local property*. Solving an equation means extending this local property to a large region. There is, of course, not just a single curve which is a solution, but rather through *any* point (x_0, y_0) there is a solution of the differential equation (see the next section for exceptions).

22.2 THE DIRECTION FIELD

The ideas of the above paragraph can be made graphic. In the x, y plane we choose various points and compute the slope $x^2 - y^2$. At each of these points we draw a short line with the computed slope. These lines indicate the local direction of the solution, and by the use of a little imagination we can easily sketch in various solutions, provided the set of points through which we draw lines is sufficiently dense.

While the above method works, it is often easier to proceed in a somewhat different fashion. We ask for the curves along which all the short-line elements have the same slope: the *isoclines*, lines of the same slope. In the particular example that we are using, when we choose the slope k, we find the isoclines are hyperbolas:

$$x^2 - y^2 = k$$

The direction field using the isocline approach is shown in Fig. 22.2.1.

A number of things should be noted. If the solution curve has a maximum, or a minimum, then it must lie on the 0 isocline. Inflection points may similarly be located by setting $y'' = 0$:

$$y'' = 0 = 2x - 2yy'$$

$$= 2[x - y(x^2 - y^2)]$$

$$x = \frac{1 \pm \sqrt{1 + 4y^4}}{2y}$$

This curve is *not* shown in Fig. 22.2.1.

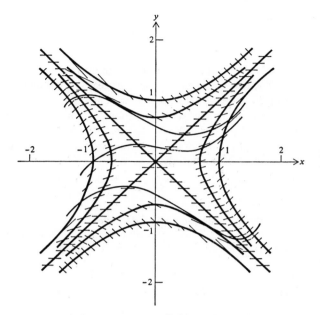

FIGURE 22.2.1
Direction field for $y' = x^2 - y^2$.

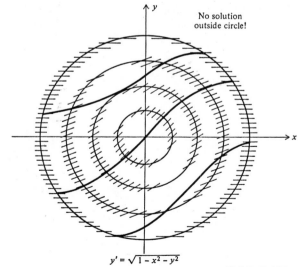

No solution
outside circle!

$y' = \sqrt{1 - x^2 - y^2}$

FIGURE 22.2.2

A direction-field picture can often shed a great deal of light on the nature of the problem.   In the following case

$$y' = \sqrt{(1 - x^2 - y^2)}$$

Fig. 22.2.2 shows that the solution is confined to the circle and cannot get outside.

The direction field is a simple idea, but it should not be scorned.   Frequently it shows the nature of the problem so that a sound approach may be made to the more accurate numerical solution.   Several times in the author's experience, the direction-field sketch has answered the original problem, and on one occasion an accurate sketch prepared on a drawing board gave sufficient accuracy to answer the problem completely.

PROBLEMS

22.2.1   Draw the direction field and solutions for $y' = -\sin y$.
22.2.2   Draw the direction field and solutions for $y' = x - y^2$.
22.2.3   Draw the direction field and solutions for $y' = x^2 + y^2$.

22.3   THE NUMERICAL SOLUTION

If all that we want is one solution through a particular point, then it would be a waste of time to draw the entire direction field.   Rather we would start at the point, compute the local direction, and then a little ahead of that compute the slopes at a few nearby points.   And so we would proceed, always drawing only those slopes which are near where we expect to find the solution (Fig. 22.3.1).

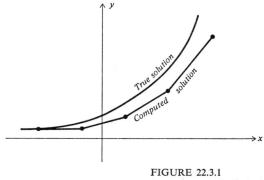

FIGURE 22.3.1
Crude numerical solution.

This process can be reduced to arithmetic. We take the starting values at point x_0, y_0, and using the slope y_0' at this point, we move forward a distance h to the next point

$$(x_0 + h, y_0 + hy_0') = (x_1, y_1)$$

Repeating this process, we go to the next point

$$(x_1 + h, y_1 + hy_1') = (x_2, y_2)$$

etc., and generate a numerical approximation to the solution of the differential equation.

For example, using the equation

$$y' = y^2 + 1 \qquad y_0 = 0 \qquad h = \tfrac{2}{10}$$

which has the known solution $y = \tan x$ for comparison purposes, we easily generate Table 22.3 which gradually drifts far from the solution. This is sometimes known as Euler's method and is clearly bad if large step sizes are used.

A simple look at Fig. 22.3.1 shows what is wrong with this method. When the true solution is concave upward, then the curve that we compute always lags below because it uses the slope from the past to get to the next point.

In view of this obvious fault we propose to look a bit ahead, *predict*, and then examine the slope at that point. Using this slope, we propose to go back and make a second, more *correct*, guess at how we should take the step. In more detail, we use the midpoint formula to *predict*

$$y_{n+1} = y_{n-1} + 2hy_n' \qquad (22.3.1)$$

Computing the slope at this point y_{n+1}' we then average the two end slopes as a more reasonable choice of slope and compute the *correct* value by the trapezoid formula

$$y_{n+1} = y_n + h\left(\frac{y_{n+1}' + y_n'}{2}\right) \qquad (22.3.2)$$

Table 22.3

x	$y_{n+1} = y_n + hy_n'$	$y' = y^2 + 1$	hy'	$y = \tan x$
0	0.000	1.00	0.200	0.000
0.2	0.200	1.04	0.208	0.203
0.4	0.408	1.17	0.234	0.423
0.6	0.642	1.41	0.282	0.684
0.8	0.724	1.52	0.304	1.030
1.0	1.028			1.557

The error term in the prediction is

$$\frac{h^3}{3} y'''(\theta_1) \qquad (22.3.3)$$

while the error term of the corrector is

$$-\frac{h^3}{12} y'''(\theta_2) \qquad (22.3.4)$$

The first question that occurs when we use this procedure is how to start. Our first prediction requires knowing one old point in addition to current slope. One answer is to use the differential equation and its derivatives to form a Taylor expansion

$$y_1 = y_0 + h y_0' + \frac{h^2}{2!} y_0'' + \frac{h^3}{3!} y_0''' + \cdots$$

about the starting point and from the series to compute the first point. Since the error term in each step depends on the third derivative, it is not necessary to extend the Taylor series beyond terms in h^3 or h^4.

This pattern of prediction and then correction has several nice properties. The difference between the predicted and the corrected values is, from Eqs. (22.3.3) and (22.3.4),

$$\frac{h^3}{12} [4y'''(\theta_1) + y'''(\theta_2)] \approx \frac{5h^3}{12} y'''(\theta)$$

If we assume that y''' does not change sign in the interval, then the two values are on opposite sides of the true value (for that single step). Thus the method provides a clue to the accuracy. If the error is too large, then the interval h can be shortened to $\bar{h} = h/2$ which multiplies the error by a factor of about $\frac{1}{8}$.

The reader should realize that we have been playing rather fast and loose with our symbols. We estimated a new value by Eq. (22.3.1) which is in error by Eq. (22.3.3). This value is used to compute the y_{n+1}', and hence while we have written y_{n+1}' as if it were the true value, it is not; it is in error by approximately the second term in

$$f\left(x, y - \frac{h^3}{3} y'''(\theta)\right) \approx f(x, y) + \frac{\partial f(x, y)}{\partial y} \left[-\frac{h^3}{3} y'''(\theta) \right]$$

It is often proposed that we first predict and then correct, and if the two are not close enough, we then repeat the correction several times, using the latest estimate, until no appreciable change occurs. As a matter of principle, it is generally better to shorten the interval than to do repeated evaluations of the corrector formula.

22.4 AN EXAMPLE

Again consider the differential equation

$$y' = y^2 + 1$$
$$y(0) = 0 \qquad 0 \le x \le 1 \qquad (22.4.1)$$

(which was chosen *because* we know that the solution is $y = \tan x$, and we can use it for a check).

To get started, we construct the first few terms of the Taylor series by differentiating the differential equation

$$y' = y^2 + 1 \qquad\qquad y'(0) = 1$$
$$y'' = 2yy' \qquad\qquad y''(0) = 0$$
$$y''' = 2yy'' + 2(y')^2 \qquad\qquad y'''(0) = 2$$
$$y^{iv} = 2y'y'' + 2yy''' + 4y'y'' \qquad y^{iv}(0) = 0$$

Hence

$$y(x) = 0 + x \cdot 1 + \frac{x^2}{2} \cdot 0 + \frac{x^3}{3!} \cdot 2 + \frac{0 \cdot x^4}{4!} + \cdots$$

$$= x + \frac{x^3}{3} + \cdots$$

We shall use the crude spacing $h = \Delta x = 0.2$. We have

$$y_1 = y(0.2) = 0.2 + \frac{0.008}{3} = 0.203$$

as the first point, and (using a slide rule)

$$y_1' = y'(0.2) = (0.203)^2 + 1 = 1.0412$$

We are now ready to start using the method. We *predict* the next y value, where we now write p for clarity

$$p_2 = y_0 + 2hy_1' = 0 + 0.4(1.0412) = 0.4165$$

and evaluate, using the differential equation (22.4.1):

$$p_2' = 1.173$$

Next we compute the corrector value, where we now write c for clarity:

$$c_2 = 0.203 + 0.1(1.0412 + 1.173) = 0.424$$

Now we know that p_2 is in error by

$$\frac{h^3}{3} y'''(\theta_1)$$

and c_2 is in error by

$$-\frac{h^3}{12} y'''(\theta_2)$$

If y''' is approximately constant in the interval, then

$$p - c = \frac{5h^3}{12} y'''$$

Thus the error in the corrector value is approximately

$$-\tfrac{1}{5}(p - c)$$

Hence we *add*

$$\tfrac{1}{5}(p - c) = -0.002$$

to obtain the final value

$$y = c + \tfrac{1}{5}(p - c) = 0.422$$

We are now ready to repeat the cycle until we reach $x = 1$.   (See Table 22.4.)
  As a check we find that we have computed $y(1) = 1.546$ while the true value
is 1.557.   The error 0.011 represents a relative error

$$\frac{0.011}{1.557} \times 100 = \frac{7}{10} \text{ percent}$$

which is fairly good, considering the spacing and the use of a slide rule.   This is
known as the modified Euler process for integrating an ordinary differential
equation.

PROBLEMS

22.4.1   Solve numerically $y' = y^2$, $(0) = 1$, $h = 0.2$ for $0 \leq x \leq 1$.
22.4.2   Solve numerically $y' = y^2 - \sin^2 x$, $y(0) = 0$, $h = \pi/6$ for $0 \leq x \leq 2\pi$.

Table 22.4   NUMERICAL SOLUTION OF $y' = y^2 + 1$, $y(0) = 0$

x	y	y'	p	p'	c	$p - c$	$\frac{p-c}{5}$
0	0	1					
0.2	0.203	1.0412					
0.4	0.422	1.178	0.41648	1.173	0.424	−0.008	−0.002
0.6	0.683	1.466	0.674	1.455	0.685	−0.011	−0.002
0.8	1.027	2.047	1.008	2.016	1.031	−0.023	−0.004
1.0	1.546		1.502	3.256	1.557	−0.055	−0.011
Correct value	1.557						
Error	0.011						

22.5 STABILITY OF THE PREDICTOR ALONE

It is sometimes proposed that the predictor formula

$$y_{n+1} = y_{n-1} + 2hy_n' \qquad (22.5.1)$$

be used with no corrector following.   Let us examine the stability of such a proposal before we examine the whole predictor-corrector process.

Let $z(x)$ be the true solution of the equation

$$z' = f(x, z) \qquad (22.5.2)$$

and let y_n be the computed solution at $x = x_n$.   If the computed solution y_n is put in the differential equation, it will fit exactly, since the computed y' value was found from the equation.   Thus we have (neglecting roundoff errors)

$$y_n' = f(x, y_n) \qquad (22.5.3)$$

Set

$$\varepsilon_n = z_n - y_n$$

Then subtracting Eq. (22.5.3) from (Eq. (22.5.2), we get

$$\varepsilon_n' = z_n' - y_n' = f(x, z_n) - f(x, y_n) \qquad (22.5.4)$$

$$= \frac{\partial f(x, \theta)}{\partial y} \varepsilon_n = A\varepsilon_n$$

where A is defined to be $A = \partial f/\partial y$ and θ lies between y_n and z_n.

When we substitute the true solution in the difference equation of the predictor, we get an error $e(x)$, since the true solution does not generally satisfy the difference equation

$$z_{n+1} = z_{n-1} + 2hz_n' + e_n \qquad (22.5.5)$$

This time subtracting the two difference equations, Eq. (22.5.1) from Eq. (22.5.5), we get

$$\varepsilon_{n+1} = \varepsilon_{n-1} + 2h\varepsilon_n' + e_n \qquad (22.5.6)$$

We are concerned with *estimating* the *local growth* of the error ε_n.   For this purpose we may assume that e_n and $\partial f/\partial y = A$ are all constants.   Unless they vary slowly, the step size of the integration is too large.   Under these assumptions, we may put Eq. (22.5.4) in Eq. (22.5.6) to get

$$\varepsilon_{n+1} = \varepsilon_{n-1} + 2hA\varepsilon_n + e_n$$

or

$$\varepsilon_{n+1} - 2Ah\varepsilon_n - \varepsilon_{n-1} = e_n \qquad (22.5.7)$$

which is a linear difference equation with constant coefficients.

The characteristic equation of this difference equation is

$$\rho^2 - 2Ah\rho - 1 = 0 \quad (22.5.8)$$

Hence the general solution is

$$\varepsilon_n = C_1(\rho_1)^n + C_2(\rho_2)^n \quad (22.5.9)$$

where

$$\rho_1 = Ah + \sqrt{A^2h^2 + 1}$$
$$\rho_2 = Ah - \sqrt{A^2h^2 + 1} \quad (22.5.10)$$

If $A = \partial f/\partial y > 0$, then $\rho_1 > 1$; while if $A < 0$, then $|\rho_2| > 1$. In either case one of the two terms $\rho_1{}^n$ or $\rho_2{}^n$ becomes large as $n \to \infty$. Hence the method is said to be unstable.

Thus we see that this predictor formula, used by itself alone, gives rise to a solution that is not stable since one of the zeros of the characteristic equation (22.5.8) will be $|\rho_i| > 1$. But see Sec. 22.6.

PROBLEMS

22.5.1 Try solving numerically $y' = -y, y(0) = 1, h = 0.1$, for $0 \leq x \leq 1$, using the predictor only, and compare the results with theory.

22.6 STABILITY OF THE CORRECTOR

The proposal that we iterate the corrector until no change occurs can be investigated by the same method we used on the predictor. The corrector formula

$$y_{n+1} = y_n + \frac{h}{2}(y'_{n+1} + y'_n) \quad (22.6.1)$$

leads, by exactly the same argument, to the characteristic equation

$$\left(1 - \frac{Ah}{2}\right)\rho - \left(1 + \frac{Ah}{2}\right) = 0$$
$$\rho = \frac{1 + Ah/2}{1 - Ah/2} = 1 + Ah + \frac{(Ah)^2}{2} + \frac{(Ah)^3}{4} + \cdots \quad (22.6.2)$$

provided that $|Ah/2| < 1$.

We first examine $Ah/2$ by examining the iteration process used on the corrector. Let c_i be the ith iteration. We compute the change in the iteration each step by writing

$$c_{i+1} - c_i = \frac{h}{2}[f(x, c_i) - f(x, c_{i-1})]$$

$$= \frac{h}{2}\frac{\partial f}{\partial c_i}(c_i - c_{i-1})$$

$$\approx \frac{Ah}{2}(c_i - c_{i-1})$$

Evidently the process will converge if and only if the *convergence factor*

$$\left|\frac{Ah}{2}\right| < 1 \qquad (22.6.3)$$

Returning to Eq. (22.6.2), we note that ρ is given by the first three terms of e^{Ah} plus a slight error in the fourth term.

In the simple equation

$$y' = Ay$$

where to be specific we assume $A > 0$, we have

$$\frac{\partial f}{\partial y} = A$$

and the true solution at x_n is

$$y_n = Ce^{Ahn}$$

which is the solution by Eq. (22.6.2) (if terms in h^3 are neglected).

The criterion that a characteristic root $|\rho_i| \approx |e^{Ah}| < 1$ means instability is thus seen to be misleading. Evidently, if the solution grows as a geometric progression, errors that grow at the same rate or less will not necessarily ruin the solution. On the other hand, if $A < 0$, then the errors must die out at least as rapidly as e^{Ahn}, or they will dominate the solution being computed. The fact that the errors are bounded is no help in the situation of a decreasing solution. Loosely speaking, the number of adjacent solution curves that we cross is the significant factor when we make an error.

This suggests that what we want for differential equations is the criterion of *relative stability*, which is defined as the rate of growth of the error *relative* to the growth of the solution. When the relative stability is less than 1 in size, then the noise due to an isolated roundoff, or other, error will not grow more rapidly than the solution.

In view of this new criterion of relative stability, it is necessary to reexamine the discussion in Sec. 22.5 of the use of the predictor alone.   If we assume that $|Ah| < 1$, then the characteristic roots ρ_i can be expanded as

$$\rho_1 = Ah + \sqrt{1 + (Ah)^2} = Ah + 1 + \frac{(Ah)^2}{2} - \frac{1}{8}(Ah)^4 + \cdots$$

$$\rho_2 = Ah - \sqrt{1 + (Ah)^2} = Ah - [1 + \frac{(Ah)^2}{2} - \frac{1}{8}(Ah)^4 + \cdots]$$

or

$$\rho_1 = 1 + Ah + \frac{(Ah)^2}{2} + \cdots \approx e^{Ah}$$

$$\rho_2 = -\left[1 - Ah + \frac{(Ah)^2}{2} - \cdots\right] \approx -e^{-Ah} \qquad (22.6.4)$$

The solution itself grows as e^{Ah}.   From this we see that if $A > 0$, then the repeated use of the predictor can be expected to give reasonable results; but if $A < 0$, then $(\rho_2)^n$ will grow in size and oscillate, while the true solution decreases.   Hence the method of repeated use of the predictor is relatively unstable only when $A < 0$. (See Prob. 22.5.1.)

Returning to the iterated corrector process, we find that the characteristic equation has only one root, Eq. (22.6.2), and the solution behaves exactly as this root indicates; thus the iterated corrector method is relatively stable.

PROBLEMS

22.6.1   Show that with the use of only the predictor, the result of numerically integrating $y' = y$, $y(0) = 1$, $h = 0.2$ for $0 \leq x \leq 1$ agrees with theory.

22.7   SOME GENERAL REMARKS

We have now given four examples of stability analysis, the first applied to indefinite integrals, the second to a particular convolution integral, and the last two to differential equations.   Note that stability for integrals and for differential equations is not the same; for differential equations there are feedback paths from the derivatives which are not present in the case of integrals.   The necessity for stability analysis arises whenever old values are used to compute new values and thus produce feedback of errors.   There is no simple rule to apply in all cases, but it should be evident that the subject of stability is closely connected with feedback, a subject on which many books have been written. We have also indicated that *for differential equations the concept of stability is misleading, and relative stability is a more reasonable criterion.*

The stability analysis that we did on the simple integration system was incomplete. We analyzed the predictor and the corrector separately, neglected any cross-effects, and ignored the final mop-up of the error. It is possible to write the complete system of difference equations and construct the corresponding characteristic equation, but the results are very involved. Rather, we prefer to make a few remarks about how they interact. The instability of the predictor has very little effect on the relative stability of the result after the corrector is applied once. The final mop-up also causes little trouble in regard to relative stability.

It may occur to the reader that the same argument that we used to mop up the error of the corrected value could be used to modify the predicted value and thus reduce the error in the evaluation of the predicted derivative which enters into the final value. To do this, we could add to the predicted value $-\frac{4}{5}(p_n - c_n)$ of the last cycle. If there is a tendency to oscillate, this will emphasize it. On the other hand, there is no doubt that if instability and roundoff are not troubling the computation, such a term has a good effect on the accuracy of the whole computation.

PROBLEMS

22.7.1 Develop the formulas for the integration system of predict, modify, correct, and final value.

22.8 SYSTEMS OF EQUATIONS

A single differential equation of nth order

$$y^{(n)} = f(x, y, y', \dots, y^{(n-1)})$$

can be reduced to a system of n first-order equations by a simple change in notation. Write

$$y = y_0 \qquad y' = y_1 \qquad y'' = y_1' = y_2$$
$$y''' = y_2' = y_3, \dots, y^{(n-1)} = y_{n-1}$$

and we have the equivalent system

$$y_0' = y_1$$
$$y_1' = y_2$$
$$y_2' = y_3$$
$$\dots\dots\dots$$
$$y_{n-1}' = f(x, y_0, y_1, y_2, \dots, y_{n-1})$$

In order to apply the method of integration to a system, we simply apply the operations to the separate equations in parallel.   The cross connections between the equations are taken care of when the derivatives are computed; otherwise the equations are treated separately, but in parallel.   However, see Chap. 24 for other methods.

The direction-field method is generally applicable to a single first-order equation only, but in the case of two first-order equations in which the independent variable x does not occur,

$$y' = f(y,z)$$
$$z' = g(y,z)$$

dividing one equation by the other produces a first-order equation without x. The direction field for this equation gives y versus z, and sometimes it is sufficient to indicate the trajectory of y versus z without specifying the x for which the point y, z on the curve occurs.

A GENERAL THEORY OF PREDICTOR-CORRECTOR METHODS

23.1 INTRODUCTION

Chapters 21 and 22 introduced ideas and techniques which we now use to develop a general theory of predictor-corrector methods for numerically integrating ordinary differential equations. This theory includes the Milne and Adams-Bashforth methods as special cases. Thus we are in a position to compare these and other widely used methods within the framework of a single theory.

Predictor-corrector methods for integrating ordinary differential equations are widely used because of the following advantages:

1 The difference between the predicted and corrected values provides one measure of the error being made at each step and hence can be used to control the step size employed in the integration.

2 Only one or two evaluations of the derivatives need to be computed at each step (as compared with four for the Runge-Kutta method of Chap. 24), and on high-order systems this can save considerable computing effort.

3 It is easy to catch many kinds of machine failures.

Some disadvantages are that the methods are complex to program and are not self-starting.   A method for starting will be discussed in Chap. 24.

There are three main sources of trouble in predictor-corrector methods for integrating ordinary differential equations.   They are:

1   Truncation errors that arise from the finite approximations for the derivatives

2   Propagation errors (instability) that arise from solutions of the approximate difference equations that do not correspond to solutions of the differential equations

3   Amplification of roundoff errors due to certain combinations of coefficients in the finite difference formulas

The plan of this chapter is to examine a broad class of formulas in a systematic fashion and indicate how the choice of a particular formula is a compromise between conflicting desires.   Since the corrector formula plays the more important role in the integration of ordinary differential equations, it is natural to examine the corrector first and more intensively.

After examining the corrector, we take up the corresponding problem for the predictor.   With families of both predictors and correctors available, we are then in a position to examine the problem of designing a system for integrating ordinary differential equations.   We conclude with a discussion of some experimental results.

The methods examined have a truncation error of order h^5; experience shows that these are the most widely used.   Chapter 22 developed a predictor-corrector method of order h^3, and the intermediate theory of order h^4 is left as exercises.

The third-order method of Chap. 22 had no free parameters; the fourth-order methods treated in the exercises introduce two new coefficients, one of which is used to increase the order of the truncation error and one for stability; the fifth-order methods in the text have still another two coefficients which are again used, one for truncation and one for stability, and so on for higher-order methods.

In many applications the truncation error of the finite approximation to the derivatives of a differential equation, or to a system of differential equations, greatly exceeds the errors due to roundoff at each step.   However, in building special-purpose computers that must operate in real time, there is great pressure to keep the length of the numbers, and hence the basic level of accuracy, as low as possible.   In such situations the truncation error may be about the same size as the roundoff error, and the latter cannot be ignored in designing the methods for integrating the ordinary differential equations of the problem. It also happens with increasing frequency these days that in some problems, such as trajectories to the moon and planets, eight- and even ten-decimal-place accuracy can produce significant roundoff errors, and some methods described here are then useful. However, whole books have been written on the topic of numerical integration of ordinary differential equations, and we can cover only a small part of the topic.

23.2 TRUNCATION ERROR

The most general linear corrector formula that uses the function and the first-derivative information at the last three points of the solution, plus an estimate of the derivative at the point being computed, is

$$y_{n+1} = a_0 y_n + a_1 y_{n-1} + a_2 y_{n-2} + h(b_{-1} y'_{n+1} + b_0 y'_n + b_1 y'_{n-1}$$
$$+ b_2 y'_{n-2}) + E_5 \frac{h^5 y^{(5)}}{5!} (\theta)$$

Other linear forms could be used, but this one includes many of the common methods and is the same as we used in Chap. 21 when examining indefinite integrals. The change from the treatment in Chap. 21 is that now we have feedback through the derivative terms, as well as from old solution values; hence the investigation of the stability is more complex.

The formula as written implies that it is exact for polynomials through degree 4; that is, it is exact for $y = 1, x, x^2, x^3, x^4$. Using these five conditions and taking a_1 and a_2 as parameters, we get, as before in Sec. 21.3,

$$a_0 = 1 - a_1 - a_2 \qquad b_0 = \tfrac{1}{24}(19 + 13a_1 + 8a_2)$$
$$a_1 = a_1 \qquad b_1 = \tfrac{1}{24}(-5 + 13a_1 + 32a_2)$$
$$a_2 = a_2 \qquad b_2 = \tfrac{1}{24}(1 - a_1 + 8a_2)$$
$$b_{-1} = \tfrac{1}{24}(9 - a_1) \qquad E_5 = \tfrac{1}{6}(-19 + 11a_1 - 8a_2)$$

Five of the seven coefficients are used to reduce the truncation error; the other two will be used to reduce the instability and roundoff errors rather than to reduce the truncation error still further. As in Chap. 21, the difference of the predictor and corrector values will be used both to provide an estimate of the error and to "mop up" the leading error terms.

The usual method of finding the error term is to use the Taylor expansion (Sec. 16.2),

$$y(x) = y(A) + (x - A)y'(A) + \frac{(x - A)^2}{2!} y''(A) + \cdots$$
$$+ \frac{(x - A)^{m-1}}{(m - 1)!} y^{(m-1)}(A) + \frac{1}{(m - 1)!} \int_A^x y^{(m)}(s)(x - s)^{m-1} \, ds$$

with $A = -2h$ and $m = 5$.   Following the steps in Chap. 21, we obtain $(B = h)$

$$\text{Error} = R(y) = \int_{-2h}^{h} y^{(5)}(s)G(s)\,ds$$

where $G(s)$ is the *influence function*

$$G(s) = (\text{LHS} - \text{RHS})\frac{(x-s)_{+}^{m-1}}{(m-1)!}$$

$$4!\,G(s) = (h-s)_{+}^{4} - a_0(-s)_{+}^{4} - a_1(-h-s)_{+}^{4} - a_2(-2h-s)_{+}^{4}$$
$$- 4h[b_{-1}(h-s)_{+}^{3} + b_0(-s)_{+}^{3} + b_1(-h-s)_{+}^{3} + b_2(-2h-s)_{+}^{3}]$$

and (see Sec. 16.4)

$$(a-s)_{+}^{k} = \begin{cases} (a-s)^{k} & \text{if } a-s \geq 0 \\ 0 & \text{if } a-s \leq 0 \end{cases} \quad k > 0$$

If $G(s)$ is of constant sign, then there exists a θ $(-2h < \theta < h)$ such that the error can be written as

$$\text{Error} = y^{(5)}(\theta)\int_{-2h}^{h} G(s)\,ds$$

If $G(s)$ is not of constant sign, then there will be functions $f(s)$ for which the error does not have this general form (Sec. 16.7).

The investigation in Sec. 21.4 produced the region of constant sign for $G(s)$ (shown in Figs. 21.4.1 and 21.4.2 of Sec. 21.4), in which the main interest lies and for which the lines of equal truncation error are shown.

23.3   STABILITY

Following the method of Sec. 21.5, we let $z = z(x)$ be the true solution, which means that z satisfies both

$$z' = f(x, z) \qquad (23.3.1)$$

and

$$z_{n+1} = a_0 z_n + a_1 z_{n+1} + a_2 z_{n-2}$$
$$+ h(b_{-1}z'_{n+1} + b_0 z'_n + b_1 z'_{n-1} + b_2 z'_{n-2}) + e_n \qquad (23.3.2)$$

and $y = y_n$ be the calculated solution

$$y'_n = f(x_n, y_n) \qquad (23.3.3)$$

$$y_{n+1} = a_0 y_n + a_1 y_{n-1} + a_2 y_{n-2} + h(b_{-1}y'_{n+1} + b_0 y'_n + b_1 y'_{n-1} + b_2 y'_{n-2})$$
$$(23.3.4)$$

We set

$$\varepsilon_n = z_n - y_n$$

and subtract Eqs. (23.3.2) and (23.3.4) to get

$$\varepsilon_{n+1} = a_0\varepsilon_n + a_1\varepsilon_{n-1} + a_2\varepsilon_{n-2} + h(b_{-1}\varepsilon'_{n+1} + b_0\varepsilon'_n + b_1\varepsilon'_{n-1} + b_2\varepsilon'_{n-2}) + e_n \tag{23.3.5}$$

while Eqs. (23.3.1) and (23.3.3) give, using the mean value theorem,

$$\varepsilon'_n = f(x_n, z_n) - f(x_n, y_n)$$
$$= \frac{\partial f(x_n, \theta)}{\partial y}\varepsilon_n = A\varepsilon_n \tag{23.3.6}$$

where θ is the usual mean value. Putting Eq. (23.3.6) in Eq. (23.3.5), we get

$$(1 - b_{-1}Ah)\varepsilon_{n+1} = (a_0 + b_0 Ah)\varepsilon_n + (a_1 + b_1 Ah)\varepsilon_{n-1} + (a_2 + b_2 Ah)\varepsilon_{n-2} + e_n \tag{23.3.7}$$

which is a linear difference equation.

In studying the growth of the error ε_n, it is reasonable to assume that e_n and $\partial f/\partial z = A$ are both constants; in practice they vary slowly from step to step. Hence we assume that Eq. (23.3.7) is a linear difference equation with constant coefficients (locally).

In Eq. (23.3.7) only the quantity Ah occurs, not h or A separately. While Ah can vary widely, it is necessary to estimate reasonable values that may be expected. To gain a little perspective, consider the equations

$$y' = \pm Ay \qquad y(0) = 1$$

The truncation error at each step will be taken as $h^5 A^5/40$ ($\frac{1}{40}$ is a typical value for $E_5/5!$; see Table 21.7). This leads to Table 23.3. The table shows that if the truncation error is to be reasonably small for these equations, then $|A| \leq 5/10$ and for accurate solutions $|Ah| \leq 2/10$.

Table 23.3 TRUNCATION ERROR
AS A FUNCTION OF $|Ah|$

$|Ah|$	$\dfrac{h^5 y^{(5)}}{40}$
0.1	2.50×10^{-7}
0.2	8.00×10^{-6}
0.3	6.07×10^{-5}
0.4	2.56×10^{-4}
0.5	7.88×10^{-4}

Of course other equations could have different patterns of higher derivatives (probably growing more rapidly than simple powers), and one cannot say anything very specific about what $|Ah|$ will be in comparison with $E_5 h^5 y^{(5)}/5!$. Nevertheless, we shall adopt the attitude of $|Ah| < 0.4$ for most equations of interest.

Figure 21.5.1 shows the stability region in case $Ah = 0$. When $Ah \neq 0$, the corresponding homogeneous difference equation leads to the characteristic equation

$$Ah = \frac{\rho^3 - a_0 \rho^2 - a_1 \rho - a_2}{b_{-1}\rho^3 + b_0 \rho^2 + b_1 \rho + b_2} \qquad (23.3.8)$$

This equation is more complex than the corresponding equation in Chap. 21 because of the extra feedback from the derivative terms. The solution is

$$\varepsilon_n = C_1(\rho_1)^n + C_2(\rho_2)^n + C_3(\rho_3)^n \qquad (23.3.9)$$

The solution of the original differential equation (23.3.1) tends to grow as

$$y_n = Ce^{Ahn}$$

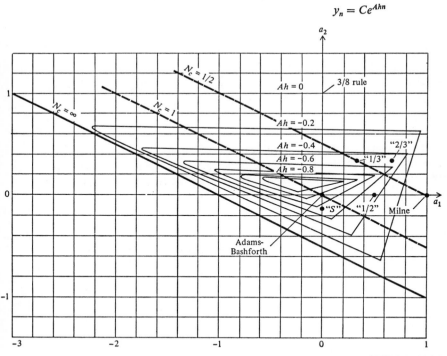

FIGURE 23.3.1
Stability region.

and generally it is not the growth of the error in itself but the growth *relative* to the local growth of the solution which concerns us. Thus it is not stability

$$|\rho_i| < 1$$

that we want, but rather *relative stability*

$$|\rho e^{-Ah}| \le 1 \quad (23.3.10)$$

One of the roots, which we shall designate as ρ_3, behaves as

$$\rho_3 = 1 + Ah + \frac{(Ah)^2}{2} + \frac{(Ah)^3}{3!} + \frac{(Ah)^4}{4!} + k\frac{(Ah)^5}{5!} \simeq e^{Ah} \quad (23.3.11)$$

for small Ah. Thus

$$\rho_3 e^{-Ah} \simeq 1$$

and ρ_1, ρ_2 are the roots which may give relative instability trouble.

The triangle along which, for $Ah = 0$, the roots take on the values

$$|\rho e^{-Ah}| = |\rho| = 1$$

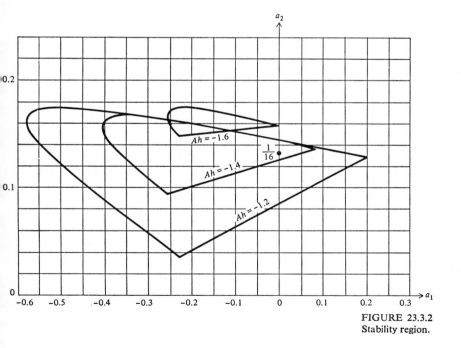

FIGURE 23.3.2
Stability region.

becomes, for nonzero Ah, somewhat more complex.   For $\rho = -e^{Ah}$ we obtain a linear equation

$$\alpha a_2 + \beta a_1 + \gamma = 0 \quad (23.3.12)$$

where α, β, and γ are combinations of Ah, e^{Ah}, e^{2Ah}, and e^{3Ah}.   But the other two sides of the triangle become a hyperbola.   In Figs. 23.3.1 and 23.3.2 we show for negative values of Ah one branch of the hyperbola until it is cut off by the line of Eq. (23.3.12).   While values inside the other branch would be stable, in the intervening gap there would be unstable values; and so they do not seem to be useful except possibly for special situations.   These charts, Figs. 23.3.1 and 23.3.2, show stability regions for the roots ρ_1 and ρ_2; for large values of $|Ah|$ the root ρ_3 is far from its required value of e^{Ah}, and the truncation error cannot be neglected.

PROBLEMS

23.3.1   Plot the line of Eq. (23.3.12) for $Ah > 0$, and discuss the region of stability.
23.3.2   Using Prob. 21.7.1, develop the corresponding theory for differential equations. Draw the stability region as a function of a_1 and Ah.

23.4   ROUNDOFF NOISE

The theory of the propagation of roundoff noise for a differential equation is very much like that for indefinite integrals treated in Sec. 21.6.   For $Ah = 0$ the line

$$a_2 = -\frac{a_1}{2}$$

is the dividing line between the region of growth and the region of decay of a random error.   Inside the stability region the noise amplification is given by (see Sec. 21.6)

$$N_c = \frac{1}{1 + a_1 + 2a_2}$$

When $Ah \neq 0$, the division between the two regions is more complex, but it is not worth exploring it here; we merely need to recall that it is better to be above the line than below it.

To control the uncorrelated noise, it is well to keep

$$N_a = (a_0{}^2 + a_1{}^2 + a_2{}^2)^{1/2}$$

reasonably small.

On machines that "chop," that is, drop the digits rather than round properly, there is a second kind of trouble with roundoff. If, for the moment, we regard the various numbers as if they were mass, then chopping a positive number decreases the mass, while chopping a negative number increases the mass. In most computations we can reasonably expect some balance between the positive and negative numbers. But for a differential equation in some interval there well may be an excess of one effect or another, and in the next interval, since the signs of the numbers involved tend to change infrequently, there will be an effect *in the same direction*. Thus step after step we will find that chopping pushes us in the same direction, and the effect accumulates linearly, rather than in a customary way such as the square root of the number of operations involved. Chopping, therefore, needs to be watched carefully when integrating differential equations.

23.5 THE THREE-POINT PREDICTOR

The most desirable predictor to go with the corrector (Sec. 23.2) would be one that used the information at the last three points, namely,

$$y_{n+1} = A_0 y_n + A_1 y_{n-1} + A_2 y_{n-2} + h(B_0 y'_n + B_1 y'_{n-1} + B_2 y'_{n-2}) + \bar{E}_5 \frac{h^5 y^{(5)}}{5!}$$

The coefficients may be found by the usual process or by setting $b_{-1} = 0$. The results are

$$A_0 = -8 - A_2 \qquad B_0 = \frac{17 + A_2}{3}$$

$$A_1 = 9 \qquad B_1 = \frac{14 + 4A_2}{3}$$

$$A_2 = A_2 \qquad B_2 = \frac{-1 + A_2}{3}$$

$$\bar{E}_5 = \frac{40 - 4A_2}{3}$$

The large value $A_1 = 9$ means that at best we shall have a large noise amplification

$$N_a = (A_0^2 + A_1^2 + A_2^2)^{1/2} > 9$$

in the predictor.

In order to keep the three-point predictor and avoid the high noise amplification, we could make the formula exact through only x^3, leaving the error term of order h^4. This leads to the family of predictors

$$A_0 = 1 - A_1 - A_2 \qquad B_0 = \frac{23 + 5A_1 + 4A_2}{12}$$

$$A_1 = A_1 \qquad\qquad B_1 = \frac{-16 + 8A_1 + 16A_2}{12}$$

$$A_2 = A_2 \qquad\qquad B_2 = \frac{5 - A_1 + 4A_2}{12}$$

$$\bar{E}_4 = 9 - A_1$$

If we now use the difference of the nth predictor minus the corrector to estimate the error of the $(n + 1)$st predictor, we would modify p_{n+1} by

$$m_{n+1} = p_{n+1} - (p_n - c_n) = c_n + (p_{n+1} - p_n)$$

While such a scheme is occasionally justified, it may also be a waste of computation, especially when computing accurate solutions of differential equations, and we shall not discuss it further, although it clearly has a fifth-order error term.

23.6   MILNE-TYPE PREDICTORS

In order to have seven parameters to use in searching for a suitable predictor, we can add either one more old value of the function, namely y_{n-3}, or one old value of the derivative, namely y'_{n-3}. The first proposal includes Milne's predictor, and we call these "Milne-type predictors," while the second contains the Adams-Bashforth predictor and is named accordingly.

We therefore try first to find a formula of the form

$$y_{n+1} = A_0 y_n + A_1 y_{n-1} + A_2 y_{n-2} + A_3 y_{n-3} + h(B_0 y'_n$$

$$+ B_1 y'_{n-1} + B_2 y'_{n-2}) + \bar{E}\frac{h^5 y^{(5)}}{5!}$$

The usual method gives

$$A_0 = -8 - A_2 + 8A_3 \qquad B_0 = \frac{17 + A_2 - 9A_3}{3}$$

$$A_1 = 9 - 9A_3 \qquad\qquad B_1 = \frac{14 + 4A_2 - 18A_3}{3}$$

$$A_2 = A_2 \qquad\qquad\quad B_2 = \frac{-1 + A_2 + 9A_3}{3}$$

$$A_3 = A_3 \qquad\qquad\quad \bar{E}_5 = \frac{40 - 4A_2 + 72A_3}{3}$$

For this formula to have the indicated error term, it is necessary that the *influence function*

$$G(s) = (h - s)_+^4 - A_0(-s)_+^4 - A_1(-h - s)_+^4 - A_2(-2h - s)_+^4$$
$$- A_3(-3h - s)_+^4 - 4h[B_0(-s)_+^3 + B_1(-h - s)_+^3 + B_2(-2h - s)_+^3]$$

be of constant sign for $h \geq s \geq -3h$. The line in the A_2, A_3 plane along which $G(s) = 0$ moves up from below to the axis $A_3 = 0$ as s goes from h to $-2h$, and it remains at $A_3 = 0$ for $-2h \geq s \geq -3h$. Thus we are restricted to $A_3 \geq 0$ if the error term is to be correct.

If we try to reduce the truncation error by choosing $A_3 = 0$, then we find $A_1 = 9$, which is uncomfortably large. On the other hand, if we try to minimize the noise amplification

$$N_a = (A_0{}^2 + A_1{}^2 + A_2{}^2 + A_3{}^2)^{1/2}$$

we get

$$A_0 = -\tfrac{4}{114} \qquad A_1 = \tfrac{9}{114} \qquad A_2 = -\tfrac{4}{114} \qquad A_3 = \tfrac{113}{114}$$

This is very close to Milne's predictor

$$y_{n+1} = y_{n-3} + \frac{4h}{3}(2y'_n - y'_{n-1} + 2y'_{n-2}) + \frac{14}{45}h^5 y^{(5)}(\theta)$$

with $A_0 = A_1 = A_2 = 0$ and $A_3 = 1$. The gain does not seem worth it.

Thus if we try this approach, we are led to Milne's predictor. As discussed in Sec. 22.4, this may be combined in the pattern of Chap. 22 with any suitable corrector. Of course other choices of the parameters A_2, A_3 are possible.

PROBLEMS

23.6.1 Examine predictors of the form

$$y_{n+1} = A_0 y_n + A_1 y_{n-1} + A_2 y_{n-2} + h(B_0 y'_n + B_1 y'_{n-1})$$
$$+ \bar{E}_4 \frac{h^4 y^{(4)}(\theta)}{4}$$

using A_2 as the parameter.

Ans. $A_0 = -4 - 5A_2$ $A_1 = 5 + 4A_2$ $A_2 = A_2$
  $B_0 = 4 + 2A_2$ $B_1 = 2 + 4A_2$ $\bar{E}_4 = 4 - 4A_2$

If $A_2 = 1$, then $\bar{E}_4 = 0$ and $\bar{E}_5 = 12$.

404   23   A GENERAL THEORY OF PREDICTOR-CORRECTOR METHODS

23.7 ADAMS-BASHFORTH-TYPE PREDICTORS

If, instead of using an extra value of the function, we use an extra value of the derivative, y'_{n+3}, then we have the form

$$y_{n+1} = A_0 y_n + A_1 y_{n-1} + A_2 y_{n-2}$$
$$+ h(B_0 y'_n + B_1 y'_{n-1} + B_2 y'_{n-2} + B_3 y'_{n-3}) + \bar{E}_5 \frac{h^5 y^{(5)}}{5!}$$

and the values

$$A_0 = 1 - A_1 - A_2 \qquad B_1 = \frac{-59 + 19A_1 + 32A_2}{24}$$

$$A_1 = A_1 \qquad B_2 = \frac{37 - 5A_1 + 8A_2}{24}$$

$$A_2 = A_2 \qquad B_3 = \frac{-9 + A_1}{24}$$

$$B_0 = \frac{55 + 9A_1 + 8A_2}{24} \qquad \bar{E}_5 = \frac{251 - 19A_1 - 8A_2}{6}$$

In this case the influence function $G(s)$ has zeros in the A_1, A_2 plane along a line which starts as $A_1 = 9$ for $s = -2h$ and rises while tilting through negative slopes as s increases, producing no serious restraints on the coefficients.

A number of cases are given in Table 23.7, including the Adams-Bashforth predictor. The noise propagation value N_c for $h \to 0$ is the same as for the corrector and need not be repeated here.

Table 23.7   ADAMS-BASHFORTH-TYPE PREDICTORS [THE VALUES OF $\bar{E}_5 - E_5$ ARE COMPUTED FOR THE CASE $a_i = A_i \, (i = 0, 1, 2)$]

	Adams-Bashforth	High accuracy	Three-eighths rule	"$\frac{1}{3}$"	"$\frac{1}{2}$"	"$\frac{2}{3}$"
A_0	1	0	0	$\frac{1}{4}$	$\frac{1}{2}$	0
A_1	0	1	0	$\frac{1}{4}$	$\frac{1}{2}$	$\frac{2}{3}$
A_2	0	0	1	$\frac{1}{4}$	0	$\frac{1}{3}$
B_0	$\frac{35}{24}$	$\frac{8}{3}$	$\frac{21}{8}$	$\frac{91}{36}$	$\frac{119}{48}$	$\frac{191}{72}$
B_1	$-\frac{39}{24}$	$-\frac{5}{3}$	$-\frac{9}{8}$	$-\frac{63}{36}$	$-\frac{99}{48}$	$-\frac{107}{72}$
B_2	$\frac{37}{24}$	$\frac{4}{3}$	$\frac{15}{8}$	$\frac{57}{36}$	$\frac{69}{48}$	$\frac{109}{72}$
B_3	$-\frac{9}{24}$	$-\frac{1}{3}$	$-\frac{3}{8}$	$-\frac{13}{36}$	$-\frac{17}{48}$	$-\frac{25}{72}$
E_5	$\frac{251}{6}$	$\frac{116}{3}$	$\frac{243}{6}$	$\frac{121}{3}$	$\frac{161}{4}$	$\frac{707}{18}$
$\bar{E}_5 - E_5$	$\frac{270}{6}$	$\frac{240}{6}$	$\frac{270}{6}$	$\frac{260}{6}$	$\frac{255}{6}$	$\frac{250}{6}$

PROBLEMS

23.7.1 Examine predictors of the form

$$y_{n+1} = A_0 y_n + A_1 y_{n-1} + h(B_0 y'_n + B_1 y'_{n-1} + B_2 y'_{n-2})$$
$$+ \bar{E}_4 \frac{h^4 y^{(4)}(\theta)}{4}$$

using A_1 as a parameter.

Ans. $A_0 = 1 - A_1$ $B_0 = \dfrac{23 + 5A_1}{12}$ $B_2 = \dfrac{5 - A_1}{12}$

$\qquad\quad A_1 = A_1$ $B_1 = \dfrac{-16 + 8A_1}{12}$ $\bar{E}_4 = 9 - A_1$

23.8 GENERAL REMARKS ON THE CHOICE OF A METHOD

In designing a specific method, the choice of the predictor and corrector involves balancing three somewhat incompatible objectives:

1 Low truncation error
2 Large margin of stability
3 Protection against roundoff troubles

It is best to begin by selecting a corrector, since it plays the more important role.   Figure 21.5.1 shows the domain from which we may select our corrector. It is clear that Milne's corrector ($a_1 = 1$, $a_2 = 0$) is about as accurate as can be found.   Figure 23.3.1 shows the stability measure.   If $A = \partial f / \partial y$ is positive in the entire interval of integration, then Milne is a good choice.   But if $A = \partial f / \partial y$ can be negative, then Milne's corrector is unstable, and we need to decide on how we shall compromise between the two troubles, instability and truncation, which are measured, respectively, by

$$\frac{\partial f}{\partial y} h \equiv Ah \qquad \text{and} \qquad \frac{E_5 h^5 y^{(5)}(\theta)}{5!}$$

No specific relationship between these two terms exists, although there is clearly a tendency for both to be large or both to be small.

We shall arbitrarily select $Ah = -\frac{4}{10}$ as a safe amount of protection against instability troubles for moderately accurate solutions for most equations.   With this choice, a comparison of Figs. 21.5.1 and 23.3.1 shows that the truncation error is about the same all along the straight side of the contour $Ah = -\frac{4}{10}$.   Round-off troubles suggest moving to the upper right and selecting the point ($a_1 = \frac{2}{3}$,

$a_2 = \frac{1}{3}$) as the appropriate corrector.   In a sense, this choice is a mixture of two-thirds of Milne with one-third of the three-eighths-rule.   Milne and Reynolds[1] have suggested the occasional use of the three-eighths-rule while using Milne's method.   The above proposed two-thirds method is a particular such mixture (at every step as it were) and provides stability in a more satisfactory manner.

While we have selected $Ah = -\frac{4}{10}$ as a generally safe amount of protection against instability, it should be noted that for computing low-accuracy solutions of equations with large negative $\partial f/\partial y$, a more stable method should be chosen. From Fig. 23.3.1 it appears that $a_1 = 0$, $a_2 = \frac{1}{16}$ is a very stable method, but it has obviously a fairly large ($E_5 = -\frac{23}{12}$) truncation error.

PROBLEMS

23.8.1   From Prob. 23.3.2, discuss the choice of a fourth-order corrector.

23.9   CHOICE OF PREDICTOR

In choosing a predictor there is a strong temptation to use the same values for the A's in the predictor as for the a's in the corrector.   For one reason, in computing trajectories to the moon or the planets, for example, we could compute the sum of the y values (positions) by a double-precision routine and probably manage to compute the y' terms, which depend on differences of position, with single precision and at the same time achieve a large increase in accuracy.   This matching of the a's and A's eliminates the Milne-type predictors, which have slightly more accuracy than the Adams-Bashforth type.   We shall, however, examine only the cases where they match, as this provides quite a range of systems.   The reader is invited to design his own method to fit his specific case.

The difference between the predicted and corrected values of y_{n+1} measures the fifth derivative.   With matching a's and A's, the y terms cancel, and we have

$$p_{n+1} - c_{n+1} = h[B_0\, y_n' + B_1 y_{n-1}' + B_2\, y_{n-2}' + B_3\, y_{n-3}'$$
$$- (b_{-1}p_{n+1}' + b_0\, y_n' + b_1 y_{n-1}' + b_2\, y_{n-2}')]$$

which can generally be computed quite accurately.

PROBLEMS

23.9.1   Discuss the corresponding fourth-order theory.

---

[1] W. E. Milne and R. R. Reynolds, Stability of Numerical Solution of Differential Equations, *J. Assoc. Computing Machinery*, vol. 6, pp. 196–203, April, 1959.

23.10   SELECTED FORMULAS

It is a common observation that $p_n - c_n$ tends to be somewhat constant from step to step so that, knowing that p_{n+1} is in error by

$$\frac{\bar{E}_5 h^5 y^{(5)}}{5!}$$

we naturally use

$$-\frac{\bar{E}_5}{\bar{E}_5 - E_5}(p_n - c_n) = -\frac{\bar{E}_5 h^5 y^{(5)}}{5!}$$

to mop up this error.   (In the first predictor step, when we lack the $p_n - c_n$, we use $p_n - c_n = 0$.)   A similar argument applies to the corrector where it is natural to use

$$\frac{E_5}{\bar{E}_5 - E_5}(p_{n+1} - c_{n+1}) \approx \frac{E_5 h^5 y^{(5)}}{5!}$$

as the correction to the corrector value.

Applying this to the two-thirds case as an example, we use the following procedure:

Predict:

$$p_{n+1} = \frac{2y_{n-1} + y_{n-2}}{3}$$

$$+ \frac{h}{72}(191y'_n - 107y'_{n-1} + 109y'_{n-2} - 25y'_{n-3}) + \frac{707}{2,160} h^5 y^{(5)}$$

Modify:

$$m_{n+1} = p_{n+1} - \tfrac{707}{750}(p_n - c_n)$$

Correct:

$$c_{n+1} = \frac{2y_{n-1} + y_{n-2}}{3}$$

$$+ \frac{h}{72}(25m'_{n+1} + 91y'_n + 43y'_{n-1} + 9y'_{n-2}) - \frac{43}{2,160} h^5 y^{(5)}$$

Final value:

$$y_{n+1} = c_{n+1} + \tfrac{43}{750}(p_{n+1} - c_{n+1})$$

In this process we get an exact answer if $y^{(5)}(x) = $ constant, so that it may be called a sixth-order method.

It should be noted that if y_{n+1} is close to m_{n+1}, then the second evaluation of the derivative to get y'_{n+1} *need not* be done. If ε is the difference, then it will result in (assuming $|Ah| = 4/10$)

$$\varepsilon \cdot h \frac{191}{72} \left| \frac{\partial f}{\partial y} \right| \le \varepsilon \cdot 2.65 |A| h \approx 1.06\varepsilon$$

in the next predicted value.

Another case worth examination is the "one-half" which is an average of the Milne and Adams-Bashforth cases. Again arbitrarily using matching coefficients,

$$p_{n+1} = \frac{y_n + y_{n-1}}{2} + \frac{h}{48}(119y'_n - 99y'_{n-1} + 69y'_{n-2} - 17y'_{n-3}) + \frac{161}{480} h^5 y^{(5)}$$

$$m_{n+1} = p_{n+1} - \tfrac{161}{170}(p_n - c_n)$$

$$c_{n+1} = \frac{y_n + y_{n-1}}{2} + \frac{h}{48}(17m'_{n+1} + 51y'_n + 3y'_{n-1} + y'_{n-2}) - \frac{9}{480} h^5 y^{(5)}$$

$$y_{n+1} = c_{n+1} + \tfrac{9}{170}(p_{n+1} - c_{n+1})$$

This has a slightly smaller truncation error than the "two-thirds" method. However, $N_c = 2/3$ and $N_a = 0.7071$, versus $N_c = 3/5$ and $N_a = 0.75$ for the two-thirds method.

PROBLEMS

23.10.1 Discuss the corresponding fourth-order theory and develop specific formulas.

23.11 DESIGNING A SYSTEM[1]

Besides the formulas for integrating each step, it is necessary to have formulas for halving and doubling the interval and criteria for when to do so. We also need a starting routine which we shall here assume is the Runge-Kutta method (see Chap. 24).

There are two possible reasons for halving (or subdividing in any other way) the step size of the integration: One is a large truncation error; the other is instability. Protection against instability during a computation depends on the

---

[1] For a good discussion see Arnold Nordsieck, On Numerical Integration of Ordinary Differential Equations, *Math. of Comp.*, vol. 16, no. 77, pp. 22–49, January, 1962, and Gear [16].

ability to estimate $A = \partial f / \partial y$. If we evaluate the derivatives for both m_{n+1} and y_{n+1}, we get

$$f(x_{n+1}, m_{n+1}) - f(x_{n+1}, y_{n+1}) \approx \frac{\partial f(x_{n+1}, y_{n+1})}{\partial y}(m_{n+1} - y_{n+1}) \approx A(m_{n+1} - y_{n+1})$$

Note that an accurate estimate of A is not possible from this formula. The occasional estimation of A by doing the two evaluations in a single step is necessary in order to be safe from instability troubles if some other method of estimating A is not available.

If A is found to be too large for the current step size h, then we can either decrease h or change to a more stable formula. If at the same time the truncation error is low, then the change of formula to a more stable one with a larger truncation error is probably better. On the other hand, if both are close to tolerance, then h should be decreased.

We have repeatedly indicated that the truncation error is measured by

$$p_n - c_n = \frac{(\bar{E}_5 - E_5)h^5 y^{(5)}}{5!}$$

Actually, this is not so in the proposed systems, since, as noted, if $y^{(5)}$ were constant, then the truncation error would be zero. The true error is measured by the change in $p_n - c_n$ from step to step. The first temptation is to compute

$$(p_{n+1} - c_{n+1}) - (p_n - c_n)$$

and use it as a guide for when to halve the interval. Unfortunately, as a few moments' reflection will reveal, such a quantity is extremely dependent upon the local "noise" in the computation (except in the case of low-accuracy computations done on a high-accuracy computer) and does not provide a satisfactory criterion for when to halve the interval size. We are, therefore, left in the unfortunate position of knowing that the truncation error per step is a good deal less than the quantity $p_n - c_n$, but we do not know how much.

To find a suitable halving formula, we have y_n, y_{n-1}, y_{n-2} and their derivatives, so that we can estimate $y_{n-1/2}$ from them while making the formula exact for $1, x, \ldots, x^5$ (which makes the error depend on $y^{(6)}$). Such a formula is

$$y_{n-1/2} = \frac{1}{128}[45y_n + 72y_{n-1} + 11y_{n-2} + h(-9y'_n + 36y'_{n-1} + 3y'_{n-2})]$$

To obtain $y_{n-3/2}$, we can reverse the formula to get

$$y_{n-3/2} = \frac{1}{128}[11y_n + 72_{n-1} + 45y_{n-2} - h(3y'_n + 36y'_{n-1} - 9y'_{n-2})]$$

When it comes to doubling, we can do one of the following:

1 Carry extra back values
2 Restart
3 Use special formulas for two steps

In addition to the extra computing involved, the difficulty with carrying back values is that special care must be taken not to double until enough back values have been developed from the last doubling or from the initial starting.

Most programs contain provisions for both doubling and halving the interval and use two constants C_d and C_h to test

$$|p_{n+1} - c_{n+1}|$$

When halving is called for, the newly computed numbers at $n + 1$ are thrown away, the interval is halved by interpolation, and the step is tried again.   If, however, halving is called for on the first predictor-corrector step, then the whole start must be redone.

Care should be taken to keep

$$C_h > 100C_d$$

Otherwise there is the risk that, because of chance fluctuations, the machine may halve and double on successive steps.

Another point should be made: If machine errors are at all likely, then a *sudden* jump in $p_n - c_n$ is probably due to machine error rather than too large a step.   The above method, which throws away the step effectively, eliminates the machine error at the cost of halving and later doubling.   A special test could be used to watch for sudden jumps, and when jumps are found, the step can merely be repeated.

23.12   NUMERICAL VERIFICATION

A popular approach to the problem of selecting a method for numerically integrating some differential equation is to use a computer to solve a number of special cases (often chosen because they have known solutions).   While this approach is sometimes necessary, it is seldom satisfactory.

In order to check the *theory* just developed, a large number of solutions to three equations

Equation	Initial condition	Analytic solution
$y' = y$	$y(0) = 1$	$y = e^x$
$y' = -y$	$y(0) = 1$	$y = e^{-x}$
$y' = -2xy^2$	$y(0) = 1$	$y = \dfrac{1}{1 + x^2}$

were computed for $0 \leq x \leq 10$ for various step sizes h. These equations were selected because the first two have essentially the "local linearized behavior" of the general equation and clearly reveal instability troubles. The third has a solution which is often troublesome to approximate by polynomials.

Only methods whose predictor and corrector had the same coefficients for the function values (y's) were used, except that the classic Milne's method was also tried.

Because the evaluations of the derivatives were so easy, flat random noise was added to simulate a higher level of roundoff noise at this point *only* for half the runs.

The results, which are too voluminous to reproduce here, can be summarized as follows:

1 The stability conditions were verified. The use of the modifier based on $p_n - c_n$ accentuates an oscillation due to instability, but use of $p_{n-1} - c_{n-1}$ helps to prevent this tendency toward instability.

2 For moderately accurate solutions the use of the modifier and final-value steps helped by reducing errors by a factor of about 10 in size (provided that stability conditions were observed); that is,

$$\tfrac{1}{10}\left|p_{n+1} - c_{n+1}\right| \approx \left|m_{n+1} - y_{n+1}\right|$$

Whenever the step size and hence the errors were too small, then these two steps did not help, since roundoff dominated the small correction effects.

3 For the equation $y' = y$ the three-eighths rule, in spite of its large truncation error, did very well.

4 If we exclude the unstable methods, then on an average, the Adams-Bashforth, two-thirds, and one-half methods appeared not to differ significantly, with perhaps some advantage to the first.

These results should not be taken as definitive, and further study of special situations will probably reveal much more. The field of numerical integration of differential equations is a very complex one, and a complete study of it does not belong in an introductory course. Hence we shall drop the topic of numerical verification without, we hope, giving the idea that it is unimportant.

24

SPECIAL METHODS OF INTEGRATING
ORDINARY DIFFERENTIAL EQUATIONS

24.1 INTRODUCTION AND OUTLINE

The predictor-corrector methods of the previous chapter require special methods until enough old values are built up to start the predictor. Probably the most widely used starting methods are those of Runge-Kutta. While these methods can also be used to compute each step, the machine efficiency is likely to be low. The typical Runge-Kutta method requires four evaluations of the derivatives for each step forward, while the predictor-corrector methods require one or two per step, and at the same time both methods have about the same truncation error. The Runge-Kutta methods often do not contain within themselves any measure of the accuracy of the solution at each step; hence it is difficult to find out when to halve or double the interval size. The idea of using two separate integrations at different step sizes has been proposed, but it is so obviously expensive of machine time that the proposal need not be taken seriously.

Whenever the system of equations of a frequently repeated problem has a special structure, it is worth examining the equations to see if one can take advantage of the special structure to reduce the amount of computation required. For example, second-order linear differential equations (with variable coefficients) have enough special structure that special methods significantly reduce the amount of computing, when compared with the time needed to write the second-order equation in the canonical form of a pair of first-order equations and then solve them in their general form.

A great many special methods have been invented, and since this is an introductory text, we can examine only a few of them here; thus we shall merely show how a couple of special methods can be constructed. The derivations illustrate the same general ideas as were used in the previous chapters and again show the power of the general methods. The reader should, at this point, be able to construct the special cases he needs to solve his own special problems.

24.2 RUNGE-KUTTA METHODS

There are many variants of the Runge-Kutta method, but the most widely used one is the following: Given

$$y' = f(x, y)$$
$$y(x_n) = y_n$$

we compute in turn

$$k_1 = hf(x_n, y_n)$$
$$k_2 = hf\left(x_n + \frac{h}{2}, y_n + \frac{k_1}{2}\right)$$
$$k_3 = hf\left(x_n + \frac{h}{2}, y_n + \frac{k_2}{2}\right)$$
$$k_4 = hf(x_n + h, y_n + k_3)$$
$$y_{n+1} = y_n + \tfrac{1}{6}(k_1 + 2k_2 + 2k_3 + k_4)$$

This process may be described in geometric terms. At the point (x_n, y_n) we compute the slope k_1/h and using it, we go one-half step forward and examine the slope there. Using this new slope (k_2/h), we again start at (x_n, y_n), go one-half step forward, and again sample the slope. Using this latest slope k_3/h, we again start at x_n, y_n, but this time we go a full step forward where we examine the slope k_4/h. The four slopes are averaged, using weights $\tfrac{1}{6}, \tfrac{2}{6}, \tfrac{2}{6}, \tfrac{1}{6}$, and using this average slope, we make the final step from x_n, y_n to x_{n+1}, y_{n+1}. If $f(x, y)$ did not depend on y, then the averaging would produce Simpson's formula. The method has an error term proportional to h^5.

There are variants on the method as to the locations where the samples are taken and hence what weights are assigned to them in the various steps. There are also lower-order methods which use fewer samples per step, as well as higher-order methods.

It is evident that the method throws away all old information and begins each complete step anew, and hence is hardly likely to be as efficient as methods which take advantage of old information.   It is also evident that there is no check on whether the step size is too small or too large, although perhaps a study of the k_i might give a clue; this is often not done.   Among other proposals to monitor the accuracy are: (1) comparing k_4 of one step with k_1 of the next step, and (2) using the quantity

$$\frac{\sum_{i=1}^{N} (k_3 - k_2)_i{}^2}{\sum_{i=1}^{N} (k_2 - k_1)_i{}^2}$$

where N is the number of equations.

The method of deriving these equations does not seem to be used elsewhere in numerical analysis, and so it is not worth giving here.[1]   The general spirit of the derivation is that the functions $f(x, y)$ which are on the right-hand sides are all expanded in series in powers of h, and the corresponding derivatives are equated to eliminate the lower powers of h.   The result is that the method is exact for polynomials of degree 4 (or less), but is *not* equivalent to a power-series expansion about the origin, as may be seen by solving the equation (using $h = 1$)

$$y' = \sin^2 2\pi x$$
$$y(0) = 0$$

since the calculated solution will be identically zero, but the true solution is

$$y = \frac{4\pi^2}{3} x^3 + \cdots$$

24.3   SECOND-ORDER-EQUATION METHODS WHEN y' IS MISSING

Many special situations may be investigated, and in this section we shall confine ourselves to a common situation where we have second-order differential equations with no first derivative present.   Frequently, if the derivative is present, a simple transformation will remove it.   We shall use the same methods for investigating this problem as we used in Chap. 23.

We begin with the corrector for integrating the differential equation

$$y'' = f(x, y)$$

---

[1] See Kopal [29] and Ralston [49].

which we shall assume in the form

$$y_{n+1} = a_0 y_n + a_1 y_{n-1} + a_2 y_{n-2} + h^2(b_{-1} y''_{n+1} + b_0 y''_n + b_1 y''_{n-1} + b_2 y''_{n-2})$$

The method of the last chapter would suggest saving two parameters, but a brief trial indicates that using a_2 as the only parameter will take us through the stability problem safely. Thus we can make the method exact for $1, x, \dots, x^5$. The results are given in Table 24.3.

The method in the last column has a zero error term; hence it has higher accuracy than the others. But the multiple characteristic root means that the propagated error ε_n will behave as

$$\varepsilon_n = C_1 + C_2 n + C_3 n^2$$

which is not pleasant to contemplate. In the other cases we have

$$\varepsilon_n = C_1 + C_2 n + C_3 (a_2)^n$$

The multiple root $\rho = 1$ is unavoidable if we make the form exact for 1 and x, and the linear growth of the error comes from an error in y' (which does not appear explicitly), propagating linearly with n.

The advantage of the two zero coefficients in the choice [39] $a_2 = 0$ makes this formula very attractive.

$$y_{n+1} = 2y_n - y_{n-1} + \frac{h^2}{12}(y''_{n+1} + 10y''_n + y''_{n-1}) - \frac{h^6 y^{(6)}(\theta)}{240}$$

This can also be written in the form

$$y_{n+1} = 2y_n - y_{n-1} + h^2\left(y''_n + \frac{\Delta^2 y''_{n-1}}{12}\right) - \frac{h^6 y^{(6)}(\theta)}{240}$$

The symmetry makes it easy to check that $G(s) \neq 0$ for $-h < s < h$.

Table 24.3

	$a_2 = -1$	$a_2 = -\frac{1}{2}$	$a_2 = 0$	$a_2 = \frac{1}{2}$	$a_2 = 1$
$a_0 = 2 + a_2$	1	$\frac{3}{2}$	2	$\frac{5}{2}$	3
$a_1 = -(1 + 2a_2)$	1	0	-1	-2	-3
$a_2 = a_2$	-1	$-\frac{1}{2}$	0	$\frac{1}{2}$	1
$b_{-1} = \frac{1}{12}$	$\frac{1}{12}$	$\frac{1}{12}$	$\frac{1}{12}$	$\frac{1}{12}$	$\frac{1}{12}$
$b_0 = (10 - a_2)/12$	$\frac{11}{12}$	$\frac{21}{24}$	$\frac{10}{12}$	$\frac{19}{24}$	$\frac{9}{12}$
$b_1 = (1 - 10a_2)/12$	$\frac{11}{12}$	$\frac{11}{24}$	$\frac{1}{12}$	$-\frac{8}{24}$	$-\frac{9}{12}$
$b_2 = -a_2/12$	$\frac{1}{12}$	$\frac{1}{24}$	0	$-\frac{1}{24}$	$-\frac{1}{12}$
$E_6 = -3 + 3a_2$	-6	$-\frac{9}{2}$	-3	$-\frac{3}{2}$	0
$\rho_1, \rho_2, \rho_3 = 1, 1, a_2$	$1, 1, -1$	$1, 1, -\frac{1}{2}$	$1, 1, 0$	$1, 1, \frac{1}{2}$	$1, 1, 1$

A convenient predictor to go with this corrector (we do not derive it here) is

$$y_{n+1} = 2y_{n-1} - y_{n-3} + \frac{4h^2}{3}(y_n'' + y_{n-1}'' + y_{n-2}'') + \frac{16}{240}h^6 y^{(6)}(\theta)$$

or

$$y_{n+1} = 2y_{n-1} - y_{n-3} + 4h^2(y_{n-1}'' + \frac{\Delta^2 y_{n-2}''}{3}) + \frac{16}{240}h^6 y^{(6)}(\theta)$$

The predictor minus corrector gives

$$\tfrac{17}{240}h^6 y^6(\theta)$$

so that the error in the corrector is about

$$-\tfrac{1}{17}(p_{n+1} - c_{n+1})$$

while the error in the predictor is

$$+\tfrac{16}{17}(p_n - c_n)$$

These can be used to improve the accuracy if desired.

The characteristic roots of the predictor are $1, -1, i, -i$ and give no trouble.

PROBLEMS

24.3.1 Derive the predictor equation.

24.3.2 Show that the predictor equation is a valid formula, by examining the G function.

24.4 LINEAR EQUATIONS

Linear differential equations occur very frequently, and in spite of the vast amount of theory associated with them it is sometimes necessary to solve one by numerical methods. The linear property makes them much easier to solve than the general equation. As an example, we consider the equation

$$y'' = f(x)y + g(x)$$

which is a special case of those in Sec. 24.3 (since y' does not occur). Using the corrector in the second form, we have

$$\Delta^2 y_{n-1} - \frac{h^2}{12}\Delta^2 y_{n-1}'' = h^2 y_n'' - \frac{h^6 y^{(6)}}{240}$$

Substituting the value for each appearance of y'', we have

$$\Delta^2 y_{n-1} - \frac{h^2}{12} \Delta^2 [f(x_{n-1})y_{n-1} + g(x_{n-1})] = h^2 [f(x_n)y_n + g(x_n)]$$

$$\Delta^2 \left\{ \left[1 - \frac{h^2}{12} f(x_{n-1}) \right] y_{n-1} \right\} = h^2 f(x_n) y_n + h^2 \left(g_n + \frac{1}{12} \Delta^2 g_{n-1} \right)$$

We now write

$$\left[1 - \frac{h^2}{12} f(x_n) \right] y_n = Y_n$$

so that

$$\Delta^2 Y_{n-1} = h^2 [f(x_n)y_n + g_n + \tfrac{1}{12}\Delta^2 g_{n-1}]$$

Thus we solve

$$Y_{n+1} = 2Y_n - Y_{n-1} + h^2 [f(x_n)y_n + g_n + \tfrac{1}{12}\Delta^2 g_{n-1}]$$

where

$$y_n = \frac{Y_n}{1 - (h^2/12)f(x_n)}$$

This method is often attributed to Numerov [19] and is quite useful and efficient.

It should be evident that similar rearrangements of various methods to eliminate the predictor can be made when the equation is linear, and we need not give all the special cases explicitly.

PROBLEMS

24.4.1 Compute the solution of

$$y'' + xy = 0 \qquad y(0) = 1 \qquad y'(0) = 0$$

for $0 \le x \le 1$ and $h = 2/10$, using Numerov's method.

24.4.2 Show that no predictor is required for the linear equation

$$y' = P(x)y + Q(x)$$

24.5 A METHOD WHICH USES y, y', AND y'' VALUES

We have pointed out several times that once a function has been computed, it is often quite economical of machine time to compute the derivative of the function. Thus, if we have

$$y' = f(x, y)$$

then

$$y'' = \frac{\partial f}{\partial y} y' + \frac{\partial f}{\partial x}$$

Problem 21.8.1 includes the formula $(b_{-1} = 1/2)$

$$y_{n+1} = y_n + \frac{h}{2}(y'_{n+1} + y'_n) + \frac{h^2}{12}(-y''_{n+1} + y''_n) + \frac{h^5}{720}y^{(5)}(\theta)$$

This appears to be an excellent choice for a corrector formula. Since only the y_n term occurs, there are no extraneous roots of the characteristic equation to produce instability. The truncation error is clearly very small, and roundoff is also under control.

We have, therefore, to find a predictor of the general form

$$y_{n+1} = A_0 y_n + A_1 y_{n-1} + h(B_0 y'_n + B_1 y'_{n-1}) + h^2(C_0 y''_n + C_1 y''_{n-1})$$

The usual process of making it exact for $1, x, x^2, x^3, x^4$ produces the results given in Table 24.5. The choice $A_1 = 1$ has particularly attractive coefficients, although $A_1 = 0$ has a slightly smaller error.

PROBLEMS

24.5.1 Develop the details of a predictor-corrector method based on the $A_1 = 1$ case, including modification and final-value equations.

24.5.2 Examine the influence function for the case of $A_1 = 1$.

24.5.3 Derive

$$y_{n+1} = y_n + \frac{h}{240}(101y'_{n+1} + 128y'_n + 11y'_{n-1})$$

$$+ \frac{h^2}{240}(-13y''_{n+1} + 40y''_n + 3y''_{n-1})$$

$$+ \frac{8h^7 y^{(7)}(\theta)}{7!15}$$

24.5.4 Derive, as in Prob. 24.5.3,

$$y_{n+1} = y_{n-1} + \frac{h}{15}(7y'_{n+1} + 16y'_n + 7y'_{n-1})$$

$$+ \frac{h^2}{15}(-y''_{n+1} + y''_{n-1}) + \frac{16h^7 y^{(7)}(\theta)}{7!15}$$

24.5.5 Find a predictor to go with Probs. 24.5.3 and 24.5.4.

Table 24.5

	$A_1 = 0$	$A_1 = \frac{1}{2}$	$A_1 = 1$
$A_0 = 1 - A_1$	1	$\frac{1}{2}$	0
$A_1 = A_1$	0	$\frac{1}{2}$	1
$B_0 = (-1 + A_1)/2$	$-\frac{1}{2}$	$-\frac{1}{4}$	0
$B_1 = (3 + A_1)/2$	$\frac{3}{2}$	$\frac{7}{4}$	2
$C_0 = (17 - A_1)/12$	$\frac{17}{12}$	$\frac{33}{24}$	$\frac{4}{3}$
$C_1 = (7 + A_1)/12$	$\frac{7}{12}$	$\frac{15}{24}$	$\frac{8}{3}$
$E_5 = (31 + A_1)/6$	$\frac{31}{6}$	$\frac{31}{4}$	$\frac{16}{3}$

24.6 WHEN THE SOLUTION IS NOT EASILY APPROXIMATED BY A POLYNOMIAL

In general, we have been assuming that the functions which we are considering can be approximated by a polynomial, if we use reasonably sized steps. The main exception was in integration where we noted that the integral could be written in the form

$$\int_b^a f(x)\, dx = \int_b^a K(x)g(x)\, dx$$

and if the moments

$$m_k = \int_a^b K(x)x^k\, dx$$

are known and if $g(x)$ is easily approximated by a polynomial, then the integral can be solved readily. Thus we approximate only part of the integrand by a polynomial.

Somewhat similar situations arise in the field of differential equations; frequently the solution we are looking for cannot be closely approximated by a polynomial of reasonable degree, with steps of reasonable size. We then look to see what part of the problem we can approximate by a polynomial and try to take advantage of that.

As a simple example, suppose that we have

$$ay'' + by' + cy = f(x) \qquad y(x_0) = y_0 \qquad y'(x_0) = y_0'$$

where a, b, and c are constants and $f(x)$ is easily approximated by, say,

$$f(x) = p_i + q_i(x - x_{i-1}) \qquad \text{for } x_{i-1} \leq x \leq x_i$$

Thus we are approximating $f(x)$ by a sequence of straight-line segments. It is easy to compute a particular solution of

$$y(x) = F_i(x) = P_i + Q_i(x - x_{i-1})$$

by direct substitution. The solution of the homogeneous differential equation leads to the characteristic equation

$$am^2 + bm + c = 0$$

with roots

$$m_1 = \frac{-b + \sqrt{b^2 - 4ac}}{2a}$$

$$m_2 = \frac{-b - \sqrt{b^2 - 4ac}}{2a}$$

Hence the solution of the differential equation in the ith interval $(x_{i-1} \leq x \leq x_i)$ starts at

$$y(x_{i-1}) = y_{i-1}$$
$$y'(x_{i-1}) = y'_{i-1}$$

and is given by

$$y(x) = C_1 e^{m_1(x-x_{i-1})} + C_2 e^{m_2(x-x_{i-1})} + F_i(x)$$

To determine C_1 and C_2, we use the initial conditions (of the interval)

$$y_{i-1} = C_1 + C_2 + F_i(x_{i-1})$$
$$y'_{i-1} = m_1 C_1 + m_2 C_2 + F'_i(x_{i-1})$$

We then compute the initial conditions for the next interval

$$y_{i-1}(x_i) = y_i$$
$$y'_{i-1}(x_i) = y'_i$$

and are ready for the next step.

If one or both of the values of m_i are large and negative, then the solution is not readily approximated by a polynomial, and by writing our solution in the form of a polynomial *plus* an exponential part, we have evaded the trouble. The approach used in no way depends on approximating $f(x)$ by linear polynomials, and generally speaking higher-order polynomials would be more economical of machine time. The approach also does not depend on the second-order equation; any order could have been used.

24.7 CONSERVATION LAWS

It often happens that the system of differential equations to be solved describes some quantities that obey conservation laws. For example, the total number of particles, molecules, or atoms is fixed (conserved). Such systems of equations have a form like

$$y'_i = f_i(t, y_1, y_2, \ldots, y_n) \qquad i = 1, 2, \ldots, n$$

and if all the equations are added (sometimes suitable multiplicative constants[1] need to be supplied before adding), then we get the equation

$$\sum_{i=1}^{n} y'_i = 0$$

which implies

$$\sum_{i=1}^{n} y_i = C$$

---

[1] If variables are used as multipliers, for example to get the total energy, then the argument does not apply.

How will predictor-corrector methods treat the conservation? If we suppose that all the steps have obeyed the conservation up to the present step, then we find that the predicted values will also show the conservation (within roundoff) *provided* the right-hand sides of the equations have been arranged so that the large roundoffs that occur are carefully matched in the corresponding equations. And so it goes, step by step, with care in the evaluation of the right-hand sides so that any large roundoff in the right-hand side of one equation is compensated for by the corresponding equation; the system will, within roundoff, obey the conservation law.

This has an important consequence; often the user of the results will *assume* that if the conservation holds, then the results must be accurate. But we have just seen that the lack of conservation is a result of roundoff and *does not* involve the truncation error, and so the solutions may be very much in error—it is in the nature of predictor-corrector methods that the solution tends to obey the conservation.

24.8 STIFF EQUATIONS

We now examine another class of problems in which we do not approximate the final answer by a polynomial but only some part of it.

Large negative characteristic roots are typical of what are sometimes called "stiff equations." The simplest form of a stiff equation is

$$y' = f(x, y) \qquad \frac{\partial f}{\partial y} = A \ll 0$$

Such equations often arise in control systems where a rapid following of the signal is desired. They also occur in problems involving space charge as well as in flame chemistry.

One approach to the equation is based on the situation when $\partial f/\partial y$, though large and negative, is slowly changing. We write (for A = constant approximately equal to $\partial f/\partial y$)

$$y' = f(x, y) + Ay - Ay$$
$$y' - Ay = f(x, y) - Ay \equiv F(x, y)$$

where, of course,

$$\frac{\partial F}{\partial y} = \frac{\partial f}{\partial y} - A \approx 0$$

Solving the equation with the usual integrating factor e^{-Ax}, we have

$$y(x) = C_1 e^{A(x - x_0)} + e^{A(x - x_0)} \int_{x_0}^{x} e^{-A(\theta - x_0)} F[\theta, y(\theta)] \, d\theta$$

We now face the problem of deriving formulas for integrals of the form

$$I(h) = \int_0^h e^{-A\theta}g(\theta)\,d\theta$$

which is easy to do, following the methods developed so far.   The predictor cannot use the value $g(h)$ but can use $g(0), g(-h), \ldots$, as well as $I(0), I(-h), \ldots$, while the corrector can use the predicted value of $g(h)$.   We leave the details to the reader, but note that we are supposing that

$$f(x, y) = f(x, y) - Ay \qquad [y = y(x)]$$

can be approximated by a polynomial using a fairly wide-spaced set of samples, and in many cases of stiff equations this is not so.

   In the first example above we *added* the nonapproximatable part; in the second it was as a *multiplicative factor*.   In both cases we abandoned the direct approximation of the solution by polynomials in favor of approximating some part of the solution.

   Stiff equations are the topic of many papers these days, but it seems that we still do not understand the basic problem very well.   Many of the proposed methods which damp out the oscillation also have a side effect of producing gross distortions in the solution being computed.[1]   Material in Chaps. 31 to 41 present a framework of ideas for what the author thinks is a better approach.

PROBLEMS

24.8.1   Suppose that $f(x)$ can be approximated by a quadratic polynomial in the interval $0 \le x \le k2\pi$, and we have the equation

$$y'' + y = f(x)$$

Develop the above theory for numerically integrating the equation in the interval.   Show how to extend it step by step to $2k\pi \le x \le 4k\pi$, $4k\pi \le x \le 6k\pi$, etc.

24.9   PROBLEMS WITH WIDELY DIFFERENT TIME CONSTANTS

The nature of problems with different time constants is perhaps best understood in terms of a concrete example.   Suppose that a new chemical plant has been designed.   The chemical engineers have carefully examined the chemical aspects of

---

[1] For a discussion of this fault, see, for example, E. E. Zajac, Note on Overly-Stable Difference Approximations, *J. Math. Phys.*, vol. 43, no. 1, pp. 51–54, 1964.   See also M. E. Fowler and R. M. Warten, A Numerical Technique for Ordinary Differential Equations with Widely Separated Eigenvalues, *IBM, J. Res. Develop.*, September, 1967, pp. 537–543.

the problem, while the control engineers have examined their part of the design. It is now necessary to fill the inevitable gap that occurs when two groups of experts work on the same project; will the combined system operate properly? For this we take the differential equations of the complete system, chemistry and control, and set out to solve them.

The first thing that we find is that the control circuits tend to have rapid time constants, meaning, to be specific, that typically they will respond in time intervals of perhaps milliseconds, while the chemical engineering time response of the flow of materials, temperature changes, etc., may take up to an hour. Thus each simulated situation, for example how the plant will start up, will have to run perhaps an hour or more. But there are 3.6×10^6 msec in the hour, and the number of steps needed to integrate the equations is unpleasant to contemplate. This makes part of the system of equations look "stiff."

It often happens in such situations, not only in chemical plants but even in solid-state physics, that the system of equations can be broken into two sets, one with the long time constants and one with the short time constants. The short-time-constant equations tend to see the quantities from the long-time-constant equations as if they were constants, and thus their changes can pretty well be ignored in the integration of the control equations. On the other side, the control equations respond so rapidly that the slow-time-constant equations see them as being very close to equilibrium. In this situation the control differential equations can have their derivatives equated to zero and treated as algebraic or transcendental equations with no derivatives. Thus the two sets can be solved separately.

But beware! The purpose of the simulation was to see if the two groups of engineers overlooked any cross-connections between their two systems. Thus a careful mathematical examination must be made to ensure that everything said about the two time constants is in fact true, that neither is close to the other for *any* set of the variables. The outside examination of this point is a valuable contribution to the whole design problem, and it can be carried out with mathematical symbols which tend to have no preconceived ideas of how the answer should come out.

24.10   TWO-POINT PROBLEMS

We have been solving initial-value problems so far, but frequently conditions on the solution are given at two points. For example, we may have

$$y'' = f(x, y) \qquad y(0) = 0 \qquad y(1) = 0$$

and want to know $y = y(x)$ for $0 \le x \le 1$.

This situation may be reduced to the previous case by a trial-and-error process. We start with

$$y'' = f(x, y) \qquad y(0) = 0 \qquad y'(0) = \lambda$$

and try to find a pair of λ values λ_1 and λ_2 such that, say,

$$\text{for } \lambda_1 \qquad y(1) < 0$$
$$\text{for } \lambda_2 \qquad y(1) > 0$$

Thus [since for most practical problems $y(1)$ is a continuous function of λ] we can use the crude process of bisecting the known range at each trial and hence in 10 trials reduce the range

$$|\lambda_1 - \lambda_2|$$

by about $2^{10} \approx 1,000$. More effective ways of locating the zero of $y(1)$ may be found easily. (See Chap. 4.)

Instead of reducing the two-point problem to a series of initial-value problems, it is more efficient to attack the problem directly. An extensive theory has been developed [13], and we shall briefly discuss only one example to show some of the ideas involved.

Suppose that we have

$$y'' = f(x)y + g(x) \qquad y(0) = A \qquad y(1) = B$$

We first approximate y'' by

$$h^2 y_n'' = \Delta^2 y_{n-1} \qquad (24.10.1)$$

where we divided the interval $0 \le x \le 1$ into N intervals of size $h = 1/N$. Thus we have

$$y_{n+1} - 2y_n + y_{n-1} = h^2 [f(x_n)y_n + g(x_n)]$$

or

$$y_n = \frac{y_{n-1} + y_{n+1} - h^2 g(x_n)}{2 + h^2 f(x_n)} \qquad n = 1, 2, \ldots, N - 1 \qquad (24.10.2)$$

We have $N - 1$ linear equations in $N - 1$ unknowns, which may be solved in many ways.

As an example, let us consider a specific case of the problem. Set $A = B = 0$, $f(x) = 1$, $g(x) = x$, and $N = 4$. We have

$$y'' = y + x \qquad y(0) = 0 \qquad y(1) = 0$$

(which has the known solution $y = \sinh x / \sinh 1 - x$ for comparison). The difference equations are

$$y_2 - 2y_1 = \tfrac{1}{16}(y_1 + \tfrac{1}{4})$$
$$y_3 - 2y_2 + y_1 = \tfrac{1}{16}(y_2 + \tfrac{2}{4})$$
$$-2y_3 + y_2 = \tfrac{1}{16}(y_3 + \tfrac{3}{4})$$

or

$$16y_2 - 33y_1 = \tfrac{1}{4}$$
$$16y_3 - 33y_2 + 16y_1 = \tfrac{2}{4}$$
$$-33y_3 + 16y_2 = \tfrac{3}{4}$$

One method of solving these particular equations is to multiply them by 16, 33, and 16, respectively, and add

$$(16^2 - 33^2 + 16^2)y_2 = \tfrac{1}{4}(16 + 2 \cdot 33 + 3 \cdot 16)$$

From y_2 it is easy to find y_1 and y_3, if we use the first and third equations in turn.

Suppose that this is not as accurate as we wish; what shall we do next? We could set $N = 8$ and using the computed solution for $N = 4$, guess at a solution for $N = 8$. Then, using the guessed solution in the right-hand side of the eight equations corresponding to Eq. (24.10.2), we could compute improved values and repeat the process until no change occurs. This process can be extended as far as we wish.

Alternatively, and probably preferably, we could use a more accurate formula in place of Eq. (24.10.1). We use the S_7 matrix of Sec. 15.6 to find such a formula. The moments of $y''(0)$ are $(0, 0, 2, 0, 0, 0, 0)$ and should be written as a column vector. This produces the column vector corresponding to twice the third column of S_7; that is,

$$\frac{1}{360}\begin{pmatrix} 4 \\ -54 \\ 540 \\ -980 \\ 540 \\ -54 \\ 4 \end{pmatrix} = \frac{1}{180}\begin{pmatrix} 2 \\ -27 \\ 270 \\ -490 \\ 270 \\ -27 \\ 2 \end{pmatrix}$$

which is

$$h^2 y_n'' = \tfrac{1}{180}(2y_{n+3} - 27y_{n+2} + 270y_{n+1} - 490y_n + 270y_{n-1} - 27y_{n-2} + 2y_{n-3})$$
$$= \Delta^2 y_{n-1} - \tfrac{1}{12}\Delta^4 y_{n-2} + \tfrac{1}{90}\Delta^6 y_{n-3}$$

Using only the first two terms, since $\tfrac{1}{90}\Delta^6 y_{n-2}$ is likely to be small, we compute

$$-1/12\,\Delta^4 y_{n-1}$$

To do this, we need some values outside the range $0 \leq x \leq 1$; in particular we need $y(-1/4)$ and $y(5/4)$. These can be found from the obvious equations

$$y_{-1} - 2y_0 + y_1 = h^2(y_0 + 0)$$

Alternatively, we note that we really need only $\Delta^2 y_{-1}$ and $\Delta^2 y_3$, which can be found from the values of y_0 and y_4 to be

$$\Delta^2 y_{-1} = 0 \qquad \Delta^2 y_4 = h^2$$

Using the table of values,

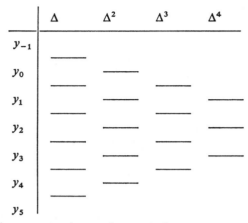

we can compute the values of $\Delta^4 y_i$, namely $\Delta^4 y_{-1}$, $\Delta^4 y_0$, and $\Delta^4 y_1$, required to improve the result.

These values of $\Delta^4 y_i$, which were neglected in the first computation, we now put in as *correction terms* on the right-hand side:

$$\Delta^2 y_{n-1} = h^2(y_n + x_n) + \tfrac{1}{12}\Delta^4 y_{n-3}$$

We solve the problem again, *without* changing the $\Delta^4 y_i$ terms. If the $\Delta^4 y_i$ values of the new solution are changed by any significant amount from the old ones, then we repeat the process *again*. This method is called the "difference-correction" method by Fox [13] who uses it with great effectiveness.

As already noted, we are not attempting to develop an elaborate theory for two-point problems but rather are content to give only a slight idea of what can be done with such problems.

PROBLEMS

24.10.1   Carry out the computation of the example in Sec. 24.10, using the difference-correction method, and compare the result with the correct answer.

25

LEAST SQUARES: THEORY

25.1 INTRODUCTION

In Sec. 14.2 we discussed the four basic questions:

1 What samples shall we use?
2 What class of functions shall we use?
3 What criterion shall we use?
4 Where shall we apply the criterion?

We have used both function and derivative values as samples (as well as using the position of the samples as parameters), but we have used only the class of polynomials and the criterion of exactly matching the samples. We now begin the examination of other criteria for picking the particular member of the class of functions being used. While we shall continue to concentrate on polynomials as the class of functions, more general classes will be mentioned from time to time.

In this chapter and in Chaps. 26 and 27, we examine the criterion of least squares, and in Chaps. 28 and 29, we examine the Chebyshev, or minimax, criterion. These are the two most widely used criteria excluding exactly matching at the sample points (within roundoff, of course), but others do arise now and then.

Having seen three different examples used in the book, hopefully the reader will be able to handle others when they arise.

The least-squares criterion is used in situations where there is much more data available than parameters so that exact matching is out of the question. Polynomials are most commonly used in least-squares matching, though any linear family will work about as well in practice.

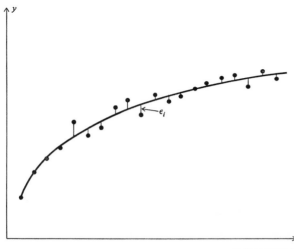

FIGURE 25.1.1

The least-squares criterion is widely used, and often believed to be the " right one " to use.   There is a saying that mathematicians believe that it is a physical principle while physicists believe that it is a mathematical principle.   Either we can assume the principle for selecting the particular member of the family of polynomials of a given degree, or we can assume some other principles and deduce that of least squares—something must be assumed in any case.

Least squares is very often regarded as *smoothing the data*, or, if you wish, *removing the noise*.   The least-squares polynomial is supposed to give the correct values (Fig. 25.1.1).   In order to understand when to use the principle and when to avoid it, we shall try to give a feeling for what it does by connecting it with other known results.   Just what this kind of smoothing does will be looked at again in Secs. 27.7 and 35.6.

25.2 THE PRINCIPLE OF LEAST SQUARES

Suppose we try to measure some quantity x and make M measurements x_i (see Fig. 25.2.1). We do not see x but only the measurements x_i with errors of measurement ε_i; that is, we see

$$x_i = x + \varepsilon_i \qquad i = 1, 2, \ldots, M$$

We will regard the residuals ε_i as "noise" and call x the true value, (whatever that may mean in a situation in which it cannot be measured directly!).

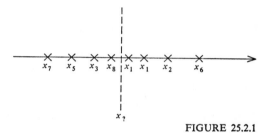

FIGURE 25.2.1

The principle of least squares states that the best estimate $\hat{x}$ of x is that number which minimizes the sum of the squares of the deviations of the data from their estimate,[1]

$$f(\hat{x}) = \sum_{i=1}^{M} \varepsilon_i{}^2 = \sum_{i=1}^{M} (x_i - \hat{x})^2$$

In the final analysis, the usefulness of this principle rests on how useful the results turn out to be in practice and how easy it is to use, not on the fanciness of the mathematical derivations.

This principle is equivalent to the assumption that the average

$$x_a = \frac{1}{M} \sum_{i=1}^{M} x_i$$

is the best estimate. To prove this equivalence, we first show that the least-squares principle leads to the average. We regard

$$f(\hat{x}) = \sum_{i=1}^{M} (x_i - \hat{x})^2$$

---

[1] The Gauss-Markov theorem states that if $\mathrm{Av}\{\varepsilon_i\} = 0$ and $\mathrm{cov}\{\varepsilon_i \varepsilon_j\} = 0$, then of all unbiassed linear estimates this estimate has the least variance. See Mood and Graybill [42].

as a function of $\hat{x}$ to be minimized.   Applying the usual calculus rule, "differentiate and set equal to zero," we get

$$\frac{df}{d\hat{x}} = -2\sum_{i=1}^{M}(x_i - \hat{x}) = 0$$

$$\sum_{i=1}^{M} x_i - \sum_{i=1}^{M} \hat{x} = 0$$

$$\sum_{i=1}^{M} x_i - M\hat{x} = 0$$

therefore

$$\hat{x} = \frac{1}{M}\sum_{i=1}^{M} x_i = x_a$$

Thus $x_a = \hat{x}$ minimizes the sum of the squares of the residuals.   We also note that

$$\frac{d^2f}{d\hat{x}^2} = 2M > 0$$

and hence we have a minimum.

We now show the converse, namely that if we pick the average x_a as the best choice, then this minimizes the sum of the squares.   We set

$$\begin{aligned}
f(x_a) &= \sum_{i=1}^{M}(x_i - x_a)^2 \\
&= \sum x_i^2 - 2x_a\sum x_i + \sum x_a^2 \\
&= \sum x_i^2 - 2x_a M x_a + M x_a^2 \\
&= \sum x_i^2 - M x_a^2
\end{aligned}$$

Now if we were to choose any other value, say $x_b \cdot$ then

$$\begin{aligned}
f(x_b) &= \sum (x_i - x_b)^2 \\
&= \sum x_i^2 - 2x_b M x_a + M x_b^2
\end{aligned}$$

Subtracting the first result from the second, we get

$$\begin{aligned}
f(x_b) - f(x_a) &= M[x_a^2 - 2x_a x_b + x_b^2] \\
&= M(x_a - x_b)^2 \geq 0
\end{aligned}$$

Thus

$$f(x_b) \geq f(x_a)$$

and they are equal only if $x_b = x_a$; that is, x_a minimizes the sum of the squares of the residuals.

We have now proved that the principle of the least squares and the choice of the average as the best value are equivalent.

The choice of the average as the best value is very common practice in many situations, but clearly is inappropriate in some. For example, the average income in a country is likely to be misleading because of the skewness of the distribution of incomes. Likewise, if the probability distribution of the errors (residuals) is believed to be highly skewed, then it may be foolish to use the principle of least squares.

Probably the major fault with least squares is that a single very wrong measurement will greatly distort the results because in the squaring process large residuals play the dominant part — one gross error 10 times larger than most of the others will have the same effect in the sum of the squares as will 100 of the others. Great care should be exercised before blindly applying any result to data (as is so often done); at least look at the residuals, either by eye or by some suitable program, to see if one or possibly a few measurements are wildly off.

PROBLEMS

25.2.1 Find the least-squares estimate of the numbers 2, 3, 2, 1, 2, 3, both ways.
Ans. $x = 13/6$
25.2.2 Find the least-squares estimate of $-5, 5, 4, -3, 7, 1, 0, 0, 4, -6$

25.3 OTHER CHOICES BESIDES LEAST SQUARES

Suppose that instead of choosing to minimize the sum of the squares of the deviations ε_i we choose x_m to minimize the sum of the absolute values

$$\sum_{i=1}^{M} |\varepsilon_i| = \sum_{i=1}^{M} |x_i - x_m| = f(x)$$

This leads to the choice of the *median* (middle) value of the x_i (if there is an even number of observations M, then we choose the average of the two middle ones). The proof is straightforward. Suppose that there is an odd number $M = 2k + 1$ of observations x_i. Consider using the middle one, x_m, as the median. Then any upward shift of this value x would increase the k terms $|x_i - x|$ that have x_i below x_m and would decrease the k values with x_i above x_m, each by the same amount, but the shift would *also* increase the term $|x_m - x|$, thus increasing the sum of all the deviations.

In place of the minimizing of the sum of the squares, another choice is to minimize the maximum deviation. This leads to

$$\frac{x_{\min} + x_{\max}}{2} = x_{\text{midrange}}$$

which is the usual *midrange* estimate of the best value. This is the Chebyshev criterion in slight disguise and will be treated in Chaps. 28 and 29.

25.4   THE NORMAL LAW OF ERRORS

There appears to be a widespread belief that the principle of least squares implies *the normal law of errors*, which states that the probability of an error ε_i in the interval $(x, x + \Delta x)$ is given by

$$\frac{k}{\sqrt{\pi}} e^{-k^2 x^2} \Delta x$$

This belief is false, and it is worth some discussion of the topic at this point to clarify matters.   Another belief sometimes met is that the normal law is a law of nature.   This situation has also been paraphrased by the remark, "mathematicians believe it to be a physical law, while physicists believe it to be a mathematical law."

One highly seductive derivation of the normal law (due to Herschel[1]) goes as follows: Consider dropping a dart from a height while aiming at some point 0 on the (horizontal) floor (see Fig. 25.4.1).   We now *assume* (1) that the errors do not depend on the coordinate system but only on the distance from 0, and (2) that larger errors are less likely than smaller ones.   These assumptions seem to be quite reasonable, since the coordinate system appears to be perfectly arbitrary.

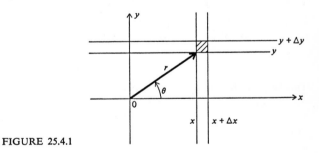

FIGURE 25.4.1

Let the probability of falling in the strip $x, x + \Delta x$ be given (approximately) by

$$f(x) \, \Delta x$$

while that of falling in the strip $y, y + \Delta y$ is

$$f(y) \, \Delta y$$

---

[1] Sir John F. W. Herschel (1792–1871).

Owing to the assumption that the errors in the two coordinates are independent, the probability that it falls in the common rectangle is then

$$f(x)f(y) \, \Delta x \, \Delta y$$

Let us now choose polar coordinates. Since the probability is assumed to be independent of direction, the probability that the dart falls in the element of area is given by

$$g(r) \, \Delta x \, \Delta y$$

The two probabilities are the same, and we have

$$g(r) = f(x)f(y)$$

The left-hand side does not depend on θ, and differentiating with respect to θ, we have

$$\frac{\partial g(r)}{\partial \theta} = 0 = f(x)\frac{\partial f(y)}{\partial \theta} + f(y)\frac{\partial f(x)}{\partial \theta} \quad (25.4.1)$$

Using the relations

$$x = r \cos \theta$$
$$y = r \sin \theta$$

we have

$$\frac{\partial f(x)}{\partial \theta} = \frac{\partial f(x)}{\partial x}\frac{\partial x}{\partial \theta} = f'(x)(-y)$$
$$\frac{\partial f(y)}{\partial \theta} = \frac{\partial f(y)}{\partial y}\frac{\partial y}{\partial \theta} = f'(y)(x) \quad (25.4.2)$$

Combining Eqs. (25.4.2) and (25.4.1), we have

$$f(x)f'(y)(x) + f(y)f'(x)(-y) = 0$$

or

$$\frac{f'(x)}{xf(x)} = \frac{f'(y)}{yf(y)}$$

Since x and y are assumed to be independent, each side must be a constant, say K, and

$$\frac{f'(x)}{xf(x)} = K = \frac{f'(y)}{yf(y)}$$

or, what is the same thing,

$$\frac{df(x)}{f(x)} = K \cdot x \cdot dx \quad \text{and} \quad \frac{df(y)}{f(y)} = K \cdot y \cdot dy$$

Integrating the first of these, we get

$$\ln f(x) = \frac{Kx^2}{2} + C$$

or

$$f(x) = Ae^{Kx^2/2}$$

We assumed that large errors were less likely than small ones; hence K must be negative, say

$$K = -2k^2$$

and so we have

$$f(x) = Ae^{-k^2x^2}$$
$$f(y) = Ae^{-k^2y^2}$$

Combining them, we get

$$g(r) = f(x)f(y) = A^2e^{-k^2(x^2+y^2)}$$

Now the dart must fall somewhere; hence, using polar coordinates, we find

$$\int_0^{2\pi} \int_0^{\infty} g(r)r \, dr \, d\theta = 1$$

$$A^2 \cdot 2\pi \int_0^{\infty} re^{-k^2r^2} \, dr = 1$$

$$A^2\pi \frac{e^{-k^2r^2}}{-k^2} \bigg|_0^{\infty} = \frac{A^2\pi}{k^2} = 1$$

and finally we get our result

$$f(x) = \frac{k}{\sqrt{\pi}} e^{-k^2x^2}$$

which is the normal (Gaussian) distribution function.

It should be noted that this derivation is only one way of obtaining the normal law. The central-limit theorem, which states, loosely, that the distribution of the sum of a large number of small errors becomes normal, provides another approach. In either case, it is not the derivation that matters as much as whether or not the model helps us understand experience.

The normal law has been found in practice to be a useful model in many applications. Deviations from it usually occur from having more values in the "tail" of the distribution when x is large than the model indicates there should be. The reason for this is that often there is a small effect which has a wide variability. In such cases, a mixture of two normal curves with different k values sometimes is useful. The theory of *quality control* is, in part, based on the observed excesses in the tails.

The *maximum-likelihood estimator* [42] of an unknown parameter maintains that we should take the value that maximizes the likelihood of observing it. If the sample is from some probability density $p(x; \theta)$, then the likelihood estimator is the product of the individual (independent) observations

$$L(\theta) = p(x_1; \theta)p(x_2; \theta) \cdots p(x_m; \theta)$$

In the case that the errors come from a normal distribution

$$\frac{k}{\sqrt{\pi}} e^{-k^2(x-\theta)^2}$$

then this leads to

$$L(\theta) = \frac{k^M}{\pi^{M/2}} e^{-k^2\Sigma(x_i-\theta)^2}$$

When this is maximized it clearly is the least-squares estimate and leads to

$$\theta = \frac{1}{M} \sum_{i=1}^{M} x_i = \bar{x}$$

Thus least squares can be derived from the normal distribution via the (assumed) maximum-likelihood estimator.

PROBLEMS

25.4.1 Show that $\sigma^2 = 1/(2k)^2$, where σ is the variance of the distribution $f(x) = (k/\sqrt{\pi})e^{-k^2x^2}$.

25.5 THE LEAST-SQUARES STRAIGHT LINE

Before examining the general case of a polynomial, it is worth becoming familiar with the special case of fitting a straight line in the least-squares sense. We fit

$$y(x) = a + bx$$

to a set of data (x_i, y_i) $(i = 1, 2, 3, \ldots, M)$. We have two parameters a and b and want to minimize the function of two variables (a and b)

$$F(a,b) = \sum_{i=1}^{M} [y(x_i) - y_i]^2 = \sum_{i=1}^{M} \varepsilon_i^2$$

$y(x_i)$ being the computed value and y_i the measured value. This is the obvious generalization of the material on a single parameter that we discussed before. We have

$$F(a, b) = \sum_{i=1}^{M} (a + bx_i - y_i)^2$$

to minimize.   Apply the calculus method

$$\frac{\partial F(a, b)}{\partial a} = 2 \sum (a + bx_i - y_i) = 0$$

$$\frac{\partial F(a, b)}{\partial b} = 2 \sum (a + bx_i - y_i)x_i = 0$$

Dropping the factor 2 and rewriting, we get

$$aM + b \sum x_i = \sum y_i$$
$$a \sum x_i + b \sum x_i^2 = \sum x_i y_i$$

These equations are easily solved for the parameters a and b.   Thus we have the least-squares straight line to the given data.   As an easily followed example, consider the data (for $M = 5$).   See Fig. 25.5.1.

x	y	x^2	xy
1	0	1	0
2	2	4	4
3	2	9	6
4	5	16	20
5	4	25	20
Sum 15	13	55	50

The equations for the parameters are:

$$\begin{array}{c|c} 5a + 15b = 13 & -3 \\ 15a + 55b = 50 & 1 \end{array}$$

$$10b = 11$$

$$\begin{cases} b = 11/10 \\ a = -7/10 \end{cases}$$

The line is

$$y = \frac{1}{10}(-7 + 11x)$$

To plot the line we use

$$y(0) = -7/10$$
$$y(5) = 48/10$$

We have

x	$y(x)$	y_i	ε_i	ε_i^2
1	0.4	0	0.4	0.16
2	1.5	2	−0.5	0.25
3	2.6	2	0.6	0.36
4	3.7	5	−1.3	1.69
5	4.8	4	0.8	0.64
			$\Sigma = 0$	3.10

check

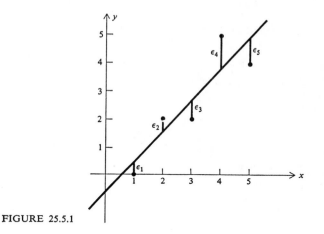

FIGURE 25.5.1

PROBLEMS

Fit a straight line to:

25.5.1

x	-3	-2	-1	0	1	2	3
y	4	4	2	2	1	0	0

25.5.2

x	2.5	1.7	3.2	0.7	0.3
y	1.7	1.4	2.5	0.6	0.1

25.5.3

x	1	2	3	4	5	6
y	0	1	4	9	16	25

25.5.4 $y = e^x$ for $0, 2/10, 4/10, \ldots, 1.0$

25.5.5 Prob. 25.5.4 for e^{-x}.

25.6 POLYNOMIAL CURVE FITTING

Fitting a polynomial of degree N to some data is a direct generalization of the previous section. Given the M measurements (observations) (x_i, y_i), we wish to fit a polynomial of degree N where N is much less than M, typically at least 10 times smaller for cubics or higher-degree polynomials. We again let the calculated data be $y(x_i)$ where

$$y(x) = a_0 + a_1 x + a_2 x^2 + \cdots + a_N x^N$$

and form the sum of the squares of the differences (residuals)

$$\sum_{i=1}^{M} [y(x_i) - y_i]^2 = F(a_0, a_1, a_2, \ldots, a_N)$$

With $N + 1$ parameters we have the $N + 1$ partial derivatives to equate to zero

$$\frac{\partial F}{\partial a_j} = 2 \sum_{i=1}^{M} [(y(x_i) - y_i]x^j = 0 \qquad j = 0, 1, \ldots, N$$

or

$$a_0 \sum x_i^j + a_1 \sum x_i^{i+1} + \cdots + a_N \sum x_i^{j+N} = \sum x_i^j y_i$$

For ease in notation we set

$$\sum_{i=1}^{M} x_i^k = S_k \qquad \sum_{i=1}^{M} x_i^k y_i = T_k$$

The equations then become

$$\sum_{n=0}^{N} a_n S_{n+j} = T_j \qquad j = 0, 1, \ldots, N$$

and are the *normal equations*.

The normal equations are $N + 1$ linear equations in $N + 1$ unknowns a_j, whose determinant we now show cannot be zero.   We do this exactly as we did in Sec. 19.4.   We consider the homogeneous equations

$$\sum_{n=0}^{N} a_n S_{n+j} = 0$$

and show that they have only the trivial solution $a_j = 0$ for all j, from which it follows that the determinant is not zero.   Multiply the jth equation by a_j and sum for all j

$$\sum_{j=0}^{N} \sum_{n=0}^{N} a_j a_n S_{n+j} = \sum_{j=0}^{N} \sum_{n=0}^{N} a_j a_n \sum_{i=0}^{M} x_i^j x_i^n$$

$$= \sum_{i=0}^{M} \left(\sum_{j=0}^{N} a_j x_i^j \right) \left(\sum_{n=0}^{N} a_n x_i^n \right)$$

$$= \sum_{i=0}^{M} y^2(x_i) = 0$$

This means that $y(x) \equiv 0$ and $a_j = 0$ for all j.   We have therefore proved that the determinant is not zero and that the nonhomogeneous equations can be solved for the coefficients of the least-squares polynomial.

In principle the problem of finding the least-squares polynomial is solved; in practice it is not easy to solve the normal equations when N is very large, say greater than 10.   To see why this is so, suppose that the x_i are more or less uniformly distributed in the interval $0 \le x \le 1$.   Then

$$S_k = \sum_{j=1}^{M} x_j^k \approx M \int_0^1 x^k \, dx = \frac{M}{k+1}$$

The resulting determinant (supressing the factors M) is the well-known Hilbert determinant

$$\left|\frac{1}{j+k+1}\right| \qquad j, k = 0, 1, \ldots, N$$

The Hilbert determinant[1] of order N has the value

$$H_N = \frac{N!(N+1)! \cdots (2N-1)!}{[1!2!3! \cdots (N-1)!]^3}$$

which approaches zero very rapidly and suggests what is true, that the system of normal equations will be difficult to solve (ill-conditioned, Sec. 7.8) when N is even moderate in size.   The following table gives some values

n	H_n
1	1
2	8.3×10^{-2}
3	4.6×10^{-4}
4	1.7×10^{-7}
5	3.7×10^{-12}
6	5.4×10^{-18}
7	4.8×10^{-25}
8	2.7×10^{-33}
9	9.7×10^{-43}

The solution to this dilemma is given in the next chapter, and it is *not* to go to multiple precision to get an answer to a set of equations which in their very formation have significant errors.

As an example of polynomial curve fitting consider fitting a quadratic

$$y(x) = a + bx + cx^2$$

to the following data.   The given data are in the two left-hand columns ($M = 7$) and are chosen to make the example easy to follow.

	x	y	xy	x^2	x^2y	x^3	x^4
	-3	4	-12	9	36	-27	81
	-2	2	-4	4	8	-8	16
	-1	3	-3	1	3	-1	1
	0	0	0	0	0	0	0
	1	-1	-1	1	-1	1	1
	2	-2	-4	4	-8	8	16
	$+3$	-5	-15	9	-45	27	81
Sums	0	1	-39	28	-7	0	196

---

[1] David Hilbert (1862–1943).

The normal equations are

$$7a + 0b + 28c = 1$$
$$0a + 28b + 0c = -39$$
$$28a + 0b + 196c = -7$$

It follows easily that $b = -39/28$.   The other two equations can be written

$$\begin{cases} a + 4c = 1/7 \\ a + 7c = -1/4 \end{cases}$$

from which

$$a = 56/84$$
$$c = -11/84$$

and

$$y = \frac{1}{84}\,(56 - 117x - 11x^2)$$

which is shown in Fig. 25.6.1.

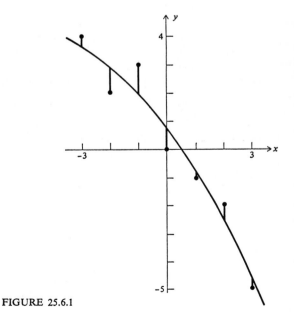

FIGURE 25.6.1

PROBLEMS

25.6.1 Fit a quadratic to

x	$-\frac{3}{2}$	-1	$-\frac{1}{2}$	0	$\frac{1}{2}$	1
y	-3	-3	-2	1	2	2

25.6.2 Fit a quadratic to

x	0	1	2	3	4	5
y	4	7	10	13	16	19

25.6.3 Draw a flow diagram for doing least squares, assuming a library program for solving the normal equations. Find the residuals.

25.7 NONPOLYNOMIAL LEAST SQUARES AND OTHER GENERALIZATIONS

If instead of fitting the data in terms of a linear combination of powers of x, suppose we are given a linearly independent set of functions $f_j(x)$ as the basis for the representation. We have

$$y(x) = a_0 f_0(x) + a_1 f_1(x) + \cdots + a_N f_N(x)$$

and the least-squares fit is given by

$$\sum_{i=1}^{M} \left[\sum_{j=0}^{N} a_j f_j(x_i) - y_i \right]^2 = \min$$

Differentiating this with respect to the parameters a_k (which still occur linearly) we are led to the system of equations

$$\sum_{i=1}^{M} \sum_{j=1}^{N} a_j f_j(x_i) f_k(x_i) = \sum_{i=1}^{M} y_i f_k(x_i)$$

Using the notation of Sec. 25.6, we set correspondingly

$$S_{j,k} = \sum_{i=k}^{M} f_j(x_i) f_k(x_i) \qquad T_k = \sum_{i=1}^{M} y_i f_k(x_i)$$

and we have similar normal equations

$$\sum_{j=1}^{N} a_j S_{j,k} = T_k \qquad k = 0, 1, \ldots, N$$

The same proof that the determinant is nonzero will go through since the family of functions was assumed to be linearly independent over the set of points x_i.

If all the data are not equally reliable and it is desirable to weight the data by nonnegative constants w_i, then again we will be led to the least-squares equation

$$\sum_{i=1}^{M} w_i [\sum a_j f_j(x_i) - y_i]^2 = \min$$

and the normal equations will include the weights w_i with nothing significantly different. The linear independence of the $f_j(x_i)$ means that the determinant will not be zero (by the same pattern of proof).

The case where the parameters do not occur linearly will be discussed in more detail in Sec. 27.5.

PROBLEMS

25.7.1 Carry out the derivation for weights $w_i \geq 0$ and prove that the normal equation can be solved.

25.8 A COMPARISON OF LEAST SQUARES AND POWER-SERIES EXPANSION

The power-series expansion of a function $y(x)$,

$$y(x) = \sum_{n=0}^{\infty} \frac{y^{(n)}(0)x^n}{n!}$$

when truncated to

$$y_N(x) = \sum_{n=0}^{N} \frac{y^{(n)}(0)x^n}{n!}$$

gives a very good approximation near $x = 0$, but as x increases, the approximation tends to get worse and worse. On the other hand, the least-squares fit tries to find a more or less uniform [depending on the weight function $w(x)$] approximation *in an interval*. And this is one of the fundamental differences: The power series fits at a point whereas the least squares fits in an interval.

When it comes to numerical computation, it is not generally possible to estimate accurately the derivatives at a point from the samples scattered in an interval; the exact-matching interpolating polynomial is frequently used in place of the truncated power series, or sometimes a least-squares polynomial.

25.9 CONCLUDING REMARKS ON LEAST SQUARES

Having been shown how to produce a least-squares polynomial, the reader probably expects the text to go on to the problems of integrals and differential equations.   While this could be done, the facts are that seldom is a least-squares approximation used as a basis for integration.   Integration itself is a "smoothing operation."   When there is a "noisy" term in a differential equation, it is customary to smooth it and then use an exact-matching polynomial for integrating the equation.

Another idea which may occur to the reader, but which has not been adequately explored, is the following: In the exact-matching process, we have used $E_0, E_1, \ldots, E_{m-1} = 0$ and let the rest, $E_m, E_{m+1}, \ldots$, fall where they may. Suppose that we were to try to minimize

$$m(a_0, a_1, \ldots) = \sum_{k=0}^{\infty} a_k E_k^2$$

The exact matching used $a_m = a_{m+1} = \cdots = 0$, but we could taper the a_k and thus allow for an improved fit for some of the higher powers at the cost of some exactness for the lower powers.

This can be viewed as a proposal to fit in error space rather than in the function space.   An adequate theory for the error term is not known at present.

An example of this proposal to sacrifice a small amount of exactness in fitting at some value for a large gain at some higher value of k is given by the idea of throwback, discussed briefly in Sec. 18.8 as applied to Everett's interpolation formula.   In the particular case, we gave up a small amount of accuracy in x^2 (modified Δ^2) for a large gain in fitting x^3 and x^4.   (See also Sec. 29.9.)

26

ORTHOGONAL FUNCTIONS

26.1 INTRODUCTION

The idea of orthogonal functions plays a central role in the development of many branches of mathematics. We shall confine our attention to the relationship of orthogonality to linear independence and least squares, and the role that it plays in some parts of computing. Thus we treat only a few aspects of the topic in this chapter.

Two vectors x and y are said to be *orthogonal* if the components x_i and y_i satisfy the equation

$$\sum_{i=1}^{n} x_i y_i = x_1 y_1 + x_2 y_2 + x_3 y_3 + \cdots + x_n y_n = 0$$

In both two and three dimensions this means that the two vectors are perpendicular, and since the algebraic relationship plays the same role in higher dimensions as in two and three dimensions, it is customary to say that they are perpendicular to each other in n-dimensional space.

Now let us imagine that the number of dimensions increases to infinity in such a way that the vectors are replaced in the limit by continuous functions, and that the sum approaches an integral. We then have

$$\int_a^b x(t)y(t)\, dt = 0$$

Thus we have the definition that two functions are orthogonal if the product of the two functions integrated over the range is zero. We have tacitly excluded the vector and the function that are identically zero, since they are perpendicular to all functions including themselves. Notice that if we are going to maintain the geometric language, then we have an infinitely nondenumerable dimensional space of functions.

Sometimes it is convenient to include a weight factor $w(t) \geq 0$, and the definition of orthogonality becomes

$$\int_a^b w(t)x(t)y(t)\, dt = 0$$

A set of functions $f_i(t)$ is said to be *orthogonal* [with respect to the weight factor $w(t)$ and over the interval $a \leq x \leq b$] if

$$\int_a^b w(t)f_i(t)f_j(t)\, dt = \begin{cases} 0 & i \neq j \\ \lambda_i > 0 & i = j \end{cases}$$

If

$$\int_a^b w(t)f_i^2(t)\, dt = 1$$

we say that the system is *orthonormal*. It is easy to make an orthogonal set orthonormal if it is not. To do this, if

$$\int_a^b w(t)f_i^2(t)\, dt = \lambda_i > 0$$

then the set $f_i(t)/\sqrt{\lambda_i}$ is orthonormal.

We have defined the idea of orthogonality over a continuous interval. There are analogous definitions over a finite set of points, although in the latter case there can be at most as many mutually orthogonal functions as there are points in the set. It is convenient to use the continuous version most of the time and only occasionally mention the discrete case.

26.2 SOME EXAMPLES OF ORTHOGONAL SYSTEMS OF FUNCTIONS

Perhaps the best-known set of orthogonal functions, and certainly the one that has given rise to the most new mathematics (including Lebesgue integration), is the family

$$1, \cos x, \sin x, \cos 2x, \sin 2x, \cos 3x, \ldots$$

We first prove that they are orthogonal over the interval $0 \le x \le 2\pi$.   We need to show that

$$\int_0^{2\pi} \cos mx \cos nx \, dx = \begin{cases} 2\pi & m = n = 0 \\ \pi & m = n \ne 0 \\ 0 & m \ne n \end{cases}$$

$$\int_0^{2\pi} \cos mx \sin nx \, dx = 0$$

$$\int_0^{2\pi} \sin mx \sin nx \, dx = \begin{cases} \pi & m = n \ne 0 \\ 0 & m \ne n \end{cases}$$

In the first case, using the standard trigonometric identity, we get

$$\int_0^{2\pi} \cos mx \cos nx \, dx = \frac{1}{2} \int_0^{2\pi} [\cos(m + n)x + \cos(m - n)x] \, dx$$

$$= \frac{1}{2} \left[\frac{\sin(m + n)x}{m + n} + \frac{\sin(m - n)x}{m - n} \right] \Big|_0^{2\pi} = 0 \qquad m - n \ne 0$$

If $m = n \ne 0$, then we have

$$\int_0^{2\pi} \cos mx \cos nx \, dx = \frac{1}{2} \int_0^{2\pi} [\cos(m + n)x + 1] \, dx = \pi$$

If $m = n = 0$, then we have

$$\int_0^{2\pi} \cos mx \cos nx \, dx = \int_0^{2\pi} 1 \, dx = 2\pi$$

The other relations can be proved similarly.

We can see a simple use for orthogonality in the formal expansion of a function; for example, suppose for some function $F(x)$ $(0 \le x \le 2\pi)$

$$F(x) = \frac{a_0}{2} + a_1 \cos x + b_1 \sin x + a_2 \cos 2x + b_2 \sin 2x + \cdots$$

$$= \frac{a_0}{2} + \sum_{k=1}^{\infty} (a_k \cos kx + b_k \sin kx)$$

Multiply both sides by $\cos mx$ and integrate $(0 \le x \le 2\pi)$

$$\int_0^{2\pi} F(x) \cos mx \, dx = \pi a_m \qquad (m = 0, 1, \ldots$$

(all other terms vanish due to the orthogonality).
Using $\sin mx$ in place of $\cos mx$, we get

$$\int_0^{2\pi} F(x) \sin mx \, dx = \pi b_m \qquad m = 1, 2, \ldots$$

Thus using orthogonality, we can formally find the individual coefficients of the expansion. The use of $a_0/2$ is for convenience.

The a_m and b_m computed this way are called *the Fourier*[1] *coefficients*, even in the case of a general system of orthogonal functions

$$\int_a^b w(x)f_i(x)f_j(x)\,dx = \begin{cases} 0 & i \neq j \\ \lambda_j & i = j \end{cases}$$

If

$$F(x) = \sum_{i=0}^{\infty} a_i f_i(x)$$

then

$$a_j = \frac{1}{\lambda_j}\int_a^b w(x)F(x)f_j(x)\,dx$$

are the Fourier coefficients.

As another example of orthogonal functions, consider the ultraspherical polynomials defined by the differential equation

$$(1 - x^2)y_n''(x) - 2(\alpha + 1)xy_n'(x) + n(n + 2\alpha + 1)y_n(x) = 0 \qquad \alpha > -1$$

If we multiply this equation by $(1 - x^2)^\alpha y_m(x), (m \neq n)$ and integrate $(-1 \leq x \leq 1)$ we get

$$\int_{-1}^1 y_m(x)\frac{d}{dx}[(1 - x^2)^{\alpha+1}y_n'(x)]\,dx + n(n + 2\alpha + 1)$$

$$\times \int_{-1}^1 (1 - x^2)^\alpha y_m(x)y_n(x)\,dx = 0$$

Integrate the first integral by parts

$$y_m(x)(1 - x^2)^{\alpha+1}y_n'(x)\Big|_{-1}^1 - \int_{-1}^1 (1 - x^2)^{\alpha+1}y_m'(x)y_n'(x)\,dx$$

$$+ n(n + 2\alpha + 1)\int_{-1}^1 (1 - x^2)^\alpha y_m(x)y_n(x)\,dx = 0$$

The integrated part vanishes at both limits. We next write the same equation with m and n interchanged and subtract the two equations to get

$$[m(m + 2\alpha + 1) - n(n + 2\alpha + 1)]\int_{-1}^1 (1 - x^2)^\alpha y_m(x)y_n(x)\,dx = 0$$

For $m \neq n$ the term inside the square bracket is not zero; hence for all $\alpha > -1$ the ultraspherical polynomials are orthogonal over the interval $-1 \leq x \leq 1$ with the weight factor $(1 - x^2)^\alpha$. Thus we have shown that the polynomial solutions of the differential equation for different n are orthogonal.

This family, depending on α, of orthogonal polynomial systems includes many special cases of importance which we will later study, $\alpha = -\frac{1}{2}, 0, \frac{1}{2}$, and 1.

[1] Jean Baptiste Joseph Fourier (1768–1830).

PROBLEMS

26.2.1 Complete the proof of the orthogonality of the trigonometric functions.

26.2.2 Prove directly the orthogonality of the set of trigonometric functions over $-\pi \leq x \leq \pi$.

26.2.3 For the set of functions $1, \cos x, \cos 2x, \ldots$ show that they are orthogonal $(0 \leq x \leq \pi)$.

26.2.4 For the set of functions $\sin x, \sin 2x, \ldots$ show that they are orthogonal $(-\pi/2 \leq x \leq \pi/2)$.

26.2.5 Show that for the case of $\alpha = -\frac{1}{2}$ in the ultraspherical polynomials the substitution $x = \cos \theta$ produces a familiar situation.

26.2.6 Normalize the Fourier-series terms.

26.3 LINEAR INDEPENDENCE AND ORTHOGONALITY

The purpose of this section is to show that the ideas of linear independence and orthogonality are closely related. First, we prove that *a set of orthogonal functions* $f_i(x)$ is linearly independent over the interval. Suppose that there is a linear relation among the $f_i(x)$ with nonzero coefficients, that is, that there exists a relation

$$c_1 f_1(x) + c_2 f_2(x) + \cdots + c_N f_N(x) \equiv 0$$

with some $c_j \neq 0$. Multiply this relation through by $w(x)f_j(x)$ and integrate over the interval. We get

$$c_1 \int_a^b w(x)f_1(x)f_j(x)\,dx + c_2 \int_a^b w(x)f_2(x)f_j(x)\,dx + \cdots$$
$$+ c_N \int_a^b w(x)f_N(x)f_j(x)\,dx = 0$$

Due to the orthogonality only one term remains $[w(x) \geq 0]$

$$c_j \int_b^a w(x)f_j^2(x)\,dx = c_j \lambda_j = 0$$

and therefore $c_j = 0$ for all j—the assumed relation does not exist.

The converse, that from a set of linearly independent functions we can construct an orthogonal set, can be shown by the Schmidt[1] process. Let the given set of linearly independent functions be $f_i(x)$. We compute

$$\int_b^a w(x)f_0^2(x)\,dx = \lambda_0 > 0 \qquad w(x) \geq 0$$

[1] Eberhart Schmidt (1850–1890).

Then

$$g_0(x) = \frac{f_0(x)}{\sqrt{\lambda_0}}$$

defines the first orthonormal function $g_0(x)$.   Using mathematical induction, we assume that we have constructed the first j orthonormal functions $g_i(x)$ $(i = 0, 1, \ldots, j - 1)$.   We set

$$F_j(x) = a_0 g_0 + a_1 g_1 + \cdots + a_{j-1} g_{j-1} + f_j(x)$$

Now $F_j(x) \not\equiv 0$, since the functions $f_i(x)$ are linearly independent and each $g_i(x)$ is a linear combination of the $f_k(x)$ for $k \leq i$.   We need

$$\int_a^b w(x) F_j(x) g_i(x)\, dx = 0 \qquad 0 \leq i \leq j - 1$$

But this is, by the definition of $F_j(x)$,

$$\lambda_i a_i + \int_a^b w(x) g_i(x) f_j(x)\, dx = 0$$

which determines the a_i and hence $F_j(x)$.   To normalize $F_j(x)$, we need to compute

$$\int_a^b w(x) F_j^2(x)\, dx = \lambda_j > 0 \qquad w(x) \geq 0$$

and then set

$$g_j(x) = \frac{F_j(x)}{\sqrt{\lambda_j}}$$

Thus we have taken one more step in the induction.

If we have only a finite number of N samples x_m, then there are at most N linearly independent functions

$$f_j(x_m)$$

That there are N follows from the particular set

$$g_j(x_m) = \begin{cases} 0 & m \neq j \\ 1 & m = j \end{cases} \qquad j = 1, \ldots, N$$

since no set of these N functions $g_j(x_m)$ can be linearly dependent.

It is worth noting that in the construction process the set that we obtain is not unique unless we choose the particular $\sqrt{\lambda_i} > 0$ and also choose the set of linearly independent functions in a fixed order.   Also, they obviously depend on the interval (a,b) and the weight function $w(x)$.

PROBLEMS

26.3.1   Given $P_0 = 1$, $P_1 = x$, and $w(x) = 1$, construct the orthogonal functions $P_2(x)$ and $P_3(x)$ over the set of points $-2, -1, 0, 1, 2$, where P_2 and P_3 are polynomials of degrees 2 and 3, respectively.

26.3.2   Given 1, $\cos x$, $\cos^2 x$, and $w \equiv 1$, construct the orthogonal functions $(0 \leq x \leq \pi)$.

26.4   LEAST-SQUARES FITS AND THE FOURIER COEFFICIENTS

In this section we show that least squares and orthogonality are closely related.

Theorem 26.4.1   The Fourier coefficients a_j give the best least-squares fit when a function $F(x)$ is expanded in terms of an orthonormal set of functions $g_j(x)$.

To prove this we want to minimize, for an arbitrary expansion,

$$m = \int_a^b w(x) \left[F(x) - \sum_{j=0}^{M} c_j g_j(x) \right]^2 dx$$

$$= \int w(x)F^2(x)\,dx - 2\sum_{j=0}^{M} c_j \int w(x)F(x)g_j(x)\,dx + \sum_{i=0}^{M} \sum_{j=0}^{M} c_j c_i \int w(x)g_i g_j\,dx$$

$$= \int w(x)F^2(x)\,dx - 2\sum_{i=0}^{M} c_i a_i + \sum_{i=0}^{M} c_i^2$$

$$= \int w(x)F^2(x)\,dx - \sum_{i=0}^{M} a_i^2 + \sum_{i=0}^{M} (a_i - c_i)^2$$

But this is minimum only if each $c_i = a_i$ as required.

A remarkable and very useful property is that in the best least-squares fit with orthogonal functions each coefficient a_i is determined independently of all the others, and that if we decide to change the number of functions $g_i(x)$ we are using, then we need not redetermine any of the coefficients that we have already found. We therefore examine the converse problem.

If the coefficients c_i in a least-squares fit of a function $F(x)$ in terms of a set $\mu_i(x)$ are not to change as we change the number of $\mu_i(x)$ that we use, then the $\mu_i(x)$ must be orthogonal.   Set

$$g(c_0, c_1, \ldots, c_n) = \int_a^b w(x) \left[F(x) - \sum_{i=0}^{M} c_i \mu_i(x) \right]^2 dx$$

Since g is to be minimized,

$$\frac{\partial g}{\partial c_j} = 0 = -2 \int_a^b w(x) \left[F(x) - \sum_{i=0}^{M} c_i \mu_i \right] \mu_j(x)\,dx$$

or

$$\int w(x)F(x)\mu_j \, dx = \sum_{i=0}^{M} c_i \int_a^b w(x)\mu_i \mu_j \, dx$$

This must also be true for $M + 1$ if the property is to be true for all M.

$$\int w(x)F(x)\mu_j(x) \, dx = \sum_{i=0}^{M+1} c_i \int_a^b w(x)\mu_i \mu_j \, dx \qquad c_{M+1} \neq 0$$

Subtracting the previous equation, we get

$$c_{M+1} \int_a^b w(x)\mu_{M+1}\mu_j \, dx = 0$$

for any j; that is, the μ_j are orthogonal to μ_{M+1} (and M was arbitrary).

Thus we see that orthogonal functions, the Fourier determination of the coefficients in an expansion, and the idea of least-squares approximation are all bound together.

26.5 BESSEL'S[1] INEQUALITY AND COMPLETENESS

The Fourier coefficients

$$a_j = \int_a^b w(x)F(x)g_j(x) \, dx$$

of an *orthonormal* set $g_j(x)$ satisfy the Bessel inequality

$$\int_a^b w(x)F^2(x) \, dx \geq \sum_{j=0}^{M} a_j^2 \qquad w \geq 0$$

The proof is straightforward.   We write, since the integrand is nonnegative,

$$\int_a^b w(x)[F(x) - \sum_{i=0}^{M} a_i g_i(x)]^2 \, dx \geq 0$$

and multiply out:

$$0 \leq \int_a^b w(x)F^2(x) \, dx - 2 \int_a^b w(x)F(x)\sum_{i=0}^{M} a_i g_i(x) \, dx$$
$$+ \int_a^b \sum_{i=0}^{M} \sum_{j=0}^{M} a_i a_j w(x)g_i g_j \, dx$$

Using the definition of the a_i and the orthogonality of the g_i, we get

$$\int_a^b w(x)F^2(x) \, dx \geq 2\sum_{i=0}^{M} a_i^2 - \sum_{i=0}^{M} a_i^2 = \sum_{i=0}^{M} a_i^2$$

for all M.

---

[1] Friedrich Wilhelm Bessel (1784–1846).

In the continuous case, if the equality holds for every continuous function $F(x)$ in the interval $[a, b]$ then the infinite set of functions is said to be "complete," and the equality

$$\int_a^b w(x)F^2(x)\,dx = \sum_{i=0}^{\infty} a_i^2$$

is called *Parseval's*[1] *equality*.

In the finite case we have, of course, no trouble in the process of interchanging limits to get Parseval's equality, and any set of N functions linearly independent over a set of N points is complete over the same set of points. Parseval's equality is an "ergodic theorem"; the sum of the squares of the function is the same as the weighted sum of the squares of the coefficients.

In Bessel's inequality the difference of the two terms is exactly the sum of the squares of the errors $\varepsilon(x)$ (residuals)

$$\sum_{i=1}^{M} w(x_i)\varepsilon_i^2 = \int_a^b w(x)F^2(x)\,dx - \sum_{j=1}^{N} a_j^2$$

In the discrete case the integrals become sums

$$\sum_{i=1}^{M} w(x_i)\varepsilon_i^2 = \sum_{i=1}^{M} w(x_i)F^2(x_i) - \sum_{j=1}^{N} a^2$$

PROBLEMS

26.5.1 If the functions are only orthogonal, then derive the corresponding formula for the sum of the squares of the ε_i.

26.6 ORTHOGONAL POLYNOMIALS

Orthogonal polynomials are a very important subclass of the class of orthogonal functions. They are defined to be such that the kth one $(k = 0, 1, 2, \ldots)$ is of degree exactly k, and they naturally have all the properties of orthogonal functions plus those peculiar to polynomials.

It is easy to show that the kth orthogonal polynomial $y_k(x)$ has exactly k real, distinct roots in the range of integration. Let us suppose that $r < k$. We form the product of the distinct, odd-multiplicity zeros

$$p(x) = (x - x_1)(x - x_2)\cdots(x - x_r) \qquad r < k$$

[1] August von Parseval (1861–1942).

Then

$$\int_a^b w(x)p(x)y_k(x)\, dx = 0 \qquad w(x) \geq 0$$

because $p(x)$ can be written as a sum of $y_0, y_1, \ldots, y_r$, and $y_k(x)$ is orthogonal to all of them. But this is impossible since the integrand does not change sign in the interval. Hence there are k real, distinct roots in (a, b).

The orthogonal polynomials $y_k(x)$ satisfy a three-term recurrence relation of the form

$$a_k y_{k+1}(x) + (b_k - x)y_k(x) + c_k y_{k-1}(x) = 0, \quad k \geq 1$$

To show this, we set

$$y_i(x) = \alpha_i x^i + \cdots \qquad \alpha_i > 0$$

(The choice of α_i as any particular positive number is a matter of convenience only.) Then

$$a_k y_{k+1} - xy_k$$

is a polynomial of degree k, provided that we choose $a_k = \alpha_k / \alpha_{k+1}$. Hence

$$a_k y_{k+1} - xy_k = \gamma_k y_k + \gamma_{k-1} y_{k-1} + \cdots + \gamma_0 y_0$$

We multiply by $w(x)y_m(x)$ and integrate. We obtain

$$\int_a^b w(x)(a_k y_{k+1} - xy_k)y_m(x)\, dx = \gamma_m \lambda_m$$

For $m = 0, 1, \ldots, k$, y_m is orthogonal to y_{k+1}; and for $m = 0, 1, \ldots, k-2$, xy_m is a polynomial of degree less than k, and hence is orthogonal to y_k. Thus $\gamma_0 = \gamma_1 = \cdots = \gamma_{k-2} = 0$. For $m = k - 1$, the above equation becomes

$$-\int_a^b wy_k(xy_{k-1})\, dx = -\int_a^b wy_k\left(\frac{\alpha_{k-1}}{\alpha_k} y_k + c_{k-1}y_{k-1} + \cdots\right) dx$$

$$= -\frac{\alpha_{k-1}}{\alpha_k} \lambda_k = \gamma_{k-1}\lambda_{k-1}$$

Using the three-term recurrence relation form, we see that

$$-c_k = \gamma_{k-1} = -\frac{\alpha_{k-1}\lambda_k}{\alpha_k \lambda_{k-1}} \neq 0$$

For $m = k$, the above equation becomes

$$\int_a^b w(a_k y_{k+1} - xy_k)y_k\, dx = -\int_a^b wxy_k^2\, dx = \gamma_k \lambda_k = -b_k\lambda_k$$

Thus we can write

$$\frac{\alpha_k}{\alpha_{k+1}} y_{k+1} - xy_k = \gamma_k y_k + \gamma_{k-1} y_{k-1}$$

or

$$\frac{\alpha_k}{\alpha_{k+1}} y_{k+1} - (\gamma_k + x)y_k + \frac{\alpha_{k-1}}{\alpha_k} \frac{\lambda_k}{\lambda_{k-1}} y_{k-1} = 0$$

Several important results follow from this formula.   We first show that the k zeros of $y_k(x)$ are separated by the $k-1$ zeros of $y_{k-1}(x)$, provided that $w(x) \geq 0$. The proof is by induction.   As a basis for the induction, we have $y_0(x) = \alpha_0 > 0$ and $y_1(x) = \alpha_1(x + \gamma_0)$.   We know that $y_1(x)$ has a real zero in the range of integration, since

$$\int_a^b w(x)y_0(x)y_1(x) \, dx = 0 \qquad \text{hence } a < -\gamma_0 < b$$

We now assume by the induction hypothesis that the zeros of $y_k(x)$ are interlaced by those of $y_{k-1}(x)$.   At the zeros x_i of $y_k(x)$, the three-term relation becomes

$$\frac{\alpha_k}{\alpha_{k+1}} y_{k+1}(x_i) = -\frac{\alpha_{k-1}}{\lambda_k} \frac{\lambda_k}{\lambda_{k-1}} y_{k-1}(x_i)$$

At the top of the range $(x = b)$ both $y_{k+1}(b)$ and $y_{k-1}(b)$ are of positive sign, since we chose the leading coefficients $\alpha_i > 0$ and since all the zeros of the functions lie inside the interval.   At the largest zero of $y_k(x_i) = 0$, $y_{k-1}(x_i)$ is still positive; hence $y_{k+1}(x_i)$ is negative.   At the next smaller zero, y_{k-1} has, owing to the induction hypothesis, changed sign; hence $y_{k+1}(x)$ is positive.   And so it proceeds; as we pass from zero to zero of $y_k(x_i)$, $y_{k-1}(x_i)$ changes sign; hence $y_{k+1}(x_i)$ changes sign.   Thus we have shown that the zeros of $y_k(x)$ interlace those of $y_{k+1}(x)$, except for the smallest zero of y_{k+1}.   But since we have just shown that $y_{k+1}(x)$ has $k + 1$ real, distinct roots in (a, b) the last zero of $y_{k+1}(x)$ is smaller than all those of $y_k(x)$, and the induction proof is complete.

There is another important consequence of the three-term recurrence relations.   Once we know the coefficients of the recurrence relation as functions of k and know $y_0(x)$ and $y_1(x)$, we can then compute the $y_k(x)$, one at a time, by this relation rather than by the cumbersome Schmidt process.   See Sec. 27.2.

Orthogonal polynomials have another rather obvious property that we have already used, namely that an arbitrary polynomial $F_N(x)$ of degree N can be written as a sum of orthogonal polynomials $P_m(x)$ of degrees 0 through N.   The process for doing this is straightforward. We merely look at the coefficients of the highest powers both of the given polynomial $F_N(x)$ and of the corresponding orthogonal polynomial $P_N(x)$ of the same degree, and we form a suitable multiple $KP_N(x)$ of the orthogonal polynomial so that when this is subtracted from the given polynomial $F_N(x)$, the highest power cancels out. We then repeat this process on what is left, $F_{N-1}(x)$, which is a polynomial of lower degree. Enough steps get us to identically zero. Thus what we have subtracted is the orthogonal polynomial representation of the given polynomial. This is illustrated in the next section by using a particular set of orthogonal polynomials.

26.7 THE LEGENDRE[1] POLYNOMIALS

One of the most commonly occurring sets of orthogonal polynomials is the Legendre polynomials $P_n(x)$, which are the ultraspherical polynomials with $\alpha = 0$ (see Sec. 26.2). Thus we have

$$\int_{-1}^{1} P_m(x)P_n(x)\, dx = \begin{cases} 0 & m \neq n \\ 2/(2n+1) & m = n \end{cases}$$

where we have not derived the normalizing factor $2/2(n+1)$. The three-term recurrence relation (Sec. 26.6) is

$$(n+1)P_{n+1}(x) - (2n+1)xP_n(x) + nP_{n-1}(x) = 0$$

This can be used to generate the Legendre polynomials (see Fig. 26.7.1), using

$$P_0 = 1$$

$$P_1 = x$$

$$P_2 = \frac{1}{2}(3x \cdot x - 1) = \frac{3x^2 - 1}{2}$$

$$P_3 = \frac{1}{3}\left(5x\,\frac{3x^2 - 1}{2} - 2x\right) = \frac{5x^3 - 3x}{2}$$

$$P_4 = \frac{1}{4}\left(7x\,\frac{5x^3 - 3x}{2} - 3\,\frac{3x^2 - 1}{2}\right) = \frac{35x^4 - 30x^2 + 3}{8}$$

etc.

[1] Adrien-Marie Legendre (1752–1833).

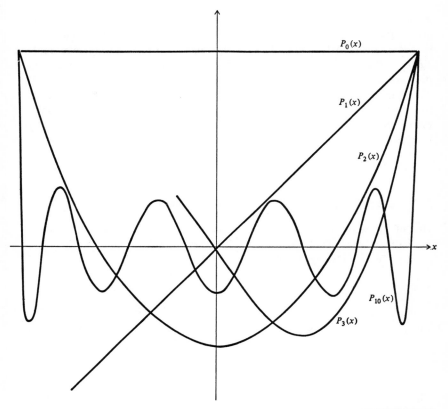

$P_0(x)$

$P_1(x)$

$P_2(x)$

$P_{10}(x)$

$P_3(x)$

x

FIGURE 26.7.1

To convert a given polynomial, say

$$F(x) = 10x^3 + 3x^2 + 7x - 9$$

to a sum of Legendre polynomials, we subtract $4P_3$

$$
\begin{array}{ll}
 & 10x^3 + 3x^2 + 7x - 9 \\
4P_3 = & 10x^3 \qquad\quad - 6x \\
\hline
 & 3x^2 + 13x - 9 \\
\text{subtract} \quad 2P_2 = & 3x^2 \qquad\quad - 1 \\
\hline
 & 13x - 8 \\
\text{subtract} \quad 13P_1 = & -13x \\
\hline
\text{and subtract} \quad -8P_0 = & \qquad\qquad\quad - 8
\end{array}
$$

Thus

$$F(x) = 10x^3 + 3x^2 + 7x - 9 \equiv 4P_3(x) + 2P_2(x) + 13P_1(x) - 8P_0$$

Other widely used sets of orthogonal polynomials are the Laguerre,[1] $L_n(x)$

$$\int_0^\infty e^{-x} L_m(x) L_n(x)\, dx = \begin{cases} 0 & m \neq n \\ n! & m = n \end{cases}$$

$$(n+1)L_n(x) - (2n+1-x)L_n(x) + nL_{n-1}(x) = 0 \qquad L_0 = 1 \qquad L_1 = 1 - x$$

$$xL_n'' + (1-x)L_n' + nL_n = 0$$

and Hermite[2] $H_n(x)$

$$\int_{-\infty}^\infty e^{-x^2} H_m(x) H_n(x)\, dx = \begin{cases} 0 & m \neq n \\ 2^n n! \sqrt{\pi} & m = n \end{cases}$$

$$H_{n+1}(x) - 2xH_n(x) + 2nH_{n-1}(x) = 0 \qquad H_0 = 1 \qquad H_1 = x$$

$$H_n'' - 2xH_n' + 2nH_n = 0$$

Many other relations can be found among these functions in standard texts.

26.8 ORTHOGONAL POLYNOMIALS AND GAUSSIAN QUADRATURE

The orthogonal polynomials are connected with Gaussian quadrature (Chap. 19). To see this, consider again the problem of finding a formula of the form

$$\int_{-1}^1 f(x)\, dx = \sum_{i=1}^N w_i f(x_i)$$

where this is to be exact for all powers of x up to $2N - 1$. Thus the formula should be true for an arbitrary polynomial $f_{2N-1}(x)$; that is,

$$\int_{-1}^1 f_{2N-1}(x)\, dx = \sum_{i=1}^N w_i f_{2N-1}(x_i)$$

Now consider dividing $f_{2N-1}(x)$ by the Legendre polynomial $P_N(x)$:

$$f_{2N-1}(x) = g(x)P_N(x) + r(x)$$

where the two functions $g(x)$ and $r(x)$ are polynomials of degree at most $N - 1$. The integration formula becomes

$$\int_{-1}^1 g(x)P_N(x)\, dx + \int_{-1}^1 r(x)\, dx = \sum_{i=1}^N w_i g(x_i)P_N(x_i) + \sum_{i=1}^N w_i r(x_i)$$

---

[1] Edmund Nicolas Laguerre (1834–1886).
[2] Charles Hermite (1822–1901).

458 26 ORTHOGONAL FUNCTIONS

But since $P_N(x)$ is orthogonal to all lower Legendre polynomials, it is also orthogonal to *any* polynomial of degree less than N, and the first of the integrals vanishes.

On the other side of the equation we can get rid of the first summation if we pick the sample points as the zeros of the Legendre polynomial $P_N(x)$. We then have left the two terms

$$\int_{-1}^{1} r(x)\, dx = \sum_{i=1}^{N} w_i\, r(x_i)$$

and we are reduced to determining the weights w_i in the standard case of known sample points x_i. From the uniqueness of the Gaussian quadrature the two results must be the same.

Notice how this parallels the earlier derivation which found the x_i so that the defining equations for $x^N, \ldots, x^{2N-1}$ were satisfied, and then having found the x_i, we were reduced to finding the weights from the first N equations.

In an exactly similar way, the zeros of the Laguerre polynomials are the sampling points for the corresponding Gaussian quadrature formula

$$\int_{0}^{\infty} e^{-x} f(x)\, dx = \sum_{i=1}^{N} w_i f(x_i)$$

while the Hermite polynomials are related to the integration formula

$$\int_{-\infty}^{\infty} e^{-x^2} f(x)\, dx = \sum_{i=1}^{N} w_i f(x_i)$$

For this approach to Gaussian quadrature, see Mineur [41].

LEAST SQUARES: PRACTICE

27.1 GENERAL REMARKS ON THE POLYNOMIAL SITUATION

The major use of the least-squares method is when the coefficients to be determined occur in a linear fashion, especially as coefficients of a polynomial. We shall therefore take up the polynomial situation first and then pass on to the other cases.

As was indicated in Sec. 25.6, the determinant of the normal equations tends to be very small; hence the solution for coefficients is likely to be uncertain. It is necessary to distinguish, however, between two things: the accuracy of the coefficients and the smallness of the sum of the squares of the errors. When the determinant is small, the coefficients may be poorly determined (that is, the curve is ill-defined by the data), but still the sum of the squares of the errors can be close to the minimum. Alternatively, it may be that the direct approach to finding the curve is at fault. Generally speaking, when there are up to five or six coefficients to be determined, the direct solution of the normal equations is usually satisfactory, but above this there is likely to be trouble.

The theory of Chap. 26 indicates that the expansion in orthogonal polynomials is a substitute for the direct solution. However, experience shows that if the orthogonalization is attempted by means of the Schmidt process, then the same difficulty arises in a different disguise. In the Schmidt process, the construction of the mth polynomial is done by first subtracting all the components of the vector x^m, that is, $(x_1{}^m, x_2{}^m, \ldots, x_N{}^m)$, which lie in the direction of the previously determined polynomials. The result is apt to be small when m is fairly large, and it is this remainder that is normalized as the last step in the Schmidt process. It is the cancellation that increases the errors, not the dividing factor which is usually much smaller than 1.

Loosely speaking, the trouble may be expressed by the remark that for large n and $x \geq 0$ the vector x^n points in about the same direction as does x^{n-1}, x^{n-2}, etc. Thus the equations leading to the Hilbert determinant are almost linearly dependent; hence the determinant is small and the accuracy low.

A novel and effective solution to this dilemma is offered in Sec. 27.3.

27.2 USE OF THE THREE-TERM RECURRENCE RELATION

If we insist on using the orthogonal polynomials, then we can apparently avoid the troubles mentioned in the last section by using the three-term recurrence relation of Sec. 26.6 to generate the orthogonal polynomials.[1]

We write the three-term recurrence relation in the form

$$p_0(x) = 1$$
$$p_1(x) = xp_0(x) - \alpha_1 p_0(x)$$
$$p_{k+1}(x) = xp_k(x) - \alpha_{k+1} p_k(x) - \beta_{k+1} p_{k-1}(x) \qquad k \geq 1$$

where α_{k+1} and β_{k+1} are to be determined.

We first determine α_1. We know that

$$\int_a^b w(x)p_0(x)p_1(x)\, dx = 0$$

Hence

$$\int_a^b w(x)x\, dx = \alpha_1 \int_a^b w(x)\, dx$$

We now suppose that we have $p_0(x)$, $p_1(x)$, $\ldots$, $p_k(x)$ and that they are orthogonal to one another. We need to compute $p_{k+1}(x)$, the next polynomial of the set.

[1] G. E. Forsythe, Generation and Use of Orthogonal Polynomials for Data-fitting with a Digital Computer, *J. Soc. Ind. Appl. Math.*, vol. 5, no. 2, June, 1957; M. Ascher and G. E. Forsythe, SWAC Experiments on the Use of Orthogonal Polynomials for Data Fitting, *J. Assoc. Computing Machinery*, vol. 5, no. 1, January, 1958.

We require both

$$\int w(x)p_{k+1}(x)p_k(x)\,dx = 0$$

and

$$\int w(x)p_{k+1}(x)p_{k-1}(x)\,dx = 0$$

These suffice to determine α_{n+1} and β_{n+1}. Using the definition of $p_{k+1}(x)$, we get after some rearrangements,

$$\int w(x)xp_k^2(x)\,dx = \alpha_{k+1}\int w(x)p_k^2(x)\,dx + \beta_{k+1}\int w(x)p_k(x)p_{k-1}(x)\,dx$$

$$\int w(x)xp_k(x)p_{k-1}(x)\,dx = \alpha_{k+1}\int w(x)p_k(x)p_{k-1}(x)\,dx + \beta_{k+1}\int w(x)p_{k-1}^2(x)\,dx$$

Since $p_k(x)$ and $p_{k-1}(x)$ are orthogonal, we have

$$\alpha_{k+1} = \frac{\int w(x)xp_k^2(x)\,dx}{\int w(x)p_k^2(x)\,dx}$$

$$\beta_{k+1} = \frac{\int w(x)xp_k(x)p_{k-1}(x)\,dx}{\int w(x)p_{k-1}^2(x)\,dx}$$

The denominator of the β_{k+1} equation was computed during the previous step when $\alpha_k(x)$ was determined; thus we have three integrals to compute on each step.

That $p_{k+1}(x)$ is orthogonal to all $p_i(x)$ $(i < k-1)$ follows from the equations

$$p_{k+1}(x) = xp_k(x) - \alpha_{k+1}p_k(x) - \beta_{k+1}p_{k-1}(x)$$

since, on multiplying by $w(x)p_i(x)$ and integrating, we have

$$\int w(x)p_k(x)xp_i(x)\,dx - \alpha_{k+1}\int w(x)p_k(x)p_i(x)\,dx - \beta_{k+1}\int w(x)p_{k-1}(x)p_i(x)\,dx = 0$$

The last two integrals are zero because of orthogonality, and in the first $xp_i(x)$ is a polynomial of degree less than k; hence the integral is also zero.

When we try to construct polynomials over a discrete set of points x_j $(j = 1, \ldots, N)$, the integrals are replaced by sums. If we attempt to construct more than N polynomials [that is, above $p_{N-1}(x)$], then the equation determining α_N fails, as it should.

PROBLEMS

27.2.1 Orthogonalize 1, x, x^2, x^3 over the interval $0 \leq x \leq 1$, using the three-term recurrence method.

27.3   THE CONSTRUCTION OF QUASI-ORTHOGONAL POLYNOMIALS

The direct approach to the least-squares fitting of polynomials leads to a system of normal equations which has large nonzero terms in places off the main diagonal and which is often hard to solve.   The use of orthogonal polynomials leads to equations for which all the terms off the main diagonal are zero; hence the solution of the equations is trivial.

If we attempt to construct some orthogonal polynomials by the Schmidt process (or by the three-term recurrence relation) and do not determine them accurately, then the resulting equations to be solved have large terms on the main diagonal, while those off the diagonal are proportional to the deviation from orthogonality.   If these are not large, then the system of equations is fairly easy to solve.

This observation suggests that we try our hand at constructing some *quasi-orthogonal polynomials* without doing much computation.   Insofar as we make them orthogonal, the terms off the main diagonal of the system of normal equations will be zero, and insofar as we miss orthogonality, they will be that far away from zero.

Further meditating on this question suggests that the fundamental property we should try to keep is the interlacing zeros.   Hence we choose $p_0(x) = 1$ as the first polynomial.   The second we take as a straight line of convenient slope but with a zero near the median of the set of points.   We next pick a quadratic whose zeros are on each side of the straight line.   And so we proceed; at each step we choose the zeros of our next polynomial so that they are separated by the preceding polynomial and also leave room for the next polynomials.   Unless we try to determine a polynomial whose degree is near that of the exact-matching polynomial through all the points (and this is rarely if ever the case) there is enough freedom to select the zeros at convenient places to make the polynomials have convenient coefficients and thus lighten the burden of subsequent calculation.

Experience with this method shows that it is very effective; usually the off-diagonal terms of the system of equations are fairly small, and the equations are easy to solve.

As a very simple example, where we have picked one unequally spaced point to avoid complete triviality (though quite commonly the data are equally spaced), consider data $x = -1, -1/2, 0, 1/2, 1, 2$.

We pick

$$P_0 = 1$$
$$P_1 = x$$
$$P_2 = (x + \tfrac{1}{2})(x - 1) = x^2 - \tfrac{1}{2}x - \tfrac{1}{2}$$

x	P_0	P_1	P_2
-1	1	-1	1
$-\frac{1}{2}$	1	$-\frac{1}{2}$	0
0	1	0	$\frac{1}{2}$
$\frac{1}{2}$	1	$\frac{1}{2}$	$\frac{1}{2}$
1	1	1	0
2	1	2	$\frac{5}{2}$

$$\sum P_0{}^2(x_i) = 6$$
$$\sum P_0(x_i)P_1(x_i) = -1 - \tfrac{1}{2} + 0 + \tfrac{1}{2} + 1 + 2 = 2$$
$$\sum P_1{}^2(x_i) = 1 + \tfrac{1}{4} + \tfrac{1}{4} + 1 + 4 = 6\tfrac{1}{2} = \tfrac{13}{2}$$
$$\sum P_0(x_i)P_2(x_i) = 1 + 0 - \tfrac{1}{2} - \tfrac{1}{2} + 0 + \tfrac{5}{2} = \tfrac{9}{2}$$
$$\sum P_1(x_i)P_2(x_i) = -1 + 0 + 0 - \tfrac{1}{4} + 0 + 5 = 3\tfrac{3}{4} = \tfrac{15}{4}$$
$$\sum P_2{}^2(x_i) = 1 + 0 + \tfrac{1}{4} + \tfrac{1}{4} + \tfrac{25}{4} = \tfrac{31}{4}$$

The matrix of the unknowns (numbers from the quasi-orthogonal polynomials) is

$$\begin{pmatrix} 6 & 2 & \frac{9}{2} \\ 2 & \frac{13}{2} & \frac{15}{4} \\ \frac{9}{2} & \frac{15}{4} & \frac{31}{4} \end{pmatrix}$$

To judge the size the off-diagonal elements, we need to normalize the quasi-orthogonal polynomials. This amounts to multiplying the ith row and the ith column by $1/\sqrt{a_{ii}}$. What amounts to the same thing, we compare $(a_{ij})^2$ with $a_{ii}a_{jj}$:

$$(a_{12})^2 = 4 \leq 6 \cdot 13/2 = 39$$
$$(a_{13})^2 = 81/4 \leq 6 \cdot 31/4 = 186/4$$
$$(a_{23})^2 = 275/16 \leq (13/2)(31/4) = 806/16$$

While we did not discuss the iterative method of solving linear equations, it should be fairly clear that if the main-diagonal terms tend to dominate the off-diagonal terms, then the equations can be solved easily—if no other way than by solving the ith equation for x_i and iterating until convergence.

PROBLEMS

27.3.1   Examine the matrix of the quasi-orthogonal polynomials $(-1 \leq x \leq 1)$ $w = 1$, $P_0 = 1, P_1 = x, P_2 = x^2 - 1/4, P_3 = x(x^2 - 9/16)$.

27.4 ON THE DEGREE OF THE POLYNOMIAL TO USE

A problem that arises regularly in least-squares fitting is what degree polynomial to use. The Bessel formula (see Fig. 27.4.1)

$$D^2(N) = \sum_{i=1}^{M} \varepsilon_i^2 = \sum_{i=1}^{M} F(x_i) - \sum_{j=1}^{N} \lambda_j a_j^2$$

gives a handy measure of the quality of the fit, and plotting this quantity as a function of the degree N of the polynomial being used gives a clue.

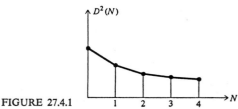

FIGURE 27.4.1

A better clue of when to stop in the process of fitting more and more terms in the least-squares approximations can be found from following the usual statistical practice of examining the residuals—in particular, examining the number of sign changes in the residuals.

Consider first the extreme condition of having only "noise" left in the residuals (see Fig. 27.4.2). Suppose that the residuals are random and independent of one another. By using simple probability we then expect that a given residual will be followed by one of the same sign half the time and one of the opposite sign half the time. Thus we expect to see about half the number of possible sign changes. If the number of sign changes is different from the number expected, either too low or too high by several times the square root of the number of possible changes, then we suspect that there was some kind of "signal" left and that our assumption of pure noise was wrong.

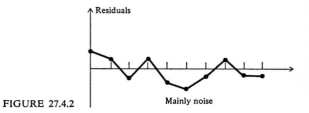

FIGURE 27.4.2

At the other extreme of no noise but all signal (see Fig. 27.4.3), then we expect (at least for the typical orthogonal system of functions) that the residuals will have about one more sign change than the last function we fitted, and a couple of extra sign changes above that will not worry us. The reason for this is that most orthogonal function systems have the property that the kth function has about k sign changes (this is exactly true of the important class of orthogonal polynomials; see Sec. 26.6).

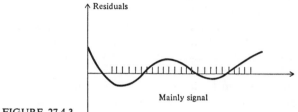

Residuals

Mainly signal

FIGURE 27.4.3

What we shall see in the residuals, of course, is probably a mixture of the two extremes, both some signal and some noise (see Fig. 27.4.4). The noise oscillations tend to increase the number of sign changes near the crossings due to the signal; and the more sign changes we see, the less we expect to reduce the sum of the squares if we fit a few more orthogonal functions.

Runs of constant sign often provide a more powerful test than the sign changes themselves—the runs reveal more structure. Furthermore, it is very easy on a computer to get the number of runs of each possible length by using an index register. If the index register is advanced one count each time the sign stays the same for the next point, and if when the sign changes the contents of the

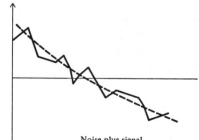

FIGURE 27.4.4

Noise plus signal

index register are used to determine the address whose contents are increased by 1, then the result is that the number of runs of length x are in location x (an initial count of 1 must be used for each run when the index register is reset after each use).

Thus we ask, What is the total length of the $k + 2$ longest runs? (We are assuming that we have fitted a polynomial of degree k and that the remainder will have $k + 1$ zeros.)   If this total length is approximately the length of the entire interval, then we know that the crossings are confined to the neighborhoods of the zeros of the relatively noise-free remainder.   If the intervals are broken up by noise, then the sum of the $k + 2$ longest intervals will not come near to the length of the interval.

This theory is a "down to earth" theory and is not given in the usual statistical courses.   The test that is currently in statistical good graces is quite complex, has some dubious assumptions (all tests must make some assumptions), and probably is of less use to the practicing scientist (who is not highly trained in statistics and is therefore not in a position to appreciate what the orthodox test means).   If you do understand advanced statistics, then by all means use it *if* it seems appropriate.

27.5   NONLINEAR PARAMETERS

It frequently happens that theory supplies the form that some data are to fit, and the data are to be used to determine the coefficients of the form, by the least-squares criterion.   For example, suppose that the given form is

$$y(x) = a + be^{cx}$$

and we are to determine a, b, and c such that for the data (x_i, y_i) $(i = 1, \ldots, N)$

$$\sum_{i=1}^{N} [y_i - (a + be^{cx_i})]^2 = \min = m \qquad (27.5.1)$$

When we differentiate with respect to a, b, and c, respectively, and set the results equal to zero, the equations are difficult to solve.

Suppose that we plot the data and make some guess for the value $c = c_1$. We now determine the a and b in the usual way.   We can then compute $m(c_1)$ from Eq. (27.5.1).

Next we try some other likely value of $c = c_2$ and again compute $m = m(c_2)$. An examination of these two values

c	$m(c)$
c_1	$m(c_1)$
c_2	$m(c_2)$

will suggest (by linear approximation) a new value of c, say c_3.

In this way we can approach the minimum value of $m(c)$.[1]   The details of the strategy for searching will be discussed in Chap. 43.   It should be evident, however, that we shall use a great deal of machine computation as compared with the situation where *all* the parameters to be determined occur linearly.   The more parameters that occur nonlinearly, the very much more the computation required to find the least-squares fit.   Experience shows that when the number of nonlinear parameters reaches four or five, the process can be exceedingly painful and slow.

27.6   LEAST SQUARES WITH RESTRAINTS: CONTINUATION OF THE EXAMPLE IN SEC. 9.10

In the example in Sec. 9.10 we fitted an exact-matching polynomial to 11 data points $x = 0(0.1)1$, using Newton's interpolation formula.   The success of the computation resulted in new laboratory equipment being built and much better and more extensive data being gathered.

A visual examination of the new data suggested about a sixth-order polynomial, and a plot of the logarithms of the values near $x = 1$ produced a slope near 6, thus tending to confirm the impression.   With no further thought, a least-squares polynomial was computed, and the results were what one would expect—stupid, thoughtless computing produced foolish results!   The polynomial began with a small positive value at $x = 0$, when the experiment clearly would give zero, and went slightly negative for a short distance, which was physically impossible.

The obvious remedy was to omit the constant term of the polynomial. Discussion with the physicist revealed that we could *not* expect the curve to be tangent to the x axis at $x = 0$, and so we could not leave out the x term.   This more carefully considered computation produced an acceptable polynomial, and the rest of the computation was straightforward, as before.

Another obvious way of approaching the problem of fitting the data at $x = 0$ would have been to give the value there a large weight, say 1,000, where the other data points had weight 1.

Many variations occur in placing extra conditions on a least-squares fit. For example, suppose that we require the solution to take on two specified values, one at a and one at b, and are using orthogonal polynomials.   We may use the classic method of Lagrange multipliers.   If the values at a and b were zero, we

---

[1] We could try fitting a quadratic and using its minimum if we wished.

could use a weight function $(x - a)^2(x - b)^2$ and construct the corresponding orthogonal polynomials $P_n(x)$.   Then the polynomials

$$Q_n(x) = (x - a)(b - x)P_n(x)$$

are orthogonal polynomials with weight factor $w(x) = 1$ passing through zero at $x = a$ and $x = b$.

PROBLEMS

27.6.1   Show that in the orthogonal-polynomial approach a straight line may be subtracted from the data first, to reduce two conditions $f(a) = A$ and $f(b) = B$ to $f(a) = f(b) = 0$.

27.7   SMOOTHING BY LEAST-SQUARES FITTING

Instead of using a single polynomial to fit all the data, it is often necessary to use a composite formula, as it were, and fit a least-squares polynomial *locally*. Typically an odd number of consecutive points is used to fit a low-order polynomial, and the value of this polynomial is taken to be the "smoothed value" at the middle point.   For example, suppose we smooth by taking a least-squares fit on five equally spaced points, using a quadratic.   We have, using the points $x = -2, -1, 0, 1,$ and 2,

$$f(a, b, c) = \sum_{i=-2}^{2} (a + bx_i + cx_i^2 - y_i)^2$$

to minimize.   The normal equations are

$$
\begin{aligned}
5a + 0b + 10c &= \sum y_i & \quad & 17 \\
0a + 10b + 0c &= \sum x_i y_i & & \\
10a + 0b + 34c &= \sum x_i^2 y_i & & -5
\end{aligned}
$$

Multiply the top equation by 17 and the bottom by -5 and add to find a, the value of the quadratic at the midpoint $x = 0$,

$$(85 - 50)a = 17 \sum y_i - 5 \sum x_i^2 y_i$$

$$a = \frac{1}{35} \sum_{i=-2}^{2} (17 - 5x_i^2) y_i$$

$$= \frac{-3y_{-2} + 12y_{-1} + 17y_0 + 12y_1 - 3y_2}{35}$$

Thus we have a formula for smoothing the data (except for the four end values). To get formulas for the end values, we have to find both b and c and evaluate the quadratic at $x = \pm 1$ and ± 2.

PROBLEMS

27.7.1 Find the formulas for the four end values.

27.7.2 Carry out the smoothing for a straight line $y = a + bx$.

$$Ans. \quad \tfrac{1}{5}(y_{-2} + y_{-1} + y_0 + y_1 + y_2)$$

27.7.3 Smooth, using seven equally spaced points and fitting a straight line, a quadratic and a quartic.

27.7.4 Discuss the fitting of an even-degree polynomial for the above method of smoothing.

27.8 ANOTHER FAULT OF LEAST-SQUARES FITTING

We have already discussed (Sec. 25.2) the fault that one or more extreme values, quite likely gross errors, play a dominant role in shaping the least-squares function that is selected from the family of functions.

Another fault with polynomial least squares is that the error curve tends to have very large errors at the ends of the interval.   This comes about because it is in the nature of polynomials to go to infinity for large values of the argument, and therefore the error is likely to be large at the ends.   Indeed, supposing for the moment that the expansion in Legendre polynomials is rapidly converging, then the error for some finite degree is approximately the first term neglected and will therefore look like the $(N + 1)$st Legendre polynomial (even if the Legendre polynomials are not used in the fitting process).   A study of the plot of the first few polynomials given in Sec. 26.7 shows this end effect clearly.   The remedy for this effect is to weight the end values relatively heavily; indeed, the correct weighting in a certain sense is given by the Chebyshev polynomials of the next two chapters.

28

CHEBYSHEV APPROXIMATION: THEORY

28.1 THE DEFINITION OF CHEBYSHEV POLYNOMIALS

The Chebyshev[1] polynomials $T_n(x)$ can be defined in many ways. Perhaps the simplest is

$$T_n(x) = \cos(n \arccos x)$$

To see that these are polynomials, recall de Moivre's[2] theorem

$$\cos n\theta + i \sin n\theta = (\cos \theta + i \sin \theta)^n$$

Expanding the binomial and taking the real part, we get

$$\cos n\theta = \sum_{k=0}^{n/2} C(n, 2k)(-1)^k \cos^{n-2k}\theta \sin^{2k}\theta$$

But using

$$\sin^{2k} \theta = (1 - \cos^2 \theta)^k$$

we get

$$\cos n\theta = \text{polynomial of degree } n \text{ in } \cos \theta$$

---

[1] The $T_n(x)$ notation comes from the spelling Chebyshev (1821–1894) used for his name in French, Tchebychef.

[2] Abraham de Moivre (1667–1754). Of him Newton late in life said "Ask M. de Moivre, he knows these things better than I do."

Now writing

$$\theta = \arccos x \qquad \text{or} \qquad x = \cos \theta$$

and using

$$\cos(\arccos x) \equiv x$$

we get the result.

Notice that

$$T_n(x) = T_{-n}(x) = T_{|n|}(x)$$

which is often useful in simplifying identities.

A second approach to the Chebyshev polynomials is via the ultraspherical polynomials defined in Sec. 26.2 with $\alpha = -\frac{1}{2}$. We have from there the orthogonality relations

$$\int_{-1}^{1} \frac{T_m(x) T_n(x)}{\sqrt{1 - x^2}} \, dx = 0 \qquad m \neq n$$

and the corresponding differential equation

$$(1 - x^2) T_n''(x) - x T_n'(x) + n^2 T_n(x) = 0$$

A third approach to the Chebyshev polynomials is to use the transformation

$$\theta = \arccos x$$

and recognize that the Chebyshev polynomials are the trigonometric cosine functions in disguise (see Fig. 28.1.1). We can show directly that the set of functions $\{\cos mx\}(0 \le x \le \pi)$ are orthogonal. We have

$$\int_0^\pi \cos m\theta \cos n\theta \, d\theta = \frac{1}{2} \int_0^\pi [\cos(m+n)\theta + \cos(m-n)\theta] \, d\theta$$

$$= \frac{1}{2} \left[\frac{\sin(m+n)\theta}{m+n} + \frac{\sin(m-n)\theta}{m-n} \right]\Bigg|_0^\pi \qquad m \neq n$$

$$= \begin{cases} 0 & m \neq n \\ \pi/2 & m = n \neq 0 \\ \pi & m = n = 0 \end{cases}$$

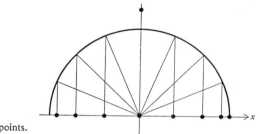

FIGURE 28.1.1
The Chebyshev sample points.

Transforming back from θ to x, we get

$$d\theta = \frac{-dx}{\sqrt{1 - x^2}}$$

$$\int_{-1}^{1} \frac{T_m(x)T_n(x)}{\sqrt{1 - x^2}} \, dx = \begin{cases} 0 & m \neq n \\ \pi/2 & m = n \neq 0 \\ \pi & m = n = 0 \end{cases}$$

Thus we see that the first definition, the ultraspherical polynomials with $\alpha = -1/2$, and the Fourier cosine series all lead to the same set of polynomials.

PROBLEMS

28.1.1 Show that $T_m\{T_n(x)\} = T_{mn}(x)$.

28.1.2 If $g(x)$ has the inverse function arc $g(x)$, then the functions $f_n(x) = g\{n \text{ arc } g(x)\}$ satisfy $f_m\{f_n(x)\} = f_{mn}(x)$.

28.2 CHEBYSHEV POLYNOMIALS OVER A DISCRETE SET OF POINTS

We have regularly faced the problem of replacing a continuous interval by some set of points.   It is an important fact, which we will prove in Chap. 31, that the Chebyshev polynomials $T_n(x)$ are orthogonal over the following discrete set of $N + 1$ points x_i, equally spaced in θ,

$$\theta_i = 0, \frac{\pi}{N}, \frac{2\pi}{N}, \ldots, \frac{(N - 1)\pi}{N}, \pi$$

where (see Fig. 28.1.1)

$$x_i = \cos \theta_i$$

We have

$$\frac{1}{2} T_m(-1)T_n(-1) + \sum_{i=2}^{N-1} T_m(x_i)T_n(x_i) + \frac{1}{2} T_m(1)T_n(1) = \begin{cases} 0 & m \neq n \\ N/2 & m = n \neq 0 \\ N & m = n = 0 \end{cases}$$

The $T_m(x)$ are also orthogonal over the following N points t_i, equally spaced in θ,

$$\theta_i = \frac{\pi}{2N}, \frac{3\pi}{2N}, \frac{5\pi}{2N}, \ldots, \frac{(2N - 1)\pi}{2N}$$

and

$$t_i = \cos \theta_i$$

$$\sum_{i=1}^{N} T_m(t_i)T_n(t_i) = \begin{cases} 0 & m \neq n \\ N/2 & m = n \neq 0 \\ N & m = n = 0 \end{cases}$$

The set of points t_i are clearly the midpoints in θ of the first case.

The unequal spacing of the points in x_i (or t_i) compensates for the weight factor

$$w(x) = \frac{1}{\sqrt{1 - x^2}}$$

in the continuous case.

28.3   FIRST PROPERTIES OF THE CHEBYSHEV POLYNOMIALS

The Chebyshev polynomials have a number of exceptional properties. First, they are orthogonal polynomials and therefore have all the properties of orthogonal polynomials. In particular, they have the three-term recurrence relation

$$T_{n+1}(x) - 2xT_n(x) + T_{n-1}(x) = 0$$

which is really the following trigonometric identity in disguise:

$$\cos(n + 1)\theta + \cos(n - 1)\theta = 2 \cos \theta \cos n\theta$$

It is important to note that the coefficients of this equation do not depend on n.

Using this relation, we construct the first few Chebyshev polynomials

$$T_0(x) = 1$$
$$T_1(x) = x$$
$$T_2(x) = 2x^2 - 1$$
$$T_3(x) = 4x^3 - 3x$$
$$T_4(x) = 8x^4 - 8x^2 + 1$$
$$T_5(x) = 16x^5 - 20x^3 + 5x$$

It is easy to see from the three-term recurrence relation that $T_k(x) = 2^{k-1}x^k + \cdots$ and that the T_k are alternately odd and even polynomials.

Second, the zeros of $T_n(x)$ interlace those of $T_{n+1}(x)$ as is also obvious from the cosine representations.

Third, it is easy to go from a general polynomial to a Chebyshev expansion and back.   Suppose we have the polynomial

$$a_0 + a_1 x + a_2 x^2 + \cdots + a_N x^N$$
$$= a_0 + x\{a_1 + \cdots [a_{N-2} + x(a_{N-1} + a_N x)] \cdots \}$$

The inner parenthesis is

$$a_{N-1} + a_N x \equiv a_{N-1} T_0 + a_N T_1$$

Multiplying by x, we get

$$x(a_{N-1} + a_N x) = a_{N-1} T_1 + a_N [\tfrac{1}{2}(T_0 + T_2)]$$
$$= \frac{a_N}{2} T_0 + a_{N-1} T_1 + \frac{a_N}{2} T_2$$

Suppose in general that we have at some stage of converting from powers of x to Chebyshev polynomials

$$\{ \cdots \cdots \} = b_0 T_0 + b_1 T_1 + b_2 T_2 + \cdots + b_k T_k$$

Multiply by x and use

$$x T_k = \tfrac{1}{2} T_{k-1} + \tfrac{1}{2} T_{k+1}$$

Pictorially (see Fig. 28.3.1), if the top line has the coefficients at one stage, then we have the lower line at the next stage.   Notice that after N stages, $a_N x^N$ goes to

$$\frac{a_N}{2^{N-1}} T_N(x) + \text{lower terms}$$

Studying this process carefully shows the basis for the claim that generally an expansion in Chebyshev polynomials converges much more rapidly than the

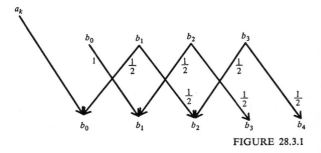

FIGURE 28.3.1

corresponding power series. It also shows that the sums of the coefficients in the two representations are the same!

To get from a Chebyshev expansion back to a polynomial, we can reverse the steps starting on the right.

PROBLEMS

28.3.1 Compute $T_6(x)$.

28.3.2 Prove that

$$T_{2k+1}(-x) = -T_{2k+1}(x)$$

and

$$T_{2k}(-x) = T_{2k}(x)$$

28.3.3 Convert

$$1 + x + x^2 + x^3 + x^4$$

to Chebyshev polynomials.

$$\text{Ans.} \quad \frac{T_4}{8} + \frac{T_3}{4} + T_2 + \frac{7T_1}{4} + \frac{15}{8} T_0$$

28.3.4 Draw a flow diagram for converting from a polynomial to Chebyshev polynomials.

28.3.5 Draw a flow diagram for the reverse of Prob. 28.3.4.

28.3.6 Show that if the three-term relation has coefficients independent of n, then they are essentially the Chebyshev polynomials.

28.4 FURTHER PROPERTIES OF THE CHEBYSHEV POLYNOMIALS

The fact that the Chebyshev polynomials are the cosines in the disguise of

$$x = \cos \theta$$

means that they have another set of remarkable properties.

First, the polynomials are equal-ripple in the sense that the alternate maxima and minima are of the same size—exactly 1 in fact (since the maximum of $\cos \theta$ is 1). See Fig. 28.4.1.

Second, the orthogonality both over the continuous range and over the sets x_i and t_i of discrete points comes from the corresponding orthogonality of the cosines, which we will prove in Chap. 31. The importance of this should be appreciated because it means that in a very natural way we can approximate the

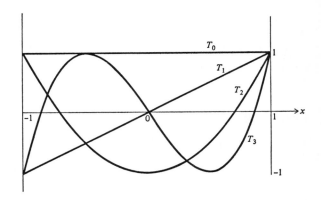

FIGURE 28.4.1

continuous case by simply doing the discrete case.   This property is not enjoyed by the other classical sets of orthogonal polynomials; one can make a set of polynomials orthogonal over a set of points in the interval $-1 \leq x \leq 1$, but they will not have exactly the same analytic form as the Legendre polynomials.

Third, at the sample points which are the zeros of the Nth Chebyshev polynomial, there is an *aliasing* effect (see Fig. 28.4.2) in that *at the sample points*, the polynomials of higher order look like lower-order Chebyshev polynomials.   We can show this directly from the Chebyshev identity that follows from the trigonometric identity

$$\cos(m + N)\theta + \cos(m - N)\theta = 2 \cos m\theta \cos N\theta$$

or

$$T_{m+N}(x) + T_{m-N}(x) = 2T_m(x)T_N(x)$$

Holding N as fixed, at the zeros of $T_N(x) = 0$, that is, at

$$\cos N\theta_k = 0 \qquad \theta_k = \frac{1}{N}\left[\frac{-\pi}{2} + k\pi\right] \qquad k = 1, 2, \ldots, N$$

we have

$$T_{m+N}(x_k) = -T_{N-m}(x_k)$$

For $m = N$ we have

$$T_{2N}(x_k) = -T_0(x_k) = -1$$

and in general it will follow that

$$T_{3N}(x_k) = 0 \qquad T_{4N}(x_k) = 1 \qquad T_{5N}(x_k) = 0 \qquad T_{6N}(x_k) = -1 \qquad \text{etc.}$$

FIGURE 28.4.2   Aliasing

It should be evident that any order Chebyshev polynomial can be reduced, *at the sample points* x_k, which are the zeros of $T_N(x)$, to an equivalent Chebyshev polynomial with order j $(0 \leq j \leq N)$. Figure 28.4.2 shows the reduction in a graphic form. Imagine the line of the index numbers of the order of the polynomial being folded like a ribbon, as shown in the figure. Then each index can be reduced to the smallest index, as shown in the figure.

The identity

$$T_m(x)T_n(x) = \tfrac{1}{2}[T_{m+n}(x) + T_{m-n}(x)]$$

has a second valuable use. It means that, similar to logs, products can be replaced by sums. Thus it is easy to multiply two series in Chebyshev polynomials —the products that arise are easily expressed as Chebyshev polynomials.

PROBLEMS

28.4.1 Draw a flow diagram for multiplying two Chebyshev expansions to get a third expansion.

28.5 THE CHEBYSHEV CRITERION

Chebyshev showed that, of all polynomials $p_n(x)$ of degree n having a leading coefficient equal to 1, *the polynomial*

$$\frac{T_n(x)}{2^{n-1}}$$

has the smallest least upper bound for its absolute value in the interval $-1 \leq x \leq 1$. Since the upper bound of $|T_n(x)|$ is 1, then this upper bound is $1/2^{n-1}$.

The proof of this remarkable property follows from an examination of the difference (see Fig. 28.5.1)

$$\phi_{n-1}(x) = \frac{T_n(x)}{2^{n-1}} - p_n(x)$$

which is a polynomial of degree $n - 1$, since x^n drops out. From the fact that $T_n(x)$ is $\cos n\theta$ in disguise, we see that in the interval $T_n(x)$ takes its extreme value $n + 1$ times, alternately positive and negative. If the extreme value of $p_n(x)$ is less than that of $T_n(x)/2^{n-1}$, then at these $n + 1$ extreme points $\phi_{n-1}(x)$ is alternately positive and negative; hence it must have n real zeros between these extremes. Since $\phi_{n-1}(x)$ is of degree $n - 1$, we conclude that $\phi_n(x) = 0$ and

$$p_n = \frac{T_n(x)}{2^{n-1}}$$

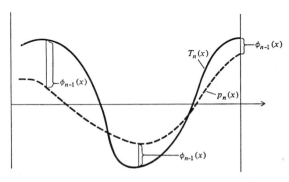

FIGURE 28.5.1

This property is of great interest in numerical computation. If some error can be expressed as a Chebyshev polynomial of degree n, then any other error which is a polynomial of degree n and which has the same leading coefficient will have, in the range $-1 \leq x \leq 1$, a greater extreme error than the Chebyshev error representation. Thus the expression *Chebyshev approximation* is associated with those approximations[1] which try to keep the maximum error to a minimum. This is sometimes called the *minimax principle*. Least-squares approximation keeps the *average* square of the error down, but in doing so isolated extreme errors are permitted; Chebyshev keeps the extreme errors down, but allows a larger average square error.

As a simple illustration, consider the problem of interpolating in the interval $-1 \leq x \leq 1$ with an exact-matching polynomial of degree n. The error term (Sec. 14.6) has the form

$$\frac{(x - x_1)(x - x_2) \cdots (x - x_{n+1})y^{(n+1)}(\bar{x})}{(n + 1)!}$$

---

[1] Chebyshev approximation should not be confused with Chebyshev integration; he had many good ideas, some of which have been named after him.

If we want to minimize the maximum error due to the factor

$$(x - x_1)(x - x_2) \cdots (x - x_{n+1})$$

we can choose as sample points the zeros of $T_{n+1}(x) = 0$. Thus we have for the part of the error that we can control easily the equal-ripple polynomial which has the least maximum error.

In a sense the Chebyshev polynomials show how the fundamental theorem of algebra collapses as the degree of the polynomial increases. One interpretation of the fundamental theorem of algebra is that the powers of x are linearly independent over any interval, and to be definite we pick the interval $-1 \le x \le 1$. By picking the nth Chebyshev polynomial divided by 2^{n-1} we get a polynomial of degree n with leading coefficient 1, which is a reasonable way to normalize the polynomials when considering linear independence. From the fact that $|T_n(x)| \le 1$ in the interval, it follows that

$$\frac{|T_n(x)|}{2^{n-1}} \le \frac{1}{2^{n-1}} \qquad -1 \le x \le 1$$

and for, say, $n = 21$, we have linear relation among the first 22 powers x^k ($k = 0$, $1, 2, \ldots, 21$) which is less than 10^{-6} in the whole interval. If the noise level of roundoff is considered to be 10^{-9}, then we can go to $n = 31$. In general for degree n we have a noise level below which a linear dependence among the powers of x occurs.

28.6 FURTHER IDENTITIES

We have already produced a number of identities among the Chebyshev polynomials. Because they are both orthogonal polynomials and the trigonometric functions cos nx in disguise, they satisfy an amazing number of relations, many of which are useful. The three-term relation shows how to multiply a Chebyshev polynomial by x (see Prob. 28.6.3 for an extension), and we have shown how to convert products of Chebyshev polynomials to sums of Chebyshev polynomials so that two Chebyshev series can be easily multiplied to get another Chebyshev series.

It is natural to ask how to differentiate and integrate. We begin with

$$T_{n+1}(x) = \cos[(n + 1)\arccos x]$$

$$\frac{1}{n+1} \frac{d[T_{n+1}(x)]}{dx} = \frac{-\sin[(n + 1)\arccos x]}{-\sqrt{1 - x^2}}$$

Write the corresponding equation for $n - 1$ and subtract to get

$$\frac{1}{n+1}\frac{d[T_{n+1}(x)]}{dx} - \frac{1}{n-1}\frac{d[T_{n-1}(x)]}{dx} = \frac{\sin(n+1)\theta - \sin(n-1)\theta}{\sin\theta}$$

or

$$\frac{T'_{n+1}(x)}{n+1} - \frac{T'_{n-1}(x)}{n-1} = \frac{2\cos n\theta \sin\theta}{\sin\theta} = 2T_n(x) \qquad n \geq 2$$

$$T'_2(x) = 4T_1$$

$$T'_1(x) = T_0$$

$$T'_0(x) = 0$$

Thus we have the formulas for the differentiation of Chebyshev polynomials.

We can use these formulas to find the formulas for integrating Chebyshev polynomials:

$$\int T_n(x)\, dx = \frac{1}{2}\left[\frac{T_{n+1}(x)}{n+1} - \frac{T_{n-1}(x)}{n-1}\right] + C \qquad n \geq 2$$

$$\int T_1(x)\, dx = \frac{T_2(x)}{4} + C$$

$$\int T_0(x)\, dx = T_1(x) + C$$

PROBLEMS

28.6.1   Show that

$$\frac{T'_{2n+1}}{2n+1} = 2(T_{2n} + T_{2n-2} + \cdots + T_2) + T_0$$

$$\frac{T'_{2n}}{2n} = 2(T_{2n-1} + T_{2n-3} + \cdots + T_1)$$

28.6.2   Show from Prob. 28.6.1 that

$$\frac{T'_k(x)}{k} = \sum_{k=0}^{n-1} T_{n-2k-1}$$

28.6.3   Show that

$$x^k T_n(x) = \frac{T_{n+k} + C(k,1)T_{n+k-2} + C(k,2)T_{n+k-4} + \cdots + T_{n-k}}{2^k}$$

28.6.4   Prove that

$$\frac{T_{2k+1}(x)}{x} = 2T_{2k}(x) - 2T_{2k-2} + \cdots \pm T_0$$

$$= (-1)^k \sum_{j=-k}^{k} (-1)^j T_{2j}(x)$$

28.6.5 Integrate

$$\int x T_n(x)\, dx = \frac{T_{m+2}}{(m+1)(m+2)} - \frac{2m^2 T_m}{m(m^2-1)} + \frac{T_{m-2}}{(m-1)(m-2)} + C$$

28.7 THE SHIFTED CHEBYSHEV POLYNOMIALS

It is often convenient to use the interval $0 \le x \le 1$ instead of the interval $-1 \le x \le 1$. For this purpose the *shifted Chebyshev polynomials*

$$T_n^*(x) = T_n(2x - 1)$$

are used. Thus we have

$$T_0^*(x) = 1$$
$$T_1^*(x) = 2x - 1$$
$$T_2^*(x) = 8x^2 - 8x + 1$$
$$T_3^*(x) = 32x^3 - 48x^2 + 18x - 1$$
$$T_4^*(x) = 128x^4 - 256x^3 + 160x^2 - 32x + 1$$
. . . .

and

$$1 = T_0^*$$
$$x = \tfrac{1}{2}(T_0^* + T_1^*)$$
$$x^2 = \tfrac{1}{8}(3T_0^* + 4T_1^* + T_2^*)$$
$$x^3 = \tfrac{1}{32}(10T_0^* + 15T_1^* + 6T_2^* + T_3^*)$$
$$x^4 = \tfrac{1}{128}(35T_0^* + 56T_1^* + 28T_2^* + 8T_3^* + T_4^*)$$
. . . .

More extensive tables are given in the National Bureau of Standards.[1]

The recurrence relation for the shifted polynomials is

$$T_{n+1}^*(x) = (4x - 2)T_n^*(x) - T_{n-1}^*(x) \qquad T_0^* = 1$$

or

or

$$x T_n^*(x) = \tfrac{1}{4}T_{n+1}^*(x) + \tfrac{1}{2}T_n^*(x) + \tfrac{1}{4}T_{n-1}^*(x)$$

where

$$T_n^*(x) = \cos[n \arccos(2x - 1)] = T_n(2x - 1)$$

[1] Tables of Chebyshev Polynomials $S_n(x)$ and $C_n(x)$, *Natl. Bur. Std. (U.S.), Appl. Math. Ser. 9*, 1952.

PROBLEMS

28.7.1   Discuss the conversion from polynomials to shifted Chebyshev polynomials and back, corresponding to Sec. 28.3.

      Find the formulas corresponding to

28.7.2   Prob. 28.6.3

28.7.3   Prob. 26.8.1

28.7.4   Differentiation

28.7.5   Integration

28.7.6 $T_m^*(x)T_n^*(x)$

CHEBYSHEV APPROXIMATION : PRACTICE

29.1 ECONOMIZATION

If a function can be represented in the form of a rapidly converging expansion in Chebyshev polynomials, and if we truncate the expansion after the term $T_{N-1}(x)$, then the error will closely resemble the first term neglected, namely, $a_N T_N(x)$. It is a folklore belief that Chebyshev expansions are among the most rapidly converging expansions there are for functions. This is clearly false in isolated cases; for example, consider an expansion in Bessel functions with coefficients of the expansion being, say, 3, 2, 1, 0, 0, 0, 0, 0. . . . If this function were expanded in Chebyshev polynomials, then arbitrarily far out there would be nonzero coefficients. On the other hand, an examination of the process that we gave (Sec. 28.3) for going from a Taylor series

$$f(x) = \sum_{n=0}^{N} a_n x^n$$

to a Chebyshev expansion

$$f(x) = \sum_{n=0}^{N} b_n T_n(x)$$

has

$$\sum_0^N a_n = \sum_0^N b_n$$

but

$$b_N = \frac{a_N}{2^{N-1}}$$

and

$$b_{N-1} = \frac{a_{N-1}}{2^{N-2}}$$

so we might *expect* the Chebyshev series to converge more rapidly than the Taylor series.

Lanczos[1] observed that if we expand a function in a series of Chebyshev polynomials, and if we truncate the series, then the remainder will look a lot like the first $T_N(x)$ that is neglected. He further observed that the error E committed in dropping the terms beyond $T_{n-1}(x)$ is bounded by

$$|E| \le \sum_{k=N}^{\infty} |b_k|$$

He called this process "economization."

Economization or processes closely allied to it are widely used in numerical methods because we tend to prefer a minimax bound to a least-squares (or any other) bound. The Chebyshev equal-ripple error is often closely approximated by the error of the economized representation, and both have a sum of squares of the error that is by definition at least as great as the least-squares fit (within roundoff, of course). On the other hand, the Chebyshev error of the least-squares fit is greater than that of the Chebyshev fit (again by definition). How do the respective errors compare? It can be shown that in a certain sense the Chebyshev approximation has a sum of squares proportional to the sum of squares of the least squares fit. On the other hand, the least-squares fit has a Chebyshev error that is a multiplier of the Chebyshev minimum fit and grows with N, the degree of the fit. Thus as a general rule when in doubt about whether to use a least-squares or a Chebyshev fit: if you choose the Chebyshev fit and are wrong, then the error is a fixed multiplier (somewhat larger than 1) times the least-squares error; but if you choose the least squares and should have picked a Chebyshev,

---

[1] Cornelius Lanczos (1893–   ).

then you can commit a large error, and the higher the order of the approximating polynomial, the (probably) higher percentage error you will make.   To see why this is so, you need only to look at the typical least-squares fit.   Assuming that the error is very like the first orthogonal polynomial dropped, say the Legendre to keep the two cases comparable, we see from Fig. 26.7.1 that the polynomial shoots up at the ends of the interval very rapidly, thus producing the large Chebyshev error.   The Chebyshev polynomials on the other hand are reasonably close to the orthogonal polynomials over most of the interval.

The field of Chebyshev approximation is now highly developed, and we refer the reader to textbooks on the subject by Fox and Parker [14] and by Snyder [53].

The obvious use for Chebyshev approximation is in the library routines for evaluating the special functions such as sine, cosine, logarithm, exponential, etc., where customarily the routines are uniformly accurate (Chebyshev) throughout the range (as measured by relative accuracy).

Another use that seldom seems to occur to users is in a problem where there is a block of heavy computation with one number as input and one number as output.   Clearly this can be viewed as a table, and a Chebyshev approximation will often greatly speed up the evaluation.   Indeed, in real time problems it may be the only way to save enough time to complete a full cycle on time.

29.2 ON FINDING A CHEBYSHEV EXPANSION (ECONOMIZATION)

Evidently the problem of finding a Chebyshev expansion is central to finding a Chebyshev approximation.   While the error in the economization process may not be exactly equal-ripple, it will generally be very close, and in any case it will furnish a very good first approximation (Sec. 29.10).

In Sec. 28.3 we showed how to go from a Taylor series of a corresponding Chebyshev series *provided* the Taylor series was truncated at some place for which the error caused by the dropped terms was negligible.   As an example of this process of conversion to Chebyshev and subsequent economization, consider the following trivial example (trivial so that it can be followed easily).   Suppose we use

$$\ln(1 + x) = x - \frac{x^2}{2} + \frac{x^3}{3} - \frac{x^4}{4} \qquad \text{(approximately)}$$

We will consider the interval $0 \leq x \leq 1$, and so we use the shifted polynomials $T_k^*(x)$ (Sec. 28.7). We have

$$x = \frac{1}{2}(T_0^* + T_1^*)$$

$$-\frac{x^2}{2} = -\frac{1}{16}(3T_0^* + 4T_1^* + T_2^*)$$

$$\frac{x^3}{3} = \frac{1}{96}(10T_0^* + 15T_1^* + 6T_2^* + T_3^*)$$

$$-\frac{x^4}{4} = \frac{-1}{512}(35T_0^* + 56T_1^* + 28T_2^* + 8T_3^* + T_4^*)$$

or

$$\ln(1 + x) = \tfrac{535}{1,536}T_0^* + \tfrac{19}{64}T_1^* - \tfrac{7}{128}T_2^* - \tfrac{1}{192}T_3^* - \tfrac{1}{512}T_4^*$$

In this approximation we know that $|T_k^*(x)| \leq 1$ $(0 \leq x \leq 1)$, and so if we drop both $T_3^*(x)$ and $T_4^*(x)$, the additional error is bounded by

$$\frac{1}{192} + \frac{1}{512} = \frac{11}{1,536}$$

Indeed, also dropping the $T_2^*(x)$ would leave the additional error less than the error $(1/3)$ due to dropping the cubic in the power-series approximation.

We see that the additional error due to the economization is indeed much as we said it would be. Thus this example, although unrealistic, shows the desired effects.

29.3   THE DIRECT EVALUATION OF THE COEFFICIENTS

The orthogonality relations enable us to compute directly the coefficients of a Chebyshev polynomial expansion,

$$f(x) = \frac{a_0}{2} + \sum_{k=1}^{\infty} a_k T_k(x)$$

We know that

$$a_k = \frac{2}{\pi} \int_{-1}^{1} f(x)\frac{T_k(x)}{\sqrt{1 - x^2}}\,dx$$

The change of variable to θ makes the relations into

$$a_k = \frac{2}{\pi} \int_{0}^{\pi} f(\cos\theta)\cos k\theta\,d\theta$$

and these can be evaluated by a variant of the Fast Fourier Transform (FFT) (Chap. 33).

For example, consider finding the Chebyshev polynomial expansion of e^{-x}. We have from the definition

$$\frac{2}{\pi}\int_0^\pi e^{-t\cos\theta}\cos k\theta\, d\theta = 2I_k(t)$$

of the modified Bessel function of order k and argument t that if we put $t = 1$, we have $a_k = 2I_k(1)$, and

$$e^{-x} = I_0(1) + 2\sum_{k=1}^\infty I_k(1)T_k(x)$$

Even if we do not know the analytic form of the integrals, for some functions they can be evaluated numerically if necessary.

For the Chebyshev expansion in terms of a discrete set of points x_i, or t_i, we have

$$f(x) = \frac{A_0}{2} + \sum_{k=1}^\infty A_k T_k(x)$$

and using, say, t_j,

$$A_k = \frac{2}{N}\sum_{j=1}^N f(t_j)T_k(t_j)$$

What is the relation between the coefficients A_k of the discrete expansion and the a_k of the continuous expansion? The aliasing discussed in Sec. 28.4 (and again in Sec. 31.6) shows that if we have for the continuous expansion

$$f(x) = \frac{a_0}{2} + \sum_{k=1}^\infty a_k T_k(x)$$

and compute the coefficients of the discrete expansion

$$A_m = \frac{1}{N}\sum_{j=1}^N f(x_j)T_m(x_j)$$
$$= \frac{1}{N}\sum_{j=1}^N \left[\frac{a_0}{2}T_m(x_j) + \sum_{k=1}^\infty a_k T_k(x_j)T_m(x_j)\right]$$

we get

$$A_m = a_m - a_{2N-m} - a_{2N+m} + a_{4N-m} + a_{4N+m} + \cdots, \qquad m = 1, 2, \ldots, N-1$$
$$\frac{A_0}{2} = \frac{a_0}{2} - a_{2N} + a_{4N} - a_{6N} + \cdots \qquad m = 0$$

Note that the formula for A_0 agrees with the one for the general m.

If the series converges rapidly, then this aliasing effect is small; if the convergence is slow, then the finite approximation may be far from the continuous one. For rapidly converging series the finite sums give ideal formulas for computing the integrals that arise in the continuous case.

29.4   A DIRECT METHOD

Another method for finding a Chebyshev expansion so that the error resembles $T_N(x)$ is to evaluate the function at the zeros x_i of

$$T_N(x_i) = 0$$

and then simply find the polynomial through these points, using perhaps divided differences (Secs. 18.2 and 18.3). The method is obvious and need not be discussed further.

29.5   THE CHEBYSHEV EXPANSION OF AN INTEGRAL

In the next case suppose we wish to represent some integral

$$\int_a^x f(t)\, dt = \sum_{k=0}^{\infty} a_k T_k(x)$$

To be specific, suppose we try

$$\arctan x = \int_0^x \frac{dt}{1 + t^2}$$

It is awkward to differentiate a Chebyshev series, but it is easy to integrate. So we begin by writing

$$\frac{1}{1 + x^2} = \sum_{k=0}^{\infty} b_k T_k(x)$$

But it is clear from symmetry that only b_{2k} terms will appear ($b_{2k+1} = 0$), and so we write (see Prob. 28.6.2)

$$\frac{1}{1 + x^2} = \sum_{k=0}^{\infty} b_{2k} T_{2k}(x)$$

$$1 = (1 + x^2) \sum_{k=0}^{\infty} b_{2k} T_{2k}(x)$$

$$= \sum_{k=0}^{\infty} b_{2k} T_{2k}(x) + b_0 x^2 T_0 + \sum_{k=1}^{\infty} b_{2k} x^2 T_{2k}(x)$$

$$= \sum_{k=0}^{\infty} b_{2k} T_{2k}(x) + \frac{b_0}{2}[T_0 + T_2] + \sum_{k=1}^{\infty} \frac{b_{2k}}{4}(T_{2k-2} + 2T_{2k} + T_{2k+2})$$

Since the $T_k(x)$ are linearly independent, we may equate coefficients. The first few equations are

$$T_0: \qquad 1 = b_0 + \frac{b_0}{2} + \frac{b_2}{4}$$

$$T_2: \qquad 0 = b_2 + \frac{b_0}{2} + \frac{b_2}{2} + \frac{b_4}{4}$$

$$T_4: \qquad 0 = b_4 + \frac{b_2}{4} + \frac{b_4}{2} + \frac{b_6}{4}$$

$$T_6: \qquad 0 = b_6 + \frac{b_4}{4} + \frac{b_6}{2} + \frac{b_8}{4}$$

Except for the first two equations, the rest have the form

$$\frac{b_{2(k-1)}}{4} + \frac{3}{2} b_{2k} + \frac{b_{2(k+1)}}{4} = 0$$

If we can satisfy these and have two parameters left, we can (probably) handle the first two equations. These difference equations have the characteristic equation

$$\rho^2 + 6\rho + 1 = 0$$

or

$$\rho_1 = -3 + 2\sqrt{2} \approx -0.172$$
$$\rho_2 = -3 - 2\sqrt{2} \approx -5.828$$

Hence

$$b_{2k} = C_1(\rho_1)^k + C_2(\rho_2)^k$$

Unless $C_2 = 0$, the original series will diverge; thus we have

$$b_{2k} = C_1(\rho_1)^k \qquad k = 1, 2, 3, \ldots$$

We have left the two equations

$$1 = \frac{3b_0}{2} + \frac{b_2}{4}$$

$$0 = \frac{b_0}{2} + \frac{3b_2}{2} + \frac{b_4}{4}$$

with b_0 and C_1 as parameters. In more detail

$$4 = 6b_0 + C_1\rho_1$$
$$0 = 2b_0 + (6\rho_1 + \rho_1{}^2)C_1$$

Thus we can find b_0 and C_1 and determine the series for $1/(1 + x^2)$. But $\rho^2 + 6\rho = -1$. It is now easy to integrate this term by term (Sec. 28.6) to get the Chebyshev expansion of arctan x.

29.6   LANCZOS' τ PROCESS

The idea behind the τ process introduced by Lanczos is that instead of approximately solving the given problem, we can perturb it slightly and solve the approximate problem exactly.   While the two processes sound different, they come down to the same thing in the long run.   Still, we shall start with Lanczos' approach and later take up the more direct process.

Given, say, a differential equation

$$y' + y = 0 \qquad y(0) = 1 \qquad 0 \le x \le 1$$

which we cannot solve in a polynomial form (the solution being e^{-x}), consider, instead, solving the problem

$$y' + y = \tau T_n^*(x)$$

where τ is to be found.   We see that if τ is small, then we have perturbed the differential equation only slightly.   To keep the computation easy to follow, we set $n = 4$ and try

$$y = a + bx + cx^2 + dx^3 + ex^4$$

We get, equating coefficients of the like power of x,

$$b + a = \tau$$
$$2c + b = -32\tau$$
$$3d + c = 160\tau$$
$$4e + d = -256\tau$$
$$e = 128\tau$$

The initial condition fixes $a = 1$.   Hence, in turn, we obtain

$$b = \tau - 1$$
$$c = \frac{1 - 33\tau}{2}$$
$$d = \frac{1}{3}\left(160\tau + \frac{33\tau - 1}{2}\right) = \frac{353\tau - 1}{6}$$
$$e = \frac{1}{4}\left(-256\tau + \frac{1 - 353\tau}{6}\right) = \frac{1}{24}(1 - 1{,}889\tau)$$
$$e = 128\tau = \tfrac{1}{24}(1 - 1{,}889\tau)$$
$$\tau = \frac{1}{4{,}961}$$

Thus the error that we have put in the original equation is

$$\left| \frac{T_4^*(x)}{4{,}961} \right| \le \frac{1}{4{,}961}$$

and for this equation we have an exact solution

$$y = 1 - \frac{4{,}960}{4{,}961} x + \frac{2{,}464}{4{,}961} x^2 - \frac{768}{4{,}961} x^3 + \frac{128}{4{,}961} x^4$$

as compared with the series solution

$$y = 1 - x + \frac{x^2}{2} - \frac{x^3}{6} + \frac{x^4}{24}$$

whose maximum error is about

$$\frac{1}{5!} = \frac{1}{120}$$

In this case it is easy to compute a bound on the error due to the addition of the Chebyshev term to the differential equation

$$y' + y = \tau T_4^*(x)$$

$$y(x) = e^{-x} \cdot 1 + \tau e^{-x} \int_0^x T_4^*(\theta) e^\theta \, d\theta$$

For $0 \le x \le 1$

$$|y - e^{-x}| = \left| \tau e^{-x} \int_0^x T_4^*(\theta) e^\theta \, d\theta \right|$$

$$\le \tau e^{-x} \int_0^x e^\theta \, d\theta = \tau(1 - e^{-x}) \le \frac{0.665}{4{,}961} \approx 1.34 \times 10^{-4}$$

We have illustrated the τ method with a simple example. Lanczos' book [33] gives a much more extensive discussion of the method. However, the main idea is the same; we slightly alter the problem and obtain an exact solution of the new problem. Since the alteration is in the domain of the physics of the original problem, we can usually understand the change more readily than some approximation made in the middle of some numerical computation. The τ method illustrates the importance of carefully answering the fourth basic question, What accuracy?

PROBLEMS

29.6.1 Apply the τ method to

$$xy'' + y = 0 \qquad \begin{cases} y\,(0) = 0 \\ y'(0) = 1 \end{cases}$$

for $0 \le x \le 1$, using $n = 4$.

29.7   THE DIRECT METHOD FOR DIFFERENTIAL EQUATIONS

Often the function we want to expand in a Chebyshev series is defined by a differential equation. Again this can often be used to get the expansion directly. For example, again suppose we have the simple differential equation

$$y' = -y \qquad y(0) = 1 \qquad -1 \le x \le 1$$

It is usually preferable to suppose that we have the expansion of the derivative of the function

$$y' = \frac{a_0}{2} + \sum_{k=1}^{\infty} a_k T_k$$

From Sec. 28.6 we can integrate

$$y(x) = \frac{a_0 T_1}{2} + a_1 \frac{T_2}{4} + \sum_{k=2}^{\infty} \frac{a_k}{2} \left[\frac{T_{k+1}}{k+1} - \frac{T_{k-1}}{k-1} \right] + C$$

We equate coefficients

$$T_0: \qquad \frac{-a_0}{2} = C$$

$$T_1: \qquad -a_1 = \frac{a_0}{2} - \frac{a_2}{2}$$

$$T_2: \qquad -a_2 = \frac{a_1}{4} - \frac{a_3}{4}$$

$$T_3: \qquad -a_3 = \frac{a_2}{6} - \frac{a_4}{6}$$

$$T_4: \qquad -a_4 = \frac{a_3}{8} - \frac{a_5}{8}$$

The general difference equation for the coefficients is

$$a_{k+1} - 2ka_k - a_{k-1} = 0$$

and has two solutions, one growing with k and one decaying. From Sec. 13.8 we see that we want to start at the top end and solve downward. If, as before, we want only through T_4, we set $a_5 = 0$ and tentatively $a_4 = 1$. As the result of computing step by step

$$a_3 = -8a_4$$

$$a_2 = 6\left(\frac{a_4}{6} - a_3 \right)$$

$$a_1 = 4\left(\frac{a_3}{4} - a_2 \right)$$

$$a_0 = \left(\frac{a_2}{2} - a_1 \right)$$

we will find that usually at $x = 0$

$$a_0 T_0 + a_2 T_2 + a_4 T_4 \neq 1$$

but that by making the obvious scale factor change an a_4 (and also a_3, a_2, a_1, a_0), we will satisfy the initial condition.

An examination of this process will reveal that it is equivalent to the τ process of rigging things so that the final equations are satisfied, since there we added the term we needed to produce an exact cancellation for the last power of x we used.

29.8 THE EVALUATION OF CHEBYSHEV EXPANSIONS

A Chebyshev polynomial expansion may be evaluated in a direct fashion. The method is based on the three-term recurrence relation

$$T_{k+1} = 2xT_k - T_{k-1}$$

If, in some fashion, we had the first two values, then each iteration would produce the next Chebyshev polynomial.

Let us examine the process more closely. Suppose we have three locations A, B, and C. What we do is multiply the contents of B by $2x$ and from this subtract the contents of C, and put this result in A. We then shift B to C and A to B. Imagine that x is fixed and that B and C both have zeros in them. Then each iteration produces more zeros. But if we were to add to B the quantity x and to C the quantity 1, then we would start to generate the Chebyshev polynomials step by step, and after k steps we would have $T_{k+1}(x)$.

If we put in $a_N x$ and a_N instead of x and 1, we would have after $N - 1$ iterations exactly $a_N T_N(x)$. If after the first iteration we added $a_{N-1}x$ and a_{N-1} in B and C, we would end up with the sum

$$a_N T_N(x) + a_{N-1}T_{N-1}(x)$$

And in general, if at each step we added in the two numbers corresponding to the term a_k that was appropriate, we would end up with the sum of the Chebyshev series. Often this process is preferable to converting to a polynomial in x and then evaluating the result. The roundoff effect in the polynomial case may be more severe than in the direct evaluation.

29.9 THROWBACK

The aliasing of the higher polynomials at the zeros of $T_N(x)$ suggests that instead of throwing away all the higher terms of a Chebyshev expansion, as is done in the economization process, they should be "thrown back" (Sec. 18.8) via the aliasing relations.   Thus while we still cannot be sure that the error curve is equal-ripple, we can be sure that the error curve has the proper zeros when we drop the $T_N(x)$ term.   We will not pursue this further.

29.10   LEVELING THE ERROR CURVE

We have seen that we can find a Chebyshev polynomial expansion for many functioñs, but that when we economize it we do not get the exact equal-ripple error curve.

There are two broad approaches, with many refinements to each.   The first looks at the positions of the zeros of the error curve and the values of the maxima between them and at the ends.   In view of these maxima the zeros are moved a bit to form the next approximation.   Evidently if a maximum or minimum is larger than the others, the adjacent (or nearest, if it is in an end interval) zeros are moved a bit together under the reasonable assumption that this will decrease the size of the peak.   Each cycle consists of carefully computing the error curve and finding the size and approximate location of the extremes.   From this the motion of the zeros is found.   Usually a few iterations suffice to level the curve at least by a factor of 10.

The second approach is more mathematical.   Having again located the extremes, both amount and location, the coefficients are perturbed one at a time to see how they alter the extremes (in the Chebyshev polynomial case these amounts can be found from the polynomials themselves).   Then a set of simultaneous linear equations are set up with arbitrary changes in the coefficients so that the result (if it were all linear) will exactly annihilate the inequalities of the peaks.   Repeated trials will again level the curve (most of the time).

Another approach to leveling the error curve is to use the sigma factors of Sec. 32.7.

*RATIONAL FUNCTION APPROXIMATION

30.1 INTRODUCTION

Up to this point in Part II we have assumed that between the sample points the function behaved as a polynomial. In this chapter we shall assume that the function behaves as the quotient of two polynomials, which is usually called a *rational function*. Rational functions have the two fundamental properties of still being a rational function under both translation and scale change of the independent variable.

Rational functions have a number of properties that are frequently needed for an approximation leading to the analytic substitution of a new function in place of the given one. Most noticeable is the fact that rational functions can approximate functions having infinite values y_i for finite values x_i. Another feature is that they can approximate straight lines, especially the x axis, for large values of x_i, which is something that nontrivial polynomials cannot do.

Inside a computer, rational functions are rapid and easy to evaluate; hence they are often used to provide a computable approximation to some of the more difficult functions. For such purposes a Chebyshev type of minimax fit is likely to be needed.

The theory of approximation by rational functions is in a confused, rapidly evolving state, and we shall give only the simplest approach.   One reason for not going farther is that in computing, the average person is not likely to do many rational approximations in a year, and more efficient methods, which save a few seconds on a fast computer, do not seem to be worth the time in an elementary course.

30.2   THE DIRECT APPROACH

Suppose we have either a function from which we can compute some samples (x_i, y_i) or the samples themselves, and we wish to fit a rational function to the data. Thus we want

$$y = f(x) \approx \frac{N(x)}{D(x)}$$

where $N(x)$ and $D(x)$ are polynomials.

The first step is to decide which polynomial forms to try.   In one particular case, plotting

$$\log y_i \text{ versus } \log x_i$$

for small $x_i > 0$ produced a slope of approximately 3, from which we deduced that we should try

$$y = f(x) \simeq \frac{ax^3 + \cdots}{1 + \cdots}$$

For large x the original data fell on what appeared to be a straight line of positive slope.   This determined that the degree of $N(x)$ should be 1 greater than the degree of $D(x)$.   Let the degree of $D(x)$ be k.   Then

$$y = f(x) = \frac{a_3 x^3 + a_4 x^4 + \cdots + a_{k+1} x^{k+1}}{1 + b_1 x + b_2 x^2 + \cdots + b_k x^k} \qquad (30.2.1)$$

It remains to choose k.   It is hard to say just how to choose k, except by noting the number of parameters that will be available and by comparing that (subjectively) with the "complexity" of the data.

When k is chosen, we then know the number of parameters M.   Note that we must fix one coefficient because only the ratio of $N(x)$ and $D(x)$ occurs; hence one coefficient (known to be nonzero) must be chosen as a fixed number.[1]

---

[1] In practice, it is usually best to select 1 as the highest power either of the numerator or of the denominator, since this saves one multiplication when evaluating the rational function.

We either select or compute the appropriate number M of samples, located where they seem to be most important. We then have

$$D(x_i)y_i = N(x_i) \qquad i = 1, \ldots, M$$

as a set of M equations to be solved. In the example we have from Eq. (30.2.1)

$$y_i + b_1 x_i y_i + b_2 x_i^2 y_i + \cdots + b_k x_i^k y_i$$
$$= a_3 x_i^3 + a_4 x_i^4 + \cdots + a_{k+1} x_i^{k+1} \qquad (30.2.2)$$

for $i = 1, 2, \ldots, 2k - 1$.

These are linear equations in the unknowns a_i and b_i and can be solved by a library routine.

The point is often raised that such a system is of high degree. As far as can be determined, seldom does the number of unknowns exceed 10 or 12, and thus the solution involves no more than a few thousand operations; this is hardly serious on a fast machine, if it is to be done once in a problem.

30.3 LEAST-SQUARES FITTING BY RATIONAL FUNCTIONS

Often there are more data points (x_i, y_i) $(i = 1, 2, \ldots, M)$ than parameters in the rational function we wish to fit. In such cases one approach is to find a least-squares fit. We want, therefore, to find the coefficients a_i and b_i such that

$$\min = \sum_{i=1}^{M} \left[y(x_i) - \frac{N(x_i)}{D(x_i)} \right]^2$$

The unknown coefficients occur nonlinearly in the normal equations, and this makes them difficult to solve. One way around this is to note that we can write the expression we wish to minimize in the form

$$\min = \sum_{i=1}^{M} \frac{1}{D^2(x_i)} [D(x_i)y(x_i) - N(x_i)]^2$$

We now set about finding an iterative solution to the following problem:

$$\min_k = \sum_{i=1}^{M} \frac{1}{D_{k-1}^2(x_i)} [D_k(x_i)y(x_i) - N(x_i)]^2 \qquad k = 1, 2, \ldots$$

where we take, if nothing else is known, $D_0 \equiv 1$. The unknowns now occur linearly in the normal equations.

In the usual course of events the convergence is rapid. However, as is always possible, the iterative loop with the feedback may cause an instability. If instability occurs, an examination of the nature of the trouble together with commonsense, will usually produce a method for dampening the instability; hence the elaborate theory of the general case is not worth going into here.

30.4 CHEBSYHEV APPROXIMATION BY RATIONAL FUNCTIONS

As indicated in Sec. 30.1, frequently a rational function is used as an easily computable approximation to a transcendental function. When such an approximation is to be used many times, either in some frequently recurring problem or in a special-purpose computer, then a Chebyshev equal-ripple (or minimax) approximation is likely to be desired.

We begin by choosing a set of samples x_i, perhaps the zeros of a suitable $T_k(x)$, and we compute the corresponding y_i. From these we find the corresponding rational function of the appropriate form (as in Sec. 30.2).

We then plot the error curve by computing the error at many points. For intervals where the error is large, we want to move the sample points closer together, and where the local extreme error is small, we want to move the samples farther apart. With this new choice of sample points, we repeat the process. Seldom are as many as 10 trials necessary to come quite close to an equal-ripple error curve. The speed of the approach to the equal-ripple error curve depends, among other things, on the method used for respacing the samples.

We offer here no proof that this heuristic method will work, but experience indicates that a number of such simple schemes work reasonably well. A little "low cunning" in arranging the computation is usually necessary to avoid the loss of too many figures.

30.5 RECIPROCAL DIFFERENCES

In Secs. 30.3 and 30.4 we used iterative schemes which require a number of trials, each of which involves the solution of the system of linear equations. This suggests that we examine the possibilities of systematizing the solution.

Following Milne [38], we consider the special case

$$y = \frac{a_0 + a_1 x + a_2 x^2 + a_3 x^3}{b_0 + b_1 x + b_2 x^2}$$

which uses six data points (x_i, y_i) $(i = 1, \ldots, 6)$. Multiplying, we get

$$a_0 - b_0 y + a_1 x - b_1 xy + a_2 x^2 - b_2 x^2 y + a_3 x^3 = 0$$

Using the method of Sec. 14.3, we can easily see that the determinant

$$
\begin{vmatrix}
1 & y & x & xy & x^2 & x^2y & x^3 \\
1 & y_1 & x_1 & x_1y_1 & x_1{}^2 & x_1{}^2y_1 & x_1{}^3 \\
1 & y_2 & x_2 & x_2y_2 & x_2{}^2 & x_2{}^2y_2 & x_2{}^2 \\
\hline
& & & & & & \\
1 & y_6 & x_6 & x_6y_6 & x_6{}^2 & x_6{}^2y_6 & x_6{}^3
\end{vmatrix} = 0
$$

is the required solution.

We propose to reduce the order of this determinant step by step in a systematic, sensible way. In theory, we first reduce the second row to the form 1, 0, 0, 0, 0, 0, 0. We multiply

1   Column 1 by y_1 and subtract from column 2
2   Column 3 by y_1 and subtract from column 4
3   Column 5 by y_1 and subtract from column 6
4   Column 5 by x_1 and subtract from column 7
5   Column 3 by x_1 and subtract from column 5
6   Column 1 by x_1 and subtract from column 3

After expanding by the second row, we divide:

7   Row 1 by $y - y_1$
8   Row 2 by $y_2 - y_1$
    Row 3 by $y_3 - y_1$

    ....
9   Row 7 by $y_6 - y_1$

This division can lead to trouble if any $y_i = y_1$ ($i \neq 1$) or even if they are close. In practice, the line to reduce to 1, 0, 0, 0, 0, 0, 0 should be chosen with this division process in mind and with the intent to avoid large cancellations in $y_i - y_1$ (and hence division by small, inaccurate numbers).

We now write

$$
\frac{x - x_1}{y - y_1} = \rho_1(x, x_1) \qquad \frac{x_i - x_1}{y_i - y_1} = \rho_1(x_i, x_1)
$$

and the determinant becomes

$$
\begin{vmatrix}
1 & \rho_1(x, x_1) & x & x\rho_1(x, x_1) & x^2 & x^2\rho_1(x, x_1) \\
1 & \rho_1(x_2, x_1) & x_2 & x_2\rho_1(x_2, x_1) & x_2{}^2 & x_2{}^2\rho_1(x_2, x_1) \\
\hline
& & & & & \\
1 & \rho_1(x_6, x_1) & x_6 & x_6\rho_1(x_6, x_1) & x_6{}^2 & x_6{}^2\rho_1(x_6, x_1)
\end{vmatrix} = 0
$$

which is of the same form as before except that it is of sixth order.

We repeat the process.   Multiply:

1 Column 1 by $\rho_1(x_2, x_1)$ and subtract from column 2
2 Column 3 by $\rho_1(x_2, x_1)$ and subtract from column 4
3 Column 5 by $\rho_1(x_2, x_1)$ and subtract from column 6
4 Column 3 by x_2 and subtract from column 5
5 Column 1 by x_2 and subtract from column 3

Reduce the determinant and divide:

6 Row 1 by $\rho_1(x, x_1) - \rho_1(x_2, x_1)$
7 Row 2 by $\rho_1(x_3, x_1) - \rho_1(x_3, x_1)$

....

Again a wise choice of the row in which we produce 1, 0, 0, 0, 0, 0 can help to keep accuracy.   If we write

$$u_x = \frac{x - x_2}{\rho_1(x, x_1) - \rho_1(x_2, x_1)}$$

we get

$$\begin{vmatrix} 1 & u_x & x & xu_x & x^2 \\ 1 & u_3 & x_3 & x_3u_3 & x_3^2 \\ 1 & u_4 & x_4 & x_4u_4 & x_4^2 \\ 1 & u_5 & x_5 & x_5u_5 & x_5^2 \\ 1 & u_6 & x_6 & x_6u_6 & x_6^2 \end{vmatrix} = 0$$

The quantity $\rho_1(x_i, x_j)$ was clearly symmetric in its variables.   But this is not true of the quantities that we have labeled u_i.   It is customary to use the symmetric quantities

$$\frac{x - x_2}{(x - x_1)/(y - y_1) - (x_2 - x_1)/(x_2 - y_1)} + y_1$$

$$\equiv \frac{x - x_1}{(x - x_1)/(y - y_1) - (x_2 - x_1)/(y_2 - y_1)} + y_1$$

$$\equiv \frac{x_2 - x_1}{(x_2 - x)/(y_2 - y) - (x - x_1)/(y - y_1)} + y$$

and call them

$$\rho_2(x, x_1, x_2)$$

but the symmetry is of little value in practice.   Continuing this process, we reduce the determinant to second order which is then trivial to handle.

We shall not go further in this highly developed field but merely refer the reader to standard books such as Milne [38], Milne-Thomson [40], and Ralston [49].

Fourier Approximation— Modern Theory

31

FOURIER SERIES:
PERIODIC FUNCTIONS

31.1 ORIENTATION

Part I was concerned with the fundamentals of computing and a number of various algorithms. Part II concentrated on approximating infinite operators like interpolation, integration, and differentiation by using polynomials. Numerous formulas were found, but the essential content of Part II was the methodology of finding formulas, as well as the rich variety of possible formulas. In Part III we use this methodology, but we use a different class of approximation functions; instead of polynomials we use sums of sines and cosines. In Chaps. 31 to 33 in Part III we use discrete sums of these functions, in the form of a Fourier series, to approximate periodic functions, while later we use a continuous, noncountable number of the functions, in the form of a Fourier integral, to approximate more general functions.

Because the Fourier series and integral *tend* to be ignored in modern mathematics (though they gave rise to a great deal of mathematics), we have to develop the mathematical manipulations we need. To keep the material within reasonable bounds the more formal mathematical rigor will be ignored [34, 35, 64].

It is recommended that Sec. 1.6 be reread at this point, and possibly all of Chap. 1, so that the material discussed in Part III will be seen in reasonable perspective.

By using Fourier approximation instead of polynomial approximation we avoid the many troubles of polynomial approximation that we have frequently mentioned.   Not only does the Fourier model provide an alternative technique for finding formulas, and hence sometimes different formulas, but it also gives an alternate way of interpreting the same formulas found in Part II.   Chapter 35 re-examines many of the results obtained in Part II and shows how the alternate view can often, but not always, shed a great deal of light on what the answers mean. Thus we have a new tool for fulfilling the purpose of the book as expressed in its motto, The Purpose of Computing Is Insight, Not Numbers.[1]

The frequency approach of Part III is very powerful in the planning stage, because it often enables us to make very accurate estimates of the spacing of the samples necessary to obtain a given accuracy.   In the computing stage, it likewise offers not only many of the old formulas, but also others that are essentially new and which may well meet the needs of the problem better.   Probably most important, in the interpretation of the results back to the physical problem, the frequency approach can interpret in a new light not only the numbers, but also the details of how the computation went, such as the physical meaning of the tight or loose spacing of the samples.   Finally, it often provides very powerful ways of verifying the physical model in many of its aspects, thus reassuring the proposer of its validity and shedding new light on the original source of the computing.   We can therefore put into practical use the theory we have learned.

The author has wondered for more than 20 years why the textbooks and mathematically inclined computer journals have almost totally neglected this fruitful and valuable approach to computing.   The answers seem to be (1) that the Fourier series and integral are seldom taught in mathematical circles except as special cases of generalizations that rob them of any physical meaning, and this ignorance tends to be followed by contempt for the knowledge; and (2) the Fourier approach puts much emphasis on the connection between the problem and the method used, and therefore the method cannot be studied in the vacuum of pure mathematics.   As a result of this neglect the reader who wishes to follow up on this approach and find out what is new is referred to the engineering journals like the *IEEE Transactions on Computers* and the *Transactions on Automatic Control,*[2] plus textbooks on topics such as automatic control [54] and digital filters [32].

---

[1] It is sometimes suggested that the motto be revised to, The Purpose of Computing Numbers Is Not Yet in Sight.
[2] IEEE, 345 East 47th Street, New York, N.Y., 10017.

31.2 THE EFFECT OF SAMPLING—ALIASING

Most computation is done with equally spaced samples. Let this spacing be chosen as the unit (of time); that is, $h = \Delta t = 1$. Section 1.6 gave one description of this effect, namely that due to the sampling, some forward rotations will appear as backward rotations. A practical application of this effect is the stroboscope which flashes a light at slightly less than the period of some rotating equipment being viewed. As a result the equipment appears to be rotating at a very slow rate; the closer the two rates, the slower the appearance of rotation.

This effect can be understood from a simple trigonometric identity. Corresponding to the sinusoid

$$\cos\left[\pi(n + \varepsilon)t + \phi\right]$$

there is another sinusoid

$$\cos\left[\pi(n - \varepsilon)t - \phi\right]$$

which has the same sample values at the equally spaced sample points as can be seen in Fig. 31.2.1. The equality of the sample values follows from

$$\cos[\pi(n + \varepsilon)t + \phi] - \cos[\pi(n - \varepsilon)t + \phi] = -2 \sin (n\pi t)\sin(n\varepsilon t + \phi) = 0$$

where both n and t are integers. Thus the sampling process confuses the two different frequencies $\pi(n + \varepsilon)$ and $\pi(n - \varepsilon)$—both appear as if they were the same frequency.

It is customary to refer to this confusion as aliasing,[1] that is, various frequencies adopting the name of one particular frequency, which is usually chosen to be the lowest positive frequency of all those which are aliased. It is obvious that *aliasing is a consequence of sampling at equally spaced intervals and cannot be undone once the sampling occurs.*

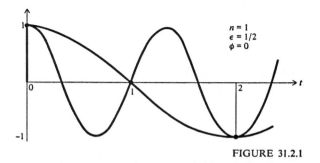

$$n = 1$$
$$\epsilon = 1/2$$
$$\phi = 0$$

FIGURE 31.2.1

[1] Apparently introduced by J. W. Tukey (1915–   ).

In a sense, the aliasing of the Chebyshev polynomials in Sec. 28.4 was that of the cosines in the disguise of

$$x = \cos \theta$$

More generally, aliasing also occurs in polynomial sampling, though it is not usually recognized as such. Consider polynomial approximation based on n sample points x_i $(i = 1, 2, \ldots, n)$. The polynomial

$$\pi(x) = (x - x_1)(x - x_2) \cdots (x - x_n)$$

plays a central role in the theory. Let $P_m(x)$ be any polynomial of degree m (we may think of m as being greater than n though it need not be). When we divide $P_m(x)$ by $\pi(x)$ we get a quotient $Q(x)$ and a remainder $R(x)$, that is,

$$P_m(x) = Q(x)\pi(x) + R(x)$$

This expression is the obvious generalization of the remainder theorem, namely *at the sample points x_i*,

$$P_m(x_i) = R(x_i) \qquad (i = 1, 2, \ldots, n)$$

and due to the sampling one function cannot be distinguished from the other *if* all we have are the sample values.

If we imagine doing this for each x^m $(m = 0, 1, \ldots)$ we have the correspondence

$$x^m \to R_m(x)$$

Now for a Taylor's series expansion of any analytic function $f(x)$

$$f(x) = \sum_{m=0}^{\infty} a_m x^m \to \sum_{m=0}^{\infty} a_m R_m(x)$$

and at the sample points

$$f(x_i) = \sum_{m=0}^{\infty} a_m R_m(x)$$

The summation is *the* interpolating polynomial of degree n which we would get if we passed an interpolating polynomial through the given samples (neglecting roundoff, of course).

31.3 THE CONTINUOUS FOURIER EXPANSION

In Sec. 26.2 we used the Fourier series as an example of a set of orthogonal functions. We showed by direct integration that

$$\int_0^{2\pi} \cos mx \cos nx \, dx = \begin{cases} 2\pi & m = n = 0 \\ \pi & m = n \neq 0 \\ 0 & m \neq n \end{cases}$$

$$\int_0^{2\pi} \cos mx \sin nx \, dx = 0$$

$$\int_0^{2\pi} \sin mx \sin nx \, dx = \begin{cases} 0 & m = n = 0 \\ \pi & m = n \neq 0 \\ 0 & m \neq n \end{cases}$$

and hence if

$$F(x) = \frac{a_0}{2} + a_1 \cos x + b_1 \sin x + a_2 \cos 2x + b_2 \sin 2x + \cdots$$

then the Fourier coefficients are given by

$$a_k = \frac{1}{\pi} \int_0^{2\pi} F(x) \cos kx \, dx$$

$$b_k = \frac{1}{\pi} \int_0^{2\pi} F(x) \sin kx \, dx$$

We also showed that (1) these Fourier coefficients give the least-squares fit to the function, and (2) that then Bessel's inequality

$$\int_0^{2\pi} F^2(x) \, dx - \pi \left[\frac{a_0^2}{2} + \sum_{k=1}^{N} (a_k^2 + b_k^2) \right] \equiv \sum_{i=1}^{M} \varepsilon_i^2 \geq 0$$

gives a measure of the goodness of fit.

It is well known that any reasonable engineering function (see Fig. 31.3.1)

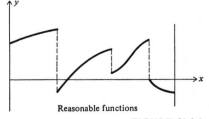

Reasonable functions

FIGURE 31.3.1

can be expanded in a Fourier series.   The rate of convergence of the series depends on the discontinuities in the function and in its derivatives (see Chap 32 for more details); but we must remember that the function is assumed to be periodic and is, as it were, wrapped around a cylinder (see Fig. 31.3.2).

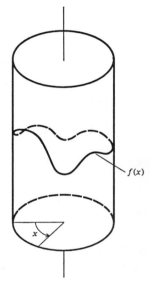

$f(x)$

FIGURE 31.3.2

Any interval of length L may be used; in particular, it is often useful to use $-L/2 \leq x \leq L/2$ instead of $0 \leq x \leq L$.

PROBLEMS

Find the Fourier expansions for:

31.3.1 $y = x$ $-1 \leq x \leq 1$

31.3.2 $y = 1 - x^2$ $-1 \leq x \leq 1$

31.3.3 $y = e^{-x}$ $0 \leq x \leq L$

31.3.4 $y = \begin{cases} 1 & 0 < x < \pi \\ 0 & \pi < x < 2\pi \end{cases}$

31.4 THE COMPLEX FORM OF THE FOURIER SERIES

It is often convenient, especially for theory, to write $\sin x$ and $\cos x$ in their equivalent complex forms. Using the Euler identity

$$e^{ix} = \cos x + i \sin x$$

we replace i with $-i$

$$e^{-ix} = \cos x - i \sin x$$

and add and subtract these two equations to get

$$\sin x = \frac{e^{ix} - e^{-ix}}{2i}$$

$$\cos x = \frac{e^{ix} + e^{-ix}}{2}$$

In this notation (k an integer) we have

$$F(x) = \sum_{k=-\infty}^{\infty} c_k e^{(2\pi i/L)kx}$$

where the discrete set of c_k is found by the simple formula

$$c_k = \frac{1}{L} \int_0^L F(x) e^{-(2\pi i/L)kx} \, dx$$

The orthogonality condition which determines the form of the c_k is then easily proved by direct integration

$$\int_0^L e^{(2\pi i/L)kx} e^{(2\pi i/L)mx} \, dx = \begin{cases} 0 & m+k \neq 0 \\ L & m+k = 0 \end{cases}$$

Comparing the two forms of representation, we find

$$c_k = \begin{cases} \dfrac{a_k - ib_k}{2} & k > 0 \\[2mm] \dfrac{a_k + ib_k}{2} & k < 0 \\[2mm] \dfrac{a_0}{2} & k = 0 \end{cases}$$

PROBLEMS

Expand in the complex Fourier-series form:

31.4.1 $y = x$ $\quad -\pi < x < \pi$

31.4.2 $y = \begin{cases} 1 & 0 \leq x \leq \frac{1}{2} \\ 0 & \frac{1}{2} < x < 1 \end{cases}$

31.5   THE FINITE FOURIER SERIES

It is a remarkable fact that the sines and cosines are orthogonal over both the continuous interval and sets of discrete, equally spaced points covering the period. *This is extremely important* because in computing we are usually given only samples of the function at a set of equally spaced points.   This means, in turn, that we can find the coefficients exactly (except for roundoff) in the discrete sampled problem, instead of approximately in the continuous model by numerical integration.

We will consider only an even number $2N$ of points.   Let the $2N$ sample points be (see Fig. 31.5.1)

$$x_p = 0, \frac{L}{2N}, \frac{2L}{2N}, \dots, \frac{(2N-1)L}{2N}$$

or more shortly,

$$x_p = \frac{Lp}{2N} \qquad p = 0, 1, \dots, 2N - 1$$

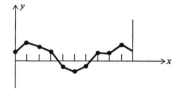

FIGURE 31.5.1

We want to show that the $2N$ Fourier functions

$$1, \cos \frac{2\pi}{L} x, \cos \frac{2\pi}{L} 2x, \dots, \cos \frac{2\pi}{L} (N-1)x, \cos \frac{2\pi}{L} Nx$$

$$\sin \frac{2\pi}{L} x, \sin \frac{2\pi}{L} 2x, \dots, \sin \frac{2\pi}{L} (N-1)x$$

form an orthogonal set of functions.   This amounts to showing that for $0 \leq k$ and $m \leq N$

$$\sum_{p=0}^{2N-1} \cos\left(\frac{2\pi}{L} k \frac{Lp}{2N}\right) \cos\left(\frac{2\pi}{L} m \frac{Lp}{2N}\right) = \begin{cases} 0 & k \neq m \\ N & k = m \neq 0, N \\ 2N & k = m = 0, N \end{cases}$$

$$\sum_{p=0}^{2N-1} \cos\left(\frac{2\pi}{L} k \frac{Lp}{2N}\right) \sin\left(\frac{2\pi}{L} m \frac{Lp}{2N}\right) = 0$$

$$\sum_{p=0}^{2N-1} \sin\left(\frac{2\pi}{L} k \frac{Lp}{2N}\right) \sin\left(\frac{2\pi}{L} m \frac{Lp}{2N}\right) = \begin{cases} 0 & k \neq m \\ N & k = m \neq 0, N \\ 0 & k = m = 0, N \end{cases}$$

In fact we shall prove more, namely that when

$$k \pm m = 0, \pm 2N, \pm 4N, \dots$$

the orthogonality conditions have special features.

We begin simply by examining the series

$$\sum_{p=0}^{2N-1} e^{(2\pi i/L)kx_p} = \sum_{p=0}^{2N-1} e^{\pi ikp/N} \qquad \text{for integer } k$$

which is a geometric progression with ratio

$$r = e^{\pi ik/N}$$

The sum of the geometric progression is

$$\begin{cases} \dfrac{1 - r^{2N}}{1 - r} = 0 & r \neq 1 \\ 2N & r = 1 \end{cases}$$

The upper sum is true because $r^{2N} = e^{2\pi ik} = 1$. The lower value is true because then $r = 1$. The condition $r = 1$ means $k = 0, \pm 2N, \pm 4N, \dots$

We next show that the set of functions

$$e^{(2\pi i/L)kx_p}$$

is orthogonal in the sense that the product of one function times the complex conjugate of another function summed over the points x_p is zero, that is, that

$$\sum_{p=0}^{2N-1} e^{(2\pi i/L)kx_p} e^{-(2\pi i/L)mx_p} = \begin{cases} 0 & |k - m| \neq 0, 2N, 4N, \dots \\ 2N & |k - m| = 0, 2N, 4N, \dots \end{cases}$$

This follows immediately from the above by writing the product as

$$\sum_{p=0}^{2N-1} e^{(2\pi i/L)(k-m)x_p}$$

and noting that $k - m$ plays the role of k.

We now return to the "real functions" by using the Euler identity

$$e^{ix} = \cos x + i \sin x$$

The condition for the single exponential summed over the points x_p becomes two equations (the real and imaginary parts separately)

$$\sum_{p=0}^{2N-1} \cos \frac{2\pi}{L} kx_p = \begin{cases} 0 & k \neq 0, \pm 2N, \pm 4N, \dots \\ 2N & k = 0, \pm 2N, \pm 4N, \dots \end{cases}$$

$$\sum_{p=0}^{2N-1} \sin \frac{2\pi}{L} kx_p = 0 \qquad \text{for all } k$$

At last we are ready to prove the orthogonality of the Fourier functions over the set of equally spaced points x_p. The first of the three orthogonality equations can be written, by using the trigonometric identity

$$\cos a \cos b = \tfrac{1}{2}\left[\cos(a+b) + \cos(a-b)\right]$$

$$\frac{1}{2}\sum_{p=0}^{2N-1}\left[\cos\pi(k+m)\frac{p}{N} + \cos\pi(k-m)\frac{p}{N}\right]$$

$$= \begin{cases} 0 & |k-m| \text{ and } |k+m| \neq 0, 2N, 4N, \ldots \\ N & |k-m| \text{ or } |k+m| = 0, 2N, 4N, \ldots \\ 2N & |k-m| \text{ and } |k+m| = 0, 2N, 4N, \ldots \end{cases}$$

For the restricted set of functions we are now using, namely,

$$1, \cos\frac{2\pi}{L}x, \ldots, \cos\frac{2\pi}{L}(N-1)x, \cos\frac{2\pi}{L}Nx$$

$$\sin\frac{2\pi}{L}x, \ldots, \sin\frac{2\pi}{L}(N-1)x$$

$k+m = 0, 2N, 4N, \ldots$ cannot occur unless $k = m = 0$ or N. Thus, we have the required orthogonality of the Fourier functions over the set x_p.

The orthogonality in turn leads directly to the expansion of an arbitrary function $F(x)$ defined on the set of points x_p,

$$F(x) = \frac{A_0}{2} + \sum_{k=1}^{N-1}\left(A_k\cos\frac{2\pi}{L}kx + B_k\sin\frac{2\pi}{L}kx\right) + \frac{A_N}{2}\cos\frac{2\pi}{L}Nx$$

where

$$A_k = \frac{1}{N}\sum_{p=0}^{2N-1}F(x_p)\cos\frac{2\pi}{L}kx_p \qquad k = 0, 1, \ldots, N$$

$$B_k = \frac{1}{N}\sum_{p=0}^{2N-1}F(x_p)\sin\frac{2\pi}{L}kx_p \qquad k = 1, 2, \ldots, N-1$$

The function $F(x)$ is often given at $2N + 1$ points, but still for $2N$ intervals, and it is assumed that

$$F(0) = F(L)$$

If this is not so, then it is customary to use

$$\frac{F(0) + F(L)}{2}$$

as the value of $F(0)$. In this case the formulas for the coefficients A_k and B_k are

$$A_k = \frac{1}{N} \left[\frac{F(0)}{2} + \sum_{p=1}^{2N-1} F(x_p) \cos \frac{2\pi}{L} k x_p + \frac{F(L)}{2} \right]$$

$$B_k = \frac{1}{N} \sum_{p=1}^{2N-1} F(x_p) \sin \frac{2\pi}{L} k x_p$$

These are effectively the trapezoid rule for numerically integrating the corresponding integrals in the continuous interval.

There is a second set of equally spaced points that we need to mention, namely,

$$t_p = x_p + \frac{L}{4N} = \frac{L}{2N} \left(p + \frac{1}{2} \right)$$

which are midway between the x_p points (recall that $x_0 = 0$ is the same as x_{2N} due to periodicity).

Repeating the steps briefly, we see that the sum

$$\sum_{p=0}^{2N-1} e^{(2\pi i k/L) t_p} = \begin{cases} 0 & r \neq 1 \\ 2N e^{\pi i k/2N} & r = 1 \end{cases}$$

as before. Hence, the rest follows as before, except for the special treatment of the end values in the summations for the coefficients.

How accurate is the finite Fourier expansion in reproducing the original data (within roundoff)? The set of functions is orthogonal; hence they are linearly independent and form a basis that spans the whole space. Thus, within roundoff, the expansion reproduces the data exactly.

PROBLEMS

Find the finite Fourier expansion for:

31.5.1 The data $x(0) = 0$; $x(1) = 1$; $x(2) = 2$; $x(3) = 3$; $N = 2$.

31.5.2 The data $x(\frac{1}{2}) = 1$; $x(\frac{3}{2}) = 1$; $x(\frac{5}{2}) = -1$; $x(\frac{7}{2}) = -1$; $N = 2$.

31.6 RELATION OF THE DISCRETE AND CONTINUOUS EXPANSIONS

It is reasonable to ask, What is the relation between the two expansions we have just found, the continuous and the discrete? Let the continuous expansion be

$$F(x) = \frac{a_0}{2} + \sum_{k=1}^{\infty} \left(a_k \cos \frac{2\pi}{L} x + b_k \sin \frac{2\pi}{L} x \right)$$

with lowercase letters for the coefficients.   We pick $x_p = Lp/(2N)$ for convenience.
If we multiply $F(x_p)$ by $\cos(2\pi/L)kx_p$ and sum, we get

$$\sum_{p=0}^{2N-1} F(x_p)\cos\frac{2\pi}{L}kx_p = NA_k = N(a_k + a_{2N-k} + a_{2N+k} + \cdots)$$

Hence, the coefficient we calculate is

$$A_k = a_k + \sum_{m=1}^{\infty}(a_{2Nm-k} + a_{2Nm+k})$$

which expresses the (uppercase) finite Fourier-series coefficients in terms of the
(lowercase) continuous Fourier-series coefficients.
Similarly,

$$B_k = b_k + \sum_{m=1}^{\infty}(-b_{2Nm-k} + b_{2Nm+k})$$

For the special constant term A_0, we have

$$A_0 = a_0 + 2\sum_{m=1}^{\infty} a_{2Nm}$$

Thus, various frequencies present in the original continuous signal $F(x)$ are
added together due to the sampling.   This effect is called "aliasing" and *is
directly attributable to the act of sampling at equally spaced points*; once the
sampling has been done, the effect cannot be undone (from the samples alone).

These relationships between the coefficients of the two expansions, the
continuous and the discrete sampled, show clearly the aliasing of the frequencies.
If we represent the frequencies as points along a line running from zero to infinity,
we can represent the aliasing by folding the line back and forth onto itself.   The
first frequency at which a fold occurs is called *the folding frequency*, or the *Nyquist*[1]
frequency, and is $k = N$.   Subsequent folds occur at this same frequency spacing.
All points on the curve (shown in Fig. 31.6.1) above the same location on the

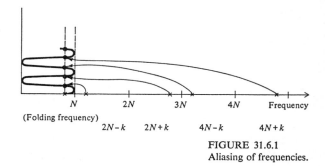

FIGURE 31.6.1
Aliasing of frequencies.

[1] Named after Harry Nyquist (1889–   ), a famous scientist in electrical engineering.

frequency axis appear as the same frequency owing to the effect of sampling, and once we have sampled, we cannot again separate the different frequencies that have been aliased together.   At the folding frequency itself, we are sampling at the crucial rate of 2 samples per cycle [and $\sin \pi t = 0$, $\cos \pi t = (-1)^t$ where $t =$ an integer].

31.7 THE POWER SPECTRUM

Up to now we have concentrated on the coefficients of the Fourier-series expansion of a periodic function.   However, the coefficients themselves are often not physically significant because they depend on the choice of the origin.   To see this, consider that

$$a_k = \frac{1}{\pi} \int_0^{2\pi} f(t)\cos kt\, dt$$

$$b_k = \frac{1}{\pi} \int_0^{2\pi} f(t)\sin kt\, dt$$

and under a shift of coordinates of a distance s [$f(t)$ is periodic]

$$t' = t + s$$

we would compute the primed coefficients a_k' and b_k'.   For a_k' we have

$$a_k' = \frac{1}{\pi} \int_0^{2\pi} f(t + s)\cos kt\, dt$$

$$= \frac{1}{\pi} \int_s^{s+2\pi} f(t')\cos k(t' - s)\, dt'$$

Due to periodicity this is the same as

$$a_k' = \frac{1}{\pi} \int_0^{2\pi} f(t')(\cos kt' \cos ks + \sin kt' \sin ks)\, dt'$$

$$= a_k \cos ks + b_k \sin ks$$

Similarly

$$b_k' = a_k \sin ks - b_k \cos ks$$

Thus the coefficients of the expansion depend on the origin we pick for the periodic function, and usually this is of no physical significance.

However, the nonlinear expression

$$(a_k')^2 + (b_k')^2 = a_k^2 \cos^2 ks + 2a_k b_k \cos ks \sin ks + b_k^2 \sin^2 ks$$
$$+ a_k^2 \sin^2 ks - 2a_k b_k \cos ks \sin ks + b_k^2 \cos^2 ks = a_k^2 + b_k^2$$

The quantity $a_k{}^2 + b_k{}^2$ is invariant under translations, and is therefore much more likely to have physical significance than the individual coefficients. This quantity (or its square root) is usually called " the power at frequency k," and a plot of

$$a_k{}^2 + b_k{}^2$$

as a function of k is called the "power spectrum."

From Sec. 31.4 we had for the complex form of the expansion

$$f(t) = \sum_{k=-1}^{\infty} c_k e^{2\pi i k t}$$

$$c_k = \frac{a_k - ib_k}{2} \qquad k > 0$$

$$c_k = \frac{a_k + ib_k}{2} \qquad k < 0$$

So that

$$4|c_k|^2 = 4|c_{-k}|^2 = a_k{}^2 + b_k{}^2$$

or for real functions $f(t)$

$$|c_k| + |c_{-k}| = \sqrt{a_k{}^2 + b_k{}^2}$$

For many purposes it is the power spectrum we care about; but let us be clear: there is a loss of information in going to it.   For each pair of terms we get $a_k \cos kt + b_k \sin kt$

$$= \sqrt{a_k{}^2 + b_k{}^2} \left(\frac{a_k}{\sqrt{a_k{}^2 + b_k{}^2}} \cos kt + \frac{b_k}{\sqrt{a_k{}^2 + b_k{}^2}} \sin kt \right)$$

$$= \sqrt{a_k{}^2 + b_k{}^2} \cos(kt - \phi)$$

where we have picked ϕ so that

$$\frac{a_k}{\sqrt{a_k{}^2 + b_k{}^2}} = \cos \phi \qquad \frac{b_k}{\sqrt{a_k{}^2 + b_k{}^2}} = \sin \phi$$

Thus the power spectrum does not include information about the phase angle ϕ.

31.8 INTERPOLATION OF PERIODIC FUNCTIONS

Part II developed the theory of polynomial approximation, and we now face the problem of developing a similar theory for the Fourier-series approximation of a continuous periodic function.

While the problem may be discussed for unequally spaced sample points, and some results are known, both theory and practice are usually based on equally spaced points. For $2N$ equally spaced sample points the interpolation problem is solved by using the finite Fourier series to estimate the function between the sample points.

As in Part II, once we have the formula, we next want to know the error of the approximation. The error can be divided into two parts: that due to taking only a finite number of terms in the series and that due to the sampling effect (aliasing). We have already studied the effects of sampling, and we now examine the effect of taking only a finite number of terms of a Fourier series.

If the continuous Fourier series is of the usual form, then substituting the formulas for the coefficients, we get for the first $2N$ terms, where we take one-half the last cosine term as we would for a discrete series, and one-half the last sine term, since it is not identically zero in the continuous case,

$$
\begin{aligned}
y_{2N}(t) &= \frac{1}{N} \int_0^{2N} y(s) \left\{ \frac{1}{2} + \sum_{k=1}^{N-1} \left[\cos \frac{\pi}{N} kt \cos \frac{\pi}{N} ks + \sin \frac{\pi}{N} kt \sin \frac{\pi}{N} ks \right] \right. \\
&\quad \left. + \frac{1}{2} \sin \pi t \sin \pi s + \frac{1}{2} \cos \pi t \cos \pi s \right\} ds \\
&= \frac{1}{N} \int_0^{2N} y(s) \left\{ \frac{1}{2} + \sum_{k=1}^{N-1} \left[\cos \frac{\pi}{N} k(t-s) \right] + \frac{1}{2} \cos \pi(t-s) \right\} ds
\end{aligned}
$$

In order to sum[1] the quantity in the braces, we multiply by

$$
\sin \frac{\pi}{2N}(t-s)
$$

and get

$$
\begin{aligned}
\frac{1}{2} \left\{ \sin \frac{\pi}{2N}(t-s) \right. & \\
&+ \left[\sin \frac{\pi}{N}\left(1+\frac{1}{2}\right)(t-s) - \sin \frac{\pi}{N}\left(1-\frac{1}{2}\right)(t-s) \right] \\
&+ \left[\sin \frac{\pi}{N}\left(2+\frac{1}{2}\right)(t-s) - \sin \frac{\pi}{N}\left(2-\frac{1}{2}\right)(t-s) \right] \\
&\quad \cdot \ \cdot \ \cdot \ \cdot \ \cdot \ \cdot \ \cdot \ \cdot \ \cdot \ \cdot \ \cdot \ \cdot \\
&+ \left[\sin \frac{\pi}{N}\left(N-1+\frac{1}{2}\right)(t-s) - \sin \frac{\pi}{N}\left(N-1-\frac{1}{2}\right)(t-s) \right] \\
&\left. + \sin \frac{\pi}{2N}(t-s)\cos \pi(t-s) \right\}
\end{aligned}
$$

---

[1] We could clearly use the methods of Chap. 11 to sum this series more elegantly, but it is likely that the reader has forgotten such details by now, and so we shall use the more clumsy but familiar trigonometric methods.

Because of cancellation of each term on the left by the term on the right one line lower, all that we have left is two terms

$$\frac{1}{2}\left\{\sin\left[\frac{\pi}{N}\left(N-\frac{1}{2}\right)(t-s)\right] + \sin\frac{\pi}{2N}(t-s)\cos\pi(t-s)\right\}$$

We now expand the first of these terms in the form

$$\sin\pi(t-s)\cos\frac{\pi}{2N}(t-s) - \cos\pi(t-s)\sin\frac{\pi}{2N}(t-s)$$

and because of cancellation of the second of these terms with the second term that we had left, for the braces in the integral we have finally

$$\frac{\sin\pi(t-s)\cos(\pi/2N)(t-s)}{2\sin(\pi/2N)(t-s)}$$

Thus

$$y_{2N}(t) = \frac{1}{N}\int_0^{2N} y(s)\frac{\sin\pi(t-s)\cos(\pi/2N)(t-s)}{2\sin(\pi/2N)(t-s)}\,ds$$

Since $y(s)$ is assumed to be periodic, we can shift the limits to any interval of length $2N$ and set $t-s=\theta$:

$$y_{2N}(t) = \frac{1}{2N}\int_{-N}^{N} y(t-\theta)\frac{\sin\pi\theta\,\cos[(\pi/2N)\theta]}{\sin[N(\pi/2)\theta]}\,d\theta$$

This form is deceptive; the more samples $2N$ we take, the longer the interval. In practice, of course, the range of periodicity is fixed, and we change the sampling interval Δt. Thus we set

$$t = 2Ns$$
$$\theta = 2N\phi$$

to get the form

$$y_{2N}(s) = \int_{1/2}^{1/2} y(s-\phi)\sin(2\pi N\phi)\frac{\cos\pi\phi}{\sin\pi\phi}\,d\phi$$

where the change in the meaning of $y_{2N}(\cdot)$ should not cause serious confusion.

If $y(t)\equiv 1$, then we have

$$1 = \int_{1/2}^{1/2} \sin(2\pi N\phi)\frac{\cos\pi\phi}{\sin\pi\phi}\,d\phi$$

Since the variable of integration is ϕ, we can write

$$y(s) = \int_{1/2}^{1/2} y(s)\sin(2\pi N\phi)\frac{\cos\pi\phi}{\sin\pi\phi}\,d\phi$$

and subtract from this $y_{2N}(s)$ to get

$$\varepsilon_{2N}(s) = y(s) - y_{2N}(s) = \int_{1/2}^{1/2} [y(s) - y(s - \phi)] \left\{ \sin(2\pi N\phi) \frac{\cos \pi\phi}{\sin \pi\phi} \right\} d\phi$$

The quantity inside the braces is often called[1] the *kernel*, $K(\phi)$. Clearly $K(0) = 2N$.
For large N, $\sin(2N\pi\phi)$ oscillates rapidly while the term $\cot \pi\phi$ dampens (modulates) the magnitude of the oscillation until at $\phi = \pm 1/2$, $K(\pm 1/2) = 0$. The
envelope of the modulation is $\cot \pi\phi$, and

$$K(\phi) = K(-\phi)$$

If the function $y(x)$ is smooth, then the major contribution to the integral, as far as
the kernel is concerned, comes from the immediate neighborhood of $\phi = 0$, say
from $-1/2N$ to $1/2N$, between the first zeros on each side. Thus the error in the
approximation tends to be a local property (see Fig. 31.8.1). As N increases, the
peak becomes higher and narrower, and the side lobe (local peaks) oscillations
tend to cancel out.

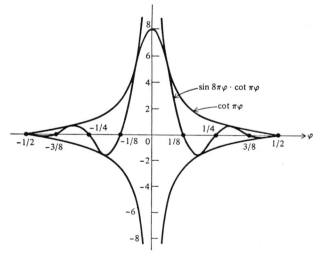

FIGURE 31.8.1

---

[1] See Ref. 64, p.50, where it is called a "modified kernel."

If, on the other hand, there are rapid changes in $y(x)$, then the tendency of the damped oscillations to approximately cancel out and thus contribute only small amounts is no longer present, and sudden changes in $y(x)$ remote from the point of interpolation can produce errors.

Note that the integrand involves the function rather than a high-order derivative as in polynomial approximation.

PROBLEMS

31.8.1 We have approximated trigonometric functions by polynomials. Approximate $y(t) = (1 - t^2)^2 \, (-1 \le t \le 1)$ by a Fourier series using $N = 3$, that is, samples at $t = -2/3, -1/3, 0, 1/3, 2/3, 1$, and interpolate for the value at $t = 1/2$.

31.8.2 In this section we have developed the equivalent of Lagrange interpolation; examine the equivalent of Hermite interpolation which matches the function and derivative at each sample point.   *Ans.* Not possible

31.9   INTEGRATION

As in Part II, having examined the question of interpolation and its errors, we next turn to the problem of integration.   If the coefficients of the continuous Fourier series of $y(t)$ fall off fairly rapidly as $n \to \infty$, then the A_0 coefficient of the finite Fourier series gives a good estimate of the integral

$$\int_0^{2N} y(t) \, dt = N a_0$$

since

$$A_0 = \frac{1}{N} \sum_{t=0}^{2N-1} y(t) = a_0 + 2 \sum_{j=1}^{\infty} a_{2Nj}$$

Thus the error involved in estimating the integral from $2N$ equally spaced, equally weighted samples is

$$2N \sum_{j=1}^{\infty} a_{2Nj}$$

and represents the inevitable aliasing due to the discrete sampling.

For periodic functions the formula for integration which uses $2N$ equally spaced, equally weighted samples corresponds to the following:

Newton-Cotes, because it is equally spaced
Chebyshev, because it is equally weighted
Gauss, because $2N$ samples give exactness for $4N$ functions

Let us illustrate the accuracy of this formula for integrating periodic functions by a pair of examples. Consider the two elliptic integrals

$$I_1 = \int_0^{\pi/2} \sqrt{1 - k^2 \sin^2 \theta} \, d\theta$$

$$I_2 = \int_0^{\pi/2} \frac{1}{\sqrt{1 - k^2 \sin^2 \theta}} \, d\theta$$

where $|k| \leq 1$ is a parameter. We can imagine expanding the square root as a binomial. If k^2 is small, then the series will converge rapidly. Converting the powers $\sin^{2n}\theta$ to multiple angles

$$\left(\frac{e^{i\theta} - e^{-i\theta}}{2i}\right)^{2n} = \frac{1}{(2^{2n})(i^{-2n})} \left[e^{2ni\theta} - C(2n, 1)e^{2(n-1)i\theta} + \cdots\right]$$

$$= \frac{1}{(2^{2n-1})(i^{-2n})} \left[\cos 2n\theta - C(2n, 1)\cos(2n-2)\theta + \cdots\right]$$

we get a rapidly converging Fourier series. Thus for small k^2 we expect accurate answers; for k^2 near 1 we do not expect accurate answers.

As an experiment, we chose sample points at 0, 30, 60, 90°, although to get periodicity we had to assume six points in the interval of periodicity, $0 \leq t \leq \pi$. Thus we actually computed only four sample points, one of which ($\theta = 0$) was trivial. The result (compared with those from a table) are given in Table 31.9.

Table 31.9 ELLIPTIC INTEGRAL $I_1(k^2)$

k^2	I_1(calc.)	I_1(table)	Error
$k^2 = \frac{1}{4}$	1.46746	1.4675	
$k^2 = \frac{1}{2}$	1.35064	1.3506	
$k^2 = \frac{3}{4}$	1.21099	1.2111	0.0001

ELLIPTIC INTEGRAL $I_2(k^2)$

k^2	I_2(calc.)	I_2(table)	Error
$k^2 = \frac{1}{4}$	1.68575	1.6858	
$k^2 = \frac{1}{2}$	1.85410	1.8541	
$k^2 = \frac{3}{4}$	2.15789	2.1565	0.0014

This example shows two things.   First, it is often possible to make reasonable estimates of the size of the coefficients of a Fourier series; hence we can estimate the error of the computation *before* starting it.   Second, the method of integration which uses $2N$ equally spaced, equally weighted samples gives very accurate results for periodic functions.   This approach has been used to compute the Bessel function $J_0(x)$ from the formula

$$J_0(x) = \frac{1}{\pi} \int_0^\pi \cos(x \sin \phi)\, d\phi$$

PROBLEMS

31.9.1   Evaluate $J_0(1/2)$ from the integral form, using six samples.

31.10   THE GENERAL-OPERATOR APPROACH

In Sec. 15.4 of Part II, we described a general method (third method) for finding formulas which were exact for $1, x, \ldots, x^n$.   In the same way we can find formulas which are exact for

$$1, \cos \frac{\pi}{N} t, \cos \frac{2\pi}{N} t, \cos \frac{3\pi}{N} t, \ldots, \cos \frac{(N-1)\pi}{N} t, \cos \pi t$$

$$\sin \frac{\pi}{N} t, \sin \frac{2\pi}{N} t, \sin \frac{3\pi}{N} t, \ldots, \sin \frac{(N-1)\pi}{N} t$$

Such conditions are the natural ones to apply to a function that is periodic and for which we use $2N$ equally spaced samples.

Let the result of the linear operator L operating on the function $f(t)$ be $L(f)$, and let it be expressed in terms of the set of $2N$ equally spaced samples of the function $t = 0, 1, 2, \ldots, 2N - 1$; that is,

$$L[f(t)] = w_0 f(0) + w_1 f(1) + \cdots + w_{2N-1} f(2N - 1)$$

For $1, \cos \dfrac{\pi}{N} m, \cos \dfrac{2\pi}{N} m, \ldots, \cos \dfrac{\pi(N-1)}{N} m, \cos \pi m, \sin \dfrac{\pi}{N} m, \ldots, \sin \dfrac{\pi(N-1)}{N} m,$

the defining equations are:

M_0
$= w_0 + w_1 \qquad\qquad\qquad + w_2 \qquad\qquad\qquad + \cdots + w_{2N-1}$

M_1
$= w_0 + w_1 \cos \dfrac{\pi}{N} \qquad + w_2 \cos \dfrac{2\pi}{N} \qquad + \cdots + w_{2N-1} \cos \dfrac{2N-1}{N}$

M_2
$= w_0 + w_1 \cos \dfrac{2\pi}{N} \qquad + w_2 \cos \dfrac{4\pi}{N} \qquad + \cdots + w_{2N-1} \cos \dfrac{2N-1}{N} 2\pi$

. . . .

M_{N-1}
$$= w_0 + w_1 \cos \frac{N-1}{N}\pi + w_2 \cos \frac{(N-1)2\pi}{N} + \cdots + w_{2N-1} \cos \frac{(2N-1)(N-1)\pi}{N}$$

M_N
$$= w_0 - w_1 \qquad\qquad + w_2 \qquad\qquad - \cdots - w_{2N-1}$$

M_{N+1}
$$= 0 + w_1 \sin \frac{\pi}{N} \qquad + w_2 \sin \frac{2\pi}{N} \qquad + \cdots + w_{2N-1} \sin \frac{(2N-1)\pi}{N}$$

$$\cdots\cdots\cdots\cdots\cdots\cdots\cdots\cdots\cdots\cdots\cdots\cdots\cdots\cdots\cdots\cdots$$

M_{2N-1}
$$= 0 + w_1 \sin \frac{(N-1)\pi}{N} + \sin \frac{2(N-1)\pi}{N} + \cdots + w_{2N-1} \sin \frac{(2N-1)(N-1)\pi}{N}$$

Thus the matrix of the unknown weights w_i is

$$
\begin{bmatrix}
1 & 1 & 1 & \cdots & 1 \\
1 & \cos \dfrac{\pi}{N} & \cos \dfrac{2\pi}{N} & \cdots & \cos \dfrac{2N-1}{N}\pi \\
\cdot & \cdot\ \ \cdot\ \ \cdot & \cdot\ \ \cdot\ \ \cdot & \cdot\ \ \cdot\ \ \cdot & \cdot\ \ \cdot \\
1 & \cos \dfrac{N-1}{N}\pi & \cos \dfrac{(N-1)2\pi}{N} & \cdots & \cos \dfrac{(N-1)(2N-1)\pi}{N} \\
1 & -1 & 1 & \cdots & -1 \\
0 & \sin \dfrac{\pi}{N} & \sin \dfrac{2\pi}{N} & \cdots & \sin \dfrac{2N-1}{N}\pi \\
\cdot & \cdot\ \ \cdot\ \ \cdot & \cdot\ \ \cdot\ \ \cdot & \cdot\ \ \cdot\ \ \cdot & \cdot\ \ \cdot \\
0 & \sin \dfrac{N-1}{N}\pi & \sin \dfrac{N-1}{N}2\pi & \cdots & \sin \dfrac{(N-1)(2N-1)\pi}{N}
\end{bmatrix}
$$

Because of the orthogonality relations

$$\sum_{k=0}^{2N-1} \cos \frac{m\pi}{N}k \cos \frac{n\pi}{N}k = N\delta_{m,n} \qquad m, n \neq 0, N$$

$$\sum_0^{2N-1} \sin \frac{m\pi}{N}k \cos \frac{n\pi}{N}k = 0$$

$$\sum_0^{2N-1} \sin \frac{m\pi}{N}k \sin \frac{n\pi}{N}k = N\delta_{m,n}$$

$$\sum_0^{2N-1} 1 = \sum_0^{2N-1} \cos^2 \pi k = 2N$$

the inverse matrix is the transpose of the matrix except that two of the columns, the first and the $(N-1)$st, corresponding to the functions 1 and $\cos \pi m$, have been divided by 2.

$$\left(\frac{1}{N}\right)\begin{pmatrix} \frac{1}{2} & 1 & \cdots & 1 & \frac{1}{2} & 0 & \cdots & 0 \\[2mm] \frac{1}{2} & \cos\dfrac{\pi}{N} & \cdots & \cos\dfrac{N-1}{N}\pi & -\frac{1}{2} & \sin\dfrac{\pi}{N} & \cdots & \sin\dfrac{N-1}{N}\pi \\[2mm] \frac{1}{2} & \cos\dfrac{2\pi}{N} & \cdots & \cos\dfrac{N-1}{N}2\pi & \frac{1}{2} & \sin\dfrac{2\pi}{N} & \cdots & \sin\dfrac{N-1}{N}2\pi \\[2mm] \cdots\cdots\cdots\cdots\cdots\cdots\cdots\cdots\cdots\cdots\cdots\cdots\cdots\cdots\cdots\cdots\cdots\cdots\cdots\cdots\cdots\cdots\cdots \\[2mm] \frac{1}{2}\cos\dfrac{2N-1}{N}\pi & \cdots & \cos\dfrac{(N-1)(2N-1)}{N}\pi & -\frac{1}{2}\sin\dfrac{2N-1}{N}\pi & \cdots & \sin\dfrac{(N-1)(2N-1)}{N}\pi \end{pmatrix}$$

Here we have produced the postmultiplier inverse, while in Sec. 15.5 we produced the premultiplier inverse; they are, of course, the same, but in each case it is harder to think about the other inverse.   Interchanging the order of the two matrices gives a different set of orthogonal relations, which, at times, can be useful.

This general approach of using the third method may be illustrated by an example.   Suppose that we wish to estimate the first derivative of the periodic function at the point $t=0$.   To be specific, suppose that we examine the case $N=3$.   The components of the " moments " M are, using $1, \cos(\pi/3)t, \cos(2\pi/3)t, \cos \pi t, \sin(\pi/3)t, \sin(2\pi/3)t,$

$$0, 0, 0, 0, \frac{\pi}{3}, \frac{2\pi}{3}$$

The inverse matrix times the moment vector gives

$$(\tfrac{1}{2})(\tfrac{1}{3})\begin{pmatrix} 1 & 2 & 2 & 1 & 0 & 0 \\ 1 & 1 & -1 & -1 & \sqrt{3} & \sqrt{3} \\ 1 & -1 & -1 & 1 & \sqrt{3} & -\sqrt{3} \\ 1 & -2 & 2 & -1 & 0 & 0 \\ 1 & -1 & -1 & 1 & -\sqrt{3} & \sqrt{3} \\ 1 & 1 & -1 & -1 & -\sqrt{3} & -\sqrt{3} \end{pmatrix}\begin{pmatrix} 0 \\ 0 \\ 0 \\ 0 \\ \dfrac{\pi}{3} \\ \dfrac{2\pi}{3} \end{pmatrix} = \tfrac{1}{6}\begin{pmatrix} 0 \\ \sqrt{3}\,\pi \\ -\dfrac{\sqrt{3}}{3}\pi \\ 0 \\ \dfrac{\sqrt{3}}{3}\pi \\ -\sqrt{3}\,\pi \end{pmatrix}$$

The last vector is the weight vector; thus the formula is

$$f'(0) = \frac{\sqrt{3}\,\pi}{18}\,[3f(1) - f(2) + f(4) - 3f(5)]$$

PROBLEMS

31.10.1   Using the matrix of the illustrative example, find a formula for $f''(0)$.

31.10.2   Using the matrix, find a formula for estimating $\int_0^{2N} \int_0^t f(t)\, dt\, dt$.

31.11   SOME REMARKS ON THE GENERAL METHOD

In the case of polynomial approximation, it was efficient to use the general matrix approach.   In the case of approximation with a Fourier series, it is not.   The difference is due to the fact that we can find the Fourier expansion of a function with much less labor than performing the matrix multiplication to get the weights (see Sec. 31.5).   Once we have the Fourier coefficients of the function, it is easy to multiply them by the moments to get the final answer.   Thus if we have the Fourier series

$$f(t) = \frac{a_0}{2} + \sum_{k=1}^{N-1} a_k \cos \frac{\pi}{N} kt + \frac{a_N}{2} \cos \pi t + \sum_{k=1}^{N-1} b_k \sin \frac{\pi}{N} kt$$

applying the operator $L(\cdot)$, we have

$$L[f(t)] = \frac{a_0}{2} L(1) + \sum_{k=1}^{N-1} a_k L\left(\cos \frac{\pi}{N} kt\right) + \frac{a_N}{2} L(\cos \pi t)$$
$$+ \sum_{k=1}^{N-1} b_k L\left(\sin \frac{\pi}{N} kt\right)$$
$$= \frac{a_0}{2} M_0 + \sum_{k=1}^{N-1} a_k M_k + \frac{a_N}{2} M_N + \sum_{k=1}^{N-1} b_k M_{N+k}$$

The reason that we can find the coefficients easily is mainly that the sines and cosines are orthogonal and can be found independently of each other.   Thus the inverse-matrix approach, while of theoretical interest for Fourier-series approximation, is not of much practical importance.

It is necessary to point out again a distinction which first arose in Sec. 21.3, which covers indefinite integration.   Whether we make a formula such as

$$y = \int_a^b f(x)\, dx$$

exact for $f(x) = 1, x, x^2, \dots$ or for $y = 1, x, x^2, \dots$ makes a difference.   As was pointed out in Sec. 21.3, exactness for $f(x) = 1, x, x^2, \dots$ is the same, for this formula, as exactness for $y = x, x^2, x^3, \dots$ since the integrals of powers of x are powers of x one higher.   We found it convenient at that time to *add* the condition of exactness for $y = 1$ and thus restore the equivalence of the two choices.

For the Fourier-series approach, if we first approximate the integrand and then integrate, the terms $\cos kx$ and $\sin kx$ go into each other, but the constant term 1 goes into x.   Thus the equivalence of approximating the integrand or the solution by a Fourier series is lost.   We must, therefore, decide *before we start* on what part of the problem we are going to approximate by the set of functions we are using—approximations at various stages of the problem are not necessarily equivalent.   In practice it is usual to know properties of the input function or the output function, and hence these are the ones usually chosen for the approximation.

This effect of nonequivalence will be seen again in Chaps. 39 and 40 on exponential functions where, upon integration, the exponential $e^{\alpha_i x}$ goes into itself (within a multiplicative constant) *except* $\alpha_i = 0$.

32

CONVERGENCE OF FOURIER SERIES

32.1 THE IMPORTANCE OF CONVERGENCE

The rate of convergence of a Fourier series is important for two reasons. First it enables us to *estimate in advance* the sampling rate we will need to use in the computation to avoid the errors that arise from the inevitable aliasing due to sampling the function at equally spaced points. As we shall show, this rate can be reasonably well estimated from the appearance of the function—the singularities along the real axis control the rate of the convergence, rather than, as in the polynomial case, the size of the higher-order derivatives, depending on the location of the singularities in the complex plane.

Second, we shall examine methods for improving the rate of convergence, so that either lower sampling rates can be used or a higher accuracy can be obtained using the same rate. Thus a great deal of machine time can often be saved. We shall give two approaches to this important topic, one applicable before numerically finding the Fourier coefficients A_k and B_k and the other after they are found.

32.2   STRAIGHT-LINE APPROXIMATION

Many times the function we wish to use can be approximated by a sequence of straight lines.   We examine two cases: first the lines are continuous (remember that the function is periodic and that the two end values are the same point), and second the function is approximated by a broken straight line.   We shall suppose that in the ith interval the function is approximated by (see Fig. 32.2.1)

$$F(x) = \alpha_i + \beta_i x \qquad x_{i-1} < x < x_i$$
$$i = 1, 2, \ldots, K$$

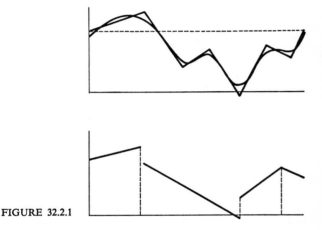

FIGURE 32.2.1

We have in the first case

$$a_k = \frac{1}{\pi} \int_0^{2\pi} F(x) \cos kx \, dx = \frac{1}{\pi} \sum_{i=1}^{K} \int_{x_{i-1}}^{x_i} (\alpha_i + \beta_i x) \cos kx \, dx$$

$$= \frac{1}{\pi} \sum_{i=1}^{K} \left[(\alpha_i + \beta_i x) \frac{\sin kx}{k} \Big|_{x_{i-1}}^{x_i} - \int_{x_{i-1}}^{x_i} \beta_i \frac{\sin kx}{k} \, dx \right]$$

Due to continuity the integrated part drops out, and we get

$$a_k = -\frac{1}{\pi} \sum_{i=1}^{K} \beta_i \frac{\cos kx}{k^2} \Big|_{x_{i-1}}^{x_i}$$

Let M_1 be the maximum of the $|\beta_i|$.   Then

$$|a_k| < \frac{M_1}{\pi} \frac{2K}{k^2} = \frac{2M_1}{k^2} \frac{K}{\pi}$$

Similarly for the b_k.   Thus the series converges like $1/k^2$.

In the second case, the broken straight line, the integrated part does *not* drop out, and we have convergence like $1/k$.

Thus we see that the behavior along the real axis tells us the rate of convergence. For practical purposes the converse is true; if the coefficients fall off like $1/k$, then there are one or more simple discontinuities; if they fall off like $1/k^2$, then there are no discontinuities.[1]

PROBLEMS

Find the rate of convergence of the Fourier series for

32.2.1 $F(t) = \begin{cases} 0 & 0 < t < \pi \\ 1 & \pi < t < 2\pi \end{cases}$

32.2.2 $F(t) = t \qquad -\pi < t < \pi$

32.2.3 $F(t) = |t| \qquad -\pi < t < \pi$

32.2.4 $F(t) = \begin{cases} 0 & 0 < t < a \\ 1 & a < t < 2\pi \end{cases}$

32.3 FUNCTIONS HAVING CONTINUOUS HIGHER DERIVATIVES

If a function has a continuous first derivative and, except for a finite number of places, a continuous second derivative, then we can proceed as in Sec. 32.2 except that we integrate by parts twice

$$a_k = \frac{1}{\pi} \int_{-\pi}^{\pi} F(t)\cos kt \, kt = -\frac{1}{\pi k} \int_{-\pi}^{\pi} F'(t)\sin kt \, dt$$

the continuity causing, as before, the integrated part to cancel out. Thus the assumption of straight lines is not essential to the argument of Sec. 32.2. We again break up the range into pieces in which $F''(t)$ is continuous

$$a_k = -\frac{1}{\pi k} \sum_{i=0}^{K} \int_{t_{i-1}}^{t_i} F'(t)\sin kt \, dt$$

$$= -\frac{1}{\pi k^2} \sum_{i=0}^{K} \int_{t_{i-1}}^{t_i} F''(t)\cos kt \, dt$$

$$|a_k| \le \frac{M_2}{\pi k^2} 2K = \frac{2M_2}{k^2} \frac{K}{\pi}$$

where M_2 is the maximum of $|F''(t)|$. We treat b_k similarly.

[1] G. Raisbeck, Order of Magnitude of Fourier Coefficients, *Am. Math. Monthly*, vol. 62, pp. 149–154, March, 1955.

If the coefficients fall off like $1/k^2$, then the Fourier series converges everywhere since

$$|a_k \cos kt + b_k \sin kt| \le \frac{4M_2}{k^2}$$

The same pattern clearly applies to functions having continuous mth derivatives, except at a finite number of points, the coefficients fall off like $1/k^m$. If it has a continuous mth derivative, the coefficients fall off like $1/k^{m+1}$ (m an integer).

PROBLEMS

Find the Fourier series for:

32.3.1 $F(t) = \begin{cases} t(\pi - t) & 0 \le t \le \pi \\ -f(-t) & 0 \ge t \ge -\pi \end{cases}$

32.3.2 $F(t) = 1 - t^2/\pi^2 \qquad -\pi \le t \le \pi$

32.3.3 $F(t) = x(1 - x^2) \qquad -1 \le x \le 1$

32.3.4   Discuss the rate of convergence of
$$y(t) = 3t^5 - 10t^3 + 7t \qquad -1 \le t \le 1$$

32.4 IMPROVING THE CONVERGENCE

Given a function to expand in a Fourier series, we examine the discontinuities in the function and in its derivatives.   Suppose, for example, we have the common case of a discontinuity in the function at the ends of the interval.   This discontinuity will control the rate of convergence and make it rather slow.   If we were to subtract the function (see Fig. 32.4.1)

$$y(x) = a + bx$$

with the value of b picked so that the difference between the given function $F(x)$ and $y(x)$ is continuous at the endpoints, then the convergence of the result would be more rapid.

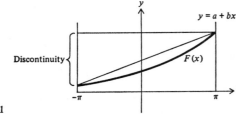

FIGURE 32.4.1

And this is a general approach.   We can subtract simple functions, typically polynomials but not necessarily only these, so that the difference has as much smoothness as we wish, and hence the resulting Fourier series will converge as rapidly as we please—thus minimizing the aliasing troubles that occur due to the sampling rate used.   For this purpose we need some standard functions with various singularities.

As an example of such a standard function, suppose we have the function

$$y(t) = t/2\pi \qquad -\pi < t < \pi$$

which has a discontinuity of unit height at $t = \pi$.   It is easy to see that

$$y(t) = -y(-t)$$

so that the $a_k = 0$.   The coefficients of the sine terms are

$$
\begin{aligned}
b_k &= \frac{1}{\pi} \int_{-\pi}^{\pi} y(t)\sin kt \, dt = \frac{2}{\pi} \int_{0}^{\pi} \frac{t}{2\pi} \sin kt \, dt \\
&= \frac{1}{\pi^2} t \left(-\frac{\cos kt}{k} \right) \Big|_{0}^{\pi} + \frac{1}{k\pi^2} \int_{0}^{\pi} \cos kt \, dt \\
&= \frac{\pi}{\pi^2 k} [(-1)\cos \pi k] = \frac{(-1)^{k-1}}{k\pi}
\end{aligned}
$$

Hence

$$y(t) = \frac{1}{\pi}\left(\sin t - \frac{\sin 2t}{2} + \frac{\sin 3t}{3} - \frac{\sin 4t}{4} + \cdots \right)$$

This series converges to $y(t)$ for all values of t *except* at the discontinuity $(t = \pi)$ where clearly it converges to zero (the average of the two values $1/2$ and $-1/2$ which are the limit values as $t \to \pi$ or $t \to -\pi$ from the origin).   See Fig. 32.4.2.

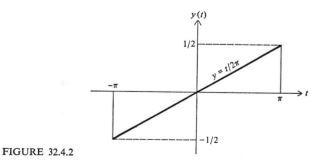

FIGURE 32.4.2

If we set

$$t = t' - a + \pi$$

we translate the discontinuity to $t' = a$. In place of the original series we have (dropping the prime on t) for a function $y(t)$ having a unit jump at $t = a$

$$y(t) = \frac{1}{\pi} \sum_{k=1}^{\infty} \frac{(-1)^{k-1}}{k} \sin[(kt - ka) + k\pi]$$

$$= \frac{1}{\pi} \sum_{k=1}^{\infty} \frac{(-1)^{k-1}}{k} \sin(kt - ka)\cos k\pi = \frac{1}{\pi} \sum_{k=1}^{\infty} \frac{\sin(ka - kt)}{k}$$

$$= \frac{1}{\pi} \sum_{k=1}^{\infty} \frac{\sin ka}{k} \cos kt - \frac{1}{\pi} \sum_{k=1}^{\infty} \frac{\cos ka}{k} \sin kt$$

In this Fourier series the coefficients again fall off as $1/k$.

If we now consider a function with a finite number M of simple discontinuities and with straight lines between the discontinuities, then if the discontinuity at $t = a_i$ is of size y_i, we take the appropriate linear combination of the expansions (including possibly an a_0 term)

$$y(t) = \frac{1}{\pi} \sum_{i=1}^{M} y_i \left(\sum_{k=1}^{\infty} \frac{\sin ka_i}{k} \cos kt - \sum_{k=1}^{\infty} \frac{\cos ka_i}{k} \sin kt \right)$$

$$= \frac{1}{\pi} \sum_{k=1}^{\infty} \left(\sum_{i=1}^{M} y_i \sin ka_i \right) \frac{\cos kt}{k} - \frac{1}{\pi} \sum_{k=1}^{\infty} \left(\sum_{i=1}^{M} y_i \cos ka_i \right) \frac{\sin kt}{k}$$

where we have formally interchanged the summation processes.

Similar expansions can easily be found for functions having other types of discontinuities and need not be given here.

PROBLEMS

32.4.1 Find an expansion for a unit discontinuity in the derivative at $t = a$.

32.4.2 Do the same as in Prob. 32.4.1 except for the second derivative.

32.5 THE GIBBS PHENOMENON[1]

In order to understand better what happens at a discontinuity, we begin with the special case of a rectangular wave $H(t)$ of period 2π (see Fig. 32.5.1). If we compute the partial sum of the first $2n$ terms, the cosine terms are all zero, and we have (in the continuous case)

$$H_{2n}(t) = \frac{1}{2} + \frac{2}{\pi} \sum_{k=1}^{n} \frac{1}{2k-1} \sin(2k-1)t$$

---

[1] Much of this is taken from a Bell Telephone Laboratories memorandum by R. G. Segers. See also Ref. 33.

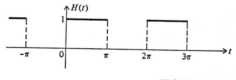

FIGURE 32.5.1

Gibbs (in 1899) pointed out that the partial sum H_{2n} overshoots the function by a certain amount (see Fig. 32.5.2). More precisely

$$H_{2n}\left(\frac{\pi}{2n}\right) \to 1.08949 \cdots \qquad \text{as } n \to \infty$$

Indeed, not only does $H_{2n}(t)$ overshoot the function $H(t)$, but it also tends to oscillate about $H(t)$, and the oscillations decrease slowly as we move away from the discontinuity.

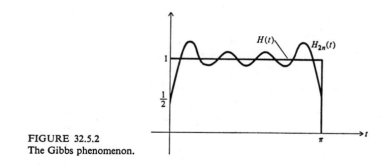

FIGURE 32.5.2
The Gibbs phenomenon.

In order to explain the phenomenon, we write the expansion as

$$H_{2n}(t) = \frac{1}{2} + \frac{2}{\pi} \sum_{k=1}^{n} \int_0^t \cos(2k-1)x \, dx$$

$$= \frac{1}{2} + \frac{2}{\pi} \int_0^t \sum_{k=1}^{n} \cos(2k-1)x \, dx$$

$$= \frac{1}{2} + \frac{1}{\pi} \int_0^t \frac{\sin 2nx}{\sin x} \, dx$$

where we have used

$$\sum_{k=1}^{n} \cos(2k-1)x = \frac{\sin 2nx}{2 \sin x}$$

From this it is clear that the maxima and minima occur (for $0 \leq t \leq \pi$) at the points where the derivative of $H_{2n}(t) = 0$, namely,

$$\frac{dH_{2n}(t)}{dt} = \frac{1}{\pi} \frac{\sin 2nt}{\sin t} = 0$$

which gives

$$t = \frac{m\pi}{2n} \qquad m = 1, 2, \ldots, 2n - 1$$

and that they alternate. The size of these has been computed by Carslaw [7].

What is true for this special function is clearly true for more general functions since the discontinuity can be regarded as coming from a rectangular wave *added* to a smooth function.

32.6 LANCZOS' σ FACTORS

As our second approach to improving the convergence, we propose to replace the rapidly oscillating function $H_{2n}(t)$ by a smoothed function

$$\overline{H_{2n}}(t) = \frac{n}{\pi} \int_{t-\pi/2n}^{t+\pi/2n} H_{2n}(s)\, ds$$

where we have averaged over one complete oscillation of $H_{2n}(t)$ centered about t. We now find the effect of the smoothing.

Using the earlier expansion of the function $H_{2n}(t)$, we get

$$
\begin{aligned}
\overline{H_{2n}}(t) &= \frac{n}{\pi} \int_{t-\pi/2n}^{t+\pi/2n} \left[\frac{1}{2} + \frac{2}{\pi} \sum_{k=1}^{n} \frac{1}{2k-1} \sin(2k-1)s \right] ds \\
&= \frac{n}{\pi} \left[\frac{\pi}{2n} + \frac{2}{\pi} \sum_{k=1}^{n} \frac{-1}{(2k-1)^2} \cos(2k-1)s \Big|_{t-\pi/2n}^{t+\pi/2n} \right] \\
&= \frac{1}{2} + \frac{2}{\pi} \sum_{k=1}^{n} \frac{1}{2k-1} \frac{\sin[(2k-1)(\pi/2n)]}{(2k-1)(\pi/2n)} \sin(2k-1)t
\end{aligned}
$$

If we compare this with $H_{2n}(t)$, we find that *we now have the extra factor in the expansion*

$$\sigma_{2k-1} = \frac{\sin(2k-1)(\pi/2n)}{(2k-1)(\pi/2n)}$$

for each term in the summation.

The effect of this factor σ_k is to reduce the limiting value of the maximum overshoot from 0.08949 to 0.01187 and the first minimum from 0.04859 to 0.00473, etc. Thus the Gibbs phenomenon has been greatly reduced by putting in the σ factors which arose from the smoothing of $H_{2n}(t)$ over a short interval of length π/n.

32.7 THE σ FACTORS IN THE GENERAL CASE

We have studied the special case of a rectangular wave; we now show that the weight factors σ_k are the same for any Fourier series.

Let $f(t)$ $(0 \leq t \leq 2\pi)$ be integrable, and let

$$a_k = \frac{1}{\pi} \int_0^{2\pi} f(t)\cos kt \, dt \qquad k = 0, 1, 2, \ldots$$

$$b_k = \frac{1}{\pi} \int_0^{2\pi} f(t)\sin kt \, dt \qquad k = 1, 2, \ldots$$

be the Fourier coefficients[1]

$$f_n(t) = \frac{a_0}{2} + \sum_{k=1}^{n-1} (a_k \cos kt + b_k \sin kt) + \frac{a_n}{2} \cos nt$$

In the previous special case we had frequencies up to $2n$; this time we have chosen those up to n only, and so averaging over the interval $(t - \pi/n, t + \pi/n)$, we get

$$\hat{f}_n(t) = \frac{1}{2\pi/n} \int_{t-\pi/n}^{t+\pi/n} f_n(s) \, ds$$

$$= \frac{n}{2\pi} \left[\frac{a_0}{2} \frac{2\pi}{n} + \sum_{k=1}^{n-1} \left(a_k \frac{\sin ks}{k} - b_k \frac{\cos ks}{k} \right) \bigg|_{t-\pi/n}^{t+\pi/n} + \frac{a_n}{2} \frac{\sin ns}{n} \bigg|_{t-\pi/n}^{t+\pi/n} \right]$$

$$= \frac{a_0}{2} + \frac{n}{2\pi} \sum_{k=1}^{n-1} \left\{ \frac{a_k}{k} \left[\sin k\left(t + \frac{\pi}{n}\right) - \sin k\left(t - \frac{\pi}{n}\right) \right] \right.$$

$$\left. - \frac{b_k}{k} \left[\cos k\left(t + \frac{\pi}{n}\right) - \cos k\left(t - \frac{\pi}{n}\right) \right] \right\}$$

$$+ \frac{a_n}{2n} \left[\sin n\left(t + \frac{\pi}{n}\right) - \sin n\left(t - \frac{\pi}{n}\right) \right]$$

$$= \frac{a_0}{2} + \frac{n}{2\pi} \sum_{k=1}^{n-1} \left(\frac{a_k}{k} 2 \cos kt \sin \frac{k\pi}{n} + \frac{b_k}{k} \cdot 2 \sin kt \sin \frac{k\pi}{n} \right)$$

$$+ \frac{a_n}{2n} 2 \cos nt \sin \frac{n\pi}{n}$$

$$= \frac{a_0}{2} + \sum_{k=1}^{n-1} \frac{\sin(k\pi/n)}{k\pi/n} (a_k \cos kt + b_k \sin kt)$$

[1] If we use the conventional form without 1/2 the last term, the result is the same.

the last term a_n having a zero coefficient.   Using the same σ factors (notice we now are using n in place of $2n$)

$$\sigma_k(n) = \frac{\sin(k\pi/n)}{k\pi/n}$$

we get, as before, the smoothed values

$$\hat{f}_n(t) = \frac{a_0}{2} + \sum_{k=1}^{n-1} \sigma_k(a_k \cos kt + b_k \sin kt)$$

Notice that $\sigma_k(n)$ is a function of n and that

$$\sigma_n(n) = \frac{\sin(n\pi/n)}{n\pi/n} = 0$$

Thus to smooth the series, we merely apply the $\sigma_k(n)$ factors (which taper from $\sigma_0 = 1$ to $\sigma_n = 0$) to both the sine and cosine terms.   The effect of the σ factors is to greatly smooth the truncated Fourier series.

32.8   A COMPARISON OF CONVERGENCE METHODS

Besides the σ-factor method of Lanczos there is the well-known Fejer method of using the arithmetic means of the partial sums.   Fejer's method completely eliminates the oscillation, while Lanczos' method greatly dampens it.   As a result, Fejer's method pays a large price in producing, for finite n, a slow-rise time for the partial sum near the discontinuity.   Fig. 32.8.1 shows the appropriate comparison of the

Fourier series
Fejer sum
Lanczos σ-factor curves

for a 12-term approximation to a rectangular wave.   The advantages of the σ-factor method are clearly apparent.

Fig. 32.8.2 shows the rapidity with which the σ-factor method approaches the rectangular wave as a function of n.   Thus even for moderate n the "rise time" of the curve is very short.

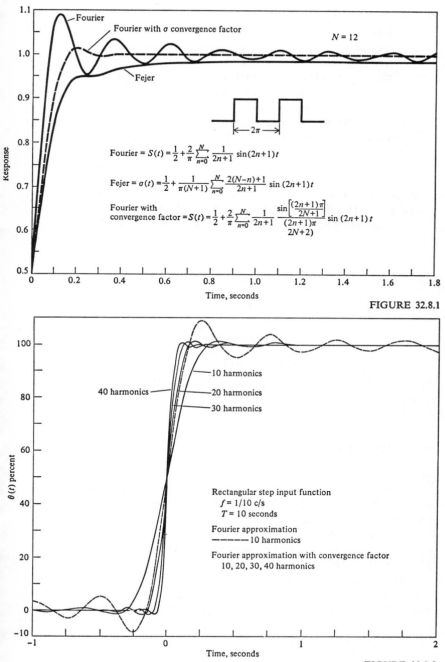

Fourier $= S(t) = \dfrac{1}{2} + \dfrac{2}{\pi} \displaystyle\sum_{n=0}^{N} \dfrac{1}{2n+1} \sin(2n+1)t$

Fejer $= \sigma(t) = \dfrac{1}{2} + \dfrac{1}{\pi(N+1)} \displaystyle\sum_{n=0}^{N} \dfrac{2(N-n)+1}{2n+1} \sin(2n+1)t$

Fourier with
convergence factor $= S(t) = \dfrac{1}{2} + \dfrac{2}{\pi} \displaystyle\sum_{n=0}^{N} \dfrac{1}{2n+1} \dfrac{\sin\left[\dfrac{(2n+1)\pi}{2N+1}\right]}{\dfrac{(2n+1)\pi}{2N+2)}} \sin(2n+1)t$

FIGURE 32.8.1

Rectangular step input function
$f = 1/10$ c/s
$T = 10$ seconds

Fourier approximation
– – – – 10 harmonics

Fourier approximation with convergence factor
10, 20, 30, 40 harmonics

FIGURE 32.8.2

32.9   LANCZOS' DIFFERENTIATION TECHNIQUE

It is sometimes necessary to differentiate a Fourier series.   But if we do so, we usually get what is not wanted in the physical problem, namely, a large high-frequency oscillation.

Lanczos [33] introduced a practical approach, namely, that when we differentiate a truncated Fourier-series approximation, we use

$$\frac{f(t + \pi/n) - f(t - \pi/n)}{2\pi/n}$$

as an estimate of the derivative of $f(t)$ at the point t (instead of the usual limit process involved in taking the derivative which has dubious physical significance). Note that the same n appears both in the formula and in the order of the truncated series

$$f(t) = \frac{a_0}{2} + \sum_{k=1}^{n} (a_k \cos kt + b_k \sin kt)$$

$$\frac{f(t + \pi/n) - f(t - \pi/n)}{2\pi/n} = \sum_{k=1}^{n} \left\{ a_k \left[\cos k\left(\frac{t + \pi}{n}\right) - \cos k\left(\frac{t - \pi}{n}\right) \right] \right.$$

$$\left. + b_k \left[\sin k\left(\frac{t + \pi}{n}\right) - \sin k\left(\frac{t - \pi}{n}\right) \right] \right\} \frac{n}{2\pi}$$

$$= \sum (-ka_k \sin kt + kb_k \cos kt)\sigma_k$$

Thus the same σ factors as before, put in the formally differentiated series, give the appropriate result.

32.10   SUMMARY

We have given two methods for improving the convergence of a Fourier series. The method of singularities subtracts out the troubles that cause the slow convergence before the series is evaluated.   Thus the numerical computation on the difference has a rapidly converging Fourier series, and the aliasing that arises from the sampling does not cause serious trouble.   The analytical results from the analytical expansions supply the other coefficients and keep the aliasing from falling into the lower frequencies.

The second method of σ_k factors applies to series *after* they are known and damps down the Gibbs phenomena at discontinuities.   It is chiefly applicable to analytical expansions, such as those we derived above and used to subtract out the singularities.

33

THE FAST FOURIER TRANSFORM

33.1 THE DIRECT CALCULATION

The Fourier coefficients of a function $F(t)$, and more often the power spectra, are used in many applications. The coefficients may be found directly from the formulas

$$a_k = \frac{1}{\pi} \int_0^{2\pi} F(t)\cos kt\, dt \qquad k = 0, 1, 2, \ldots$$

$$b_k = \frac{1}{\pi} \int_0^{2\pi} F(t)\sin kt\, dt \qquad k = 1, 2, 3, \ldots$$

From Sec. 31.8 we see that the equally spaced, equally weighted samples provide an excellent way of evaluating these integrals in case they cannot be done in closed form. They are effectively the Gaussian quadrature formulas and have, therefore, the corresponding accuracy associated with such formulas.

Another approach to finding the power spectrum of a function is to take the equally spaced samples (with the consequent aliasing) and fit a finite Fourier series to them. A little thought shows that these two, apparently alternate, approaches are equivalent.

The problem of aliasing appears more complicated than it is in reality. Suppose, for example, that the function has a discontinuity at the ends of the interval. This could be removed by subtracting a straight line. If then we do the sampling, there will be less aliasing due to the sampling than before because the series now converges more rapidly.

On the other hand, suppose we sample first and then from the discrete samples remove the (same) straight line.   The numbers we operate upon are the same in the two cases.   How do we reconcile this with the claim that once the aliasing occurs it cannot be undone?   The answer lies in understanding properly "the aliasing cannot be undone."   Clearly, if we have a frequency above the folding frequency and then sample, it will appear as if it were in the band, and we cannot *know* that it is not the frequency inside the band.   On the other hand, we may at times do operations that cause further aliasing, and it can happen that sometimes these operations undo an earlier aliasing—it is simply that without further information there are aliasings that we cannot *know* will be undone.

33.2   INTRODUCTION TO THE FAST FOURIER TRANSFORM (FFT)

The fast Fourier transform was rediscovered and adequately publicized by Cooley and Tukey,[1] though it had been discovered a number of times before, and was to some extent understood in the general literature.   The explicit recognition of its importance is one of the few significant advances in numerical analysis in the past few decades.   Much has been learned about various aspects of computation, but no other recent discovery has so profoundly affected so many different fields as has the fast Fourier transform.

For N samples the direct calculation, as indicated in the previous section, would involve approximately N^2 multiplications and N^2 additions.   The fast Fourier transform has, under suitable circumstances, reduced the number of operations to something like $N \log N$.   For N small, say around a hundred or so, the change is not tremendous, but when N is around 10,000 as it is in many applications, the effect is to reduce the computing to below 1 percent of the direct method and to make what were totally impractical computations now a matter of routine processing.

The fast Fourier transform is still as of 1971 in a state of rapid evolution with numerous papers appearing each year.   Thus the treatment we give must be limited to the basics, and cannot be definitive.   Before committing oneself to a

---

[1] J. W. Cooley and J. W. Tukey, An Algorithm for the Machine Calculation of Complex Fourier Series, *Math. of Comp.*, vol. 19, pp. 297–301, April, 1965.

large project using the fast Fourier transform, an examination of the current literature should be made (probably the *IEEE Transactions on Electronic Computers* should be examined first).

33.3 THE CENTRAL IDEA OF THE FAST FOURIER TRANSFORM

The fast Fourier transform is based on a method of factoring the transform into the product of two transforms—a product in the sense that one transform *follows* the other. As a result of the factoring, if we have M data points and if M can be factored into

$$M = GH$$

then in place of M^2 multiplications and additions we get approximately

$$M(G + H)$$

operations of each type. Repeated applications of the factoring leads to the following result. If

$$M = m_1 m_2 \cdots m_k$$

then we will have approximately

$$M(m_1 + m_2 + \cdots + m_k)$$

operations. In the most favorable case, when M is a power of 2, say 2^k, we have

$$M(2k)$$

operations, where $k = \log_2 M$. Thus we have approximately $M \log M$ operations in place of M^2 operations. In other cases, where M has many small factors, somewhat the same effect of greatly decreasing the number of operations happens.

It is easiest to demonstrate the way the factorization occurs in terms of complex notation, where $M = GH$ and the coefficient is given by

$$C_k \equiv C(k) = \frac{1}{GH} \sum_{p=0}^{M-1} F(x_p) e^{-2\pi i k x_p}$$

$$= \frac{1}{GH} \sum_{p=0}^{GH-1} F\left(\frac{Lp}{M}\right) e^{-2\pi i k L p/M}$$

using $x_p = Lp/M (p = 0, 1, \ldots, M - 1)$. For convenience we set $L = 1$.

The heart of the method is writing the two indices k and p in a suitable way. Divide k by G to get

$$k = k_1 G + k_0$$

Similarly divide p by H to get

$$p = p_1 H + p_0$$

where, of course, $k_0 < G$, $k_1 < H$, $p_0 < H$, and $p_1 < G$. With this notation the formula becomes (the double sum covers the *same numbers* as the original single sum)

$$C(k_1 G + k_0) = \frac{1}{GH} \sum_{p_0=0}^{H-1} \sum_{p_1=0}^{G-1} F\left(\frac{p_1 H + p_0}{GH}\right) e^{-2\pi i (k_1 G + k_0)(p_1 H + p_0)/(GH)}$$

$$= \frac{1}{GH} \sum_{p_0=0}^{H-1} e^{-2\pi i k_0 p_0/GH} e^{-2\pi i k_1 p_0/H} \left[\sum_{p_1=0}^{G-1} F\left(\frac{p_1}{G} + \frac{p_0}{GH}\right) e^{-2\pi i k_0 p_1]G}\right]$$

where we have used the fact that

$$e^{-2\pi i k_1 p_1} \equiv 1$$

and suppressed those terms.

The square brackets effectively contain the Fourier coefficient of the expansion of $1/H$ of the samples, phase shifted $p_0/(GH)$. There are H such sums to be done, one for each p_0. Let us label these sums

$$\hat{C}(k_0, p_0) = \sum_{p_1=0}^{G-1} F\left(\frac{p_1}{G} + \frac{p_0}{GH}\right) e^{-2\pi i k_0 p_1/G}$$

and we have

$$C(k_1 G + k_0) = \frac{1}{GH} \sum_{p_0=0}^{H-1} \hat{C}(k_0, p_0) e^{-2\pi i [k_0/(GH) + k_1/H] p_0}$$

which are the coefficients of a second Fourier expansion, this time of the $\hat{C}(k_0, p_0)$, and there are H terms to be evaluated in the sum.

A count of the operations shows that they are proportional to

$$GH(G + H)$$

as claimed.

33.4   THE FAST FOURIER TRANSFORM IN PRACTICE

At first it was believed that using the factors of 2 as often as possible was the best way of implementing the fast Fourier transform. Then it was realized that when there was a pair of factors of 2, it was best handled as a factor of 4, because of the fact that the values of the trigonometric functions were all 0s and 1s. Thus it was advocated that the G (or H) be taken as 4 whenever possible, and if there was an odd number of 2s, then only one stage should use a factor of 2. Clearly the factor of 4 uses only additions and subtractions.

More recently a radix-8 fast Fourier transform has been recommended. However, the situation is still in a state of flux and the final word is not in. For further references see the sources cited below.[1]

Because of the importance of taking Fourier transforms, the fast Fourier transform has been implemented in hardware directly in a number of cases, and is reasonably available. The hardware implementation makes further spectacular improvements in speed (and cost) of the computation.

33.5 DANGERS OF THE FOURIER TRANSFORM

The fast Fourier transform calculates (within roundoff) the same quantities as does the direct method, so that the troubles we shall discuss are not functions of the fast Fourier transform but rather of the Fourier transform itself.

First, there is the ever present problem of *aliasing* (Secs. 31.2 and 31.6).

Second, there is the problem of *leakage* due to the use of a finite length of record. The effect on a single spike is that it is replaced by a $(\sin x)/x$ ripple, and on a continuous spectrum, a large peak tends to "leak out" into the nearby parts of the spectrum (see Sec. 34.8).

Finally, there is the *picket-fence effect*. Each value of the sample spectrum acts like a spike, and the sum of the $(\sin x)/x$ terms, each shifted to its proper place, gives the spectrum the appearance of a sort of picket-fence effect of equally spaced ripples. This is noticeable if other than the properly spaced frequencies are computed.

These effects will not be discussed in the text at greater length because of the theoretical difficulties involved. We mention them here only as a matter of caution.

33.6 FOURIER ANALYSIS USING 12 POINTS

Because the fast Fourier transform is often so obscure when first seen, we give the classic, closely related 12-point case for the Fourier series. In this example the reduction is made to follow from the periodicity of the trigonometric functions, which is in fact what the fast Fourier transform actually uses. But since it is in the real domain, it may be easier for some to follow—for others because of the clumsy symbolism it may be harder!

---

[1] G. D. Bergland, A Guided Tour of the Fast Fourier Transform, *IEEE Spectrum*, July, 1969, pp. 41–52. W. M. Gentleman and G. Sande, Fast Fourier Transforms—For Fun and Profit, *Proc. FJCC*, 1966, pp. 563–578.

The special case of 12 points ($N = 6$) is of interest because it occurs frequently and is easy to do, even by hand!   We have

$$6a_k = \sum_{x=0}^{11} f(x)\cos\frac{\pi}{6}kx$$

$$6b_k = \sum_{x=0}^{11} f(x)\sin\frac{\pi}{6}kx$$

If we break up each sum into the ranges 0 to 6 and 7 to 11 in x, and in the second range put $x = 12 - x'$, we get

$$6a_k = \sum_{x=0}^{6} f(x)\cos\frac{\pi}{6}kx + \sum_{x'=1}^{5} f(12 - x')\cos\frac{\pi}{6}kx'$$

$$6b_k = \sum_{x=0}^{6} f(x)\sin\frac{\pi}{6}kx - \sum_{x'=1}^{5} f(12 - x')\sin\frac{\pi}{6}kx'$$

This suggests writing

	$f(0)$	$f(1)$	$f(2)$	$f(3)$	$f(4)$	$f(5)$	$f(6)$
		$f(11)$	$f(10)$	$f(9)$	$f(8)$	$f(7)$	
Add	$s(0)$	$s(1)$	$s(2)$	$s(3)$	$s(4)$	$s(5)$	$s(6)$
Subtract		$t(1)$	$t(2)$	$t(3)$	$t(4)$	$t(5)$	

$$6a_k = \sum_{x=0}^{6} s(x)\cos\frac{\pi}{6}kx$$

$$6b_k = \sum_{x=1}^{5} t(x)\sin\frac{\pi}{6}kx$$

We again break up the range into two parts, this time 0 to 3 and 4 to 6, setting $x = 6 - x'$ in the second range.

$$6a_k = \sum_{x=0}^{3} s(x)\cos\frac{\pi}{6}kx + (-1)^k\sum_{x'=0}^{2} s(6 - x')\cos\frac{\pi}{6}kx'$$

$$6b_k = \sum_{x=1}^{3} t(x)\sin\frac{\pi}{6}kx - (-1)^k\sum_{x'=1}^{2} t(6 - x')\sin\frac{\pi}{6}kx'$$

This suggests writing

	$s(0)$	$s(1)$	$s(2)$	$s(3)$	$t(1)$	$t(2)$	$t(3)$
	$s(6)$	$s(5)$	$s(4)$		$t(5)$	$t(4)$	
Add	$u(0)$	$u(1)$	$u(2)$	$u(3)$	$p(1)$	$p(2)$	$p(3)$
Subtract	$v(0)$	$v(1)$	$v(2)$		$q(1)$	$q(2)$	

The result, when written out, is

$$6a_0 = u(0) + u(1) + u(2) + u(3) \quad = [u(0) + u(3)] + [u(1) + u(2)]$$

$$6a_1 = v(0) + \frac{\sqrt{3}}{2}v(1) + \frac{1}{2}v(2) \quad = \left[v(0) + \frac{1}{2}v(2)\right] + \frac{\sqrt{3}}{2}v(1)$$

$$6a_2 = u(0) + \tfrac{1}{2}[u(1) - u(2)] - u(3) = [u(0) - u(3)] + \tfrac{1}{2}[u(1) - u(2)]$$

$$6a_3 = v(0) - v(2)$$

$$6a_4 = u(0) - \tfrac{1}{2}u(1) - \tfrac{1}{2}u(2) + u(3) = [u(0) + u(3)] - \tfrac{1}{2}[u(1) + u(2)]$$

$$6a_5 = v(0) - \frac{\sqrt{3}}{2}v(1) + \frac{1}{2}v(2) \quad = \left[v(0) + \frac{1}{2}v(2)\right] - \frac{\sqrt{3}}{2}v(1)$$

$$6a_6 = u(0) - [u(1) - u(2)] - u(3) = [u(0) - u(3)] - [u(1) - u(2)]$$

$$6b_1 = \frac{1}{2}p(1) + \frac{\sqrt{3}}{2}p(2) + p(3) \quad = \left[\frac{1}{2}p(1) + p(3)\right] + \frac{\sqrt{3}}{2}p(2)$$

$$6b_2 = \frac{\sqrt{3}}{2}[q(1) + q(2)]$$

$$6b_3 = p(1) - p(3)$$

$$6b_4 = \frac{\sqrt{3}}{2}[q(1) - q(2)]$$

$$6b_6 = \frac{1}{2}p(1) - \frac{\sqrt{3}}{2}p(2) + p(3) \quad = \left[\frac{1}{2}p(1) + p(3)\right] - \frac{\sqrt{3}}{2}p(2)$$

In the process we do less than 60 arithmetic operations, and most of them are simple additions. This compares favorably with the $6N^2 = 6^3 = 216$ needed by the direct method of Sec. 33.1. The programming, of course, is a bit longer to write, but is very easy since there is no complex logic and since only one difficult number $\sqrt{3}/2 = 0.8660254075$ is required, and this only four times. It is remarkably easy to do by hand.

These general methods can be, and have been, applied in cases other than $N = 6$.

The central idea in both Secs. 33.3 and 33.6 is to cause the additions to occur *before* the multiplications whenever possible.

PROBLEMS

33.6.1 If $f(x) = x(12 - x)$ for $x = 0, 1, \ldots, 11$, find, by hand, the Fourier expansion of $f(x)$.

33.6.2 Do the same for $f(x) = \begin{cases} 0 & 0 \leq x < 6 \\ 1 & 7 \leq x < 12 \end{cases}$

33.6.3 Carry out an analysis using the midpoints t_p.

33.7 COSINE EXPANSIONS

If we assume that the function $f(x)$ is periodic of period $2N$, then whether we use $x = 0, 1, \ldots, 2N - 1$ or the sequence $-(N - 1), -(N - 2), \ldots, 0, 1, \ldots, N$, the result is the same.   Now suppose that we have a function defined for $x = 0, 1, \ldots, N$ and we wish to expand it in a series of cosine terms alone.   Let us *define* $f(-x) \equiv f(x)$ for $x = 1, 2, \ldots, N - 1$.   The function $f(x)$ is now an *even function*.   When we make the Fourier expansion, we find that all the $b_k = 0$, since

$$b_k = \frac{1}{N} \sum_{x=-N+1}^{N} f(x)\sin \frac{\pi k}{N} x$$

and the sine is an odd function.   Thus we have a cosine expansion.   The formulas for the a_k can be simplified:

$$a_k = \frac{1}{N} \sum_{x=-N+1}^{N-1} f(x)\cos \frac{\pi k}{N} x$$

$$= \frac{2}{N} \left[\frac{1}{2}f_0 + \sum_{x=1}^{N-1} f(x)\cos \frac{\pi k}{N} x + \frac{1}{2}f_N(-1)^k \right]$$

By defining $f(x)$ to be an odd function, we can get a sine expansion in a similar fashion.

We also note that if we have a range $0 \leq x \leq R$, the substitution

$$x = \frac{R}{2}(1 + \cos t)$$

makes the function periodic in t as well as an even function of t.   This substitution has a number of advantages in matters of speed of convergence.

PROBLEMS

33.7.1 If $f(x)$ is defined for the interval $0 \leq x \leq \pi/2$, show how the function may be extended to $-\pi < x \leq \pi$ so that only odd cosine harmonics occur; only even cosine harmonics.

33.7.2 Discuss in detail the problem of finding a Fourier series in sines only; in odd (or even) sine harmonics.

33.8 LOCAL FOURIER SERIES

In many problems the phenomena change slowly, and the idea of a "local Fourier series" whose coefficients change slowly with time is quite natural.   It is also natural to center the range about the origin and use the values from $-N + 1$ to N

rather than from 0 to $2N - 1$. Thus expanding about the value t, we have the coefficients

$$a_m(t) = \frac{1}{N} \sum_{x=-N+1}^{N} f(x + t)\cos\frac{\pi}{N} mx \qquad m = 0, 1, \ldots, N$$

$$b_m(t) = \frac{1}{N} \sum_{x=-N+1}^{N} f(x + t)\sin\frac{\pi}{N} mx \qquad m = 1, 2, \ldots, N - 1$$

It is easy to obtain $a(t + 1)$ and $b(t + 1)$ from $a(t)$ and $b(t)$. For example,

$$a_m(t + 1) = \frac{1}{N} \sum_{x=-N+1}^{N} f(x + t + 1)\cos\frac{\pi}{N} mx$$

Set $1 + x = x'$, or $x = x' - 1$, and

$$a_m(t + 1) = \frac{1}{N} \sum_{x'=-N+2}^{N+1} f(x' + t)\cos\frac{\pi}{N} m(x' - 1)$$

$$= \frac{1}{N} \left[\sum_{x'=-N+1}^{N} f(x' + t)\cos\frac{\pi}{N} m(x' - 1) + f(N + 1 + t)\cos\frac{\pi}{N} m(N) \right.$$

$$\left. - f(-N + 1 + t)\cos\frac{\pi}{N} m(-N) \right]$$

$$= \frac{1}{N} \left\{ \sum_{x'=-N+1}^{N} f(x' + t)\cos\frac{\pi}{N} m(x' - 1) \right.$$

$$\left. + (-1)^m [f(N + 1 + t) - f(-N + 1 + t)] \right\}$$

Expanding the cosine term, we get

$$a_m(t + 1) = a_m(t)\cos\frac{\pi}{N} m + b_m(t)\sin\frac{\pi}{N} m$$

$$+ \frac{(-1)^m}{N} [f(N - 1 + t) - f(-N + 1 + t)]$$

Note that we use fixed (for a given frequency) multipliers $\cos \pi m/N$ and $\sin \pi m/N$ and add or subtract the two new values in the range.

Similarly for the sine term,

$$b_m(t + 1) = -a_m(t)\sin\frac{\pi}{\pi} m + b(t)\cos\frac{\pi}{\pi} m$$

PROBLEMS

33.8.1 Derive the formula for $b_m(t + 1)$.

34

THE FOURIER INTEGRAL : NONPERIODIC
FUNCTIONS

34.1 OUTLINE AND PURPOSE OF CHAPTER

In Chaps. 31 to 33 we approximated periodic functions by linear combinations of periodic functions—sines and cosines. Periodic functions are fairly rare in practice, and we now turn to the study of more general functions, functions which are not necessarily periodic. A very simple example of a nonperiodic function which is the sum of two periodic functions is

$$y(t) = \cos t + \cos(\sqrt{2}t)$$

Since 1 and $\sqrt{2}$ are not commensurate, this function cannot repeat itself exactly.

The idea that we can approximate a general (nonperiodic) function by a sum of periodic functions (sines and cosines) is no more unreasonable than the Taylor-series representation of a periodic function by means of nonperiodic terms, namely $1, x, x^2, \ldots$.

In the Fourier series we used an infinite but countable number of discrete frequencies to represent a periodic function. In the Fourier integral we shall use a noncountable number of frequencies to represent the function.   In the formal definition of the Fourier-integral representation of $f(t)$

$$f(t) = \int_{-\infty}^{\infty} F(\sigma)e^{2\pi i \sigma t}\, d\sigma$$

we have frequencies σ corresponding to *all* possible real numbers (not just integers).   In practice we will often find that the frequencies are confined to an interval and that $F(\sigma) = 0$ outside of the interval.   Such a function will be called *band-limited*.   As an example of a band-limited function, consider the usual hi-fi system which can transmit all the frequencies from some fairly low frequency to some upper limit around 20,000 Hz (cycles per second).[1]

The purpose of this chapter is to introduce the Fourier integral and a number of associated results that will enable us to answer questions like:

1 What is the effect of sampling at equally spaced points?

2 How can we reconstruct a band-limited function from its samples (the so-called sampling theorem) so that we can interpolate for the missing values and use it in the analytic substitution method?

3 What is the effect of taking a short run of samples from an infinitely long function?

These three questions are typical of those that cannot be intelligently discussed within the framework of polynomial approximation, and they serve to point out some of the advantages of the frequency approach when trying to understand what is happening in an experiment or a computation.

34.2 NOTATION

This section and the next are devoted to introducing the notation we shall use. They do not contain proofs of most of the statements made, and intuitive, but non-rigorous proofs will be supplied in later sections.[2]

The continuous Fourier series in the interval $-N < t \leq N$ can be written (Sec. 31.3 in a slightly changed notation)

$$f(t) = \frac{a_0}{2} + \sum_{k=1}^{\infty} a_k \cos \frac{\pi}{N} kt + \sum_{k=1}^{\infty} b_k \sin \frac{\pi}{N} kt$$

---

[1] It has been recently decided to use the name of a scientist in place of "cycles per second," in analogy with amperes being "coulombs per second," and Hertz was so honored.   Some of us consider it unfortunate that they did not choose Steinmetz instead of Hertz so that we could use his initials (Charles Proteus Steinmetz).

[2] We again refer the reader to Lighthill [37], and Lanczos' books [33–35] for more details; we cannot supply a course in the Fourier integral in one short chapter.

where

$$a_k = \frac{1}{N} \int_{-N}^{N} f(t)\cos\frac{\pi}{N}t\,dt \qquad k = 0, 1, 2, \ldots$$

$$b_k = \frac{1}{N} \int_{-N}^{N} f(t)\sin\frac{\pi}{N}t\,dt \qquad k = 1, 2, \ldots$$

or it can be written in the complex notation (Sec. 31.4)

$$f(t) = \sum_{k=-\infty}^{\infty} c_k e^{(\pi i/N)kt}$$

where

$$c_k = \frac{1}{2N} \int_{-N}^{N} f(t')e^{-(\pi i/N)kt'}\,dt'$$

and

$$c_k = \begin{cases} \dfrac{a_k - ib_k}{2} & k > 0 \\ a_0 & k = 0 \\ \dfrac{a_k + ib_k}{2} & k < 0 \end{cases}$$

The complex form of the Fourier series is, as we have already seen, much easier to handle in theoretical discussions, but of course we still compute with real numbers, and the complex notation is apt to conceal a great deal of computing. Since our use of the Fourier integral is mainly for understanding what is going on in a computation and for formal derivations of results, we shall use the complex form rather than the equivalent real form.

In the complex form both positive and negative frequencies occur, and this is apt to bother the beginner. The two arise formally from the fact that in the real case we have two functions at each positive frequency, the sine and the cosine, and we need two functions to maintain the linear independence at each frequency.

34.3   SUMMARY OF RESULTS

This section summarizes the results of the rest of the chapter and in general does not try to prove things.   First, we shall show that corresponding to the Fourier-series representation of a periodic function there is the Fourier-integral representation of a general function

$$f(t) = \int_{-\infty}^{\infty} F(\sigma)e^{2\pi i\sigma t}\,d\sigma$$

where

$$F(\sigma) = \int_{-\infty}^{\infty} f(t)e^{-2\pi i\sigma t}\, dt$$

The quantity $F(\sigma)\, d\sigma$ corresponds, loosely, to the c_k in the Fourier series.

$F(\sigma)$ is the *density function* which describes how much a frequency σ is present in the function $f(t)$. The density times the interval $\Delta\sigma$ gives the amount in measurable units. The function $F(\sigma)$ is called the *transform* of $f(t)$. It is customary to use capital and lowercase letters to represent the transform and the original function. The only difference is the change in the sign of the exponent (change from i to $-i$) between the formulas for each of the two functions in terms of the other. We will also systematically use Latin letters in the time domain (or domain of real measurements) and Greek in the frequency domain.

Second, we must again examine *aliasing*. The discussion for periodic functions in Sec. 31.6 was in no way dependent upon the discrete frequencies of the function being sampled, but referred to the confusion due to the sampling of *any* frequency. Thus the results apply to the general function, not just to periodic functions, and the details need not be repeated here.

This aliasing is the reason why band-limited functions (functions whose frequencies all fall in a limited band) play such a leading role in the theory. Provided the frequencies are limited to the band

$$-\Omega < \sigma < \Omega$$

of width 2Ω, and provided we sample at an interval Δt between samples, then we must have (see Sec. 31.6)

$$2\Omega\, \Delta t < 1$$

to avoid aliasing. The maximum sampling interval $1/(2\Omega)$ is related to the folding, or Nyquist, frequency shown in Fig. 31.6.1. A common way of expressing this relation is to say that if we are to avoid aliasing, we must have at least two samples in the highest frequency present.

Third, the question naturally arises, Can we reconstruct a band-limited function from its samples if we obey the sampling restriction? While we will later (Sec. 34.6) give a somewhat more rigorous proof of the result, we now give an intuitive argument that it is possible.

In the Lagrange interpolation (Sec. 14.5) we introduced the $\pi_i(x)$ functions which were used to form (remember the ith factor is omitted)

$$\frac{\pi_i(x)}{\pi_i(x_i)} = \frac{(x - x_1)(x - x_2) \cdots (x - x_n)}{(x_i - x_1)(x_i - x_2) \cdots (x_i - x_n)}$$

This expression has the property that it takes on the value 1 at the ith sample point x_i and is 0 at all the other sample points. It was then used to form an interpolating polynomial

$$f(x) = \sum_{i=1}^{N} y_i \frac{\pi_i(x)}{\pi_i(x_i)}$$

through the given data.

The function

$$\frac{\sin \pi(t - k)}{\pi(t - k)}$$

which for equally spaced data has exactly the same properties, leads to the interpolating function

$$f(t) = \sum_{k=-\infty}^{\infty} f(k) \frac{\sin \pi(t - k)}{\pi(t - k)}$$

This is a formal expansion which will pass through all the sample points $(k, f(k))$ $(-\infty < k < \infty)$, and we have ignored convergence questions.

We want to show that this function is band-limited and in particular has frequencies limited to the band $-1/2 < \sigma < 1/2$. It is easy to see by direct integration that

$$\int_{-1/2}^{1/2} e^{2\pi i\sigma t}\, d\sigma = \frac{e^{2\pi i t/2} - e^{-2\pi i t/2}}{2\pi i t} = \frac{\sin \pi t}{\pi t}$$

In this integral the function being integrated, $F(\sigma)$, is as shown in Fig. 34.3.1 and is band-limited $(-1/2 < \sigma < 1/2)$.

Thus the function (see Fig. 34.5.2)

$$\frac{\sin \pi(k - t)}{\pi(k - t)}$$

is band-limited to the proper interval, using, as we have, $\Delta t = 1$. This result is the essence of the sampling theorem; given the proper equally spaced samples of a band-limited function $f(t)$, we can reconstruct the function from the samples by the above formula for $f(t)$.

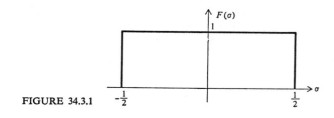

FIGURE 34.3.1

In analogy with the power spectrum of the Fourier-series power spectrum, the quantity

$$|F(\sigma)|^2 = F(\sigma)\bar{F}(\sigma)$$

is often called the *power spectrum* of the function $f(t)$. Instead of power spectrum the simple word *spectrum* is often used. One can regard a function $f(t)$ as if it were a light beam being broken up by a prism into a spectrum of colors—the Fourier integral breaks up the function into a spectrum of frequencies $F(\sigma)$. For a single frequency we often speak of *a spectral line*. Of course physically a pure spectral line does not occur, though very narrow lines are often produced by careful laboratory arrangements.

Fourth, as we shall show in Sec. 34.8, the effect of taking a finite-length sample of the function, in place of the true infinitely long function, is that the individual spectral lines are broadened in a specific way; and the longer the sample, the less the broadening.

The purpose of the rest of the chapter is to develop a little of the theory of the Fourier integral in order to give a more adequate presentation of the above results. We do not intend to become involved in the mathematical rigor since the class of functions that one is apt to use in computing is restricted to "well behaved" and excludes the pathological functions that are so dear to the hearts of the mathematically inclined.

For those whose skill in manipulating complex numbers is a bit rusty, the following problems are given for practice. If they cause any trouble, it would be well to study the topic before going on to the Fourier-integral theory since it makes use of them.

PROBLEMS

34.3.1 Prove $|e^{ix}| = 1$ (for real x).

34.3.2 Prove that the addition formulas of trigonometry follow from $e^{i\alpha}e^{i\beta} = e^{i(\alpha+\beta)}$.

34.3.3 If $1/(a + ib) = u + iv$, find u, v in terms of a and b.

34.3.4 If $u + iv = (a + ib)^{1/2}$, find u, v. (Watch the sign of the u term.)

34.3.5 If $c_k = \bar{c}_{-k}$, prove that $\sum_{k=-\infty}^{\infty} c_k e^{ikt}$ is real.

34.3.6 If m and k are integers, prove that

$$\int_{-\pi}^{\pi} e^{imx}e^{-ikx}\,dx = \begin{cases} 2\pi & m \neq k \\ 0 & m = k \end{cases}$$

34.3.7 If $f(t) = t(-\pi < t < \pi)$, find the Fourier coefficients in $f(t) = \sum_{k=-\infty}^{\infty} c_k e^{ikt}$.

34.3.8 If $f(t) = \int_{-\infty}^{\infty} F(\sigma)e^{2\pi i\sigma t}\,d\sigma$, find the transform of $f'(t), f''(t), f^{(k)}(t)$.

34.3.9 If

$$f(t) = \begin{cases} t & -\pi < t < \pi \\ 0 & \text{elsewhere} \end{cases}$$

find $F(\sigma)$.

34.3.10 If

$$F(\sigma) = \begin{cases} \sigma & -\pi < \sigma < \pi \\ 0 & \text{elsewhere} \end{cases}$$

find $f(t)$.

34.4   THE FOURIER INTEGRAL

The Fourier series for a periodic function in the interval $-N < t \le N$ was given in Sec. 34.2, and if we eliminate the coefficients c_k, we will have, for the complex form

$$f(t) = \sum_{k=-\infty}^{\infty} \left[\int_{-N}^{N} f(t')e^{-(\pi i/N)kt'}\, dt' \right] e^{(\pi i/N)kt} \frac{1}{2N}$$

In order to approach the approximation of a nonperiodic function we will let the interval get longer and longer, that is, let $N \to \infty$.   In the limit, the function will no longer be periodic since the interval of periodicity is the whole axis.

For the purpose of following what happens as $N \to \infty$ we set

$$\frac{1}{2N} = \Delta\sigma$$

and hence

$$\frac{k}{2N} = k\,\Delta\sigma = \sigma$$

We have therefore

$$f(t) = \sum_{k=-\infty}^{\infty} \left[\int_{-N}^{N} f(t')e^{-2\pi i\sigma t'}\, dt' \right] e^{2\pi i\sigma t}\,\Delta\sigma$$

Note that as N gets larger and larger the successive terms in the summation get closer and closer to each other; the exponentials are being packed closer and closer as $N \to \infty$.   It is reasonable to suppose that in the limit the summation will go over into an integral, *provided* the function $f(t)$ is reasonably well behaved

$$f(t) = \int_{-\infty}^{\infty} \left[\int_{-\infty}^{\infty} f(t')e^{-2\pi i\sigma t'}\, dt' \right] e^{2\pi i\sigma t}\, d\sigma$$

We now set

$$F(\sigma) = \int_{-\infty}^{\infty} f(t')e^{-2\pi i\sigma t'}\, dt'$$

and we have

$$f(t) = \int_{-\infty}^{\infty} F(\sigma)e^{2\pi i\sigma t}\, d\sigma$$

The function $F(\sigma)$ is said to be the Fourier transform of the function $f(t)$. The two functions $f(t)$ and $F(\sigma)$ have almost exactly reciprocal relationships to each other. The exception is in the sign of the exponent, $i \to -i$. Both functions contain the same information since each can be uniquely determined from the other. They present the same information in different forms; $f(t)$ is in the time domain, and $F(\sigma)$ is in the (complex) frequency domain. These two alternate, equivalent, views of the same information account for much of the value of the Fourier integral representation of the function.

34.5 SOME TRANSFORM PAIRS

The important relationship between the function $f(t)$ and its Fourier transform $F(\sigma)$ indicates that a table of the correspondences would be useful, and such tables have been made [6, 9]. We propose to develop only a very few elementary transforms that we will need for further work, and leave the more general results to books which concentrate on the Fourier integral.

As a first example consider the band-limited function (see Fig. 34.5.1)

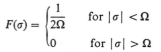

$$F(\sigma) = \begin{cases} \dfrac{1}{2\Omega} & \text{for } |\sigma| < \Omega \\ 0 & \text{for } |\sigma| > \Omega \end{cases}$$

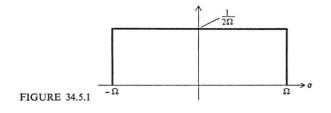

FIGURE 34.5.1

Then, as before,

$$f(t) = \int_{-\infty}^{\infty} F(\sigma)e^{2\pi i\sigma t}\,d\sigma = \frac{1}{2\Omega}\int_{-\Omega}^{\Omega} e^{2\pi i\sigma t}\,d\sigma$$

$$= \frac{1}{2\Omega}\frac{e^{2\pi i\sigma t}}{2\pi it}\bigg|_{-\Omega}^{\Omega} = \frac{e^{2\pi i\Omega t} - e^{-2\pi i\Omega t}}{2i}\frac{1}{2\Omega\pi t}$$

$$= \frac{\sin 2\pi\Omega t}{2\pi\Omega t}$$

From this it follows immediately that

$$F(\sigma) = \int_{-\infty}^{\infty} \frac{\sin 2\pi\Omega t}{2\pi\Omega t} e^{-2\pi i\sigma t} \, dt$$

The function

$$\frac{\sin 2\pi\Omega t}{2\pi\Omega t} = 1 - \frac{(2\pi\Omega t)^2}{3!} + \frac{(2\pi\Omega t)^4}{5!} - \cdots$$

occurs frequently and Fig. 34.5.2 shows how it looks.

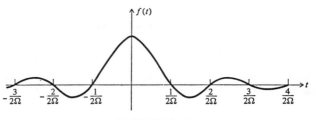

FIGURE 34.5.2
The band-limited function $\sin (2\pi\Omega t/2\pi\Omega t)$.

We see that there is a main lobe extending from $-1/2(\Omega)$ to $1/(2\Omega)$ and that beyond this the function oscillates with the zeros equally spaced while the amplitude of the oscillation slowly decays toward zero.   The larger the Ω, the narrower the spike at the origin.

As a second illustration of a relationship between Fourier transforms, suppose that we already know a transform pair $f(t)$ and $F(\sigma)$; that is, we know

$$f(t) = \int_{-\infty}^{\infty} F(\sigma) e^{2\pi i\sigma t} \, d\sigma$$

We ask, "What function $f_1(t)$ corresponds to $F(\sigma)e^{2\pi i\sigma y}$?"   We have from the definition

$$f_1(t) = \int_{-\infty}^{\infty} F(\sigma) e^{2\pi i\sigma y} \cdot e^{2\pi i\sigma t} \, d\sigma$$

$$= \int_{-\infty}^{\infty} F(\sigma) e^{2\pi i\sigma(y+t)} \, d\sigma$$

Clearly $y + t$ here plays the role of t above, so that

$$f_1(t) \equiv f(y + t)$$

Thus the effect of the exponential multiplier $e^{2\pi i\sigma y}$ is to *shift* the argument of the transform y units.   This result is often known as the "shifting theorem."

PROBLEMS

34.5.1   Find the transform of

$$f(t) = \begin{cases} |1 - t| & |t| \leq 1 \\ 0 & |t| \geq 1 \end{cases}$$

34.5.2   Find the transform of

$$f(t) = \begin{cases} t & \text{for } |t| \leq 1 \\ 0 & \text{for } |t| > 1 \end{cases}$$

34.5.3   What are the transforms of Probs. 34.5.1 and 34.5.2 if the origin is shifted k units?

34.5.4   Note that the product σy does not change if $\sigma \to k\sigma$ and $y \to y/k$. Deduce the "stretch theorem" for Fourier integrals.

34.6   BAND-LIMITED FUNCTIONS AND THE SAMPLING THEOREM

In Sec. 34.3 we gave a formal intuitive justification of the sampling theorem. Because of the importance of this theorem in computing we shall give another, somewhat more mathematical, derivation though it is still not mathematically rigorous with all the pathological details examined carefully.

The central idea of the sampling theorem is that a band-limited function $f(t)$, extending from $t = -\infty$ to $t = +\infty$, is sampled at equally spaced points with a spacing such that at least two samples occur in the highest frequency present. The restriction to band-limited functions is very reasonable since many physical situations are approximately band-limited. However, it is necessary to include a few words of caution. If $f(t)$ is band-limited, then the mathematical model says that it cannot be "time-limited" in the sense that outside some interval, no matter how large, $f(t)$ cannot vanish identically. In particular, if the band-limited function represents an electrical current, then the current must have been flowing for all past time and will flow for all future time. Correspondingly, if $f(t)$ is time-limited, then it cannot be band-limited, and there must be arbitrarily high frequencies present. Evidently the mathematical model should not be pushed too hard when applying it to the real world; it is only a very useful approximation to reality.

In the mathematical treatment, the band-limited aspect arises from the effect of sampling we have called aliasing. If aliasing is to be avoided, then we must limit ourselves to frequencies inside the Nyquist, or folding, frequency (for equally spaced data).

To derive the sampling theorem, suppose we are given the function $F(\sigma)$ in the band $-\Omega < \sigma < \Omega$ and outside the interval $F(\sigma)$ is zero. We first replace it by

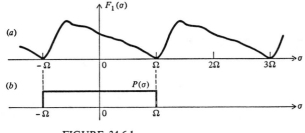

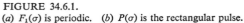

FIGURE 34.6.1.
(a) $F_1(\sigma)$ is periodic.   (b) $P(\sigma)$ is the rectangular pulse.

the function $F_1(\sigma)$ which coincides with $F(\sigma)$ inside the interval and is periodic outside as shown in Fig. 34.6.1.

We have for $F(\sigma)$

$$F(\sigma) = \int_{-\infty}^{\infty} f(t)e^{-2\pi i\sigma t}\, dt$$

$$f(t) = \int_{-\infty}^{\infty} F(\sigma)e^{2\pi i\sigma t}\, d\sigma = \int_{-\Omega}^{\Omega} F(\sigma)e^{2\pi i\sigma t}\, d\sigma$$

For $F_1(\sigma)$ we have, since it is periodic, the Fourier-series expansion

$$F_1(\sigma) = \sum_{k=-\infty}^{\infty} c_k e^{(\pi i/\Omega)k\sigma}$$

where

$$c_k = \frac{1}{2\Omega} \int_{-\Omega}^{\Omega} F_1(\sigma)e^{-(\pi i/\Omega)k\sigma}\, d\sigma$$

From the equation for $f(t)$ this is exactly

$$c_k = \frac{1}{2\Omega} f\!\left(\frac{-k}{2\Omega}\right)$$

Now consider $P(\sigma)$ a rectangular pulse (see Fig. 34.6.1b)

$$P(\sigma) = \begin{cases} \dfrac{1}{2\Omega} & \text{for} \qquad |\sigma| < \Omega \\[2mm] 0 & \text{for} \qquad |\sigma| > \Omega \end{cases}$$

By our previous result (Sec. 34.5) the transform of $P(\sigma)$ is the band-limited function

$$p(t) = \frac{\sin 2\pi\Omega t}{2\pi\Omega t}$$

From the definitions of $F_1(\sigma)$ and $P(\sigma)$ it is clear that the original band-limited function $F(\sigma)$ is

$$F(\sigma) = F_1(\sigma)P(\sigma)2\Omega$$

Substituting for $F_1(\sigma)$, we get

$$F(\sigma) = \sum_{k=-\infty}^{\infty} c_k\, e^{(\pi i/\Omega)k}P(\sigma)2\Omega$$

Using the values of the c_k, we find

$$F(\sigma) = \sum_{k=-\infty}^{\infty} f\!\left(\frac{-k}{2\Omega}\right)P(\sigma)e^{(\pi i/\Omega)k\sigma}$$

We now transform back to the time domain and apply the shifting theorem

$$f(t) = \int_{-\infty}^{\infty} F(\sigma)e^{2\pi i\sigma t}\, d\sigma$$

$$= \sum_{k=-\infty}^{\infty} f\!\left(\frac{-k}{2\Omega}\right)\int_{-\infty}^{\infty} P(\sigma)e^{\pi i/\Omega k}e^{2\pi i\sigma t}\, d\sigma$$

$$= \sum_{k=-\infty}^{\infty} f\!\left(\frac{-k}{2\Omega}\right)\frac{\sin 2\pi\Omega(t + k/2\Omega)}{2\pi\Omega(t + k/2\Omega)}$$

Now replace k by $-k$

$$f(t) = \sum_{k=-\infty}^{\infty} f\!\left(\frac{k}{2\Omega}\right)\frac{\sin \pi(2\Omega t - k)}{\pi(2\Omega t - k)}$$

and we have the sampling theorem.

At the folding frequency itself we cannot reconstruct the function from the samples, since if $\Delta t = 1$ and $f(t) = \sin \pi t$, then all the samples would be zero and the sampling theorem would give $f(t) \equiv 0$.

PROBLEMS

34.6.1 Write the function that is band-limited $-1/2 < \sigma < 1/2$ and has $f(0) = f(1) = f(2) = 1$ and all other $f(k) = 0$.

34.7 THE CONVOLUTION THEOREM

A fundamental relation involving Fourier transforms is known as *the convolution theorem.* Suppose that we have two functions $f(t)$ and $g(t)$. The convolution $h(t)$ of $f(t)$ with $g(t)$ is defined as

$$h(t) = \int_{-\infty}^{\infty} f(s)g(t - s)\, ds$$

Note that for a fixed t we can write $t - s = s'$, and the above equation becomes

$$h(t) = \int_{-\infty}^{\infty} f(t - s')g(s') \, ds'$$

Thus the convolution of $f(t)$ with $g(t)$ is the same as the convolution of $g(t)$ with $f(t)$.

We now ask, "What is the Fourier transform of the convolution which we have labeled $h(t)$?" By definition, the transform of $h(t)$ is

$$H(\sigma) = \int_{-\infty}^{\infty} h(t)e^{-2\pi i \sigma t} \, dt$$

Substituting for $h(t)$, we get

$$H(\sigma) = \int_{-\infty}^{\infty} f(s)e^{-2\pi i s \sigma} \left[\int_{\infty}^{\infty} g(t - s)e^{-2\pi i (t-s)\sigma} \, dt \right] ds$$

$$= \int_{-\infty}^{\infty} f(s)e^{-2\pi i s \sigma} G(\sigma) \, ds$$

$$= F(\sigma)G(\sigma)$$

Thus we have the result: *the transform of the convolution of two functions is the product of their transforms.*

The above formal derivation proves the result for two time functions, and Prob. 34.7.1 covers the case of the convolution of two frequency functions.

Frequently the convolution of a function with itself is of interest. We have

$$h(t) = \int_{-\infty}^{\infty} f(s)f(t - s) \, ds$$

Applying the Fourier transform twice, we get

$$h(t) = \int_{-\infty}^{\infty} F^2(\sigma)e^{-2\pi i \sigma t} d\sigma$$

and for $t = 0$

$$h(0) = \int_{-\infty}^{\infty} f(s)f(-s) \, ds = \int_{-\infty}^{\infty} F^2(\sigma) \, d\sigma$$

PROBLEMS

34.7.1   Prove the formula for the convolution of two frequency functions.
34.7.2   Derive the last formula starting with the frequency functions.
34.7.3   Calculate the convolution of a rectangular pulse of unit area with itself.

34.8   THE EFFECT OF A FINITE SAMPLE SIZE

Many time functions $f(t)$ can be regarded as infinitely long signals (in time). Of necessity we must take a finite length sample. Thus in astronomy we can observe a pulsar or a Cepheid variable for a finite length of time only. What does this limitation do to the original signal? We can regard the limitation as being equivalent to multiplication by the rectangular pulse

$$p(t) = \begin{cases} \dfrac{1}{2T} & |t| < T \\[2mm] 0 & |t| > T \end{cases}$$

Thus what we see is not $f(t)$ but rather $f_1(t)$, defined as

$$f_1(t) = p(t)f(t)2T$$

The Fourier transform is, by the convolution theorem,

$$F_1(\sigma) = 2T \int_{-\infty}^{\infty} F(\sigma_1) \frac{\sin 2\pi T(\sigma_1 - \sigma)}{2\pi T(\sigma_1 - \sigma)} \, d\sigma_1$$

Thus the transform of what we see is the convolution of the true transform $F(\sigma)$ with the function

$$2T \frac{\sin 2\pi T(\sigma_1 - \sigma)}{2\pi T(\sigma_1 - \sigma)}$$

To understand this, consider the situation in which the signal is a single frequency, a spike in the spectrum. This spike is convolved with the above function, and therefore we will see the above function. This is the usual $(\sin x)/x$ function with the amplitude growing like $2T$ and the half-width to the nearest zero contracting like $1/2T$. If instead of a single spike in the spectrum we have many lines, each will be smeared out by the same $(\sin x)/x$, and in the limit of a continuous spectrum we will get the above convolution integral. In an optical analogy this smearing out corresponds to the idea of resolving power—the longer the time signal is observed, the better we can resolve adjacent spectral lines. While it is intuitively evident that something like this should happen, we have exhibited the actual dependence on T. Thus we can now understand what happens to a function when we chop out a piece of it to observe closely, and this limits what we can hope to find out from the sample.

35

A SECOND LOOK AT POLYNOMIAL APPROXIMATION—FILTERS

35.1 PURPOSE OF CHAPTER

The purpose of this chapter is to look at a number of things we did earlier in Part II, using the tools we have just developed, so that we may see the results in a different light. We shall look at only the formulas which used equally spaced points, since this is an essential part of the Fourier-approximation approach. It is more convenient to use the notation

$$y(t) = e^{2\pi i\sigma t} \equiv e^{i\omega t} \qquad \omega = 2\pi\sigma$$

than it is to use the sines and cosines.

It is important to notice that since the formulas are linear, the result of substituting $e^{i\omega t}$ into a formula is that the complex exponential depending on t will factor out and leave a fixed multiplier independent of t. This is often called *filtering*, and the formula is called a *filter*. The word *filter* arose in engineering practice and, like *power spectrum*, persists in the language even when the method is applied outside the original field.

35.2 ROUNDOFF NOISE

Roundoff is an essential, inescapable effect in scientific computing, and a roundoff theory was developed in Chap. 10. The use of the difference table there was justified on the basis of local polynomial approximation. Let us now ask what the difference process does to a single complex sinusoid (where for convenience we have scaled σ, or ω, to make $\Delta t = h = 1$)

$$y(t) = e^{2\pi i \sigma t} = e^{i\omega t}$$

We have the linear formula

$$\Delta y_n = y_{n+1} - y_n = e^{i\omega(t+1)} - e^{i\omega t}$$

hence

$$\Delta y = e^{i\omega t} \cdot e^{i\omega/2}[e^{i\omega/2} - e^{-i\omega/2}] = \left[2i \sin \frac{\omega}{2}\right](e^{i\omega/2})e^{i\omega t}$$

Thus differencing multiplies the function by a factor of size

$$2 \sin \frac{\omega}{2} = 2 \sin \pi\sigma$$

See Fig. 35.2.1. The second difference naturally has a factor of size

$$\left(2 \sin \frac{\omega}{2}\right)^2$$

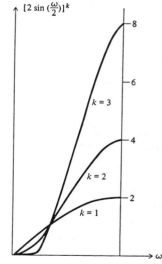

FIGURE 35.2.1

and the kth difference has a factor of size

$$\left(2 \sin \frac{\omega}{2}\right)^k = (2 \sin \pi\sigma)^k$$

As a check at $\sigma = 1/2$, which is the Nyquist sampling rate,

$$y(t) = e^{\pi i t} = (-1)^t$$

and the kth difference is, as it should be, of size 2^k.

t	$(-1)^t$	Δ	Δ^2	Δ^3	Δ^4	Δ^5	Δ^6	Δ^7
0	+1							
		−2						
1	−1		+4					
		+2		−8				
2	+1		−4		+16			
		−2		+8		−32		
3	−1		+4		−16		+64	
		+2		−8		+32		−128
4	+1		−4		+16		−64	
		−2		+8		−32		
5	−1		+4		−16			
		+2		−8				
6	+1		−4					
		−2						
7	−1							

An examination of the figure shows that at 1/3 the Nyquist sampling rate the multiplying factor is exactly 1. Thus below 1/3 the Nyquist rate, the process of repeated differencing (the usual difference table) tends to dampen (decrease the amplitude of) the frequency, while above this rate it amplifies the amplitude of the frequency. We are led, in this fashion, *to identify noise with high frequency*, and signal with low frequency, with 1/3 the Nyquist rate being the boundary between them.

And this is not unreasonable. Roundoff at a point resembles suddenly adding a small square pulse to the function, and we have already seen that the coefficients of a square pulse drop off like $1/n$. The difference table will rapidly amplify the higher frequencies and dampen the lower ones of each roundoff that occurs.

Thus we see that exactly the same computation, the difference table in this case, can be seen in a different light when we ask what the process of differencing does to the various frequencies that may be in the function being differenced. Although the computing is the same, the roundoff model is different and often more suggestive of physical reality.

35.3 DERIVATIVES

Let us look at the simple formula for finding a derivative from the differences of a function. This resembles differencing. We use the central difference formula

$$f_k' = \frac{f_{k+1} - f_{k-1}}{2h} \qquad h = \text{spacing}$$

and again ask what it does to a single frequency

$$e^{2\pi i \sigma t} = e^{i\omega t}$$

At the kth sample point $t_k = hk = k$ (since we are assuming $h = 1$)

$$f_k = e^{i\omega k}$$

We compute

$$f_k' = \frac{e^{i\omega(k+1)} - e^{i\omega(k-1)}}{2} = e^{i\omega k}(i \sin \omega)$$

On the other hand, the correct value is

$$f_k' = i\omega e^{i\omega k}$$

The ratio of the computed answer to the true answer is

$$\frac{i \sin \omega e^{i\omega k}}{i\omega e^{i\omega k}} = \frac{\sin \omega}{\omega}$$

From the plot of $(\sin \omega)/\omega$ in Sec. 34.5 we see that the difference formula clearly underestimates all frequencies except $\omega = 0$ (which is often called dc for direct current). As one would expect, when $\omega = \pi$, that is, when $\sigma = 1/2$, we get an estimate of zero for the derivative ($\sigma = 1/2$ is the folding, or Nyquist, frequency).

Similarly for the second derivative,

$$f_k'' = \frac{f_{k+1} - 2f_k + f_{k-1}}{h^2} \qquad h = 1$$

we get the estimate

$$f_k'' = e^{i\omega k}(e^{i\omega} - 2 + e^{-i\omega}) = -e^{i\omega k}[2(1 - \cos \omega)]$$

The correct answer is

$$f_k'' = -\omega^2 e^{i\omega k}$$

so that the ratio of the computed to true is

$$\frac{2(1 - \cos \omega)}{\omega^2} = \frac{\sin^2 \omega/2}{(\omega/2)^2} = \left[1 - \frac{\omega^2}{24} + \cdots\right]^2$$

which again underestimates the quantity except at dc. At the folding frequency the ratio is $4/\pi^2$.

Since, as we already knew, the estimation of derivatives from computed or tabulated values is dangerous, we usually try to avoid them when possible.

Many times the problem of estimating a derivative can be avoided by proper analysis.   As an example, suppose we have data f_k and we know that theoretically $f(t)$ satisfies a second-order differential equation of the form

$$f'' = H(f, t)$$

Using this, we can go from values of $f(t)$ to values of $f''(t)$ and can integrate the latter to obtain $f'(t)$.   We frequently prefer integration to differentiation, provided that we are not operating on so long a run of data that a slow, steady drift in the integration process would ultimately produce a large error.

Numerous other tricks have been found in special cases, but there appears to be no general theory of when and how derivatives can be avoided.

For differentiation we have, also, Lanczos' method of differentiation, Sec. 32.9, but this is based on Fourier approximation, not on polynomial approximation.

35.4   INTEGRATION—A FIRST LOOK

We will take up a detailed examination of integration along with the treatment of differential equations in the next chapter, but it is worthwhile to first look at what the frequency approach tells us about simple formulas such as the composite trapezoid and midpoint formulas.   We have for the trapezoid rule the formula

$$\frac{1}{2} f_0 + f_1 + f_2 + \cdots + f_{n-1} + \frac{1}{2} f_n$$

($h = 1$) and again suppose that

$$f_k = e^{2\pi i \sigma k} = e^{i\omega k}$$

We get immediately

$$\frac{1}{2} + e^{i\omega} + e^{2i\omega} + \cdots + e^{i(n-1)\omega} + \frac{1}{2} e^{i\omega n} = \frac{1 - e^{i(n+1)\omega}}{1 - e^{i\omega}} - \frac{1}{2}(1 + e^{i\omega n})$$

$$= \frac{(1 + e^{i\omega})(1 - e^{i\omega n})}{2(1 - e^{i\omega})}$$

The true answer is, of course,

$$\int_0^n e^{i\omega t} \, dt = \frac{e^{i\omega n} - 1}{i\omega}$$

The ratio of the computed to true is therefore

$$\frac{-(1 + e^{i\omega}) \cdot i\omega}{2(1 - e^{i\omega})} = \frac{e^{i\omega/2} + e^{-i\omega/2}}{e^{i\omega/2} - e^{-i\omega/2}} \frac{i\omega}{2}$$

$$= \frac{\cos \omega/2}{\sin \omega/2} \frac{\omega}{2} = \frac{\omega}{2} \cot \frac{\omega}{2}$$

$$= 1 - \frac{\omega^2}{12} - \frac{\omega^4}{720} - \cdots$$

We see that the error behaves like ω^2, and for low frequencies the formula is quite accurate.

For the midpoint formula we get, correspondingly,

$$f_{1/2} + f_{3/2} + \cdots + f_{(2n-1)/2} = e^{i\omega/2} + e^{3i\omega/2} + \cdots + e^{(2n-1)i\omega/2}$$

$$= e^{i\omega/2} \frac{1 - e^{i\omega n}}{1 - e^{i\omega}}$$

The ratio of computed to true is

$$\frac{e^{i\omega n} - 1}{e^{i\omega/2} - e^{-i\omega/2}} \frac{i\omega}{e^{i\omega n} - 1} = \frac{\omega/2}{\sin(\omega/2)}$$

$$\approx 1 + \frac{\omega^2}{24} + \frac{7\omega^4}{5760} + \cdots$$

which is by now a familiar function. This differs from the trapezoid rule by the omission of the factor $\cos(\omega/2)$ and already does better at the lower frequencies by a factor of 2.

35.5   SMOOTHING, AN EXAMPLE OF DESIGN

In the polynomial treatment of formulas we tried to avoid the topics of both differentiation and smoothing, though a little of each crept in. Least-squares smoothing, by fitting a local polynomial and using the value from the polynomial as the smoothed value, was an example we gave of smoothing. In the simplest case we fit by a constant value, and found in Sec. 27.7 (Prob. 27.7.2) that this led to the average value.

For three points we get

$$\frac{f_{k-1} + f_k + f_{k+1}}{3} = g_k = \text{smoothed value}$$

What happens to our complex sinusoid? Let

$$f_k = e^{2\pi i \sigma k}$$

$$g_k = e^{i\omega k} \frac{e^{-i\omega} + 1 + e^{i\omega}}{3}$$

$$= e^{i\omega k} \left[\frac{1 + 2\cos\omega}{3} \right]$$

The ratio of smoothed to true is

$$\frac{1 + 2\cos\omega}{3} = \frac{\sin(3\omega/2)}{3\sin(\omega/2)}$$

In the spectrum, this factor is

$$\frac{\sin^2(3\omega/2)}{9\sin^2(\omega/2)}$$

We have already learned that the sampling frequency is directly connected with the folding frequency in the spectrum. Suppose we have a signal that is band-limited and has a maximum frequency of 10 Hz. This forces us to 20 samples per second to avoid aliasing.

But suppose that we are interested only in frequencies less than 5 Hz. We dare not sample at 10 Hz, since then the part of the spectrum above 5 Hz would be aliased down into the interval of interest.

The filter in this section suggests a way of eliminating the upper half of the spectrum. As the filter stands, it reduces the frequencies by at least a factor of 1/9.

We can do better by following the first filter with a second smoothing filter, smoothing by 4s,

$$\frac{g_k + g_{k+1} + g_{k+2} + g_{k+3}}{4} \equiv h_k$$

Then h_k has this effect on $e^{2\pi i \sigma t} = e^{i\omega t}$

$$h_k = \frac{e^{i\omega k}}{4} (1 + e^{i\omega} + e^{2i\omega} + e^{3i\omega})$$

$$= \frac{e^{i\omega k}}{4} \frac{1 - e^{4i\omega}}{1 - e^{i\omega}} = \frac{e^{i\omega(k+2)}}{4e^{i\omega/2}} \frac{e^{2i\omega} - e^{-2i\omega}}{e^{i\omega/2} - e^{-i\omega/2}}$$

$$= \frac{e^{i(k+3/2)\omega}}{4} \frac{\sin 2\omega}{\sin \omega/2}$$

Hence the spectrum is affected in magnitude by

$$\left(\frac{\sin 2\omega}{4 \sin \omega/2} \right)^2 = \left(\frac{\sin 4\pi\sigma}{4 \sin \pi\sigma} \right)^2$$

which has zeros at $\sigma = 1/4, 1/2, \ldots$.

The result of the two filters, the smoothing by 3s followed by smoothing by 4s, is the filter

$$h_k = \frac{f_k + 2f_{k+1} + 3f_{k+2} + 3f_{k+3} + 2f_{k+4} + f_{k+5}}{12}$$

and its ability to reject frequencies in the upper half of the spectrum is very good. See Fig. 35.5.1 where we have plotted the log of the spectrum versus σ.

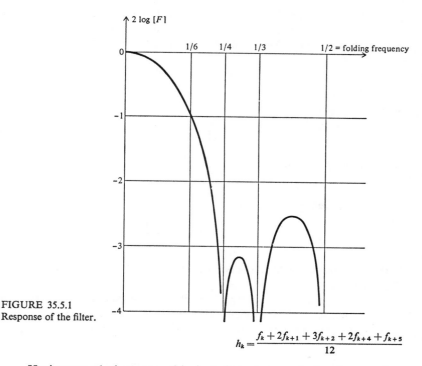

FIGURE 35.5.1
Response of the filter.

$$h_k = \frac{f_k + 2f_{k+1} + 3f_{k+2} + 2f_{k+4} + f_{k+5}}{12}$$

Having smoothed, we can safely drop alternate terms h_k, keeping, say, h_{2k}, and analyze the result without any serious fear of the resulting aliasing (the folding frequency is now at 5 Hz) *unless* there was a very large amount of "power" in the upper half of the spectrum compared to that in the lower half. When we finally use the results, we must make due allowance for the effect of filtering on the frequencies up to 5 Hz. Thus the filtering action of the filters allows us to sub-sample the function without having aliasing troubles. The accuracy for frequencies between 1/3 and 1/2 of the original folding frequency is not very good.

It is easy to compute the effect of smoothing by averaging m consecutive values, namely

$$\frac{\sin(\pi m \sigma)}{m \sin(\pi \sigma)}$$

35.6   LEAST-SQUARES SMOOTHING

In Sec. 27.7 we examined the smoothing of data by passing a least-squares quadratic through five points, and then using the value of the quadratic as the smoothed value at the midpoint.   This resulted in the smoothing formula

$$g = \frac{-3y_{-2} + 12y_{-1} + 17y_0 + 12y_1 - 3y_2}{35}$$

or

$$g = y_0 - \frac{3}{35} \Delta^4 y_{-2}$$

Using the usual substitution of

$$y = e^{i\omega t}$$

we find in the frequency domain

$$G(\omega) = \frac{17 + 24 \cos \omega - 6 \cos 2\omega}{35}$$

In the spectrum we would, of course, use the squares of the tabled values of $G(\omega)$ (Table 35.6).

If we smoothed more drastically by fitting a constant, then we would have the least-squares smoothing by 5s.   These two curves are shown in Fig. 35.6.1, along with a curve from the next section on Chebyshev smoothing.

Table 35.6

ω	$G(\omega)$
0	1.000
$\pi/6$	0.995
$2\pi/6$	0.915
$3\pi/6$	0.658
$4\pi/6$	0.228
$5\pi/6$	−0.194
$6\pi/6$	−0.371

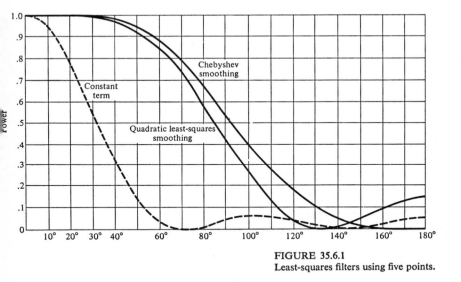

FIGURE 35.6.1
Least-squares filters using five points.

PROBLEMS

35.6.1 Plot the curve for smoothing by 3s on Fig. 35.6.1. Discuss the results.

35.7 CHEBYSHEV SMOOTHING

After examining least-squares smoothing on 5 points $(-2, -1, 0, 1, 2)$ from the frequency point of view, it is natural to ask the same question about Chebyshev smoothing. What does it do to the signal? We proceed as follows:

1 Find the quartic through the 5 points. To do this, assume that the quartic is

$$y(x) = a + bx + cx^2 + dx^3 + ex^4$$

and impose the five conditions. We easily see that

$$a = y_0$$
$$c = \tfrac{1}{24}(-y_{-2} + 16y_{-1} - 30y_0 + 16y_1 - y_2)$$
$$e = \tfrac{1}{24}(y_{-2} - 4y_{-1} + 6y_0 - 4y_1 + y_2)$$

2 We next observe that the particular quartic

$$T(x) = 2x^4 - 8x^2 + 3$$

is equal-ripple *at the sample points* since

$$T(-2) = 3$$
$$T(-1) = -3$$
$$T(0)\ \ = 3$$
$$T(1)\ \ = -3$$
$$T(2)\ \ = 3$$

3 From the quartic fitting of the data we subtract a suitable multiple of $T(x)$, so that we have a cubic

$$g(x) = y(x) - \frac{e}{2} T(x) = \text{cubic in } x$$

and therefore we have a Chebyshev fit to the data.   The smoothed result is

$$g(0) = a - \frac{e}{2} T(0) = \text{smoothed value}$$

$$= \frac{1}{16}(-y_{-2} + 4y_{-1} + 10y_0 + 4y_1 - y_2)$$

$$= y_0 - \frac{1}{16}\Delta^4 y_{-2}$$

4 In the smoothing formula we put $y_n = e^{i\omega n}$ to get the effect on the frequencies, the "transfer function"

$$\frac{1}{8}(5 + 4\cos\omega - \cos 2\omega)$$

5 To find the zeros of this function we set $\cos\omega = z$

$$5 + 4z - (2z^2 - 1) = 0$$
$$z^2 - 2z - 3 = 0$$
$$(z + 1)(z - 3) = 0$$
$$z = -1, 3$$

Only $z = \cos\omega = -1$ leads to real frequencies, and this is $\omega = \pi$.

6 We tabulate the transfer function (Table 35.7)

Table 35.7

ω	Transfer function
0	1.0000
/6	0.9955
$2\pi/6$	0.9375
$3\pi/6$	0.7500
$4\pi/6$	0.4375
$5\pi/6$	0.1295
$6\pi/6$	0.0000

and plot it on the chart with the least-squares smoothing functions for comparison purposes.

The comparisons of the smoothing formulas as shown in Fig. 35.6.1 speak for themselves.

PROBLEMS

35.7.1 Discuss smoothing by the family of formulas $g_n = y_n + \alpha\Delta^4 y_{n-2}$

35.8 THE FOURIER INTEGRAL

We have derived a number of expressions for the ratio of the computed to the true value for a sinusoid of a given frequency. Since the frequency was arbitrary, we have, in effect, the result for all frequencies. Thus we are in a position to apply the Fourier integral.

Suppose the function to which we wish to apply the formula has the Fourier transform

$$f(t) = \int_{-\infty}^{\infty} F(\sigma)e^{2\pi i\sigma t}\, d\sigma$$

Thus $F(\sigma)$ gives the amplitude of the frequency σ. Let the ratio of the calculated to true values be labeled

$$G(\sigma)$$

Then the result of amplitude $F(\sigma)$ is

$$F(\sigma)G(\sigma)$$

and we have that the output, or computed value for $f(t)$, is

$$\hat{f}(t) = \int_{-\infty}^{\infty} F(\sigma)G(\sigma)e^{2\pi i\sigma t}\, d\sigma$$

The process then is: given the function $f(t)$, we find the transform $F(\sigma)$ and then transform back the product $F(\sigma)G(\sigma)$ to get the effect of the formula, where $G(\sigma)$ is the ratio we have been computing.

35.9   SUMMARY

In summary, we have looked at a number of formulas in Part II that we first found by polynomial approximation.   In each case the frequency approach to the same formula has given new understanding of what the formula does.   It is important to realize that the errors in polynomial approximation are usually expressed in terms of high-order derivatives which are very rarely available before the computation, and often not even during the computation (consider, for example, Gaussian quadrature).   Fourier approximation, on the other hand, uses frequencies which can often be estimated from the physical source of the problem.   They can also be estimated from the rate of convergence of the Fourier representation of the function, which is controlled by the singularities of the function along the real axis where the measurements are made.   The size of the derivatives in the polynomial case are controlled by the singularities in the complex part of the plane, where there are no measurements and little intuition to be used.   Furthermore, the step size used in the computation can be related to the frequencies present in the problem itself.   Thus the Fourier approach lives up to the motto of the book, The Purpose of Computing Is Insight, Not Numbers.

36

*INTEGRALS AND DIFFERENTIAL EQUATIONS

36.1 INTRODUCTION

Now that we have developed suitable tools, we can look more closely at the important computing problems of integration and the numerical solution of ordinary differential equations.   The main new tool we have is the Fourier integral.   It is important to notice that in the polynomial case we rated a formula by being exactly true, within roundoff, for $1, x, x^2, \ldots, x^{m-1}$ and measured the error by the next power E_m.   Similarly, in the Fourier series we made the formula exact for a finite sequence of functions $1, \cos x, \sin x, \ldots, \cos(n-1)x, \sin(n-1)x, \cos nx$.   But with the Fourier integral we look at the error as a function of the noncountable number of frequencies in the Nyquist interval, and we have been plotting the ratio of the computed result to the true answer as a function of the (continuous) frequency σ (or ω).

For a while we will use this new method of evaluating old formulas, and variants of the method, but any new method of evaluation gradually brings new methods of design.   These we will look at only briefly.

There is a second important idea that we have to discuss.   It goes under the fancy names of *recursive* and *nonrecursive* filter design.   The difference is simple.

In a nonrecursive design only the function (input) values are used, while for a recursive design both input and output can be used (which means a feedback loop and possible instability). Clearly, we have used a recursive design in the past when we studied integration because there we used both the integrand and the solution values to predict the next value. And in any case the distinction is not all that clear. In Sec. 35.4 we analyzed the trapezoid rule as a nonrecursive formula, and in the next section we will analyze it as a recursive formula. Because recursive formulas use more information, generally they give better results—but because of feedback they also require more care in their design.

36.2 SIMPLE RECURSIVE INTEGRATION FORMULAS

The trapezoid rule may be looked at as an open-loop nonrecursive formula, as we did in Sec. 35.4, or else as a recursive formula in the form

$$y_{n+1} = y_n + \frac{h}{2}[f_n + f_{n+1}] \qquad \text{Set } h = 1$$

We set

$$f_k = A_I e^{i\omega k}$$
$$y_k = A_0 e^{i\omega k}$$

where A_I is the input amplitude and A_0 is the output amplitude. We get

$$A_0(e^{i\omega} - 1) = \frac{A_I}{2}(e^{i\omega} + 1)$$

Hence

$$\frac{A_0}{A_I} = \frac{1}{2}\frac{e^{i\omega} + 1}{e^{i\omega} - 1} = \frac{1}{2i}\frac{\cos(\omega/2)}{\sin(\omega/2)}$$

The true ratio is $1/(i\omega)$; hence the ratio of the computed to true answers is, as before,

$$\frac{\omega}{2}\cot\frac{\omega}{2}$$

Similarly, for the midpoint formula the recursive form is

$$y_{n+1} = y_n + f_{n+1/2}$$

We get

$$A_0(e^{i\omega} - 1) = A_I e^{i\omega/2}$$
$$\frac{A_0}{A_I} = \frac{1}{e^{i\omega/2} - e^{-i\omega/2}} = \frac{1}{2i\sin(\omega/2)}$$

and the usual ratio of computed (ratio) to true (ratio) is

$$\frac{\omega/2}{\sin(\omega/2)}$$

For these two cases the recursive form is easier to handle in the analyses of what they do.

Let us pause to comment of this method in more detail so that we will be clear about what we are doing and what we are getting from the method. Suppose we were operating on a real sinusoid $A \sin 2\pi\sigma t$. If we were to measure the quality of the formula by the size of the error itself, then we would need to know whether $A = 1$ or $A = 1,000$ before stating how serious an error really is. It is much more reasonable to measure, as we have done repeatedly in the book, the *relative error* rather than the absolute error. But as we have noted in the past, near a zero the relative error is very misleading. However, when we use the complex sinusoid $Ae^{2\pi i\sigma t}$, as we do, for real σ and t, the modulus of the complex sinusoid is fixed in size

$$|Ae^{2\pi i\sigma t}| = |A|$$

and there are no zeros; hence the relative error now provides a very satisfactory measure of accuracy.

We have been cavalier about the phase angle. We can always suppose that it is incorporated into the factor A, and for our current method of taking the ratio of the output amplitude to the input amplitude, we need only worry about the *relative phase* of one to the other. This relative phase cannot be neglected in the general case, although it is often zero in important special cases.

36.3 THE TRANSFER-FUNCTION APPROACH TO INTEGRATION FORMULAS

In Chap. 21 we used a general approach to a class of formulas for computing indefinite integrals. The basic formula for integration was

$$y_{n+1} = a_0 y_n + a_1 y_{n-1} + a_2 y_{n-2} + h(b_{-1} y'_{n+1} + b_0 y'_n + b_1 y'_{n-1} + b_2 y'_{n-2})$$

where the y'_k are the integrand values and the y_k are the computed answers to the problem of computing

$$y(t) = y(0) + \int_0^t y'(x)\, dx$$

We propose to reexamine these formulas in the light of our present approach. Suppose, again, that the integrand (the input) is a pure sinusoid

$$y'(t) = A_I e^{2\pi i\sigma t} = A_I e^{i\omega t} \qquad \omega = 2\pi\sigma$$

where the subscript I refers to input.   The true answer is

$$y(t) = -\frac{iA_I}{\omega}\,e^{i\omega t} + C \qquad C = y(0) + \frac{iA_I}{\omega}$$

Since the basic integrated formula is linear, we expect the computed values (the output) to have the same frequency, but perhaps a different phase and amplitude. Let A_0 be the output amplitude, where we allow A_0 to be complex in order to include the phase angle.   Thus we can assume that the computed $y(t)$, except for roundoff, are of the form

$$y(t) = A_0\,e^{i\omega t}$$

We now put these two functions into the basic formula and solve for the ratio A_0/A_I:

$$\frac{A_0}{A_I} = \frac{h(b_{-1} + b_0 e^{-i\omega h} + b_1 e^{-2i\omega h} + b_2 e^{-3i\omega h})}{1 - a_0 e^{-i\omega h} - a_1 e^{-2i\omega h} - a_2 e^{-3i\omega h}}$$

This ratio we call the *transfer function.*   It is the multiplicative factor by which an input function at frequency ω is transformed to obtain the output function at the same frequency.   This idea of a transfer function is a more formal statement of what we did frequently in earlier chapters when we examined what happened to a single frequency.

Using this, let us examine the special case of Simpson's formula:

$$y_{n+1} = y_{n-1} + \frac{h}{3}(y'_{n+1} + 4y'_n + y'_{n-1})$$

Here we have for our basic formula

$$a_0 = 0 \qquad a_1 = 1 \qquad a_2 = 0$$
$$b_{-1} = 1/3 \qquad b_0 = 4/3 \qquad b_1 = 1/3 \qquad b_2 = 0$$

and the transfer function is

$$\frac{A_0}{A_I} = \frac{h(1 + 4e^{-i\omega h} + e^{-2i\omega h})}{3(1 - e^{-2i\omega h})} = \frac{-ih}{3}\,\frac{\cos\omega h + 2}{\sin\omega h}$$

which again has exactly the correct phase (the factor $-i$) for all ω and for small ωh has an amplitude like $1/\omega$.

Next, let us consider the three-eighths rule.   The transfer function is

$$\frac{3h}{8}\frac{1 + 3e^{-i\omega h} + 3e^{-2i\omega h} + e^{-3i\omega h}}{1 - e^{-3i\omega h}} = (-i)\frac{3h}{8}\frac{\cos(3\omega h/2) + 3\cos(\omega h/2)}{\sin(3\omega h/2)}$$

In order to standardize these formulas, we use the *ratio of the computed to the true answer* and scale ω so that $h = 1$. We have as results:

Trapezoid rule: $\qquad H_1(\omega) = \dfrac{\omega}{2} \cot \dfrac{\omega}{2}$

Simpson's formula: $\qquad H_2(\omega) = \dfrac{\omega}{3} \dfrac{2 + \cos \omega}{\sin \omega}$

Three-eighths rule: $\qquad H_3(\omega) = \dfrac{3\omega}{8} \dfrac{\cos(3\omega/2) + 3 \cos(\omega/2)}{\sin(3\omega/2)}$

These three formulas are plotted in Fig. 36.3.1. We have used the $\log_{10}$ of the square of the ratio as the measure, since being twice too large is about as serious as being twice too small, and this makes logarithms the natural scale to use. We have shown these formulas plotted against both $\sigma = \omega/(2\pi)$ and the sampling rate. It is easy to see that Simpson's formula is the best of the three (for reasonably high sampling rates) as far as the size of the error is concerned. The table in Fig. 36.3.1 shows some values which are too small to read conveniently from the graph but which are important for applications.

Figure 36.3.1 also shows the curve for the formula

$$y_{n+1} = y_{n-1} + h(0.3584y'_{n+1} + 1.2832y'_n + 0.3584y'_{n-1})$$

The coefficients were determined experimentally by Leo Tick so that in the interval $0 \le \omega \le \pi/2$ the error would be equal-ripple (Chebyshev) and also exact for $\omega = 0$ (dc). Tick clearly used twice the Nyquist sampling rate. This formula will be discussed in more detail in Sec. 36.7.

The significance of the trapezoid rule going down while Simpson's formula goes up requires some explanation. Simpson's formula increases the amplitude of the higher frequencies whereas the trapezoid tends to smother them. Although not enough is actually known about roundoff effects, we know that sudden jumps in a function due to roundoff tend to produce some high frequencies; hence Simpson's formula amplifies these whereas the trapezoid rule tends to smother them.

If we examine Simpson's formula as a function of the sampling rate, we see that at 5 samples per cycle the error per step is about 1.5 percent, which would almost always be too big. At 7 samples per cycle we approach crude but sometimes useful accuracy, and at 10 samples per cycle we are committing an error of less than 1 part per 1,000, which in many cases is good enough.

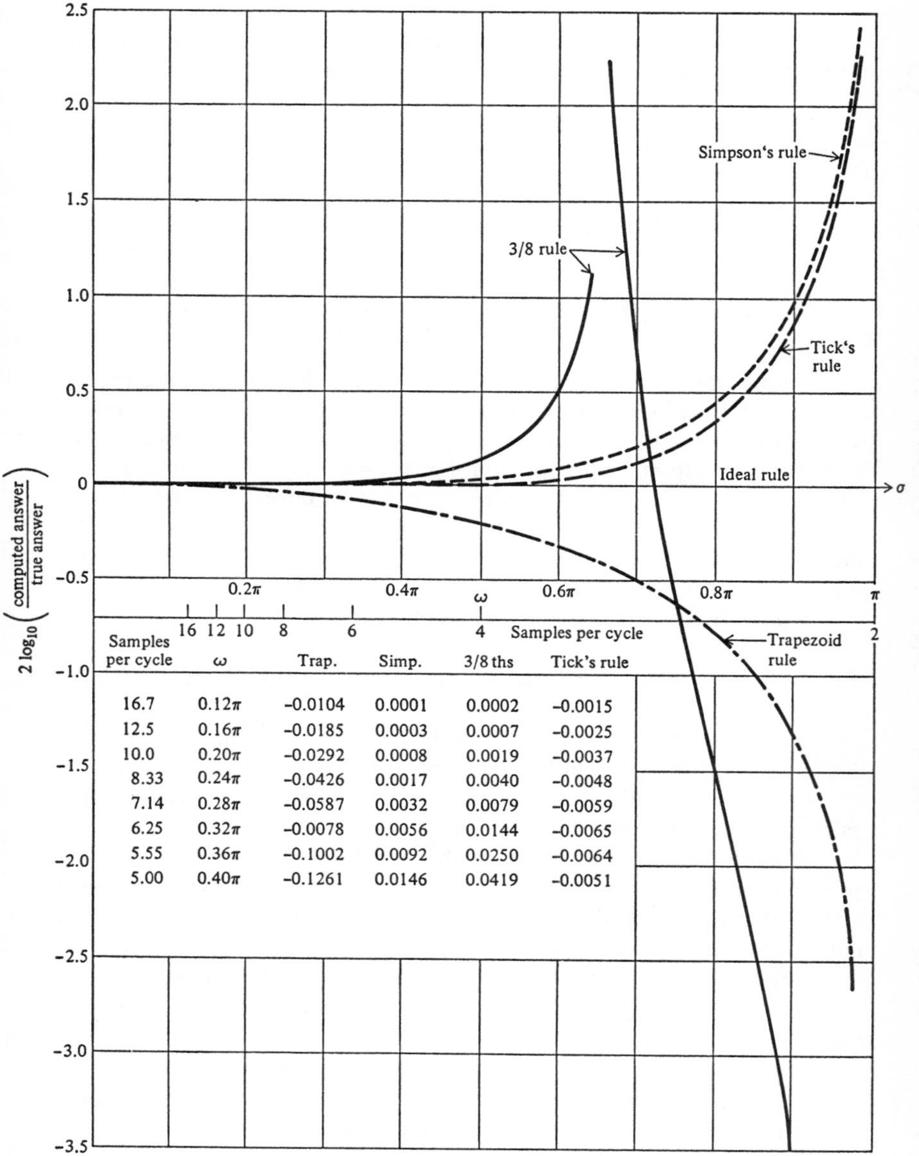

Samples per cycle	ω	Trap.	Simp.	3/8 ths	Tick's rule
16.7	0.12π	−0.0104	0.0001	0.0002	−0.0015
12.5	0.16π	−0.0185	0.0003	0.0007	−0.0025
10.0	0.20π	−0.0292	0.0008	0.0019	−0.0037
8.33	0.24π	−0.0426	0.0017	0.0040	−0.0048
7.14	0.28π	−0.0587	0.0032	0.0079	−0.0059
6.25	0.32π	−0.0078	0.0056	0.0144	−0.0065
5.55	0.36π	−0.1002	0.0092	0.0250	−0.0064
5.00	0.40π	−0.1261	0.0146	0.0419	−0.0051

FIGURE 36.3.1
Frequency response to some integration formulas.

So far we have, in principle, been examining a single frequency $\omega = 2\pi\sigma$. Suppose that we have a general function (input)

$$f_I(t) = \int_{-\infty}^{\infty} F(\sigma)e^{2\pi i\sigma t}\, d\sigma$$

Each term $e^{2\pi i\sigma t}$ is transformed by the transfer function

$$G(\omega) = G(2\pi\sigma) = G_1(\sigma)$$

and we obtain for the output

$$f_0(t) = \int_{-\infty}^{\infty} F(\sigma)G_1(\sigma)e^{2\pi i\sigma t}\, d\sigma$$

We have on the right-hand side the transform of a product, and applying the convolution theorem (Sec. 34.7), we have

$$f_0(t) = \int_{-\infty}^{\infty} f_I(\tau)g_1(t - \tau)\, d\tau$$

Thus, the transform $g_1(t)$ of the transfer function $G_1(\sigma)$ convolved with the input function $f_I(t)$ gives the output function $f_0(t)$.

The transfer function $G_1(\sigma)$ contains all the information about the formula and in a sense is equivalent to the formula of integration being used.

PROBLEMS

36.3.1   Discuss the conditions necessary for an integration formula to have exactly the right phase $(-i)$.

36.4   GENERAL INTEGRATION FORMULAS

The transfer-function approach gave the exact phase for the three formulas examined in Sec. 36.3 as well as those examined earlier.   When we examine the general form in the cases listed in Table 21.7 (Sec. 21.7), matters do not go so smoothly.   For example, using the Adams-Bashforth method, we get (for $h = 1$)

$$\frac{9 + 19e^{-i\omega} - 5e^{-2i\omega} + e^{-3i\omega}}{24(1 - e^{-i\omega})} = -i\frac{(9e^{i\omega/2} + 19e^{-i\omega/2} - 5e^{-3i\omega/2} + e^{-5i\omega/2})}{48\sin(\omega/2)}$$

and the numerator is not purely imaginary since the imaginary part of the parentheses is for small ω,

$$\left(9\sin\frac{\omega}{2} - 19\sin\frac{\omega}{2} + 5\sin\frac{3\omega}{2} - \sin\frac{5\omega}{2}\right) \approx \omega^5$$

The real part of the parentheses is

$$\left(9\cos\frac{\omega}{2} + 19\cos\frac{\omega}{2} - 5\cos\frac{3\omega}{2} + \cos\frac{5\omega}{2}\right) \approx 24$$

so that the deviation from a purely imaginary transfer number is quite small for small ω.

This incorrect phase should not be taken too seriously. While both the amplitude and phase can be studied separately, we are concerned with the failure to have the value 1, and so we now consider

$$\left| \frac{i\omega A_0}{A_I} - 1 \right|$$

which gives a good single measure of the error.[1] The ratio we used before— $|\omega A_0/A_I|$—can be close to 1 and still have a large phase error; this did not occur in the cases studied previously, since the phase was exactly correct.

Figure 36.4.1 shows this quantity as a function of frequency.

Table 36.4 gives the values of the quantity for the usual range of sampling used. An examination of the table shows that of the stable methods (which excludes Simpson's) the two-thirds has the least error for low and moderate sampling rates.

---

[1] In some problems, such as in many acoustical ones, phase errors are of minor importance; in some problems such as in the estimation of cospectra, they can be of prime importance.

Table 36.4 $\left| \dfrac{i\omega A_0}{A_1} - 1 \right|$ FOR VARIOUS INTEGRATION METHODS

Sampling rate	Angle	Trapezoid	Adams-Bashforth	Simpson	Three-eighths
25.00	0.08π	5.27×10^{-2}	1.05×10^{-4}	2.23×10^{-5}	5.06×10^{-5}
16.67	0.12π	1.19×10^{-2}	5.29×10^{-4}	1.14×10^{-4}	2.61×10^{-4}
12.50	0.16π	2.11×10^{-2}	1.66×10^{-3}	3.66×10^{-4}	8.49×10^{-4}
10.00	0.20π	3.31×10^{-2}	4.02×10^{-3}	9.08×10^{-4}	2.15×10^{-3}
8.33	0.24π	4.78×10^{-2}	8.25×10^{-3}	1.92×10^{-3}	4.67×10^{-3}
7.14	0.28π	6.53×10^{-2}	1.51×10^{-2}	3.66×10^{-3}	9.16×10^{-3}
6.25	0.32π	8.57×10^{-2}	2.54×10^{-2}	6.44×10^{-3}	1.68×10^{-2}

Sampling rate	Angle	One-third	One-half	Two-thirds	"S"
25.00	0.08π	5.05×10^{-5}	5.00×10^{-5}	3.44×10^{-5}	1.28×10^{-4}
16.67	0.12π	2.60×10^{-4}	2.55×10^{-4}	1.76×10^{-4}	6.24×10^{-4}
12.50	0.16π	8.39×10^{-4}	8.10×10^{-4}	5.66×10^{-4}	1.88×10^{-3}
10.00	0.20π	2.11×10^{-3}	2.00×10^{-3}	1.41×10^{-3}	4.34×10^{-3}
8.33	0.24π	4.53×10^{-3}	4.18×10^{-3}	3.02×10^{-3}	8.51×10^{-3}
7.14	0.28π	8.75×10^{-3}	7.85×10^{-3}	5.78×10^{-3}	1.49×10^{-2}
6.25	0.32π	1.57×10^{-2}	1.36×10^{-2}	1.03×10^{-2}	2.41×10^{-2}

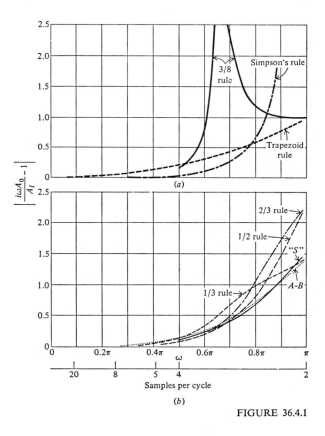

FIGURE 36.4.1

36.5 DIFFERENTIAL EQUATIONS

As we have observed before, there is an essential difference between computing an indefinite integral and solving an ordinary differential equation. For the purpose of clarity, suppose that the differential equation is

$$y' = Ay + f(t)$$

and the indefinite integral is

$$y' = f(t)$$

We have for the integral the schematic diagram

$$f(t) \rightarrow \boxed{\int} \rightarrow y(t)$$

and for the differential equation

$$f(t) \rightarrow \boxed{\int} \rightarrow y(t)$$
$$\boxed{A}$$

where we have a feedback loop, and need to worry even more about stability.

This difference makes a different approach more suitable, although the transfer-function approach can be used if desired. We shall use the approach of studying the size of the error in a step, rather than the ratio of output to input, since it is usually of more interest for differential equations. We are, in fact, giving different answers to the question of What accuracy? since different situations require the accuracy criterion to be applied in different ways.

We adopt the general form for integration that we have been using all along, which looks the same, as far as notation is concerned, as in Secs. 36.3 and 36.4, but is different in meaning since the y' values now come from the y values via the differential equation:

$$y_{n+1} = a_0 y_n + a_1 y_{n-1} + a_2 y_{n-2} + h(b_{-1}y'_{n+1} + b_0 y'_n + b_1 y'_{n-1} + b_2 y'_{n-2})$$

In order to study the frequency response of this formula, we substitute the function

$$y(t) = e^{2\pi i \sigma t} = e^{i\omega t} \qquad \omega = 2\pi\sigma$$

and compute the error as a function of frequency.

$$G(\sigma) = G_1(\omega) = e^{i\omega(t+h)} - a_0 e^{i\omega t} - a_1 e^{i\omega(t-h)} - a_2 e^{i\omega(t-2h)}$$
$$- hi\omega(b_1 e^{i\omega(t+h)} + b_0 e^{i\omega t} + b_1 e^{i\omega(t-h)} + b_2 e^{i\omega(t-2h)})$$
$$|G_1(\omega)| = |1 - a_0 e^{-i\omega h} - a_1 e^{-2i\omega h} - a_2 e^{-3i\omega h}$$
$$- ih\omega(b_{-1} + b_0 e^{-i\omega h} + b e^{-2i\omega h} + b_2 e^{-3i\omega h})|$$

If now we use a general function

$$y(t) = f(t) = \int_{-\infty}^{\infty} F(\sigma)e^{2\pi i \sigma t} \, d\sigma$$

we find that the error is

$$R[f(t)] = \int_{-\infty}^{\infty} F(\sigma)G(\sigma)e^{2\pi i \sigma t} \, d\sigma$$

Using the convolution theorem, we get

$$R[f(t)] = \int_{-\infty}^{\infty} f(s)g(t-s)\,ds = \int_{-\infty}^{\infty} g(t-s)y(s)\,ds$$

In this form the error may be compared with the error form of Chap. 16.   In particular, we had

$$R[f(t)] = \int_{A}^{B} f^{(m)}(s)G(s)\,ds$$

where $G(s)$ was the influence function and was composed of terms of the form

$$(x_k - s)_+^j$$

We again see a significant difference between the two theories: The polynomial approach leads us to express the error as an integral of the mth *derivative* of the function times an influence function, whereas the Fourier-integral approach leads us to use *the function itself* or, alternatively, its transform rather than the mth derivative and of course a different influence function.   However, this last form is deceptive; if the function $g(t - s)$ is examined closely, then it is seen to be merely the original formula in disguise.

36.6   CHEBYSHEV DESIGN OF INTEGRATION FORMULAS: THEORY

The availability of a method of evaluation leads sooner or later to a new method of design.   So many different possibilities arise in practice that we cannot hope to cover them all; rather we shall content ourselves with sketching an approach to Chebyshev design, as this type is becoming increasingly important.   One of the aims of this book has been to gradually bring the reader to the position where he is able to use the principles given in the text to design his own formulas, and we believe that this point has now been reached.

Suppose, then, that we wish to design a method for integrating differential equations such that the error in the frequency domain has a Chebyshev form rather than the form that we have had so far—very good at the origin and increasingly poor as the frequency increases.   For some purposes the Chebyshev error form is clearly preferable to the power-series error form.

We begin by transforming our range of frequencies in ω so that they lie between -1 and $+1$; hence we can now use the standard Chebyshev polynomials. If we adopt the general form of the previous section, we are led to the G function for the error at frequency ω

$$G(\sigma) = e^{t\omega(t+h)} - a_0 e^{i\omega t} - a_1 e^{i\omega(t-h)} - a_2 e^{i\omega(t-2h)}$$
$$- ih(b_{-1} e^{i\omega(t+h)} + b_0 e^{i\omega t} + b_1 e^{i\omega(t-h)} + b_2 e^{i\omega(t-2h)})$$

We now recall the observation that if we wish to have the error in the form of a Chebyshev polynomial, then we should consider getting the whole expression in Chebyshev polynomials.   For this we will need to express $e^{i\omega z}$ in terms of Chebyshev polynomials.   A search of Watson's " Bessel Functions " [62] reveals the generating function

$$e^{(z/2)(t - 1/t)} = \sum_{n = -\infty}^{\infty} t^n J_n(z)$$

for the Bessel functions $J_n(z)$.   Set $t = ie^{i\theta}$ and use the obvious relation (from the generating function)

$$J_{-n}(z) = (-1)^n J_n(z)$$

which is the same as

$$i^{-n} J_{-n}(z) = i^n J_n(z)$$

to get

$$e^{iz \cos \theta} = J_0(z) + 2 \sum_{n=1}^{\infty} i^n J_n(z) \cos n\theta$$

We set $\cos \theta = \omega$, and we have the desired expansion of $e^{i\omega z}$ in terms of Chebyshev polynomials

$$e^{iz\omega} = J_0(z) + 2 \sum_{n=1}^{\infty} i^n J_n(z) T_n(\omega) \qquad -1 \le \omega \le 1$$

In $G(\sigma)$ we factor out the term $e^{i\omega t}$, and we have

$$|G(\sigma)| = |e^{i\omega h} - a_0 - a_1 e^{-i\omega h} - a_2 e^{-2i\omega h}$$
$$- ih\omega(b_{-1}e^{i\omega h} + b_0 + b_1 e^{-i\omega h} + b_2 e^{-2i\omega h})|$$

Using the expansion of the exponential with $z = h, 0, -h, -2h$, in turn, and the appropriate identities involving $T_k(\omega)$, we can arrange everything in an expansion in Chebyshev polynomials with complex coefficients (involving Bessel functions).

If we now start equating coefficients to zero, we get one equation for each complex coefficient, and we could proceed as far as we wished, but we would have to save a parameter or two for stability and other desirable properties.   The result would be an error curve that did not pass through zero at zero (dc) frequency, and this would probably, though not always, cause us some embarrassment; the simple equation

$$y' = 0 \qquad y(0) = A \ne 0$$

would have a solution that was not a constant.

To remedy this defect, we start by requiring that zero frequency fit exactly, that is, require that

$$1 = a_0 + a_1 + a_2$$

We then begin equating coefficients to zero.   Thus we are sure that (within round-off) dc is computed correctly.   Note that if we keep the same frequency band and change the sampling rate by changing h, then *all* the coefficients change, rather than a factor h merely changing as in the polynomial case.

Thus we see the main outlines of the Chebyshev approach.   We see how to find the desired formula, provided we are willing to do some work to obtain it. And this is true over a wide range of formulas—the general approach frequently works provided we do some hard work to obtain it, and at times use a little imagination to overcome a few messy spots.   The art of finding formulas has been practically reduced to a science; the methods developed so far suffice to provide more formulas than there is room to list in a book many times thicker than this one, since there are many combinations of ideas that can be used in different circumstances.   It is hoped that the reader will now feel that he can, with some labor, design formulas to fit the situation rather than force the situation to fit the classical formulas.

36.7   SOME DETAILS OF CHEBYSHEV DESIGN

Since the Chebyshev design methods are not readily available in the literature, we indicate some of the details for integration formulas.   Let the integration formula be of the standard form

$$y_{n+1} = a_0 y_n + a_1 y_{n-1} + a_2 y_{n-2} + h(b_{-1} y'_{n+1} + b_0 y'_n + b_1 y'_{n-1} + b_2 y'_{n-2})$$

We assume, as in the Fourier approach, that

$$y(t) = e^{i\omega t}$$

and using the Chebyshev expansion of $e^{i\omega t}$, we have

$$y(t) = J_0(t) + 2 \sum_{n=1}^{\infty} i^n J_n(t) T_n(\omega)$$

The derivative is

$$y'(t) = J'_0(t) + 2 \sum_{n=1}^{\infty} i^n J'_n(t) T_n(\omega)$$

We now substitute these two expressions into our basic integration formula, using the appropriate values of t, namely, $t = h, 0, -h, -2h$, and factor out the term $e^{i\omega nh}$. We get immediately

$$J_0(h) + 2\sum_{n=1}^{\infty} i^n J_n(h)T_n(\omega) = a_0 J_0(0)$$

$$+ a_1\left[J_0(-h) + 2\sum_{n=1}^{\infty} i^n J_n(-h)T_n(\omega)\right]$$

$$+ a_2\left[J_0(-2h) + 2\sum_{n=1}^{\infty} i^n J_n(-2h)T_n(\omega)\right]$$

$$+ b_{-1}h\left[J_0'(h) + 2\sum_{n=1}^{\infty} i^n J_n'(h)T_n(\omega)\right]$$

$$+ b_0 h[2iJ_1'(0)T_1(\omega)]$$

$$+ b_1 h\left[J_0'(-h) + 2\sum_{n=1}^{\infty} i^n J_n'(-h)T_n(\omega)\right]$$

$$+ b_2 h\left[J_0'(-2h) + 2\sum_{n=1}^{\infty} i^n J_n'(-2h)T_n(\omega)\right]$$

We now rearrange this to have the form of an expansion in Chebyshev polynomials, and we equate coefficients of T_n [since the $T_n(\omega)$ are linearly independent]

$T_0(\omega)$:
$$J_0(h) = a_0 + a_1 J_0(h) + a_2 J_0(2h) + h[b_{-1}J_0'(h) - b_1 J_0'(h) - b_2 J_0'(2h)]$$

$\dfrac{T_1(\omega)}{2i}$:
$$J_1(h) = -a_1 J_1(h) - a_2 J_1(2h) + h[b_{-1}J_1'(h) + b_0 J_1'(0) + b_1 J_1'(h) + b_2 J_1'(2h)]$$

$\dfrac{T_k(\omega)}{2(i)^k}$:
$$J_k(h) = (-1)^k[a_1 J_k(h) + a_2 J_k(2h)] + h[b_{-1}J_k'(h) - (-1)^k b_1 J_k'(h)]$$
$$-(-1)^k b_2 J_k'(2h)]$$

As a first (and degenerate) case, consider the equivalent of the trapezoid rule; that is, set $a_1 = a_2 = b_1 = b_2 = 0$. We shall also require that the formula be exact for $y \equiv 1$ (the dc case), which amounts to setting

$$a_0 = 1$$

We now take the coefficients of $T_0(\omega)$ and $T_1(\omega)$

$$J_0(h) = 1 + hb_{-1}J_0'(n)$$
$$J_1(h) = h[b_{-1}J_1'(n) + b_0 J_1'(0)]$$

From these we get [where $J_0'(h) = -J_1(h)$]

$$b_{-1} = \frac{1 - J_0(h)}{hJ_1(h)}$$

$$b_0 = 2\frac{J_1(h) - 2hb_{-1}J_1'(h)}{h}$$

as the coefficients of the integration formula.

In order to understand (and check) these equations, we examine what happens as h approaches zero. We know [62] that

$$J_n(z) = \frac{1}{n!}\left(\frac{z}{2}\right)^n\left[1 - \frac{(z/2)^2}{n+1} + \frac{(z/2)^4}{2!(n+1)(n+2)}\right.$$
$$\left. - \frac{(z/2)^6}{3!(n+1)(n+2)(n+3)} + \cdots\right]$$

In particular

$$J_0(h) = 1 - \frac{h^2}{4} + \frac{h^4}{64} - \cdots$$

$$J_1(h) = \frac{h}{2} - \frac{h^3}{16} + \frac{h^5}{384} - \cdots$$

$$J_2(h) = \frac{h^2}{8} - \frac{h^4}{96} + \frac{h^6}{3,072} - \cdots$$

Hence as $h \to 0$,

$$b_{-1} = \frac{h^2/4 - h^4/64 + \cdots}{h^2/2 - h^4/16 + \cdots} \simeq \frac{1 - h^2/16}{2(1 - h^2/8)} \to \frac{1}{2}$$

$$b_0 = \frac{h - h^3/8 + \cdots - 2b_{-1}h(\frac{1}{2} - 3h^2/16 + \cdots)}{h} \to \frac{1}{2}$$

At first it may seem surprising that the two coefficients b_{-1} and b_0 are not equal, but a little thought shows that there was nothing else to be expected. The difficulty lies in the fact that we made the error curve go through zero at $\omega = 0$ and then tried to make the dominant term in the error expansion proportional to $T_2(\omega)$, which clearly does not have a zero at $\omega = 0$. The case is really too trivial to be meaningful.

The next case uses two old values of the function and the derivative, setting only $a_2 = b_2 = 0$. In this case we save one parameter for stability. Thus we have the equations

dc: $\qquad 1 = a_0 + a_1$

$T_0(\omega)$: $\qquad J_0(h) = a_0 + a_1 J_0(h) + h[b_{-1}J_0'(h) - b_1 J_0'(h)]$

$T_1(\omega)$: $\qquad J_1(h) = -a_1 J_1(h) + h[b_{-1}J_1'(h) + b_0 J_1'(0) + b_1 J_1'(h)]$

$T_2(\omega)$: $\qquad J_2(h) = a_1 J_2(h) + h[b_{-1}J_2'(h) - b_1 J_2'(h)]$

Eliminating a_0, we can write the second equation as

$$0 = (1 - J_0)(1 - a_1) + hJ_0'(b_{-1} - b_1)$$

and the fourth equation as

$$0 = - J_2(1 - a_1) + hJ_2'(b_{-1} - b_1)$$

From these we see that

$$a_1 = 1$$
$$b_{-1} = b_1$$

and we are on the edge of stability. From the first equation

$$a_0 = 0$$

The formula is, therefore, of the form

$$y_{n+1} = y_{n-1} + b_1(y_{n+1}' + y_{n-1}') + b_0 y_n'$$

The third equation is now [where we use the fact that $J_1'(0) = 1/2$]

$$2J_1 = 2hb_{-1}J_1' + \frac{hb_0}{2}$$

As $h \to 0$, this approaches

$$2b_{-1} + b_0 = 2$$

as it should since for a straight line (dc) it is correct and it includes Simpson's formula.

This result brings us back to Tick's method mentioned earlier in Sec. 34.6 Tick asked himself the question, "What integration formula should I use if I want the error curve in the frequency domain to be as small as possible over the lower half of the Nyquist interval?" He added the condition that the formula be exact if $y' = $ constant or (what is the same thing) if $y(x)$ is a straight line. Thus he required the symmetric formula to be of the form

$$y_{n+1} = y_{n-1} + h(b_1 y_{n+1}' + b_0 y_n' + b_1 y_{n-1}')$$

and to satisfy

$$2 = b_0 + 2b_1$$

He then proceeded to use a computer to compute the error curve for various choices of the parameter b_0. By a sequence of trials he came to the value

$$b_0 = 1.2832$$

and the formula we analyzed in Sec. 36.3.

By our analytical method we use both

$$2J_1 = 2hb_{-1}J_1' + \frac{hb_0}{2}$$

and the condition Tick used, namely

$$2 = b_0 + 2b_1$$

to get

$$b_{-1} = \frac{2J_1(h) - h}{[2J_1'(h) - 1]h}$$

Tick used half the Nyquist interval, that is, 4 samples per cycle, or what is the same thing $h = \pi/2$. For this value of h

$$b_{-1} = 0.35785 \cdots$$

Compared to Tick's empirical result of $b_{-1} = 0.3584$ (difference = 0.00055) this is very good since we have not done the final leveling of the Chebyshev error curve as he did. On the other hand, we have a result for *all* step sizes, or what is the same thing, for all fractions of the whole Nyquist interval.

36.8 SUMMARY

In this chapter we have examined a number of integration formulas from the point of view of what they do to all the different frequencies. In practice it is often possible to connect such information with the physical problem. Thus, before an integration is begun, the proposer often has some idea of the frequency content of the solution that he thinks is important, and after the problem is done, he can relate the step size used to the highest frequency that is present to any large extent in the solution. Of course there are problems in which the frequency content has no physical meaning and this approach is not relevant.

From the new method of looking at the characteristics of a formula we have passed to one type of formula design, namely the requirement that the error curve in the frequency domain be exactly zero at dc and have a Chebyshev characteristic in some subpart of the total Nyquist interval. This is a very natural requirement in many applications.

37

*DESIGN OF DIGITAL FILTERS

37.1 BACKGROUND

In processing analog signals, especially those in the form of electrical currents, it is usually necessary to change the amount of various frequencies present in the signal. In particular, it is often necessary to pass certain frequencies and to block others, to filter out certain frequencies. As a result of the importance of this problem there is a large, highly developed field of analog filter design.

When we digitize a signal (for machine processing or for more reliable transmission over noisy communication lines), we inherit the same filtering problems, but now we must filter out frequencies from a stream of digits. At first this problem of digital filtering was approached via the earlier field of analog filters, and if you already know a great deal about analog filters, this is a reasonable approach. But if you know almost nothing about this field, then the direct approach is more practical, and it is the one we will adopt as an introduction to digital filters. However, much can be learned from the older analog methods.

Digital filtering is not only easily done on a general-purpose computer, but special high-speed multipliers[1] and adders make it easy and economical in special applications.

37.2 A NONRECURSIVE CLASS OF DIGITAL SMOOTHING FILTERS

We begin the study of digital filters by recalling those in Sec. 35.5.   There we studied smoothing and designed one simple filter that smoothed by 3s and then by 4s and which did a very good job at blocking out the upper half of the frequency spectrum (a low-pass filter).   We also examined the filtering action of two least-squares smoothing formulas, and one Chebyshev formula.

These examples suggest that we study the general family of 5-term linear formulas (filters) of the form

$$g_k = af_{k-2} + bf_{k-1} + cf_k + df_{k+1} + ef_{k+2}$$

However, in this section we shall limit ourselves to *symmetric* formulas ($e = a$, $d = b$) both because they are the ones that are widely used in practice and because for smoothing they have no phase distortion (no imaginary terms) and this makes them easier to understand.   Later (Sec. 37.4) we will examine differentiation formulas which have a pure imaginary phase component.   We therefore first examine the restricted class of symmetric smoothing formulas

$$g_k = af_{k-2} + bf_{k-1} + cf_k + bf_{k+1} + af_{k+2}$$

As usual, we start by asking what happens to a single frequency by assuming that

$$f(t) = e^{2\pi i \sigma t} = e^{i\omega t}$$

or

$$f_k = e^{i\omega k}$$

Thus we get

$$g_k = (2a \cos 2\omega + 2b \cos \omega + c)f_k$$

The transfer function $H(\omega)$ which multiplies the amplitude of the input signal of frequency to get the output amplitude is

$$H(\omega) = 2a \cos 2\omega + 2b \cos \omega + c$$

Without the even symmetry we would have had some sine terms with imaginary coefficients.   In some of the previous work we used $H(\omega)$ as the ratio of the computed to true, and for smoothing g_k/f_k is this ratio, and so our notation is consistent.

[1] 40 ns, 17-bit multipliers in 1971.

How shall we choose the filter coefficients a, b, and c?   That depends, of course, on what we want to do.   Suppose that again we want to block out the high frequencies and pass the low frequencies (a low-pass filter).   To convert this to passing the high and blocking the low frequencies (a high-pass filter), we have only to compute $f_k - g_k$ as a new filter.

We will start by arbitrarily imposing two conditions on the frequency curve, namely that at the low end

$$H(0) = 1 \qquad \text{full dc (lowest frequency)}$$

and at the upper end

$$H(\pi) = 0 \qquad \text{no highest frequency}$$

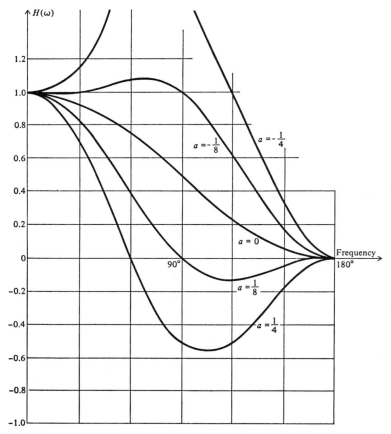

FIGURE 37.2.1

These two conditions mean

$$H(0) = 2a + 2b + c = 1$$
$$H(\pi) = 2a - 2b + c = 0$$

From these we get

$$b = 1/4$$
$$c = 1/2 - 2a$$

and we have a one-parameter family of filters (see Fig. 37.2.1).

$$H(\omega) = 2a \cos 2\omega + 1/2 \cos \omega + (1/2 - 2a) = 4a(1 + \cos \omega)\left[\cos \omega - \left(1 - \frac{1}{8a}\right)\right]$$

Note that $H(\omega)$ is a periodic even function of ω and that it is also symmetric about $\omega = \pi$.

From this figure we can pick out a filter that approximately meets our needs. Alternately, we could impose one more condition on the transfer function $H(\omega)$ and directly determine the filter.

As a first example of directly determining the filter, suppose we require

$$H_1\left(\frac{\pi}{2}\right) = 1$$

Thus we get

$$H_1\left(\frac{\pi}{2}\right) = -2a + 1/2 - 2a = 1$$
$$a = -1/8$$
$$c = \quad 3/4$$

Therefore

$$g_k = 1/8(-f_{k-2} + 2f_{k-1} + 6f_k + 2f_{k+1} - f_{k+2})$$

and

$$H_1(\omega) = -\frac{\cos 2\omega}{4} + \frac{\cos \omega}{2} + \frac{3}{4}$$

As a second example of the design of a filter, suppose we balance the two halves of the filter by setting

$$H_2\left(\frac{\pi}{2}\right) = 1/2$$

Then we get

$$H_2\left(\frac{\pi}{2}\right) = -2a + 1/2 - 2a = 1/2$$

$$a = 0$$

$$c = 1/2$$

$$g_k = 1/4(f_{k-1} + 2f_k + f_{k+1})$$

$$H_2(\omega) = \frac{\cos \omega}{2} + \frac{1}{2}$$

As a third example, suppose we try to do as well as we can in the neighborhood of dc (zero frequency). We already have both

$$H_3(0) = 1$$

and

$$\frac{dH_3(0)}{d\omega} = 0$$

hence we impose the further condition that

$$\frac{d^2 H_3(0)}{d\omega^2} = 0 = -8a - 1/2$$

$$a = -1/16$$

$$c = \quad 5/8$$

$$g_k = \frac{1}{16}\left(-f_{k-2} + 4f_{k-1} + 10f_k + 4f_{k+1} - f_{k+2}\right)$$

$$H_3(\omega) = -\frac{\cos 2\omega}{8} + \frac{\cos \omega}{2} + \frac{5}{8}$$

As a final example, suppose we ask that in the lower half of the Nyquist interval $0 \le \omega \le \pi/2$, $H(\omega)$ have the shape of a Chebyshev approximation to unity, while still keeping the two conditions $H_4(0) = 1$ and $H_4(\pi) = 0$. The position of the extreme value of $H_4(\omega)$ is defined by

$$\frac{dH_4(\omega)}{d\omega} = -4a \sin 2\omega - \frac{1}{2} \sin \omega = 0$$

This defines the value ω_0 at which the extreme value occurs. To find ω_0, we use the double-angle formula for $\sin 2\omega$ to get

$$-8a \sin \omega \cos \omega - \frac{1}{2} \sin \omega = 0$$

$$\sin \omega = 0 \qquad \omega = 0, \pi$$

$$\cos \omega = -\frac{1}{16a}$$

Hence the extreme value is

$$H_4(\omega_0) = 2a \cos 2\omega_0 + \frac{1}{2} \cos \omega_0 + \frac{1}{2} - 2a$$

$$= \frac{1}{2} - 4a - \frac{1}{64a}$$

The value at $\omega = \pi/2$ is

$$H_4\left(\frac{\pi}{2}\right) = 1/2 - 4a$$

The excess above 1 at ω_0 is to be equated to the amount below at $\pi/2$, since the Chebyshev condition means that

$$H_4(\omega_0) - 1 = 1 - H_4\left(\frac{\pi}{2}\right)$$

This leads to the equation

$$2^9 a^2 + 2^6 a + 1 = 0$$

whose solution is

$$a = \frac{-2 \pm \sqrt{2}}{32}$$

Thus

$$\cos \omega_0 = \frac{-1}{16a} = \frac{-2}{-2 \pm \sqrt{2}} = 2 \pm \sqrt{2}$$

Evidently we must use the minus sign (if we want a real ω_0), and

$$a = \frac{-2 - \sqrt{2}}{32} = \frac{-3.414}{32} = -0.1067 \cdots$$

$$c = 0.7134$$

Looking at the figure we see that this is about the expected value.

We have thus shown a number of simple ways we can design a low-pass filter.

37.3 AN ESSAY ON SMOOTHING

We have just examined some low-pass digital filters. We again note that by merely taking $f_k - g_k$ we can get a high-pass filter that accepts what the previous one rejected and rejects what the previous one accepted. Thus we have a high-pass filter corresponding to each low-pass filter in the previous section.

We earlier identified smoothing with the removal of high frequencies on the grounds that random roundoffs are like small jumps in the function which give rise to high frequencies as well as low frequencies, while the signal we are usually dealing with has mainly low frequencies. We further showed (Sec. 35.2) that the difference-table method of measuring the noise amplified the upper two-thirds of the Nyquist interval and damped out the lower one-third.

It is not easy to state in a vacuum exactly what filter to use to smooth a given function. It depends on the spectra of *both* the signal and noise. Evidently we can build a filter to reject the frequencies that we think are noise and not signal, but we cannot, by any linear method of smoothing, reject the frequencies of the noise in the range of the signal *unless* we are prepared to lose some of the signal as well. The practical art of filtering to remove noise often includes this difficult compromise of removing a small amount of the signal to get rid of a lot of the noise as well. And to do this we need to have an accurate estimate of the spectrum of the signal as well as the spectrum of the noise.

In practical measurements it is often possible to form some judgments as to the respective spectra, but in computing all too little is known of the spectrum of the roundoff noise—more studies are required in this area. It is generally believed to be flat (white noise).

Viewed as smoothing filters, some of the formulas we have given violate a very natural condition, namely that if one function is uniformly larger than another, then the smoothed values should also be larger. This condition is formalized in the following theorem.

Theorem 37.3. If $u_k \geq v_k$ for all k, then a necessary and sufficient condition for the smoothed value

$$\bar{u}_k = \sum_{i=k-N}^{k+N} a_i u_i$$

to be greater than or equal to the corresponding smoothed value $\bar{v}_k$ is that all the

$$a_i \geq 0$$

PROOF   If the a_i are all positive, then the smoothed values are clearly in the right order (we have only to look at the smoothed difference of $u_k - v_k$). On the other side of the argument, if one of the $a_i < 0$, then consider $u_k = 0$ for all k and $v_k = 0$ except for $v_i < 0$. Then

$$\bar{v}_i > \bar{u}_i = 0$$

Hence the theorem is proved.   ////

For the family in Sec. 37.2

$$a = a$$
$$b = 1/4$$
$$c = 1/2 - 2a$$

This means that $0 \le a \le 1/4$ ($1/2 \ge c \ge 0$), and many of the interesting filters are eliminated by this condition.

This theorem should not be taken as gospel. Consider Fig. 37.3.1 where $u_k \equiv 0$ (circles) and where v_k is as shown (crosses). It is *not* unreasonable to have the smoothed middle value of $v_k > 0$ as shown by the curve.

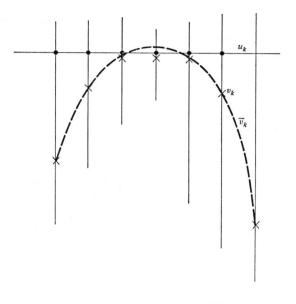

FIGURE 37.3.1

37.4 DIFFERENTIATION FILTERS

Differentiation, even more than smoothing, is a delicate process, since it tends to amplify the high frequencies (the noise in the signal). For differentiation we use a filter with odd symmetry ($c = -a$, $d = -b$, $c = 0$)

$$g_k = af_{k+2} + bf_{k+1} - bf_{k-1} - af_{k-2}$$

The usual substitution

$$f_k = e^{i\omega k}$$

gives

$$g_k = i(2a \sin 2\omega + 2b \sin \omega)f_k$$

where of course the derivative (with respect to k) is

$$f_k' = i\omega f_k \simeq g_k$$

Thus the parentheses should be an approximation to ω since the transfer function is the ratio of the computed to the true

$$H(\omega) = \frac{2a \sin 2\omega + 2b \sin \omega}{\omega}$$

We cannot approximate it in the whole interval, for it is clear that the term in the parentheses is zero at $\omega = \pi$. We shall impose the condition

$$H(0) = 1$$

This means that as $\omega \to 0$,

$$4a \cos 2\omega + 2b \cos \omega \to 1$$
$$4a + 2b = 1$$

or

$$b = \frac{1}{2}(1 - 4a)$$

and

$$g_k = a(f_{k+2} - f_{k-2}) + \frac{1 - 4a}{2}(f_{k+1} - f_{k-1})$$

$$H(\omega) = \frac{2a \sin 2\omega + (1 - 4a) \sin \omega}{\omega}$$

$$= \frac{\sin \omega}{\omega}[1 - 4a(1 - \cos \omega)]$$

Thus we have a one-parameter family of differentiation formulas to explore.
As our first example, we ask that

$$H\left(\frac{\pi}{2}\right) = 1 = \frac{1 - 4a}{\pi/2}$$

hence

$$a = \frac{1}{4}\left(1 - \frac{\pi}{2}\right)$$

$$b = \frac{\pi}{4}$$

As a second example, suppose we try to do as well as we can in the neighborhood of $\omega = 0$, that is, $H'(0) = 0$. This becomes

$$H(\omega) = \frac{1}{\omega} \left\{ 2a \left[2\omega - \frac{(2\omega)^3}{3!} + \frac{(2\omega)^5}{5!} - \cdots \right] + (1 - 4a) \left[\omega - \frac{\omega^3}{3!} + \frac{\omega^5}{5!} - \cdots \right] \right\}$$

$$= 1 + \omega^2 \left(\frac{-8a}{3} - \frac{1}{6} + \frac{2a}{3} \right) + \omega^4 \left(\frac{8a}{15} + \frac{1}{120} - \frac{a}{30} \right) + \cdots$$

We equate the coefficient of ω^2 to zero to get $a = -1/12$, which leads to

$$H_2(\omega) = 1 - \frac{\omega^4}{30} + \cdots$$

$$g_k = -\frac{1}{12} [f_{k+2} - f_{k-2}] + \frac{2}{3} [f_{k+1} - f_{k-1}]$$

We can similarly design other filters to meet other conditions.

PROBLEMS

37.4.1 Design a filter for the second derivative.

37.4.2 Design a Chebyshev in $0 \leq \omega \leq \pi/2$ filter for a first derivative. [Use $H(\omega)$.]

37.5 RECURSIVE FILTERS

Nonrecursive filters use only the input function values and do not make use of already computed output numbers, while recursive filters use both.   There are occasions when, as in the two integration formulas for the trapezoid and midpoint rules (Secs. 35.4 and 36.2), both methods give the same result.   For a given span (number of back terms used in the formula) the nonrecursive formula has a limited memory of the past, while a recursive formula, in a certain sense, remembers all the past.

For a given span the recursive formula will have almost twice as many parameters as the nonrecursive formula, and hence it can be made to fit a desired frequency response far better.   However, it should be remembered that the recursive formula will have to be watched for stability problems.

The predictor-corrector methods for integrating ordinary differential equations that we developed using polynomial approximation, and later in Chap. 36 analyzed in the frequency domain, are good examples of recursive formulas. They also illustrate the situation where we tend to use recursive filters, namely when we have information only up to the present and we need to take some action or make some prediction of what is going to happen next (extrapolation). Non-recursive filters tend to be used when all the data are available at the time we are processing them.

The three classical operations of filters are integration, smoothing (interpolation, extrapolation), and differentiation, but in principle filters apply to any linear operation on the data.

The recursive filters have another property worth mentioning. In the integration of differential equations, when we want great accuracy, we use at least 10 samples in the highest frequency present, and even for rough work we use at least 5 samples. But in many data processing problems, where the noise is high, we cannot afford that luxury (because among other reasons the spectrum of the noise for high frequencies is too large). Thus we find that we are occasionally working around the Nyquist rate of 2 samples in the highest frequency present (to any extent).

The topic of filters is of great importance in many digital problems of signal processing, such as the Moon and Mars pictures, and whole books[1] are currently devoted to the topic of digital filters, so we must (reluctantly) refer the reader to them.

---

[1] The classic reference is F. F. Kuo and J. F. Kaiser [32], especially the chapter by Kaiser. A more modern reference is B. Gold and C. M. Rader [17]. A further important reference with respect to practical computing is V. V. Solodovnikov [54]. Also *IEEE Transactions on Audio and Electroacoustics*, vol. AU-16, no. 3, September, 1968.

38

*QUANTIZATION OF SIGNALS

38.1 INTRODUCTION

We have already examined the effects of sampling an analog signal in the time (independent) variable. The process of sampling also replaces the continuous analog signal by a digital output that can be only one of comparatively few numbers. Thus the signal is not only sampled in the independent variable, it is also *quantized* in the dependent variable. Since digital computing involves a great deal of processing of quantized physical data, it is necessary to look carefully at this topic. (See Fig. 38.1.1.)

The quantization is similar to roundoff *except* that we normally think of roundoff as occurring in relatively high precision numbers, while the quantization effect in measured signals may often be severe. The process of quantization occurs in what is usually called analog-to-digital conversion, or *A*-to-*D* conversion. This process may have precision anywhere from 1 to 17 bits (possibly more in a few cases), so that quantization may be very important or may be a relatively small effect. In many situations a 1-bit digitizer can give a surprising amount of information (Sec. 38.6).

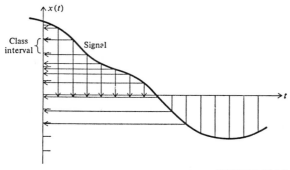

FIGURE 38.1.1

The topic of quantization is a large one and we can give only a brief introduction, referring the interested reader to the literature [17, 20, 58] for more details; in a sense we are only trying to create an awareness of the consequences of these widely ocurring effects.

38.2 THE GRAY CODE

The analog to digital converter usually does not go from the analog signal directly to the binary representation of the quantized number, rather there is an intermediary step of representation in a form known as *the Gray code*.[1] The reason for this is easily understood if we try to imagine how the converter works. It is convenient to think of a rotating wheel with sectors, and we are trying to read optically the encoding on the sector under the reading heads. If there is a sector directly under the heads at the moment of reading, there will be little trouble, but suppose the reading heads are partly over one sector and partly over another sector, in particular between sectors 7 and 8. (See Fig. 38.2.1. In binary form, this is:

$$0 \ 111$$
$$1 \ 000$$

Clearly, almost anything between 0 000 and 1 111 (0 and 15) might be read out.

[1] Technically *a* Gray code, but this one is so universally used that it is *the* Gray code.

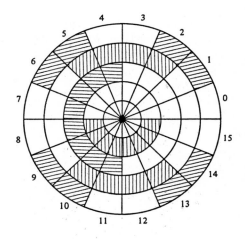

FIGURE 38.2.1

The Gray code has the important property that in going from one number to the next only one digit changes, thus eliminating this awkward problem of reading parts of two sectors at the same time and then getting almost any number. Because only one digit changed, one of the two possible numbers must come out.

We shall first show how to construct a Gray code. The code using 1 bit is clearly

$$\begin{array}{c|c} 0 & 0 \\ 1 & 1 \end{array}$$

and note that it is really circular and the bottom is followed by the top of the list. To form the 2-bit code we put a 0 in front of each to get

$$\begin{array}{c} 00 \\ 01 \end{array}$$

We also copy the original code below this in the *reverse* order and put 1s in front of this, getting, finally, the 2-bit Gray code

$$\begin{array}{c|c} 0 & 00 \\ 1 & 01 \\ 2 & 11 \\ 3 & 10 \end{array}$$

At each stage we get the next larger code by first copying the earlier code with an added 0 in front and then copying the original code in reverse order with 1s in front. For example, in going from a 3-bit to a 4-bit code we get

0	0 000
1	0 001
2	0 011
3	0 010
4	0 110
5	0 111
6	0 101
7	0 100
8	1 100
9	1 101
10	1 111
⋮	⋮
15	1 000

The characteristic property of the Gray code follows immediately from the method of construction.

It sometimes happens that the conversion from the Gray code to binary number code is not done in the converter, and therefore it must be done in the computer. We now describe how this is done. A study of the method of construction shows that:

1 The left-hand digit is the correct binary digit.

2 If the leftmost digit is 0, then the next digit is correct; but if it is a 1, then the next digit should be complemented.

3 In general, if at some stage in the decoding there has been an odd number of 1s in the Gray-code representation, then the next digit must be complemented; but if there has been an even number of 1s, then the next digit is correct.

The rare case of having to convert the other way (it can arise in a simulation of a system which has converters) is left as an exercise.

PROBLEMS

38.2.1 Give the equations for determining the digits of the Gray-code representation of a binary number.

38.2.2 Sketch a 5-bit Gray-code wheel.

38.3 THE STATISTICAL DISTRIBUTION OF VALUES

Since quantization is very nonlinear, it is not possible to give a simple analysis of what happens, and we will examine only the *statistics* of the process.

We begin with an examination of the distribution of the values of the function. We want to know how many times a given value occurs. To get this we need to project the curve horizontally onto the vertical axis and note how many values end up at a given point along this axis, for both the discrete and the continuous cases. See Fig. 38.3.1.

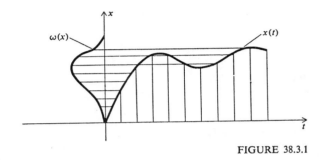

FIGURE 38.3.1

As an example of this process, consider the distribution of the values of a single sinusoid $x = \sin t$. See Fig. 38.3.2. Clearly we need to examine only the interval $-\pi/2$ to $\pi/2$. The area under the density curve $x(w)$ between x and $x + dx$ is the fraction of the time the signal $x(t) = \sin t$ is between x and $x + dx$, that is,

$$dt = w(x)\, dx$$

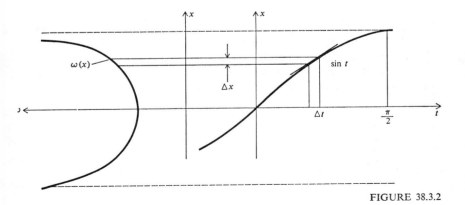

FIGURE 38.3.2

Since $x = \sin t$, then

$$\frac{dx}{dt} = \cos t$$

or

$$dt = \frac{dx}{\cos t}$$

Comparing the two expressions for dt, we find

$$w(x) = \frac{1}{\cos t} = \frac{1}{\sqrt{1 - \sin^2 t}} = \frac{1}{\sqrt{1 - x^2}}$$

To convert this distribution to a probability distribution, we need to normalize it so that

$$\int_{-1}^{1} w(x)\, dx = 1$$

But we have

$$\int_{-1}^{1} \frac{dx}{\sqrt{1 - x^2}} = \arcsin x \, \Big|_{-1}^{1} = \pi$$

Hence we must use

$$w(x) = \begin{cases} \dfrac{1}{\pi}\dfrac{1}{\sqrt{1 - x^2}} & -1 \le x \le 1 \\ 0 & \text{elsewhere} \end{cases}$$

in place of the old $w(x)$.

In the general case of $x = x(t)$ we use

$$\frac{dx}{d\,[x(t)]/dt}$$

We can either normalize the probability range before we start (note in the above example the range is π, the normalizing factor) or else normalize when we are done.

It is this function $w(x)$ that leads to the statistics of the analog-to-digital conversion process. Notice in the case of $x = \sin t$ that the "favorite values" are at the extremes of the range and not in the middle.

PROBLEMS

Find $w(x)$ for:

38.3.1 $x = e^{-t}$ $0 \le t \le N$, and 0 elsewhere

38.3.2 A circle

38.3.3 $x(t) = \tan t$ $(0 \le t \le \pi/4)$, 0 elsewhere

38.4 NOISE DUE TO QUANTIZATION

When we studied the roundoff noise, the precision of the numbers was such that it was reasonable to assume that the roundoff errors were uniformly distributed in the interval, and any skewness of the distribution of the numbers (see Secs. 2.8 and 2.9) did not reach that far down. But when the quantization is imprecise, then the skewness of the distribution of the values can affect the statistics of the quantization (roundoff). We therefore examine this effect, which turns out to be the well-known Sheppard's corrections (which we now derive).

We *assume* that the distribution function of the values of $w(x)$ has the property that the derivatives

$$w'(x), w''(x), w'''(x), \ldots$$

all approach zero as the value of x gets large in either direction, and also that

$$x^k w'(x), x^k w''(x), \ldots$$

approach zero for finite k. Usually this is true for distribution functions, and so it is not an unreasonable assumption.

In our quantization problem the range can be reduced to $a \leq x \leq b$ and divided into n quantization (class) intervals of equal size that correspond to the equal-size steps of the usual analog-to-digital converter. The intervals will be

$$(x_i - q/2, x_i + q/2)$$

where $b - a = qn$.

Under these assumptions when we apply the Euler-Maclaurin, Gregory, or Hamming-Pinkham equal-weight integration formulas (Sec. 18.9), the correction terms will all vanish, and we will have

$$\int_{-\infty}^{\infty} x^r w^{(s)}(x)\, dx = q \sum_{i=1}^{n} x_j^{\,r} w^{(s)}(x_j)$$

The probability p_j that we are in the jth quantization interval is

$$p_j = \int_{x_j - q/2}^{x_j + q/2} w(x)\, dx$$

and the rth moment is calculated from the quantized (grouped) data by

$$\mu_r'' = \sum_{j=1}^{n} x_j^{\,r} p_j$$

where the double primes mean calculated values, not derivatives. On the other hand the true value is, using a single prime for it,

$$\mu_r' = \int_a^b x^r w(x)\, dx$$

We can now expand the expression for the probability p_j

$$p_j = \int_{x_j - q/2}^{x_j + q/2} w(x)\, dx = \int_{-q/2}^{q/2} w(x_j + x)\, dx$$

$$= \int_{-q/2}^{q/2} \left[w(x_j) + xw'(x_j) + \frac{x^2}{2} w''(x_j) + \frac{x^3}{3!} w'''(x_j) + \cdots \right] dx$$

$$= qw(x_j) + \frac{q^3}{24} w''(x_j) + \frac{q^5}{1,920} w^{iv}(x_j) + \cdots$$

The calculated rth moment is therefore, using our assumptions that led to the equivalence of the sum and integral,

$$\mu_r'' = \sum_{j=1}^{n} x_j^r p_j$$

$$= \int_a^b x^r w(x)\, dx + \frac{q^2}{24} \int_a^b x^r w''(x)\, dx + \frac{q^4}{1,920} \int_a^b x^r w^{iv}(x)\, dx + \cdots$$

When integrating by parts, the integrated terms vanish, and we get

$$\mu_r'' = \mu_r' + \frac{q^2}{24} r(r-1) + \frac{q^4}{1,920} r(r-1)(r-2)(r-3) + \cdots$$

If the moments about the mean are μ (no primes), we get

$$\mu_0'' = \mu_0 = 1$$

$$\mu_1'' = \mu_1 = 0$$

$$\mu_2'' = \mu_2 + \frac{1}{12} q^2 \mu_0 = \mu_2 + \frac{q^2}{12}$$

$$\mu_3'' = \mu_3 + \frac{1}{4} q^2 \mu_1 = \mu_3$$

$$\mu_4'' = \mu_4 + \frac{q^2}{2} \mu_2 + \frac{q^4}{80}$$

Solving the other way, we get the true (unprimed) moments in terms of the calculated (grouped, double-primed) moments

$$\mu_1 = \mu_1'' = 0$$

$$\mu_2 = \mu_2'' - \frac{q^2}{12}$$

$$\mu_3 = \mu_3''$$

$$\mu_4 = \mu_4'' - \frac{q^2}{2} \mu_2'' + \frac{7}{240} q^4$$

. . . .

Evidently the correction in the second moment calculated from the quantized (grouped) data is particularly simple, and it is the main effect due to quantization that is used in practice—rarely do we go to the fourth moments.

38.5 THE QUANTIZATION THEOREM

Although we have expressed the quantized moments μ_r'' in terms of the true moments μ_r, and vice versa, there is still the question of when we can be sure that from the quantized data we can recover the true moments—the formalism may fail us if the series, for example, does not converge. This question is much like that answered by the sampling theorem which gives the conditions under which we can reconstruct the original function from the samples. There is a very similar theorem for quantization, based in the same way on the Fourier transform, but the physical meaning of the theorem is difficult to apply in practice, so we shall ignore it and proceed with an intuitive approach.

At first one might reason as follows. Let's suppose that the distribution of values $w(x)$ is reasonably normal. This assumption seems to fly in the face of the special case of $x = \sin t$ that we used as an example and where we found that the distribution of the values tended to pile up at the extremes. But for the moment we proceed anyway, holding on to our doubts. We could explain it, for example, by assuming that the extremes do *not* occur at the same level each time, and the result gives the normal distribution. One would be tempted in the case of, say, 8 levels of quantization to pick the step size to be σ. Thus most of the cases will fall in the range of the middle two intervals, and almost every value will fall in the 3σ range, and we still have a fourth to cover the rest, though there will be a little "clipping" when the value exceeds 4σ.

It is clear what is wrong with the plan. We are using most of the intervals of quantization for almost none of the values, and this is stupid. But then probably so was the assumption of a normal distribution.

Thus it is necessary to look more closely at the distribution of the values $w(x)$. It has a "dynamic" range from the least to the maximum that must be covered, or else we must accept some clipping of the extreme cases. Once we have decided this problem, if the rest of the values do not fall rather uniformly in the range, then we must consider what is known as "companding"—compressing and later expanding. We look for a transformation of the dependent variable that will transform the distribution of values to make it fairly uniform. Then, using this distribution, we quantize into equal steps and get the numbers into the computer. Inside we have to first undo the compression (by using the inverse transformation to expand it). In this way we get the most information through the narrow bottleneck of the analog-to-digital converter with its limited range of numbers. Once inside the computer, we almost always have a far greater range available for the expanded range.

"Companding" in one form or another is widely used; for example, in both phonograph records and tapes the playing unit has "compensation circuits" (expanding) to correct for the deliberate recording distortion to keep a reasonable dynamic range.

38.6   THE POOR MAN'S FOURIER SERIES

When most people who work in Fourier series hear that a 1-bit quantizer can give a lot of information, they immediately think along the following lines. Suppose I were to quantize the trigonometric functions that occur in the computation of the coefficients of a Fourier series, how well would I do using only additions and subtractions of the original data? (Carried further, this idea leads to Walsh functions, etc.)

It is easy to analyze. We get the calculated coefficient values[1]

$$A_m = \frac{1}{\pi} \int_0^{2\pi} f(x) \, \text{sign}(\cos mx) \, dx \qquad B_m = \frac{1}{\pi} \int_0^{2\pi} f(x) \, \text{sign}(\sin mx) \, dx$$

In order to analyze exactly what these numbers are, we begin by analyzing the square wave

$$\text{sign}(\sin x) = \begin{cases} 1 & 0 \leq x < \pi \\ -1 & \pi \leq x < 2\pi \end{cases}$$

This function has the Fourier-series coefficients (following the sin x only)

$$b_{1,k} = \frac{1}{\pi} \int_0^{2\pi} \text{sign}(\sin x) \sin kx \, dx$$

which is zero for even frequencies and for odd frequencies is

$$b_{1,2k+1} = \frac{2}{\pi} \int_0^{\pi} \sin(2k+1) \, x \, dx = \frac{2}{\pi} \frac{\cos(2k+1) \, x}{-(2k+1)} \bigg|_0^{\pi} = \frac{4}{\pi(2k+1)}$$

Thus we have

$$\text{sign}(\sin x) = \frac{4}{\pi} \sum_{k=0}^{\infty} \frac{\sin(2k+1) \, x}{2k+1}$$

Replacing x by mx, we get

$$\text{sign}(\sin mx) = \frac{4}{\pi} \sum_{k=0}^{\infty} \frac{\sin(2k+1) \, mx}{2k+1}$$

[1] The sign function is often written *sgn* (signum).

Thus

$$B_m = \frac{1}{\pi} \int_0^{2\pi} f(x) \frac{4}{\pi} \sum_{k=0}^{\infty} \frac{\sin(2k+1)\,mx}{2k+1}\,dx$$

$$= \frac{4}{\pi} \sum_{k=0}^{\infty} \frac{1}{2k+1} \frac{1}{\pi} \int_0^{2\pi} f(x) \sin(2k+1)\,mx\,dx$$

$$= \frac{4}{\pi} \sum_{k=0}^{\infty} \frac{1}{2k+1} b_{m(2k+1)}$$

where the b_j are the true Fourier-series coefficients of $\sin jx$. These equations are triangular in form

$$B_1 = \frac{4}{\pi} \left\{ b_1 + \frac{b_3}{3} + \frac{b_5}{5} + \cdots \right\}$$

$$B_2 = \frac{4}{\pi} \left\{ b_2 + \frac{b_6}{3} + \frac{b_{10}}{5} + \right\}$$

$$B_3 = \frac{4}{\pi} \left\{ b_3 + \frac{b_9}{3} + \cdot\cdot \right\}$$

etc.

If the original series converges rapidly, then we can solve these equations approximately for the b_j in terms of the B_m, and the solution is quite stable. Similarly for the cosine-term coefficients a_j in terms of the A_m.

In the case of a finite number of equally spaced points (and a finite number of terms in the Fourier expansion), we need clearly to use the midpoints t_p and avoid the endpoints of the intervals x_p. Confusion begins if you try to go directly from the continuous to the quantized discrete; the two stages done separately limits the confusion.

38.7 SOME GENERAL REMARKS ON QUANTIZATION EFFECTS

Quantization effects are often not as serious as the beginner is apt to think. It depends a great deal on what the quantized signal is going to be used for. The reader has probably seen the effects of quantization of photographs into two, four, eight, and sixteen levels. Using 1 bit is simply going to the extreme of contrast in the developing stage of the photograph. Sec. 38.6 shows reasonably clearly how much two or four levels would affect a Fourier expansion. Thus simple two-level quantization can at times be used successfully.

On the other hand, for some purposes quantization can be very serious. Much depends on the consumer of the information. If the consumer is a human, who to some extent has been engineered by evolution to be able to comprehend signals in the presence of severe noise, quantization can leave the message still intelligible. But it is also true that if you want high fidelity, the human is again well engineered to detect very small differences.

If the consumer is not a human, but further formulas, then we need to consider the two extreme conditions of the previous paragraph. The mere detection of the presence of a signal may be easy, but the detailed information carried by the signal may be very hard to recapture if the quantization has been at all serious. Thus the *use* of the quantized signal is central to the question of how serious the quantization is.

PART IV

Exponential
Approximation

39

SUMS OF EXPONENTIALS

39.1 INTRODUCTION

When considering a basis for the representation of a function there are three classes of functions that occur naturally: polynomials, sines and cosines, and exponentials. All these have the property that if a function can be represented in terms of them, then it can also be so represented if the origin is shifted. This is not true, for example, of the class $1, 1/x, 1/x^2, \ldots, 1/x^n$. It is only true for the above three classes and combinations of them. The numerical integration of differential equations shows why this property is relevant.

We studied the first class, polynomials, in Part II (Chaps. 14 through 30) and the second class, sinusoids, in Part III (Chaps. 31 through 38), and now we come to the exponentials in Part IV (three short chapters). The reasons for this disparity of treatment are several. First, polynomials are indeed more useful in many situations because they are also invariant under a stretch of $x \to kx'$ (because a polynomial of degree n in x is still a polynomial of degree n in kx), but for the sinusoids the frequencies will be changed, and hence they are not invariant under the transformation. Thus when the units of the independent variable are not reasonably well known, the polynomial approximation is the natural one to use.

Second, the polynomial treatment was arranged to introduce many of the ideas and methods necessary for a careful examination of the problems of computing. The Fourier approach further developed many of the ideas we need because in a very real sense they are simply exponentials with a pure imaginary exponent. Thus much of the material needed for exponential approximation has already been presented.

Third, exponentials are much less used in practice than the other two, and indeed they have played a much smaller role in the history of mathematics. Thus there is less necessity to present material on this topic in a reasonably balanced presentation of numerical methods involving one independent variable.

Finally, it is part of the plan of this book to deliberately pass from giving many of the details to giving fewer as the reader matures; it should no longer be necessary to give as many details as in the earlier parts of the book.

The problems of exponential approximation using a finite number of terms can be divided as follows: first, problems in which the exponents are known in advance; second, those in which the exponents are to be determined. We examine these two in this chapter. In the following chapter we examine the use of a continuum of exponentials in the Laplace transform (which corresponds to the Fourier transform).

39.2 LINEAR INDEPENDENCE

Before plunging into the problem of exponential approximation, we need to prove that the functions we are going to use are linearly independent over the sample points. For polynomials this was proved by showing that the Vandermonde determinant (Sec. 14.4) was not zero if the sample points were distinct. For the Fourier series we appealed to the orthogonality, and hence the linear independence, of the functions.

Consider, first, the following sum of exponential terms.

$$f(x) = C_1 e^{\alpha_1 x} + C_2 e^{\alpha_2 x} + C_3 e^{\alpha_3 x} + \cdots + C_n e^{\alpha_n x} \qquad \alpha_i \neq \alpha_j, \ i \neq j$$

To show that they are linearly independent in any continuous interval, suppose they are dependent and that the interval for which $f(x) \equiv 0$ includes the origin. Then on successive differentiations we get

$$f'(x) = C_1 \alpha_1 e^{\alpha_1 x} + C_2 \alpha_2 e^{\alpha_2 x} + \cdots + C_n \alpha_n e^{\alpha_n x} = 0$$
$$f''(x) = C_1 \alpha_1^2 e^{\alpha_1 x} + C_2 \alpha_2^2 e^{\alpha_2 x} + \cdots + C_n \alpha_n^2 e^{\alpha_n x} = 0$$
$$\cdot \quad \cdot \quad \cdot \quad \cdot \quad \cdot \quad \cdot \quad \cdot \quad \cdot \quad \cdot \quad \cdot \quad \cdot \quad \cdot \quad \cdot$$
$$f^{(n-1)}(x) = C_1 \alpha_1^{n-1} e^{\alpha_1 x} + C_2 \alpha_2^{n-1} e^{\alpha_2 x} + \cdots + C_n \alpha_n^{n-1} e^{\alpha_n x} = 0$$

At $x = 0$ the determinant of the equations for determining the coefficients C_k is exactly the Vandermonde, which cannot be zero if $\alpha_i \neq \alpha_j$; hence all the C_k are zero, contrary to the assumption of linear dependence.

For the case of discrete samples we use the fact that $e^{\alpha x} \neq 0$ for any finite x. This is the basis for an inductive proof. We therefore assume any $n - 1$ exponentials are linearly independent over any $n - 1$ distinct points. Consider now n exponentials at n points.

$$\sum_{k=1}^{n} C_k e^{\alpha_k x} = \left(\sum_{k=1}^{n-1} C_k e^{(\alpha_k - \alpha_n)x} + C_n \right) e^{\alpha_n x}$$

The term in parentheses is zero for n values, and by Rolle's theorem the derivative

$$\sum_{k=1}^{n-1} C_k(\alpha_k - \alpha_n)e^{(\alpha_k - \alpha_n)x} = 0$$

for at least $(n - 1)$ values, contrary to the induction hypothesis. Thus n exponentials are linearly independent over any n distinct points.

39.3 KNOWN EXPONENTS

In Chap. 15 we developed a uniform method for finding formulas based on polynomial approximation. We use this same method for finding formulas which are to be accurate for a set of exponentials having known exponents.

Suppose we have some linear operator $L(f)$, such as integration, differentiation, or interpolation, and we want the answer to be exact for a set of exponentials with known coefficients (given a corresponding set of samples). For convenience we will exclude the use of derivatives. Thus we are trying to estimate $L(f)$ as a linear combination of samples of the function $f(x_i)$ such that the formula

$$L(f) = w_0 f(x_0) + w_1 f(x_1) + \cdots + w_{n-1} f(x_{n-1})$$

will be exact for the n functions

$$e^{a_0 x}, e^{a_1 x}, \ldots, e^{a_{n-1} x}$$

(Often $a_0 = 0$, so that we have a constant term in the family of exponentials.)

The defining equations are, naturally,

$$m_k = \sum_{i=0}^{n-1} w_i e^{a_k x_i} \qquad k = 0, 1, \ldots, n - 1$$

where we have now written

$$m_k = L(e^{a_k x})$$

We have shown in Sec. 39.2 that the functions $e^{a_k x}$ are linearly independent over any set of n distinct points x_i, and so the determinant of these equations is not zero, and the equations can be solved for the unknown weights w_i of the formula.

As an experiment, consider the function

$$y = \frac{1}{4} + \frac{e^{-x}}{2} + \frac{e^{-2x}}{4}$$

It takes the values

x	y
0	1.00
1	0.468
2	0.322

where we are using 3-significant-digit arithmetic.   Given these data, we now try to fit it by 1, e^{-x}, e^{-2x}.   The equations are:

$$1.00 \; = a + 1.00 \; b + 1.00c$$
$$0.468 = a + 0.368b + 0.135c$$
$$0.322 = a + 0.135b + 0.183c$$

We eliminate a, using the top equation as the pivot,

$$0.532 = 0.632b + 0.865c$$
$$0.678 = 0.865b + 0.982c$$

Eliminating b, we find that back substitution leads to

$$c = \frac{0.032}{0.127} = 0.252$$
$$b = 0.497$$
$$a = 0.251$$

which is pretty good for 3-digit arithmetic.

PROBLEMS

39.3.1   Discuss the case of the given data including both function values and derivatives.

39.4   UNKNOWN EXPONENTS

The problem of interpolation (representation) of a function using sums of exponentials with unknown exponents is important since it arises frequently in practice and is the basis for the usual analytic-substitution approach.   Prony gave a simple method for finding the exponents *when* the given data are equally spaced. The essence of Prony's method is the separation of the problem into that of first finding the exponents and then finding the coefficients. In this respect it closely resembles the method used in Gaussian quadrature (Chap. 19).

We are assuming that the desired function $f(x)$ can be written in the form

$$f(x) = A_0 e^{\alpha_0 x} + A_1 e^{\alpha_1 x} + \cdots + A_{k-1} e^{\alpha_{k-1} x}$$

for some set of given values $x = x_j$ ($j = 1, 2, \ldots, n$) which are equally spaced. It is no real restraint to suppose that $x_j = j - 1$.

Prony observed that each of the exponentials

$$e^{\alpha_i j} = (e^{\alpha_i})^j \equiv \rho_i{}^j \qquad i = 0, 1, \ldots, k - 1$$

satisfies some fixed kth-order linear difference equation

$$y(j + k) + C_{k-1} y(j + k - 1) + C_{k-2} y(j + k - 2) + \cdots + C_0 y(j) = 0$$

with constant coefficients. The characteristic equation of this difference equation is

$$\rho^k + C_{k-1}\rho^{k-1} + C_{k-2}\rho^{k-2} + \cdots + C_0 = 0$$

and has the roots ρ_i. Since the individual terms satisfy the kth-order, linear, homogeneous difference equation, then any linear combination of the terms also satisfies the equation. In particular the original function satisfies the equation, that is,

$$f(j + k) + C_{k-1} f(j + k - 1) + \cdots + C_0 f(j) = 0 \qquad j = 1, 2, \ldots, n - k$$

The assumed form of the function has $2k$ parameters, and for the unique determination of them we need $n = 2k$ samples of data. In this case we have exactly k equations to determine the C_i ($i = 0, 1, \ldots, k - 1$). If we are to find the C_i, then the persymmetric determinant (Sec. 19.3)

$$\Delta = |f(j + k)|$$

must not be zero. Knowing the C_i, we can put them into the characteristic equation and from this determine the roots ρ_i, which in turn give the exponents α_i. If the exponents are to be real, then the roots must be nonnegative. Once we have the exponents we can then use the first k of the defining equations to determine the coefficients A_j. Just as in the Gaussian integration we are assured that these will fit the last k equations also.

An examination of this process shows that it is very similar to the Gaussian integration formulas.

As an easily followed example, consider the data $k(= 2)$

x	y
0	32
1	20
2	14
3	11

Working our way up the displayed equations, we get, in turn, the equations for the C_i

$$14 + 20C_1 + 32C_0 = 0$$
$$11 + 14C_1 + 20C_0 = 0$$

whose solutions are

$$C_1 = -3/2$$
$$C_0 = 1/2$$

The characteristic equation is therefore

$$\rho^2 - \frac{3}{2}\rho + \frac{1}{2} = 0$$

from which we get

$$(\rho - 1)\left(\rho - \frac{1}{2}\right) = 0$$
$$\rho = 1, \tfrac{1}{2}$$

and

$$y_n = f(n) = A_0(1)^n + A_1(\tfrac{1}{2})^n$$

The first two defining equations are

$$32 = A_0 + A_1$$
$$20 = A_0 + A_1/2$$

whose solutions are

$$A_1 = 24$$
$$A_0 = 8$$

and we have, finally,

$$y_n = f(n) = 8 + \frac{24}{2^n} = 8 + 3 x 2^{3-n}$$

PROBLEMS

39.4.1 Fit

$$y = A_1 e^{\alpha_1 x} + A_2 e^{\alpha_2 x}$$

to the data

x	0	1	2	3
y	128	48	10	9

39.4.2 Fit

$$y = A_1 e^{\alpha_1 x} + A_2 e^{\alpha_2 x}$$

to the data

x	0	1	2	3
y	0	$\frac{1}{2}$	$\frac{3}{4}$	$\frac{7}{8}$

39.4.3 Discuss the case when $\Delta = 0$ and we cannot determine the C_i.

39.5 LEAST-SQUARES FITTING

Many times we have more data than parameters, and we wish to find a least-squares fit to the data. Again, supposing that the data points x_j are equally spaced, we will come down to more than k equations to determine the C_i. Thus we are in a position to apply the methods of least-squares solution (Chaps. 25 and 27) of the simultaneous linear equations. Having found the coefficients of the characteristic equation, we solve them for the characteristic roots and hence the exponents. Now with these known, we can again resort to a least-squares fitting of the data by the coefficients A_j.

The details are straightforward, though a bit messy, and need not be given here; we need to note only that there are two stages of least-squares fitting, those for the coefficients C_i of the characteristic equation and those for the coefficients A_j of the exponentials.

What relation this two-stage least-squares fitting bears to the direct fitting (which is often difficult to do) is open for discussion, but it is likely to be reasonably close as measured by the resulting sum of squares of the residuals.

39.6 PRONY'S METHOD WITH CONSTRAINTS

Just as we examined integration formulas when various, often naturally occurring, constraints were applied, we now look at Prony's method when there are constraints.

A very common constraint is that one or more of the exponents α_i are known (frequently $\alpha_0 = 0$).

Suppose that we have the special case (5 parameters)

$$f(x) = A_0 + A_1 e^{\alpha_1 x} + A_2 e^{\alpha_2 x}$$

When we come to the characteristic equation, which will be a cubic

$$\rho^3 + C_2 \rho^2 + C_1 \rho + C_0 = 0$$

we know that one of the roots is $\rho = 1$. This means that

$$1 + C_2 + C_1 + C_0 = 0$$

From the data we get the equations

$$f(3) + C_2 f(2) + C_1 f(1) + C_0 f(0) = 0$$
$$f(4) + C_2 f(3) + C_2 f(2) + C_0 f(1) = 0$$

and adding the equation

$$1 + C_2 + C_1 + C_0 = 0$$

we have the three equations necessary to determine C_0, C_1, and C_2. Each known exponent reduces the necessary given data by 1, thus reducing the standard set of equations by 1, but by substituting each known exponent in the characteristic equation, we gain one corresponding equation. As a result, we still have the necessary number of equations.

Conditions on the coefficients tend to resemble that of Chebyshev integration if they are on unknown exponents and Ralston integration if they are on known exponents. The constraints may lead to a linear programming problem (Sec. 43.10).

39.7 WARNINGS

We have supposed that all will go properly in Prony's method, but surprises await the unwary. Consider, for example, the problem of fitting

$$y(x) = A_1 e^{\alpha_1 x} + A_2 e^{\alpha_2 x}$$

to the data

x	0	1	2	3
y	16	12	4	3

The equations for determining the coefficients of the characteristic polynomial are

$$4 + 12C_1 + 16C_2 = 0$$
$$3 + \ 4C_1 + 12C_2 = 0$$

Solving these, we get

$$C_1 = 0$$
$$C_2 = -\tfrac{1}{4}$$

The characteristic equation is therefore

$$\rho^2 - \tfrac{1}{4} = 0$$

and it has the roots

$$\rho = \pm\tfrac{1}{2}$$

This produces the exponents

$$\alpha_1 = -ln\,2$$
$$\alpha_2 = \pi i + \ln\tfrac{1}{2} = \pi i - \ln 2$$

Thus, while the approximating sum of exponentials will assume the (real) given values at the sample points, between the sample points the approximating function will be a complex number! Hence, it will normally be dangerous to use for analytic substitution.

On the other hand, complex conjugates ρ_i with the real part greater than zero produce complex conjugate exponents, and with the corresponding coefficients A_i conjugates of each other the sum is real.

Our trouble is that we can solve the difference equation easily in the conventional form, and this leads to complex values for the interpolating function. The solution proceeds

$$y(n) = A_1(\tfrac{1}{2})^n + A_2(-\tfrac{1}{2})^n$$
$$n = 0: \quad 16 = A_1 + A_2$$
$$n = 1: \quad 12 = A_1/2 - A_2/2$$
$$A_1 = 20$$
$$A_2 = -4$$

Solution:

$$y(n) = 20(\tfrac{1}{2})^n - 4(-\tfrac{1}{2})^n$$

However, this form gives trouble in interpolating in the whole interval. A dodge is to write,[1] since $\cos \pi n = (-1)^n$,

$$y(n) = (\tfrac{1}{2})^n(20 - 4\cos \pi n)$$

[1] We could also include an arbitrary term $\sin \pi n$.

Then it is easy to get a form for interpolating all x values, namely

$$y(x) = (\tfrac{1}{2})^x(20 - 4 \cos \pi x)$$

While in principle we have shown how to find the exponents, things still do not always go so well in practice; sometimes it happens that the number of terms to use is not known but is to be found.   An illustration of this is radioactive decay, where the terms correspond to various half-lives in the decay chain being investigated.

Let us examine the simple case of trying to distinguish between

$$Ae^{-\alpha t} \qquad \text{and} \qquad \frac{A}{2}e^{-(\alpha+\varepsilon)t} + \frac{A}{2}e^{-(\alpha-\varepsilon)t} = Ae^{-\alpha t}\left(\frac{e^{-\varepsilon t} + e^{\varepsilon t}}{2}\right) \qquad \alpha > 0$$

The expression in the parentheses is

$$1 + \frac{\varepsilon^2 t^2}{2} + \frac{\varepsilon^4 t^4}{24} + \cdots$$

The difference in the two forms depends on ε^2, and only for large t can we hope to detect this amid the background noise of measurement.   But for large t, $e^{-\alpha t}$ is small!   A similar situation applies to Laplace transforms (Chap. 40).

$$f(t) = \int_0^\infty F(\sigma)e^{-\sigma t}\, d\sigma$$

Given $F(\sigma)$, it is easy to compute $f(t)$, but given $f(t)$ in the form of data, the problem of finding $F(\sigma)$ is more difficult.   One of the difficulties is that the values of $F(\sigma)$ for large σ determine the values of $f(t)$ for small t, and vice versa.   With proper care and known limitations on $f(t)$ or with some extra information about $F(\sigma)$, the inverse transform can sometimes be found.

39.8 EXPONENTIALS AND POLYNOMIALS

When the overall behavior of a problem is exponential in character, the use of a polynomial times a suitable exponent often is more manageable than a sum of exponentials.   Gauss-Laguerre integration

$$\int_0^\infty e^{-x}f(x)\, dx = \sum_{i=1}^n w_i f(x_1)$$

further illustrates this point.   Since the general techniques necessary to carry out this suggestion have been developed in Part II, we do not need to discuss them further.

39.9   ERROR TERMS

When the formulas have been found, it is natural to ask for the error terms.   The approach in Chap. 16 of Part II started with an expansion of the arbitrary function $f(x)$ in terms of the particular function $1, x, \ldots, x^{m-1}$ for which the formula was made exact.   This expansion has been generalized to arbitary sets of functions by Peterson, and a treatment of it can be found in A. S. Householder's excellent (but difficult-to-read) book " Principles of Numerical Analysis " [21].

In practice the usefulness of such error terms is open to debate, and so far they have seldom been used.   The error term corresponding to that developed for band-limited functions has apparently not been investigated.

40

*THE LAPLACE TRANSFORM

40.1 WHAT IS THE LAPLACE TRANSFORM?

Given a function $F(t)$, we say that the Laplace transform $f(s)$ is

$$f(s) = \int_0^\infty e^{-ts} F(t)\, dt$$

where for the moment we think of the variable s as being real, though later we will let it take on complex values. The Laplace transform is closely related to the Fourier transform, especially the so-called two-sided Laplace transform [61]

$$\int_{-\infty}^\infty e^{-ts} F(t)\, dt$$

For functions which vanish identically for all negative time, the two-sided transform reduces to the one-sided, or plain, Laplace transform. This class of functions is very natural to consider in many situations, especially in experiments where before starting it is reasonable to suppose that nothing happened and that the functions of interest are identically zero. Thus we regard $F(t) = 0$ for $t < 0$.

The Laplace transform has many uses, and we cannot in one short chapter develop more than a small feeling for what it is all about. Thus we must refer the reader to the many standard texts on the topic.[1]

To each function $F(t)$ there corresponds a transform $f(s)$. It is an important theorem (Lerch's theorem) that the reverse is true within functions that are zero *except* for a set of measure zero—and such exceptions are hardly likely to be relevant in practical numerical computation. Tables of Laplace transforms are often made and are widely used.

40.2 SOME EXAMPLES OF LAPLACE TRANSFORMS

In order to get a feel for Laplace transforms we shall find a number of simple ones.

EXAMPLE 40.2.1 $F(t) = t^n$
Then

$$f(s) = \int_0^\infty e^{-st} t^n \, dt$$

Integration by parts gives

$$f(s) = t^n \frac{e^{-st}}{-s} \bigg|_0^\infty + \frac{n}{s} \int_0^\infty t^{n-1} e^{-st} \, ds$$

$$= \frac{n}{s} \int_0^\infty e^{-st} t^{n-1} \, ds$$

and by repetition we get

$$f(s) = \frac{n!}{s^{n+1}}$$

EXAMPLE 40.2.2 $F(t) = \sin at$
Then

$$f(s) = \int_0^\infty e^{-st} \sin at \, dt$$

Standard integration tables give

$$f(s) = e^{-st} \left(\frac{-s \sin at - a \cos at}{s^2 + a^2} \right) \bigg|_0^\infty$$

$$= \frac{a}{s^2 + a^2} \qquad s > 0$$

[1] An excellent introduction is M. R. Spiegel [55]; see also W. R. LePage [36].

EXAMPLE 40.2.3　$F(t) = \cos at$
As in Example 40.2.2 we get

$$f(s) = \int_0^\infty e^{-st} \cos at \, dt$$

$$= e^{-st} \left(\frac{-s \cos at + a \sin at}{s^2 + a^2} \right) \bigg|_0^\infty$$

$$= \frac{s}{s^2 + a^2} \qquad s > 0$$

EXAMPLE 40.2.4　$F(t) = \begin{cases} 0 & t < a \\ 1 & t > a \end{cases}$

$$f(s) = \int_0^\infty e^{-st} F(t) \, dt$$

$$= \int_a^\infty e^{-st} \, dt = \frac{e^{-st}}{-s} \bigg|_a^\infty = \frac{e^{-as}}{s}$$

PROBLEMS

Find the Laplace transform $F(t) \leftrightarrow f(s)$.

40.2.1　$\sinh at \leftrightarrow \dfrac{a}{s^2 - a^2} \qquad s > |a|$

40.2.2　$\cosh at \leftrightarrow \dfrac{s}{s^2 - a^2} \qquad s > |a|$

40.2.3　$c_1 F_1(t) + c_2 F_2(t) \leftrightarrow c_1 f_1(s) + c_2 f_2(s)$

40.2.4　$F(t) = \begin{cases} 1 & 0 < a \leq t \leq b \\ 0 & \text{elsewhere} \end{cases}$

40.2.5　$F(t) = \begin{cases} 1 - t & 0 \leq t \leq 1 \\ 0 & \text{elsewhere} \end{cases}$

40.3　SOME GENERAL PROPERTIES OF LAPLACE TRANSFORMS

Differentiation and integration of Laplace transforms are very common, and the rules are given as follows, where we need to take the limit from the right for a discontinuity at the origin and appropriate directions at other points of discontinuity.
　　If

$$f(s) = \int_0^\infty e^{-st} F(t) \, dt$$

then

$$sf(s) - F(0) = \int_0^\infty e^{-st} F'(t) \, dt$$

$$s^2 f(s) - sF(0) - F'(0) = \int_0^\infty e^{-st} F''(t) \, dt$$

$$\frac{f(s)}{s} = \int_0^\infty e^{-st} \left[\int_0^t F(u) \, du \right] dt$$

Using these we can find

$$\int_0^\infty e^{-st} [t^n F(t)] \, dt = (-1)^n \frac{d^n}{ds^n} f(s)$$

$$\int_0^\infty e^{-st} \left[\frac{F(t)}{t} \right] dt = \int_s^\infty f(u) \, du$$

EXAMPLE 40.3.1   Given $F(t) = \sin t$, then

$$\int_0^\infty e^{-st} \left(\frac{\sin t}{t} \right) dt = \int_0^\infty \frac{du}{u^2 + 1} = \arctan 1/s$$

There are two shifting theorems for the Laplace transforms.
Given

$$f(s) = \int_0^\infty e^{-st} F(t) \, dt$$

then

$$f(s - a) = \int_0^\infty e^{-st} \{ e^{at} F(t) \} \, dt$$

and if

$$G(t) = \begin{cases} F(t) - a & t > a \\ 0 & t < a \end{cases}$$

then

$$e^{-as} f(s) = \int_0^\infty e^{-st} G(t) \, dt$$

Both are immediately verifiable.
    We defined (Sec. 34.7) the convolution of two functions which are 0 for $x < 0$ as

$$H(t) = \int_0^t F(u) G(t - u) \, du$$

We have, similar to the Fourier transform,

$$f(s) g(s) = \int_0^\infty e^{-st} H(t) \, dt$$

All these properties are useful in practice.

PROBLEMS

Give the details for:
40.3.1   The first shifting theorem
40.3.2   The second shifting theorem
40.3.3   The convolution theorem

40.4   PERIODIC FUNCTIONS

We examined periodic functions in Chaps. 31 to 33 in terms of the Fourier transform.   In view of their importance we again examine this class of functions, this time in terms of the Laplace transform.

If $F(t)$ has a period $T > 0$, that is, $F(t + T) = F(t)$, then we can write

$$\int_0^\infty e^{-st}F(t)\,dt = \sum_{k=0}^\infty \int_{kT}^{k(T+1)} e^{-st}F(t)\,dt$$

$$= \sum_{k=0}^\infty \int_0^T e^{-s(t-kT)}F(t - kT)\,dt$$

$$= \int_0^T F(t)\sum_{k=0}^\infty e^{-st}\cdot e^{skT}\,dt$$

$$= \int_0^T \frac{e^{-st}F(t)}{1 - e^{-sT}}\,dt \qquad s > 0$$

Thus we have the extra factor

$$\frac{1}{1 - e^{-sT}}$$

in the integrand to compensate for the reduced range $0 \le t \le T$.

40.5   APPROXIMATION OF LAPLACE TRANSFORMS

The method of analytic substitution, which we will discuss again in Chap. 42, is quite useful in the evaluation of Laplace transforms.   Suppose we have some function $F(t)$ for which there is no value in the tables of Laplace transforms that we consult, or suppose they are experimentally determined data.   What can we do? We merely examine a table of transforms for family of functions $F_i(t)$ that look suitable for approximating our given function $F(t)$.   If we can find a linear representation of $F(t)$ in terms of this family

$$F(t) \simeq c_1 F_1(t) + c_2 F_2(t) + \cdots + c_N F_N(t)$$

then using analytic substitution and the linearity property of the Laplace transform, we get our answer.   (See Secs. 9.10 and 12.5 for similar examples.)

Everything seems fine.   We can apply any method we please to find the approximation, exact matching at selected points, least squares, Chebyshev, or anything else that seems appropriate.   Thus we have apparently thrown the problem back onto the approximation problem.   And this is what we have been doing all along.   We first used the class of functions $1, x, x^2, \ldots, x^{n-1}$; later we used the class of sines and cosines; and now we are looking at exponentials.   But any family of functions for which we know the analytic answers can be used provided the process is linear (and the functions are linearly independent).

Our wording, however, raises the question of how it will go in practice, or in other words, how should you match the function?   An examination of the Laplace transform suggests that this is not a trivial question.   Consider, for example, two functions which differ far out, but are close to each other for moderate values of t.   How will the Laplace transform "see" the difference?   We weight the difference by the factor e^{-st}, and for moderate values of $s > 0$ this difference simply will not be seen in the accumulated sum of integrand values.   What happens far out in the t variable can only be seen in the transform near the origin (s small).   And conversely, what is near the origin will be seen far out in the transform since even when s is large, it will still enter significantly in the sum that the integral computes.

We now see the fundamental trouble with the Laplace transforms.   The neighborhood of infinity is transformed into the neighborhood of the origin, and vice versa.   But in the time domain, where we usually make the measurements, we have a limit of how fine a spacing we can take near the origin, and we also have a limit on how long a time we can make measurements.   Thus the transform is not as well determined as we could wish.   Indeed, one can say with Lanczos [33] that the inversion cannot be done accurately.

However, we find in practice that both the direct and inverse processes are done.   This is accomplished by making assumptions about the behavior of the function.   For example, it is often convenient to assume that the transform is, or can be approximated by, a rational function with a known degree for the numerator and the denominator.   With this assumption we fit (somehow) the proper rational function to the transform, and from a table we can get the original function.   This is one example of what we said in the first paragraph in this section: "We merely examine a table of transforms for a family of functions $F_i(t)$ that look suitable for approximating our given function $F(t)$."   We can also apply the method for $f(s)$ if we wish.

PROBLEMS

40.5.1   Discuss approximating $f(s) \simeq \sum_{i=1}^{N} c_i f_i(s)$.

40.6   COMPLEX FREQUENCIES

In practice the variable s is allowed to be complex, usually with a limit on the size of the real part so that the integral

$$f(s) = \int_0^\infty e^{-st} F(t)\, dt$$

will converge.   Thinking back on the underlying meaning of the integral as a limit of a sum, we are now summing terms like (for $s = x + iy$)

$$F(t) e^{-xt} e^{-iyt}$$

which can also be written as

$$F(t)\, e^{-xt} \cos yt \qquad F(t)\, e^{-xt} \sin yt$$

Thus the Laplace transform handles complex frequencies, frequencies that are sinusoidal but with an exponential decaying factor, while the Fourier transform handles only sinusoidal terms.   In this sense the Laplace transform includes the Fourier transform as a special case.

The Fourier transform considers values only along a line, much as real-variable theory considers values along the real axis.   The theory of complex variables, by considering values in the whole of the complex plane, both greatly limits the class of functions and at the same time greatly expands the power of the methods used.   The Laplace transform with its use of the complex plane similarly gains in power.   As an example, the Fourier transform of a constant 1, (or even the function sin t) does not technically exist for lack of convergence, but the Laplace transform, by using a slightly negative real part of s, has all the necessary convergence and provides a more powerful analytical tool.   (One may question, however, the increased physical power.)

When we get into the complex plane, we get an inversion formula corresponding to the Fourier inversion formula.   This is

$$F(t) = \frac{1}{2\pi i} \int_{\gamma - i\infty}^{\gamma + i\infty} e^{st} f(s)\, ds \qquad t > 0$$

and is sometimes called "Bromwich's integral formula."[1]   The value of γ is chosen so that the integral converges for s values on this line, and hence the Cauchy integral can be closed on the left.

---

[1] Oliver Heaviside (1850–1925) by his formal approach initiated research on the mathematical foundations by others and is justly regarded as the "father of the field."

The complex inversion formula has great practical limitations.   Generally speaking, we cannot expect to make the measurements necessary to evaluate this integral—they can only come from some theoretical assumptions and hence are not likely to lead to computing problems.

40.7   A FORMULA FOR NUMERICAL INTEGRATION[1]

The numerical evaluation of the Laplace transform, given equally spaced data, is often necessary.   We shall find a formula, like the Hamming-Pinkham (Sec. 18.9), that uses end corrections in the form of differences.

We begin by examining integrals, over a finite range, of the form

$$I(\lambda) = \int_0^n e^{-\lambda x} f(x)\, dx$$

Given equally spaced data, which we assume is at unit spacing, we approximate the integral by the form

$$\int_0^n e^{-\lambda x} f(x)\, dx = \sum_{k=0}^n e^{-\lambda k} f(k) + \sum_{s=0}^\infty [h_s(\lambda) e^{-\lambda n} \Delta^s f(n-s) + (-1)^s m_s(\lambda) \Delta^s f(0)]$$

where the $h_s(\lambda)$ and $m_s(\lambda)$ play the role of the q_s in the earlier formula.   The dependence of the coefficients $h_s(\lambda)$ and $m_s(\lambda)$ on the parameter is to be expected.

As before, we begin by requiring the formula to be exact for the integrand $f(x) = 1, x, x^2, \ldots$ and get the defining equations

$$\int_0^n e^{-\lambda x} x^p\, dx = \sum_{k=0}^n e^{-\lambda k} k^p + \sum_{s=0}^\infty [h_s(\lambda) e^{-\lambda n} \Delta^s (n-s)^p + (-1)^s m_s(\lambda) \Delta^s (0)^p]$$

$$p = 0, 1, 2, \ldots$$

To get the exponential generating function, we multiply the pth equation by $t^p/p!$ and sum over all the equations

$$\int_0^n e^{-\lambda x} e^{tx}\, dx = \sum_{k=0}^n e^{-\lambda k} e^{kt} + \sum_{s=0}^\infty [h_s(\lambda) e^{-\lambda n} \Delta^s e^{(n-s)t} + (-1)^s m_s(\lambda) \Delta^s e^{0t}]$$

Why do we use the exponential generating function?   Simply because for the operations we are going to do, integration, summation of a series, and differencing, the exponential function is left unchanged.   As a result we will be able to arrange the various terms into exponential and nonexponential terms, and the terms with the exponential will carry all the dependence on n.

---

[1] See also V. I. Krylov and N. S. Skoblya [31].

We now do the indicated operations

1    $\int_0^n e^{-(\lambda-t)x}\,dx = \dfrac{1 - e^{-(\lambda-t)n}}{\lambda - t}$

2    $\sum_{k=0}^n e^{-(\lambda-t)k} = \dfrac{e^{-(\lambda-t)(n+1)} - 1}{e^{-(\lambda-t)} - 1} = \dfrac{e^{-(\lambda-t)n} - e^{(\lambda-t)}}{1 - e^{(\lambda-t)}}$

3    $\Delta^s e^{(n-s)t} = e^{nt}(1 - e^{-t})^s$

4    $\Delta^s e^{0t} = (e^t - 1)^s$

and assemble the parts

$$e^{-(\lambda-t)n}\left[-\frac{1}{\lambda-t} - \frac{1}{1 - e^{(\lambda-t)}} - \sum_{s=0}^{\infty} h_s(\lambda)(1 - e^{-t})^s\right]$$
$$+ \left[\frac{1}{\lambda-t} + \frac{e^{(\lambda-t)}}{1 - e^{(\lambda-t)}} - \sum_{s=0}^{\infty}(-1)^s m_s(\lambda)(e^t - 1)^s\right] = 0$$

If we now define

$$Q(t, \lambda) = -\frac{1}{\lambda-t} + \frac{1}{e^{(\lambda-t)} - 1} - \sum_{s=0}^{\infty} h_s(\lambda)(1 - e^{-t})^s$$

$$P(t, \lambda) = \frac{1}{\lambda-t} + \frac{1}{e^{-(\lambda-t)} - 1} - \sum_{s=0}^{\infty}(-1)^s m_s(\lambda)(e^t - 1)^s$$

we have

$$e^{-(\lambda-t)n}Q(t, \lambda) + P(t, \lambda) = 0$$

Comparing $Q(t, \lambda)$ with $P(t, \lambda)$ we have

$$Q(t, \lambda) = P(-t, -\lambda)$$

provided

$$m_s(-\lambda) \equiv h_s(\lambda)$$

Thus we need only use $Q(t, \lambda) = 0$ to find the coefficients $h_s(\lambda)$ of the **differences**, where

$$-\sum_{s=0}^{\infty} h_s(\lambda)(1 - e^{-t})^s = \frac{1}{\lambda-t} - \frac{1}{e^{(\lambda-t)} - 1}$$

Again we set

$$1 - e^{-t} = v \qquad t = -\ln(1 - v)$$

to get the power series in v

$$\sum_{s=0}^{\infty} h_s(\lambda) v^s = \frac{-1}{\lambda + \ln(1 - v)} + \frac{1}{(1 - v)e^{\lambda} - 1}$$

The expansion of the right-hand side in powers of v will yield the coefficients $h_s(\lambda)$. As a result of a lot of tedious algebra we find Table 40.7[1] for $A_s(\lambda)$ where

$$h_s(\lambda) = (-1)^s A_s(\lambda) + \frac{e^{\lambda s}}{(e^{\lambda} - 1)^{s+1}}$$

$$m_s(\lambda) = h_s(-\lambda)$$

Thus we have the integration formula we sought.

[1] See M. M. Kaplan, "Numerical Integration with Complex Exponential Kernels," doctoral thesis, Stevens Institute of Technology, 1969, for tables and details.

Table 40.7

s	$A_s(\lambda)$
0	$-\dfrac{1}{\lambda}$
1	$\dfrac{1}{\lambda^2}$
2	$-\left(\dfrac{1}{\lambda^3} + \dfrac{1}{2\lambda^2}\right)$
3	$\dfrac{1}{\lambda^4} + \dfrac{1}{\lambda^3} + \dfrac{1}{3\lambda^2}$
4	$-\left(\dfrac{1}{\lambda^5} + \dfrac{1}{2\lambda^4} + \dfrac{11}{12\lambda^3} + \dfrac{1}{4\lambda^2}\right)$
5	$\dfrac{1}{\lambda^6} + \dfrac{2}{\lambda^5} + \dfrac{7}{4\lambda^4} + \dfrac{5}{6\lambda^3} + \dfrac{1}{5\lambda^2}$
6	$-\left(\dfrac{1}{\lambda^7} + \dfrac{5}{2\lambda^6} + \dfrac{2}{\lambda^5} + \dfrac{15}{8\lambda^4} + \dfrac{137}{180\lambda^3} + \dfrac{1}{6\lambda^2}\right)$
7	$\dfrac{1}{\lambda^8} + \dfrac{3}{\lambda^7} + \dfrac{25}{7\lambda^6} + \dfrac{39}{14\lambda^5} + \dfrac{29}{15\lambda^4} + \dfrac{7}{10\lambda^3} + \dfrac{1}{7\lambda^2}$

40.8   MIDPOINT FORMULAS

Corresponding to the midpoint integration formula we have a midpoint formula

$$\int_0^n e^{-\lambda x} f(x)\, dx = \sum_{k=0}^{n-1} e^{-\lambda(k+1/2)} f(k+1/2)$$

$$+ \sum_{s=0}^{\infty} e^{-\lambda n} h_s^*(\lambda) \Delta^s f(n-s-1/2) + (-1)^s m_s^*(\lambda) \Delta^s f(1/2)$$

and Table 40.8.

Table 40.8

s	$A_s^*(\lambda)$
0	$-\dfrac{1}{\lambda}$
1	$\dfrac{1}{\lambda^2} + \dfrac{1}{2\lambda}$
2	$-\left(\dfrac{1}{\lambda^3} + \dfrac{1}{\lambda^2} + \dfrac{3}{8\lambda}\right)$
3	$\dfrac{1}{\lambda^4} + \dfrac{3}{2\lambda^3} + \dfrac{23}{24\lambda^2} + \dfrac{5}{16\lambda}$
4	$-\left(\dfrac{1}{\lambda^5} + \dfrac{2}{\lambda^4} + \dfrac{43}{14\lambda^3} + \dfrac{11}{12\lambda^2} + \dfrac{35}{128\lambda}\right)$
5	$\dfrac{1}{\lambda^6} + \dfrac{5}{2\lambda^5} + \dfrac{51}{14\lambda^4} + \dfrac{1,303}{420\lambda^3} + \dfrac{563}{640\lambda^2} + \dfrac{105}{4\lambda}$
6	$-\left(\dfrac{1}{\lambda^7} + \dfrac{77}{30\lambda^6} + \dfrac{111}{28\lambda^5} + \dfrac{4,057}{840\lambda^4} + \dfrac{69,079}{22,400\lambda^3} + \dfrac{10,459}{12,800\lambda^2} + \dfrac{567}{2,560\lambda}\right)$
7	$\dfrac{1}{\lambda^8} + \dfrac{219}{70\lambda^7} + \dfrac{61,328}{11,760\lambda^6} + \dfrac{3,711,554}{419,440\lambda^5} + \dfrac{207,237}{156,800\lambda^4} + \dfrac{52,741}{12,260\lambda^4}$ $+ \dfrac{10,459}{44,800\lambda^3} + \dfrac{898,027}{313,600\lambda^3} + \dfrac{81}{2,560\lambda^2} + \dfrac{135,967}{179,200\lambda^2} + \dfrac{1,053}{5,120\lambda}$

40.9   EXPERIMENTAL RESULTS

Numerical experiments (by Kaplan) indicate that the formula works about as expected.

For the Laplace transform the upper limit is infinity, and for most practical problems the corresponding differences all approach zero so that the m_s terms drop out.

40.10   FOURIER TRANSFORMS

It is immediately evident that if in the preceding formula we write $\lambda = i\omega$, and if we do sufficient algebra, then we will have formulas for both

$$\int_0^n \cos \omega t f(t)\, dt \qquad \text{and} \qquad \int_0^n \sin \omega t f(t)\, dt$$

Alternatively, we could start from scratch and use a generating-function type of derivation technique to obtain the corresponding formulas directly.

For the Fourier transform we let both limits go to infinity, and, of course, the end correction terms will drop out, leading to the trapezoid rule for integration.

The experimental comparison of the above type of formulas with the classical Filon formula (based on local quadratic approximation of the integrand, similar to Simpson's formula) shows that the above formula is superior in accuracy as well as in the more important properties of providing an estimate of both the roundoff and truncation errors.   For details the reader is referred to Kaplan's thesis.[1]

---

[1] Op. cit.

41

*SIMULATION AND THE METHOD OF ZEROS AND POLES

41.1 INTRODUCTION

Simulation consumes a great deal of time on many computing machines; indeed one can look at a computer as a large, flexible general-purpose device for simulating various processes. Many simulations depend mainly on the integration of ordinary differential equations, and this is the reason we have devoted a great deal of space to the subject.

On the other hand it is difficult to discuss simulation in neat classifications with nice definite formulas strung across the page. What to do in a simulation depends very much on the source of the problem, the money available, and what is going to be done with the results. Consequently, the subject apparently is not susceptible to the usual mathematical treatment.

These two conflicting aspects of the problem cause most texts to ignore the topic after giving a few methods for integrating ordinary differential equations. However, we shall not dismiss it, even though we are reduced to working around the topic with many words and few formulas.

41.2 SIMULATION LANGUAGES

There are many, and undoubtedly there will be many more, simulation languages applicable to various fields.   Usually the descriptions of these specialized languages give a great deal of space to how to use the language, but very little to what method of integration is used, and almost no space as to *why* the particular method was chosen.

Given a short problem where the machine time is not important, most people will grab a handy simulation language and get an answer.   The answers they get will often conceal any interesting aspect of the computation that might shed light on the meaning of the solution.   Consider, for example, a method of integration that actually does halve and double as the accuracy requirements demand (many methods do not, and you will never know from the output in those cases that the answer is apt to be wrong).   For purposes of comparison of successive runs, the output is almost always given at a regular spacing assigned by the user.   This naturally conceals the detailed behavior of the solution at exactly those interesting parts where there was a great deal of structure.   Yes, we wanted numbers to study, but the motto of the book also states that we want insight—insight that can often arise from the *way* the computation proceeds, rather than from the numbers. Why at this place was a small step necessary?   What regions had very large steps, and what does that mean?   Does this suggest some analytic approximation in this region, and does the behavior in the region of small steps suggest some very nonpolynomial-like behavior compatible with the theory?   Perhaps we can find other analytic approximations that will more clearly reveal what is going on.

This is not to say that simulation languages should not be used; often their ease of use outweighs the probable gain in insight of a more careful, and initially more costly, approach.

Most simulation languages use a method of integration at least as good as Euler's modified predictor-corrector method (Sec. 22.3), though some are as crude as the simple Euler predictor method.   Usually on these latter ones, the user must specify the step size to be used throughout the whole run.

Runge-Kutta methods (Sec. 24.2) are very popular because they are both accurate (hence reasonably efficient) and self-starting (hence compact coding). They have the weakness of not doing too well on the matter of monitoring their own accuracy (unless further evaluations or other features are incorporated into the method, and these are seldom done).

The better predictor-corrector methods are occasionally used, but the rumored instabilities, no self-starting properties, and general untidiness tend to make them be passed over by programmers.

When using a simulation language with its packaged routine, or indeed any packaged routine for differential equations, one should watch carefully to see how to avoid the loss of insight that these approaches tend to produce (along with their great ease of use).

When a simulation is extremely important, or when the machine time used is going to amount to enough money that it becomes a significant part of the budget, then we need to turn to the design of special methods of integration that fit the needs of the problem. As Fowler observes,[1] the methods of design and simulation should be done together, not separately. Also, we sometimes build whole simulators in hardware, with the digital computer as one component buried inside. In those cases we also want to adapt our methods of integration to fit the situation.

There are numerous published comparisons of the various methods used in a simulation.[2] Theory says that the predictor-corrector methods should do about half as many evaluations as a Runge-Kutta method of comparable accuracy, but many reports do not seem to bear this out. One can only suspect why. One possible explanation is that the predictor-corrector methods spent a lot of their time halving and doubling, because the halving and doubling constants were too close and did not allow for the simulation of noise that is so often a part of the total problem. Another obvious explanation is that the author of the comparison does not understand the newer methods and evaluates them for what they are not supposed to do rather than for what they do in fact do. Many fail to realize that step-size changes can require changing many of the coefficients (see Sec. 36.7).

Stability of the predictor-corrector method is of course another source of trouble, but if the step size has been shortened to take care of it, and if the Runge-Kutta method did not correspondingly shorten, then it is likely that the second method is in error too, but the error is not so obvious.

41.3 SPECIAL METHODS[3]

Usually the simulation package reduces all the equations to a single system of first-order equations. This can be very costly in machine time and include some loss of accuracy. In the very common case of second-order equations with no first derivative present there is considerable advantage in time and accuracy in going to the appropriate method (Sec. 24.3).

[1] M. E. Fowler, Numerical Methods for Use in the Study of Spacecraft Guidance and Control Systems, *IBM Scientific Computing Symposium on Digital Simulation of Continuous Systems*, pp. 47–66, 1966. See, in particular, p. 62.
[2] P. R. Benyon, A Review of Numerical Methods for Digital Simulation, *Simulation*, November, 1968, pp. 219–237, gives 75 references. H. R. Martens, A Comparative Study of Digital Integration Methods, *Simulation*, February, 1969, pp. 87–94.
[3] M. E. Fowler, A New Numerical Method for Simulation, *Simulation*, May, 1961, pp. 324–330.

The case of stiff equations,[1] which is related to the problem of widely separated time constants, is also of common occurrence.   The solution to this difficulty depends on the source.   Often the very short time constant that is producing the stiff part is a parasitic one that could be left out of the simulation with no serious loss in the modeling.   Or, it can be replaced by an implicit equation whose derivative is taken to be zero as far as the other questions of the system are concerned. (Sec. 24.8).

Sometimes some of the equations in the simulation have solutions which are not easily approximated by polynomials (something like the stiff-equation problem), and we are forced to devise appropriate methods for integrating those particular equations in the system (Sec. 24.6).

We looked briefly at the design of integration methods from the frequency approach (Chap. 36), and we now turn to a more complete discussion (not derivation) of the approach.

41.4   THE FREQUENCY APPROACH AGAIN

The traditional approach of the mathematically inclined to the problem of the numerical integration of ordinary differential equations is through polynomial approximation and measuring the accuracy of the solution by a comparison of the calculated and true solutions—a comparison in the solution space.   The usual approach of the electrical engineer to the same problem is through his knowledge of the frequency approach, and he measures his success by how close the Fourier, or more often the Laplace, transforms of the calculated and true solutions agree.

These two approaches are not the same, and in practice can give very different results.   It is true, of course, that if one is exact, then the other is; but it is not true that if the two solutions are close to each other in the solution space, then the frequencies in each solution are close.   This can be seen in Fig. 41.4.1.   The first example clearly has a high-frequency component that the other function does not have, while in the second example there will be jerks at the corners of the straight-line solution that are not in the true solution, again a different frequency content.

When the solution is being used, for example, in a training simulator for supersonic flight or a space vehicle, the difference in frequency content can be serious indeed.   What is wanted is a trainer that "feels" like the real thing.   This is much more like the Fourier transforms being close than the solutions being

---

[1] M. E. Fowler and R. M. Warten, A Numerical Integration Technique for Ordinary Differential Equations with Widely Separated Eigenvalues, *IBM Journal*, September, 1967, pp. 537–543. See also Ref. 16.

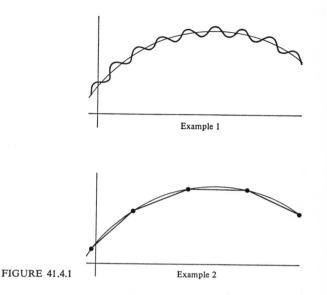

Example 1

FIGURE 41.4.1   Example 2

close.   Indeed, it is a custom to match, not the pure sinusoid content of the solution à la Fourier, but rather the Laplace transforms over the complex plane so that the damped sinusoids will "feel proper" to the trainee.

This discussion should show why the method of zeros and poles which tries to match the Laplace transforms is in many cases much more appropriate than the conventional method of matching solutions in the solution space.

41.5 THE z TRANSFORM

The z transform is a formal analogy with other transforms and is closely related to generating functions   It is very useful when dealing with sampled data systems (meaning that equally spaced samples of a function are all that are available).   Let the data be (at spacing T)

$$f(nT) \qquad n = 0, 1, \ldots$$

where the assumption is that for negative t the function is identically zero, much like the Laplace transform.   The z transform is defined as

$$F(z) = \sum_{n=0}^{\infty} f(nT)z^{-n} \qquad |z| > R$$

It is customary to use negative powers of z, though obviously a reciprocal transformation

$$z = 1/z'$$

would yield positive powers.

Like other transforms the z transform is linear

$$\sum_{n=0}^{\infty} [c_1 f_1(nT) + c_2 f_2(nT)] z^{-n} = c_1 \sum_{n=0}^{\infty} f_1(nT) z^{-n} + c_2 \sum_{n=0}^{\infty} f_2(nT)$$

There is a shifting theorem

$$z \left[\sum f(nT) z^{-n} - f(0^+) \right] = \sum_{n=0}^{\infty} [f(n+1)] z^{-n}$$

The convolution is clearly the Cauchy power-series multiplication, and we have a corresponding convolution theorem

$$F_1(z) F_2(z) = \sum_{n=0}^{\infty} \sum_{n=0}^{n} f_1(kT) f_2[(n-k)T] z^{-n}$$

Thus it is easy to see that we will have all the formal properties of most transforms. We can also construct tables corresponding to functions that we can sum in closed form. The formalities need not be gone into here.[1]

The z transform has been popular in some circles, and it is often used for the design of integration methods for systems of differential equations. The derivations are easier than those of the method of zeros and poles, but they lack the flexibility of the latter. Therefore, we shall not go farther into this topic, but be content to alert the reader to its existence and relation to the rest of the material.

Many of the comparisons of various methods seem to confuse the z transform methods and the more general zero and poles method, so that this confusion needs to be watched carefully.

---

[1] See, for example, E. I. Jury [26].

PART V

Miscellaneous

APPROXIMATIONS TO SINGULARITIES

42.1 INTRODUCTION

Problems that arise in physics very often have a singularity at one end of the interval of interest, and sometimes at both ends. We have been approximating functions using (1) polynomials, (2) sines and cosines, and (3) exponentials, all of which are finite at all finite values of the argument. Thus essentially we have been solving problems with no singularities. The exceptions to this remark are that in integration we have occasionally incorporated the singularity into the kernel $K(x)$, and in differential equations we have made (by implication) transformations on the variables to eliminate the singularities; and also rational functions (Chap. 30).

What is true about problems in physics having singularities is perhaps less true in other fields of science, and even less true in engineering, but singularities do occur in practice. We therefore need to examine this important topic.

We are going to apply analytic substitution, and approximate the behavior of functions by terms which resemble the singularity. The methods we use can be adapted to any problem where we have knowledge of the nature of the solution and the appropriate family of approximating functions. When there is an appropriate family for a problem, then the use of the more general families we have studied is foolish and a waste of both machine time and insight.   Since the appropriate family to use is so difficult to discuss in the general case, we shall confine the chapter to approximating particular cases and assume that the reader realizes that in other cases the same techniques apply and should be used.

Much of mathematics involves the art of studying the behavior of functions in the neighborhood of a singularity, and we cannot in this book go into this important topic—we must simply assume that it is known, especially the difficult art of handling "movable singularities."

In mathematics the singularities are usually subtracted out, but in computing it is usually better to arrange for the behavior in the neighborhood of the singularity to be written as a product of the known general behavior times an unknown function that will, if the choice is good, remain finite and vary slowly in the region of interest.   We must, therefore, resist using many of the standard tricks taught in mathematics and learn to proceed somewhat differently, if we are to avoid having to deal with differences of large numbers and the consequent heavy cancellations.

42.2   SOME EXAMPLES OF INTEGRALS WITH SINGULARITIES

As a first example, consider the integration of

$$I = \int_0^{\pi/2} \ln(\sin x)\, dx$$

If we write it as

$$I = \int_0^{\pi/2} \left[\ln x + \ln\left(\frac{\sin x}{x}\right) \right] dx$$

we see that it can be written as a piece that we can integrate analytically and which contains the singularity plus a second piece which we can hope to do numerically since the integrand is now well behaved

$$I = \frac{\pi}{2}\left(\ln \frac{\pi}{2} - 1 \right) + \int_0^{\pi/2} \ln\left(\frac{\sin x}{x}\right) dx$$

Using $h = \pi/4$ and Simpson's formula, we get

$$I = -0.862 + \frac{\pi}{12}(0 - 0.420 - 0.045)$$

$$= -0.862 - 0.228 = -1.090$$

$$\left(\text{Known answer} = -\frac{\pi}{2}\ln 2 = -1.089 \cdots \right)$$

Here we have used a mixture of subtracting out the singularity and writing it as a product.

Second, it is often possible to find a transformation that eliminates the singularity. Thus in the problem in Sec. 9.10

$$g(y) = \frac{d}{dy} \int_0^y \frac{f(x)}{\sqrt{y-x}} \, dx$$

the transformation

$$x = y \sin^2 \theta$$

removed the singularity. This again falls in the domain of formal mathematical manipulation and outside that of numerical methods.

As a third example, consider

$$f(x) = \int_a^x e^{x^2} \, dx = \int_a^x \frac{1}{2x} (2xe^{x^2}) \, dx \qquad a > 0$$

$$= \frac{e^{x^2}}{2x} \Big|_a^x + \int_a^x \frac{e^{x^2}}{2x^2} \, dx$$

so that the asymptotic expansion of $f(x)$ is

$$f(x) \simeq \frac{e^{x^2}}{2x} + \cdots$$

Near $x = 0$ this form is not useful and we are led to try (now $a = 0$)

$$f(x) = \int_0^x e^{x^2} \, dx = e^{x^2} D(x)$$

It is a common trick to get a differential equation from an integral, and so we differentiate to get

$$e^{x^2} D'(x) + 2xe^{x^2} D(x) = e^{x^2}$$

or

$$D'(x) + 2x D(x) = 1$$
$$D(0) = 0$$

We now have a simple differential equation and know that as $x \to \infty$, $D(x) \to 1/(2x)$. This suggests that the function $D(x)$ (for the whole range $0 \le x \le \infty$), known as Dawson's integral, is more suitable for computing than the original function.

As a fourth example, consider Bode's integral for the phase in terms of the amplitude $A(\omega)$:

$$\frac{\omega_0}{\pi} \int_{-\infty}^{\infty} \frac{A(\omega)}{\omega^2 - \omega_0^2} \, d\omega$$

Here we have singularities at

$$\omega = \pm \omega_0$$

A numerical computation from experimental data used the trapezoid rule for the intervals outside of the singularities, and near the singularities the function $A(\omega)$ was approximated by a cubic through four adjacent, centrally located values. This led to integrals of the form

$$\int_{\omega_0 - h/2}^{\omega_0 + h/2} \left[\frac{a + b(\omega - \omega_0) + c(\omega - \omega_0)^2 + d(\omega - \omega_0)^3}{(\omega - \omega_0)(\omega + \omega_0)} \right] d\omega$$

which can be handled analytically.

The use of the structure of the singularity has been illustrated to some extent by these examples, and further examples occur in the next section.

PROBLEMS

42.2.1 Writing

$$I = \int_{-1}^{1} \frac{dx}{\sqrt{1 - x^4}} = \int_{-1}^{1} \frac{1}{\sqrt{1 - x^2}} \frac{1}{\sqrt{1 + x^2}} \, dx$$

apply Chebyshev integration.

42.2.2 In Prob. 42.2.1, set $x = \cos \theta$ and integrate numerically. Compare to Prob. 42.2.1.

42.3 A SINGULARITY IN A LINEAR DIFFERENTIAL EQUATION

It occasionally happens that a numerical solution is required for a linear differential equation which has a singularity[1] in the range of integration. One way of doing this, which was first suggested by Prof. J. W. Tukey, is the following, and it again illustrates the technique of using the structure of the singularity in a multiplicative fashion.

Suppose we are given

$$y'' + P(x)y = 0 \qquad \begin{array}{l} y(x_0) = A \\ y'(x_0) = B \end{array}$$

---

[1] By singularity we mean that the function does not behave as a power series about the point.

where $P(x)$ has a singularity in the range of integration, say at $x = 0$.   We choose as a comparison equation

$$u'' + Q(x)u = 0$$

which has the known solutions $u_1(x)$ and $u_2(x)$ and has the same kind of singularity in $Q(x)$ at the same place as $P(x)$.   We shall later examine how this can be done.

We next assume that

$$y(x) = \alpha(x)u_1(x) + \beta(x)u_2(x)$$

where $\alpha(x)$ and $\beta(x)$ are functions to be determined but which we expect will vary smoothly.   The method now goes along the same lines as the classical method of "variation of parameters."   Having introduced two unknown functions $\alpha(x)$ and $\beta(x)$, we may require that $y(x)$ satisfy both the differential equation and one more condition.   We choose this second condition as

$$\alpha'u_1 + \beta'u_2 = 0$$

From this we have

$$y = \alpha u_1 + \beta u_2$$
$$y' = \alpha u_1' + \beta u_2'$$
$$y'' = \alpha u_1'' + \alpha'u_1' + \beta u_2'' + \beta'u_2'$$

Putting these in the original equation, we get

$$\alpha(u_1'' + Pu_1) + \beta(u_2'' + Pu_2) + \alpha'u_1' + \beta'u_2' = 0$$

But using the fact that u_1 and u_2 are solutions of the comparison equation, we have

$$u_1'' + Pu_1 \equiv (P - Q)u_1$$
$$u_2'' + Pu_2 \equiv (P - Q)u_2$$

Thus the differential equation becomes

$$\alpha'u_1' + \beta'u_2' = (Q - P)(\alpha u_1 + \beta u_2) = (Q - P)y$$

Solving this with the definition of $y(x)$, we get the two equations

$$\alpha'(u_1u_2' - u_2u_1') = -u_2(Q - P)y$$
$$\beta'(u_1u_2' - u_2u_1') = u_1(Q - P)y$$

But $u_1u_2' - u_2u_1'$ is the Wronskian of u_1 and u_2, and for this case (no y' term) it is a constant W_0, where

$$W_0 = u_1(x_0)u_2'(x_0) - u_2(x_0)u_1'(x_0)$$

W_0 is readily determined from the known solutions $u_1(x)$ and $u_2(x)$.

We thus have to solve

$$\alpha' = -\frac{u_2(Q - P)y}{W_0}$$

$$\beta' = \frac{u_1(Q - P)y}{W_0} \qquad \begin{matrix} y(x_0) = A \\ y'(x_0) = B \end{matrix}$$

$$y = \alpha u_1 + \beta u_2$$

The problem is how we can choose a $Q(x)$ so that the factor $Q - P$ vanishes sufficiently strongly at the singularity to cover the peculiar behavior of the products $u_2\, y$ and $u_1 y$ (or, equivalently, the products $u_2{}^2$, $u_1 u_2$, and $u_1{}^2$). If this can be done, then the equations determining $\alpha(x)$ and $\beta(x)$ will have a much less peculiar behavior at the point where there was a singularity.

We now show how $u_1(x)$ and $u_2(x)$ may be chosen in one particular example. Suppose that

$$P(x) = \frac{a_{-1}}{x} + a_0 + a_1 x + \cdots \qquad a_{-1} \neq 0$$

Thus we know that if $y(0) \neq 0$, then $y''(0)$ is infinite, and it is apparent that we could not approximate the solution by a polynomial with any fidelity near $x = 0$. We try

$$u_1 = \frac{x}{1 + ax + bx^2} \equiv \frac{x}{D(x)} \equiv xD^{-1}(x)$$

Then

$$u_1'' = [2(D^{-1})' + x(D^{-1})''] \left(\frac{D}{x} u_1\right) \qquad \frac{D}{x} u_1 \equiv 1$$

$$= \frac{x[2(D')^2 - DD''] - 2DD'}{xD^2} u_1$$

But

$$D = 1 + ax + bx^2$$

$$D' = a + 2bx$$

$$D'' = 2b$$

Hence for small x

$$u_1'' = \frac{-2a - 6bx + 2b^2 x^2}{xD^2} u_1 = \frac{-2a - 6bx + 2b^2 x^2}{x[1 + 2ax + (a^2 + 2b)x^2 + \cdots]} u_1$$

$$= -2\left[\frac{a}{x} + (3b - 2a^2) + \cdots\right] u_1$$

Thus

$$Q = -\frac{2a}{x} + (-6b + 4a^2) + (\cdots)x + \cdots$$

and we want to choose (comparing Q and P)

$$-2a = a_{-1} \qquad a = \frac{a_{-1}}{2}$$

or

$$-6b + 4a^2 = a_0 \qquad b = \frac{a_{-1}^2 - a_0}{6}$$

With this choice of a and b in our trial function $Q(x)$, we have from the structure of $P(x)$

$$Q - P \simeq Cx + \cdots$$

and $Q - P$ vanishes sufficiently rapidly. We now determine u_2 in the usual fashion.

$$
\begin{aligned}
u_2(x) &= u_1(x) \int^x \frac{d\theta}{u_1^2(\theta)} \\
&= u_1(x) \int^x \frac{1 + 2a\theta + (a^2 + 2b)\theta^2 + 2ab\theta^3 + b^2\theta^4}{\theta^2} \, d\theta \\
&= u_1(x) \left[-\frac{1}{x} + 2a \ln|x| + (a^2 + 2b)x + abx^2 + \frac{b^2 x^3}{3} \right]
\end{aligned}
$$

and the rest is easy. We have $u_1(x)$ with no singularity at $x = 0$, and $u_2(0) = -1$. We also have a $u_2(x)$ whose second derivative is infinite at $x = 0$. The products

$$(Q - P)u_1 y \qquad \text{and} \qquad (Q - P)u_2 y$$

vanish at $x = 0$, allowing the smooth integration of the differential equations that determine $\alpha(x)$ and $\beta(x)$, since the peculiar behavior is contained in the u_2 and is partially masked by the zero of $Q - P$.

In this example we again see the advantage of handling the singularities by first examining the dominant behavior of the solution near the singularity and then using a known function to approximate this behavior.

42.4 GENERAL REMARKS

We have given some examples of how to handle singularities in a few specific cases. We have no detailed rules for the general case, especially for singularities in nonlinear problems. Much depends on the mathematical ability of the person planning the solution.

With modern floating-point machines it is sometimes possible to integrate close enough to the singularity so that an analytic approximation can be fitted and used to get over the singularity. Once past the singularity, numerical values can be computed from the analytic expression in order to resume the computation. However, this should be used only when better methods fail, as the accuracy is often hard to estimate and control.

When the singularity is a so-called "movable singularity," which can occur in nonlinear problems, then it is particularly important to be careful how the numerical data are used to fit the function approximating the singularity. Yet such things can be and have been done successfully; it is a matter of courage and careful computation.

Experience shows that frequently the singularities were not in the original physical problem but got there by some standard mathematical transformation. Thus there is a high probability that a transformation can be found which removes the singularity, and that you are undoing what the mathematician did before you got the problem. It is therefore highly recommended that if a problem turns up with a singularity in it, then a careful examination of the source should be made to see how it arose. Occasionally it is in the original formulation, but an equally satisfactory approximation theory without the singularity can be made—often more satisfactory from the computing point of view. In the author's experience once after solving, with great labor, a problem having a singularity in it, the proposer asked if it could be done with some viscous terms in it—terms that effectively eliminated the trouble! What appears to the proposer to be a more complicated and more realistic model often is easier for the computer expert to handle.

43

OPTIMIZATION

43.1 INTRODUCTION

Simulations are usually undertaken with the goal of ultimately optimizing something about the behavior of the system.   Thus the usual steps in a simulation are:

1 Construct the model of the situation (or process)
2 Decide on the objective function that combines *all* the various goals into one single numerical value
3 Optimize

Each of these three steps is much harder than is usually realized when a study is first proposed.

First, the situation to be simulated is usually not as well defined as was supposed.   Many parts have not been thought through, and some parts do not have their proper physical constants for complete specification, since usually the simulation is supposed to help in the design.   In the ideal situation the simulation grows into the design, and that in turn flows into the evaluation of the system; it is wrong to separate the three phases.

Second, the so-called "objective function," which is supposed to combine all the various goals and desirable properties of the system into one single real number, is not known at the start.   The question of "trade-offs" of this for that has not been thought through, and will probably not be until some of the results are available so that people can begin to think about those that will matter in the final design.   At the beginning of the study the idea of a single number measuring the quality of the whole complex system does violence to the engineer's feelings of what is important.   Yet, in the final analysis, somehow the noncomparable aspects must be combined and a single decision reached as to which one of the possible designs to use.

Third, while it is the simulation that takes so much machine time, both in the programming and in the actual running, it is the optimization process which often controls the success or failure.   Inefficient searches of the space of possible designs can require more time and energy than are available.   Thus the part that tends to get neglected in the original planning often dominates the question of success or failure.

It is necessary, therefore, to look at the optimization process.   There are, broadly speaking, two situations.   First, the classical one that arose in agricultural experimentation where, because of the length of time to do one experiment, many cases must be run at the same time (in parallel).   From them a conclusion is drawn.   Second, modern situations that generally allow sequential decisions; one case is run, and then a partial analysis of the results is made which is used to help select the next trial.   It should be intuitively clear that the sequential process is generally very much more powerful in arriving at the optimum in a few steps than is the parallel case.   These two approaches have given rise to two distinct fields; the parallel one that arose from agricultural experiments gave rise to the field called "design of experiments," while the sequential approach has given rise to the more recent "sequential design" and the still rapidly developing field of optimization.

The problem of finding the optimum has a number of traps for the unwary person who believes everything he reads.   For example, consider the two situations in Fig. 43.1.1.   In the first, with a broad minimum, the mathematician will say "it is ill-conditioned and difficult to find," while in the second he will say "it is well-conditioned and easy to find."   The thoughtful engineer who is going to use the results will soon realize that in the first case he can safely proceed, because small changes in the position of the x value, or in the mathematical model, or again in the physical system that is ultimately constructed, will make small differences in the quality (y) of the performance.   On the other hand, for the sharp minimum the opposite applies, and small changes will greatly affect the behavior (y) of the resulting system.   Thus the mathematician's view is almost diametrically opposite to that of the engineer's, and it is the engineer who is often right in the matter.

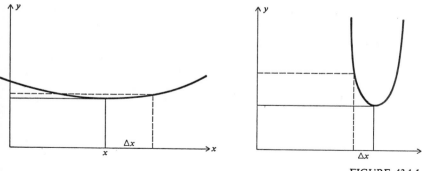

FIGURE 43.1.1

43.2 REVIEW OF CALCULUS RESULTS

Optimization is a standard topic in the calculus, but it is necessary to recall some vital details that tend to be forgotten. The main method for functions of a single variable is to obtain the first derivative, set it equal to zero, and then find the roots of this equation. These roots give the positions of the extremes (maxima,

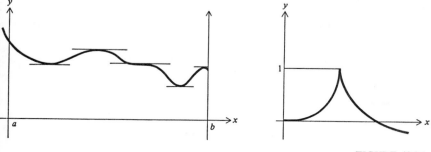

FIGURE 43.2.1

minima, and horizontal tangents). In the textbook examples these steps are usually fairly easy to do, but in practice the real zeros (Chap. 4) can be difficult to find.

What is usually glossed over are the assumptions that not only the function, but also the derivative, is continuous—that there are no cusps—and that the extreme value does not occur on the boundary. Both are significant in practice.

The test for a maximum is that the second derivative at the extreme point be negative, while for a minimum the second derivative must be positive (see Fig. 43.2.2). If the second derivative is zero at the point, then higher derivatives must be examined. If the first nonzero derivative at the point is of odd order, then it is an inflection point, while if it is of even order, then a sign test like that for the second derivative applies.

FIGURE 43.2.2   Min   Max

For functions of two variables the matter is much more complex. The usual test for two variables is

$$\frac{\partial^2 f}{\partial x^2}\frac{\partial^2 f}{\partial y^2} - \left(\frac{\partial^2 f}{\partial x \partial y}\right)^2 \begin{cases} > 0 & \dfrac{\partial^2 f}{\partial x^2}\begin{cases} < 0 & \text{maximum} \\ > 0 & \text{minimum} \end{cases} \\ = 0 & \text{undecided} \\ < 0 & \text{neither max. nor min.} \end{cases}$$

(again recall that this applies only on the inside of the region and that the boundaries require special attention). The textbooks rarely go beyond two variables.

The classic example of the peculiar behavior of a function of two variables near what seems to be a minimum is

$$z = (y - x^2)(y - 2x^2)$$
$$= y^2 - 3x^2 y + 2x^4$$

At the origin

$$\frac{\partial z}{\partial x} = -6xy + 8x^3 = 0$$

$$\frac{\partial z}{\partial y} = 2y - 3x^2 = 0$$

$$\frac{\partial^2 f}{\partial x^2}\frac{\partial^2 f}{\partial y^2} - \left(\frac{\partial^2 f}{\partial x\,\partial y}\right)^2 = 0$$

Now consider planes that pass through the z axis, $y = \lambda x$. These planes cut the surface in the curves

$$z = \lambda^2 x^2 - 3\lambda x^3 + 2x^4$$

and for them

$$\frac{dz}{dx} = 0$$

$$\frac{d^2z}{dx^2} = 2\lambda^2 > 0 \qquad \text{a minimum at } x = 0, y = 0 \text{ in the plane}$$

hence we have a minimum in each plane. The two extreme cases (planes that also contain the x axis or the y axis) also show a minimum at the origin. Thus we are inclined to think that the origin is a true minimum. See Fig. 43.2.3. However, in every neighborhood of the origin the function takes on negative values! If functions of two variables can be this tricky, consider those of many more variables!

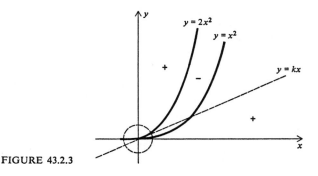

FIGURE 43.2.3

It is also worth noting that many apparently simple problems have no mathematically defined minimum value. For example, suppose that there are two towns A and B, with B due east of A, and suppose that you are required to depart from A in a northerly direction and travel the minimum path to B. See Fig. 43.2.4. Evidently in a mathematical sense this problem is not really properly stated, though at first glance it might appear to be. In practical situations the foolishness is generally better disguised, but may be there just the same to trap the unwary.

FIGURE 43.2.4 A B

43.3 LAGRANGE MULTIPLIERS

Frequently the problem we are given to minimize has n variables, but there are k constraints among the variables. Thus we have to minimize

$$y = f(x_1, x_2, \ldots, x_n)$$

when

$$\begin{cases} \phi_1(x_1, x_2, \ldots, x_n) = 0 \\ \cdots\cdots\cdots\cdots\cdots\cdots \\ \phi_k(x_1, x_2, \ldots, x_n) = 0 \end{cases} \quad k < n$$

For example, the values of the points $(x_1, x_2, \ldots, x_n)$ may be required to lie on the surface of a sphere

$$\phi = x_1^2 + x_2^2 + \cdots + x_n^2 - a^2 = 0$$

In principle the k constraint equations can be solved for k of the variables x_i, and these can be used to eliminate these variables from the original function $y = f(x_1, x_2, \ldots, x_n)$. In practice this is apt to be impossible to carry out, and a trick known as the method of Lagrange multipliers can be used instead. We will illustrate it with one constraint. Using differentials, we have

$$dy = \frac{\partial f}{\partial x_1} \Delta x_1 + \frac{\partial f}{\partial x_2} \Delta x_2 + \cdots + \frac{\partial f}{\partial x_n} \Delta x_n = 0$$

and

$$0 = \frac{\partial \phi}{\partial x_1} \Delta x_1 + \frac{\partial \phi}{\partial x_2} \Delta x_2 + \cdots + \frac{\partial f}{\partial x_n} \Delta x_n = 0$$

Using λ times the second equation added to the first, we get

$$\left(\frac{\partial f}{\partial x_1} + \lambda \frac{\partial \phi}{\partial x_1} \right) \Delta x_1 + \cdots + \left(\frac{\partial f}{\partial x_n} + \lambda \frac{\partial \phi}{\partial x_n} \right) \Delta x_n = 0$$

If each of these parentheses vanishes, then for any variation of the Δx_i we will have a horizontal tangent; thus the n equations

$$\frac{\partial f}{\partial x_i} + \lambda \frac{\partial \phi}{\partial x_i} = 0 \qquad i = 1, 2, \ldots, n$$

plus the original constraint

$$\phi(x_1, x_2, \ldots, x_n) = 0$$

gives us $n + 1$ equations in the $n + 1$ variables $x_1, x_2, \ldots, x_n$ and λ. These equations define the solution we are looking for. Thus in place of the original function we wanted to minimize we now consider the function

$$L \equiv f + \lambda \phi$$

and optimize this as a function of the x_i with λ as an additional variable.

As an illustration of the Lagrange multiplier method, consider the problem of finding the maximum rectangular block that will fit inside the ellipsoid

$$\phi = \frac{x^2}{a^2} + \frac{y^2}{b^2} + \frac{z^2}{c^2} - 1 = 0$$

The volume to be maximized is

$$V = 8xyz$$

subject to the constraint of the point x, y, z lying on the ellipsoid. We therefore consider the function

$$L = V + \lambda\phi$$

and have to solve the equations

$$\frac{\partial L}{\partial x} = 8yz + \lambda\frac{2x}{a^2} = 0$$

$$\frac{\partial L}{\partial y} = 8xz + \lambda\frac{2y}{b^2} = 0$$

$$\frac{\partial L}{\partial z} = 8xy + \lambda\frac{2z}{c^2} = 0$$

together with the original constraint, the ellipsoid. To solve these equations, multiply the first equation by x, the second by y, and the third by z, and add to get

$$24xyz + 2\lambda\left(\frac{x^2}{a^2} + \frac{y^2}{b^2} + \frac{z^2}{c^2}\right) = 0$$

or

$$3V + 2\lambda = 0$$

Again multiply the top equation by x to get

$$8xyz + 2\frac{x^2}{a^2} = 0$$

or

$$V + 2\lambda\frac{x^2}{a^2} = 0$$

Using this with the above equation connecting V and λ, we get

$$V - 3V\frac{x^2}{a^2} = 0$$

Excluding $V = 0$ as a minimum, we get

$$x^2 = \frac{a^2}{3}$$

$$x = \frac{a}{\sqrt{3}}$$

By symmetry we also get

$$y = \frac{b}{\sqrt{3}}$$

$$z = \frac{c}{\sqrt{3}}$$

and the maximum volume is

$$V = \frac{8abc}{3\sqrt{3}}$$

To generalize the Lagrange multiplier method when there are k constraints, we merely use k Lagrange multipliers λ_i and proceed accordingly.

PROBLEMS

43.3.3   Find the minimum distance from the plane $Ax + By + Cz - D = 0$ to the origin.

$$\textit{Ans.}\quad \frac{|D|}{\sqrt{A^2 + B^2 + C^2}}$$

43.4   THE CURSE OF DIMENSION

We have looked at, and will continue to use, problems in two independent variables, but in practice the number of parameters in the problem being optimized may run much higher, and it is necessary to discuss the difference between problems in a few dimensions and those in many. The difference is known as "the curse of dimension," as the following discussion will show.

Consider a unit cube in n dimensions. It will have volume $1^n = 1$. Now consider all the volume in the cube that is within a distance of ε from the surface. This volume is the difference between two cubes, the original cube of side 1 and the inner cube of side $1 - 2\varepsilon$. This difference is

$$1 - (1 - 2\varepsilon)^n$$

and we see that for a fixed ε, when we let n get larger and larger, the difference approaches 1. Thus in a high-dimension space almost all the volume of a cube is near the surface!

For a sphere we have a similar result. The volume is proportional to the nth power of the radius.

$$V = \frac{\pi^{n/2} r^n}{\Gamma(n/2 + 1)}$$

Thus as we search high-dimension spaces for optimal values, we find that almost all the volume is on the surface and that what in low-dimension spaces was a nuisance is now a necessity: we must pay close attention to the values on the surface of the region, since we can expect that frequently the optimum will be there.

In many cases that are casually talked about, the volume of the high-dimension space is so great that it is impracticable to explore it in any detail at all. Problems with 50 and 100 parameters are hopeless *unless* a great deal more is known about where to look. For example, if it is known that there is only one local maximum, then it is a reasonable problem. Usually only local extreme values can be found, with a high probability of significantly higher values being somewhere in the region where you have never looked. The methods we shall give are practical only for comparatively few parameter problems, or else in controlled situations where few extremes can occur.

43.5 THE GRADIENT

The methods of the calculus attempt to find the extreme points by direct methods in that they produce the answer without looking at other places. These methods fail when there is no nice, simple, explicit formula for the objective function which measures the goodness of the trial.

We now turn to methods which resemble hill climbing or valley descending. They picture the surface generated by the objective function as a topographic map with the level lines drawn as shown in Fig. 43.5.1. The level lines are defined by

$$f(x_1, x_2, \ldots, x_n) = \text{constant}$$

where of course $f(x_1, \ldots, x_n)$ is the objective function.

Since the maxima of f are the minima of $-f$, we are not limiting ourselves if we search for minima. The method consists, crudely, in taking the next step so as to go downhill rapidly—it follows the local topography and ignores any attempt to see the surface as a whole. The main weakness of this method of finding the minima of a function by valley descending is that it is apt to miss very narrow, deep, pits, such as a mine opening. If the search of the map does not

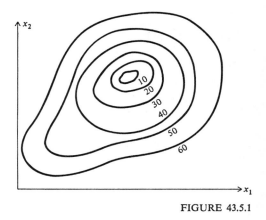

FIGURE 43.5.1

by chance run into the opening, then it will not be found by the method of searching the surface step by step.   What the method does find is the minimum of the local (smooth) valley that you start in.

The level curves (surfaces and hypersurfaces if $n > 2$) are defined by equating the objective function to the appropriate constant value.   If we differentiate this function, holding the value constant (along the level curve), we get

$$\frac{dy}{ds} = 0 = \frac{\partial f}{\partial x_1} \frac{\partial x_1}{\partial s} + \frac{\partial x}{\partial x_2} \frac{\partial x_2}{\partial s} + \cdots + \frac{\partial f}{\partial x_n} \frac{\partial x_n}{\partial s}$$

where s is the arc length along the level curve.   The n functions

$$\frac{\partial x_j}{\partial s} \qquad j = 1, 2, \ldots, n$$

form a vector which gives the direction cosines of the level curve.   It follows that the functions

$$\frac{\partial f}{\partial x_j} \qquad j = 1, 2, \ldots, n$$

also define a vector of direction numbers.   We shall indicate a vector by boldface type as

$$\left(\overline{\frac{\partial \mathbf{f}}{\partial \mathbf{x}_j}} \right)$$

The original direction cosines $\partial \mathbf{x}_j / \partial \mathbf{s}$ define the local direction of the level curve.   The condition for orthogonality of two vectors is that the sum of the products of the corresponding components be zero (see Sec. 26.1). Thus the above equation, which we found by differentiating the surface along the level curves, shows that the two vectors are perpendicular. One is the direction numbers of the level curve, so that the other

$$\left(\overline{\frac{\partial f}{\partial x_j}}\right) \equiv \begin{pmatrix} \dfrac{\partial f}{\partial x_1} \\[2mm] \dfrac{\partial f}{\partial x_2} \\[1mm] \vdots \\[1mm] \dfrac{\partial f}{\partial x_n} \end{pmatrix}$$

which is called *the gradient*, is perpendicular to the level curve. It is easy to see that the negative gradient is the path of steepest descent (locally). The gradient changes its direction from place to place, but if we follow the local gradient downward, we will get to the (local) minimum.

The path of steepest descent is not necessarily the shortest path to the minimum. Consider a meandering river. Except for the inertia of the flowing water, the river follows the local steepest descent, but clearly there are paths that descend much more rapidly without taking all the meanders that the river does. Nevertheless, the path of steepest descent is, on the average, a good way to go to the local minimum.

Since nearby paths all go to the same minimum, it is not necessary to follow the path of steepest descent closely—if we do not wander too far from it on any one step, we will still end up at the same place. Thus we have a reasonable amount of robustness built into the method of steepest descent.

The method of steepest descent has another weakness, however. Near a minimum the negative gradient points only weakly to the minimum (because of rounding errors in the computation, if for no other reason). But this should not be exaggerated—if the minimum is broad and shallow, then almost any point nearby will give almost as good a response for the function being examined.

43.6 FOLLOWING THE GRADIENT

The negative gradient gives the differential direction of the path of steepest descent—thus the problem of following the path resembles that of integrating a system of n ordinary differential equations. In principle all we need to do is follow the local negative gradient. The crudest method would be compute the vector whose components are

$$x_j{}^{(i+1)} = x_j{}^{(i)} - h\frac{\partial f^{(i)}}{\partial x_j} \qquad j = 1, 2, \ldots, n$$

where we have used superscripts to indicate the step number and subscripts to indicate the variables, and h is the step size. The method may seem crude, but remember that we do not have to be accurate in our individual steps; all the local paths will end at the same place *independent* of the local errors (provided the errors are not too big).

What step size shall we use? One way is to use the quantity which measures the size of each step

$$\sum_{j=1}^{n}(x_j{}^{(i+1)} - x_j{}^{(i)})^2$$

and if this is too small, then increase the step size. Near the minimum we cannot expect to gain much, even for big steps.

The fault of this method, and all others that use frequent reevaluations of the gradient, is simply that it usually costs a great deal of computing time to evaluate the n components of the gradient. If you insist on using this crude approach that reevaluates the gradient each step, then probably a modified Euler method (Sec. 22.3) would be better.

Rather than following the local gradient each step, it is much more efficient to find a local gradient and then follow that direction down until you meet a minimum on that path (on the surface). At this point you find a new local negative gradient and follow that direction to a local minimum on that line. It is comparatively easy to see that the successive directions you take will be perpendicular to each other. See Fig. 43.6.1.

In more detail, we find the direction of the negative gradient and take a step size h in that direction. We expect to find a smaller value of the objective func-

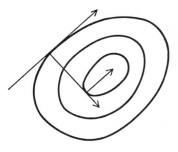

FIGURE 43.6.1

tion at that point, and if we do, then we keep on taking steps until we find an increase in the successive steps. We could then use something like the bisection method to try to find the lowest point on the line. If the first step gives us a value too high, then we have to resort to the bisection method immediately.

Alternately, once we find three points with the middle one lower than the other two, we could use quadratic interpolation and find the minimum of the quadratic as the next choice, rather than use the bisection method (which is guaranteed to be slow, but reliable). This speeds up the search, so that the first phase is to find three points with the middle one lower than the other two. If we find after a number of initial steps that we have not found the three points, then we are tempted to increase our search step size, say double it, and if still we are taking too many steps, then we double again. The process is much like that for doubling the step size for differential equations. Similarly halving can be used if we overshoot the minimum on the first step.

43.7 ESTIMATING THE GRADIENT

In many problems it is totally impractical to find expressions for the local gradient, let alone evaluate them. Instead it is necessary to examine the surface of the objective function and from it to guess at the gradient. To do this we simply make small changes in the coordinate we are interested in, and from the difference of the two values on the surface we estimate the derivative in that direction. For n independent variables we will have to evaluate the surface at n nearby points to get the estimates of the n partial derivatives. Furthermore, the choice of nearby points means that the estimates will be mainly roundoff, while placing the points far apart means that we will not be getting the local derivative. Either way you can lose!

The method and its problems are so obvious that the details of what can happen to you and how to cope with them need not be discussed here. For example, if the point where you are is a saddle point, then you can be badly fooled. Consider the surface

$$z = xy \qquad \text{at } x = 0, y = 0$$

We have

$$\frac{\partial z}{\partial x} = 0 \qquad \text{and} \qquad (\Delta z)_x = 0$$

$$\frac{\partial z}{\partial y} = 0 \qquad \text{and} \qquad (\Delta z)_y = 0$$

so that you would not have a reliable indication of the (local) surface and where to go next.

43.8   SOME PRACTICAL OBSERVATIONS

The valleys that seem to cause the most trouble when hunting for a local minimum are long, thin, curving ones that occur much more often than seems reasonable at first glance.   As Fig. 43.8.1 shows, for such valleys the search path is not efficient, and it runs back and forth.   Thus for such valleys the direct search can be expensive.

Our trouble is that we do not have an invariant algorithm (Sec. 4.8) for our search.   To see this most clearly, consider an elliptical valley where most of the time the gradient does not point to the minimum.   If we could find the transformation that opened up the valley into a circle, then we would get to the minimum

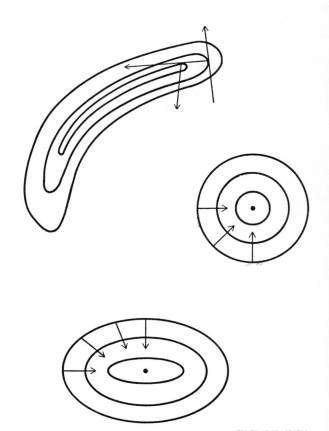

FIGURE 43.8.1

in the search along the first gradient line rather than having to shift from gradient line to gradient line. We shall look briefly at this in the next section on the variable metric methods. These methods attempt to find the shape of the valley in the large by keeping information about the shape, using where the path has been rather than depending completely on the local topography (as does the method of steepest descent).

A valley (ridge) that is not parallel to a coordinate axis and is not straight is very common and difficult to follow. The so-called "ridge-following methods" seem to work sometimes, but the better they work on a given type of ridge, the worse they are apt to be in finding a ridge in the first place. They should be used only with caution.

43.9 THE FLETCHER-POWELL METHOD

As we observed, Newton's method may be very good when the zero is successfully isolated (though there is no reasonable test to know when this happens), but it is apt to be slow when far from a zero. Hence methods which attempt to fit the surface with a quadratic and then find the minimum of the quadratic work fine when near the extreme value, but they are not very good when far away. Thus the variable metric methods tend to be used near the end of the search rather than at the beginning.

Newton's method in one variable uses the function and the first derivative to get the next value. Since the minimum is a zero of the derivative, a quadratic search method will involve the derivative and the second derivative which is the Hessian. In fact it is the reciprocal of the matrix of second partial derivatives that is needed. If we had this inverse, and if the surface were exactly a quadratic, then we would be able to find the minimum in one step. In a sense it is an attempt to produce an invariant algorithm that will take the correspondingly scaled steps for any of the multiplicative transformations that leave floating-point numbers with the same accuracy.

The Fletcher-Powell method[1] seems to be the most popular of the quadratic convergence methods. In order to follow the process, we introduce the notation

$$(x, y) = x_1 y_1 + x_2 y_2 + \cdots + x_n y_n = \text{scalar}$$

and

$$\langle x, y \rangle = (x_i y_j) = \text{matrix}$$

where x and y are vectors having n components.

[1] R. Fletcher, and M. J. D. Powell, A Rapidly Convergent Descent Method for Minimization, *Computer Journal*, vol. 6, pp. 163–168, 1963.

We suppose that the surface is a quadratic in n variables

$$f = f_0 + \sum_{i=1}^{n} a_i x_i + \frac{1}{2} \sum_{i=1}^{n} \sum_{j=1}^{n} G_{i,j} x_i x_j$$

$$= f_0 + (a, x) + \frac{1}{2}(x, Gx)$$

where, of course, G is the matrix (G_{ij}) and is taken to be symmetric. We want G^{-1} and start with the positive definite[1] matrix

$$H^{(0)} = I$$

where I is the identity matrix.

Let the current point be $x^{(i)}$, with gradient $g^{(i)}$ and current matrix $H^{(i)}$. Set

$$S^{(i)} = -H^{(i)}g^{(i)}$$

and search in this direction until you find a minimum which defines α by

$$f(x^i + \alpha s^{(i)}) = \text{minimum}$$

Set

$$\sigma^{(i)} = \alpha^{(i)} S^{(i)}$$
$$x^{(i+1)} = x^{(i)} + \sigma^{(i)}$$

Evaluate

$$f(x^{(i+1)}) \quad \text{and} \quad g^{(i+1)}$$

Set

$$y^{(i)} = g^{(i+1)} - g^{(i)}$$
$$H^{(i+1)} = H^{(i)} + A^{(i)} + B^{(i)}$$

where

$$A^{(i)} = \frac{\langle \sigma^{(i)}, \sigma^{(i)} \rangle}{(\sigma^{(i)}, y^{(i)})}$$

$$B^{(i)} = \frac{\langle -H^{(i)}y^{(i)}, y^{(i)}H^{(i)} \rangle}{(y^{(i)}, H^{(i)}y^{(i)})}$$

It can be shown that the matrix $H^{(i)}$ remains positive definite and that the convergence is quadratic. We shall not go into the details since they involve a good deal of linear algebra and are available in the original paper.

In practice it is well to restart the matrix H after n or possibly $2n$ steps to get rid of the old estimates of the surface that were far from the minimum.

---

[1] Positive definite matrix means that for all $(x, x) \neq 0$ then $(x, Hx) > 0$.

43.10 OPTIMIZATION SUBJECT TO LINEAR CONSTRAINTS

Up to now the problems in this chapter have been problems of unconstrained optimization. This has meant that in our search for a minimum of $f(x_1, \ldots, x_n)$, any value of the variables $x_1, \ldots, x_n$ was a permissible one. There are many practical cases, however, in which physical or mathematical reasoning forces us to restrict our search to values of the variables which satisfy certain conditions.

These conditions, or constraints as we will call them from now on, can take on many forms. In this section and the next we shall examine some the the forms in which constraints appear and how they affect the solution, and we shall suggest methods for handling them.

Some constraints are easier to handle than others. Consider, for example, nonnegative constraints: that is, each of the variables $x_1, \ldots, x_n$ is restricted to nonnegative values. Under these conditions, the statement of the optimization problem would read: Minimize the function

$$f(x_1, \ldots, x_n)$$

subject to the constraints

$$x_i \geq 0 \qquad i = 1, \ldots, n$$

The easiest way to solve this problem is through a transformation of variables. If we let

$$x_i = y_i^2 \qquad i = 1, \ldots, n$$

then we may minimize f as an unconstrained function of the $y_1, \ldots, y_n$ and be certain that the nonnegative constraints will be satisfied at the optimum. An alternative way is to apply the gradient or the quasi-Newton technique of Sec. 43.9 with the following modification: Whenever a variable, say x_k, becomes zero and at the same time the xth component of the gradient is positive, then stop updating x_k until $\partial f / \partial x_k$ changes sign. This is another way in which we can be sure that the search will be restricted to nonnegative values.

The above tricks apply equally well when the variables are restricted to lying between lower and upper bounds, that is, when the constraints are of the form

$$\begin{aligned} x_i - l_i &\geq 0 \\ u_i - x_i &\geq 0 \end{aligned} \qquad i = 1, \ldots, n$$

where the l_i and u_i are given and, of course, $u_i > l_i$. Here, we may again use a transformation on the variables. We may, for example, let

$$x_i = (u_i - l_i)\sin^2 y_i + l_i$$

and solve the problem as unconstrained in terms of the y's. Other transformations are possible; however, one has to be careful about introducing new local minima. This is because the minimization of $f(y_1, \ldots, y_n)$ will stop whenever $\partial f/\partial y_i = 0 (i = 1, \ldots, n)$. But by the chain rule,

$$\frac{\partial f}{\partial y_i} = \frac{\partial f}{\partial x_i} \frac{\partial x_i}{\partial y_i}$$

Therefore, any transformation on the x's which causes $\partial x_i/\partial y_i$ to be zero at a point which meets this constraint and is not on the boundary will introduce new minima.

The constraint types discussed so far are linear and are special cases of the general linear constraints of the form

$$\sum_{i=1}^{n} a_{ij} x_i - b_j \geq 0 \qquad j = 1, \ldots, m$$

Linear constraints are much easier to handle than the nonlinear ones, which we discuss in the next section. The main reason for this is that the boundary of the feasible region is composed of straight lines in two dimensions, planes in three dimensions, and hyperplanes in higher dimensions. It is, consequently, easy to travel along a boundary during the search for the optimum. Several methods, developed especially for linearly constrained problems, take full advantage of this fact. Best known among these is the so-called gradient-projection method. This method has been designed primarily for linearly constrained problems with a nonlinear objective function; it is essentially an extension of the steepest-descent method to linear constraints. This presentation is too lengthy to be included in this book and the reader is referred to the literature.[1]

A very important class of optimization problems is the one in which both the objective function and all the constraints are linear. Problems of this type which minimize

$$f(x) = \sum_{i=1}^{n} c_i x_i$$

subject to

$$\sum_{i=1}^{n} a_{ij} x_i - b_j \geq 0 \qquad j = 1, \ldots, m$$

are called *linear programming* problems, and they are important because of the enormous variety of applications which they have found in practical situations.[2] A little thought will make it intuitively clear that the solution of a linear program

---

[1] Read, for example, the section beginning on p. 133 of T. L. Saaty and J. Bram [50].
[2] See, for example, S. I. Gass [15].

lies on the boundary of the feasible region, usually on a corner, but sometimes along one side of a constraint. This observation provides an efficient technique for solving linear programs. More details can, again, be found in the literature.

43.11 OTHER METHODS

The Lagrange multiplier approach is a classical one and has endured a long time because of its practical and theoretical importance. There are occasions, however, when the solution of the nonlinear equations resulting from this approach is quite difficult. In such cases, other methods are preferred in practice.

One of the best-known alternative techniques for handling equality constraints was originated by R. Courant. In this method we minimize the function

$$F(x_1, \ldots, x_n) = f(x_1, \ldots, x_n) + r \sum_{i=1}^{m} [g_i(x_1, \ldots, x_n)]^2 \qquad r > 0$$

as an unconstrained problem for a sequence of monotonically increasing values of r. As r becomes infinitely large, the sum of the squares of the constraints is forced to zero. In this way, the sequence of the successive minima of F converges toward the solution of the given problem.

We conclude this chapter with a few brief comments about nonlinear inequality constraints of the form

$$g_i(x_1, \ldots, x_n) \geq 0 \qquad i, \ldots, 1, \ldots, m$$

First let us examine the necessary conditions for the solution. It turns out that here, as for the equality constraints, necessary conditions can be expressed through a Lagrange function. If a point $x_1, \ldots, x_n$ is a minimum of the given objective function $f(x_1, \ldots, x_n)$ subject to the above constraints, then it must be true that the negative gradient vector of f can be expressed as a linear combination of the negative gradients of the constraints which are binding. Geometrically this means that if the minimum point is on the boundary of the feasible region, then the negative gradient of f must fall within the cone which is formed by the gradients of the constraints which are binding, Fig. 43.11.1.

A little thought will show that if this were not true, then we should be able to move a little farther along the boundary to obtain a point at which f is smaller.

There are two common philosophies in the various solution approaches suggested for problems with nonlinear inequality constraints: "the boundary-following approaches" and the "penalty-function techniques."

As their name implies, the boundary-following approaches suggest that when a constraint is (or is about to be) violated, then follow the boundary of the

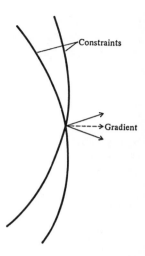

FIGURE 43.11.1

feasible region defined by that constraint until a point which satisfies the necessary conditions for a minimum is found. If the boundary is highly nonlinear, then the convergence of such approaches will be slow. In such cases, we prefer the penalty-function techniques.

The idea of a penalty function was introduced earlier when we added the sum of the squares of equality constraints to f in order to "penalize" the minimization of the resulting function whenever a constraint was violated. A similar technique can be used for inequality constraints. Consider, for example, the functions

$$F(x_1, \ldots, x_n) = f(x_1, \ldots, x_n) + r \sum_{i=1}^{m} \frac{1}{g_1(x_1, \ldots, x_n)} \qquad \begin{matrix} r > 0 \\ r \to 0 \end{matrix}$$

or

$$F(x_1, \ldots x_n) = f(\ \ ) - r \sum_{i=1}^{m} \ln[g_i(x_1, \ldots, x_n)] \qquad r \to 0$$

or

$$F(\ \ ) = f(\ \ ) + r \sum_{i=1}^{m} [\tilde{g}_i(x, \ldots, x_n)]^2 \qquad r \to \infty$$

where $\tilde{g}_i = 0$ if $g_i \geq 0$, and $\tilde{g}_i = g_i$ if $g_i < 0$.

If we minimize these functions sequentially for a series of positive values for r (monotonically decreasing for the first two functions and increasing for the third one), then we approach a solution to the constrained problem.

44

LINEAR INDEPENDENCE

44.1 INTRODUCTION

Linear independence plays a central role in much of mathematics. In mathematics it is a yes/no thing, and there are no measures of the degree of linear independence.

A little thought shows that linear independence is also very important in numerical methods, but as is so often the case, there are probably a number of different ways to measure the degree of linear independence. Unfortunately none of these possible ways has been developed to any extent, let alone the whole organized into a reasonable classification. Thus the best we can do in this chapter is to go through the material presented so far and indicate where various possible ideas might be used, if only we could develop them adequately!

The fact that the material in this chapter, like much of the more recent chapters, is vague and not of the theorem-proof form is not an indication of unimportance, but rather of being near the frontier of the field. In time it is probable that the material will become organized as more definite bodies of knowledge. But the material being at present vague, it is conventional to skip over it in the standard mathematically inclined texts, and therefore references are sometimes hard to give.

As simple examples of the basic role of linear independence in mathematics, consider the following definitions.

1 A number is *transcendental* if the powers of it are linearly independent over the integers (or fractions), that is, if

$$\sum_{k=0}^{N} a_k x^k = 0$$

implies all the $a_k = 0$; otherwise it is algebraic.

2 A function $y = y(x)$ is transcendental if the powers are linearly independent over the ring of polynomials, that is,

$$\sum_{k=0}^{N} P_k(x) y^k(x) \equiv 0$$

implies all the polynomials $P_k(x) \equiv 0$.

44.2   LINEAR EQUATIONS

In Chap. 7 we discussed the solution of a system of linear algebraic equations, and we went to considerable trouble to point out some of the faults of the widely accepted methods of solution.   The methods tend to avoid mentioning the linear dependence (except for rank) of the equations.   An ill-conditioned system is recognized as being somewhat dependent, but the direct approach to creating a measure of independence is avoided (condition numbers are hardly a measure of it).

We would like to have a measure of independence for a set of equations. The values could run from 1 to ∞ (dependent) much as entropy does, or they could run from 1 to 0 as probability does.   One intuitively feels that dependence is like entropy: every operation that is done on the system can only increase the dependence.   This is consistent with our basic feelings on the matter—it should not be possible by any mathematical operations to increase the independence of the system, but it should be possible to decrease it.   Reformulation of the problem should be the only way to increase the independence.

The use of the angle between the lines or planes is often thought to be relevant, but Prob. 7.9.2 discusses the case for just two equations.   The solution shows that a floating-point transformation by a power of 2 in one variable changes the angle but clearly leaves the equations as independent as before.   For two lines with slopes of opposite sign there is always a transformation that makes the lines perpendicular; but when the slopes are the same sign, there is a maximum angle that can be achieved. Whether this is relevant or not and how it works in n dimensions are not clear to the author.

If we keep the vague general idea of linear dependence in our minds and ask how to make it as large as we can in one step of elimination of simultaneous linear equations, then we decide that using *the same equation* to eliminate from all the others would do the best job—but this is pivoting, the most widely used method! Thus one can only wonder about the wisdom of pivoting—yes, at the current step it seems to have some nice roundoff properties, but it also seems to be designed to increase the linear dependence (assuming we do have an intuitive idea of its meaning), and hence at a later stage we may come to regret it. The truth of this observation is shown in the example of Sec. 7.8 of solving the 3×3 symmetric system that is apparently well scaled, and for which Gaussian elimination produces dependence while another method keeps the independence and permits the accurate solution of the system.

How could we proceed other than by pivoting? We could try eliminating by pairs, thus not favoring one particular equation which runs the risk that later everywhere in the derived system we will "see" the same pivot equation staring at us. This method still tends to favor the two end equations which will enter into only one equation each, while all the other equations will enter into two derived equations. This could be coped with by imagining the equations on a cylinder, and eliminating by pairs to get n equations in one less variable at each stage. At the end we would have n equations in one unknown which could be solved by least squares (the mean) for the most probable value. Putting this in the earlier system we again have n equations in the next unknown, and so we go "least squaring" ourselves up the back solution equations one at a time. Almost nothing is known about such a method except the obvious fact that it will take about twice as long in machine time to do as the conventional Gaussian elimination method and a good deal of storage as well.

Frankly the author does not know how to solve linear equations ideally, but feels strongly that any method of analysis that does not come to terms with the linear independence of the system, and the loss per step, is likely to be superficial. Linear independence is too basic in such a problem to be ignored. The difficulty, of course, is to find an adequate, relevant measure of dependence of a system of equations.

One final word of suggestion for doing what the author does not know how to do. When framing definitions of linear independence (dependence), keep in mind that we use floating-point arithmetic of a fixed precision (number of digits), and are not dealing in the usual mathematical fixed-point arithmetic. If this is kept in mind, it is soon seen that several of the arguments usually given about the size of the roundoff do not apply.   Thus the idea of taking the largest coefficient as a pivot in order to reduce the roundoff does not hold.   For suppose that with only two equations we use the smaller to eliminate from the larger.   The result is, of course, an equation with larger coefficients (hence larger fixed-point roundoff) but the same floating-point roundoff as the other way round.   It is the way the equations are combined, not the coefficients, that matters.   Of course, in the usual elimination process the final equation for each x_i is the same linear combination of the original equations *regardless* of the process along the way, but the roundoff effects of the recombinations of the same equation can be considerably different.

44.3   SAMPLING AND LINEAR INDEPENDENCE

For each family of functions, polynomials (the Vandermonde determinant in Sec. 14.4), the Fourier series (orthogonality in Sec. 31.5), and the exponentials (Sec. 39.2), we had to prove the very important result that the mathematical linear independence of the functions over any continuous interval could be translated into linear independence of n functions over the set of n distinct samples.   *This is the basis for the validity of the numerical approach.*

If we ask, as we do in this chapter, for the *amount* of linear independence, then we have to describe a measure of it.   For polynomials it is natural to ask where we can put the sample points x_i so that the crucial Vandermonde determinant has the largest possible value for its square.   The farther apart the points, of course, the larger the value, and it is necessary to normalize the question to get a reasonable result.   It is natural, again, to ask that the sample points x_i be confined to some fixed interval, say $-1 \leq x \leq 1$.

Given this problem, we can easily see that for two points the ends of the interval are best, for three the ends and the middle.   For four points, using symmetry, we want to maximize the product

$$V_4 = \sum_{i \geq j = 1}^{4} (x_i - x_j)^2 \qquad x_1 = -1$$

$$= [4x_2(1 - x_2^2)^2] \qquad \begin{aligned} x_4 &= +1 \\ x_2 &= -x_3 \end{aligned}$$

Equating the derivative to zero and finding the maximum, we get

$$5x_2^2 - 1 = 0$$

$$x_2 = 1/\sqrt{5} = -x_3$$

The next case is clearly

$$x_1 = -1$$
$$x_2 = -x_4$$
$$x_3 = 0$$
$$x_5 = 1$$
$$V_5 = 4x_2{}^3(1 - x_2{}^2)^2$$

which leads to

$$7x_2{}^2 - 3 = 0 \quad \text{or} \quad x_2 = \pm\sqrt{3/7}$$

It is futile to go on case by case, so we ask how could the samples be found in general. Clearly they can come from some family of polynomials. We have, so far

Vandermonde	Polynomial [less $(1 - x^2)$]
$V_2 = (1 - x^2)$	$p_0 = 1$
$V_3 = x(1 - x^2)$	$p_1 = x$
$V_4 = 4x(1 - x^2)^2$	$p_2 = 5x^2 - 1$
$V_5 = 4x^3(1 - x^2)^2$	$p_3 = x(7x^2 - 3)$

Where we can expect to get the polynomials from? Probably from some orthogonal system. The first try, using weight factor $1 - x^2$ and, of course, range $-1 \le x \le 1$, is

$$\int_{-1}^{1} (1 - x^2)p_i(x)p_j(x)\, dx = 0 \qquad i \ne j$$

and shows that they are indeed orthogonal (we need only do $p_0 p_2$ and $p_1 p_3$) Thus we have ultraspherical polynomials (Sec. 26.2 with $\alpha = 1$). A check of Szego[1] shows that we are correct in our heuristic approach; the zeros of the ultraspherical polynomials are indeed the sample points that maximize the Vandermonde, and hence in this sense give the maximum linear independence. Of course, the Gaussian quadrature points give another version of where to pick the sample points when integrating.

PROBLEMS

44.3.1 Show that the next polynomial is $p_4 = 21x^4 - 14x^2 + 1$ and check the orthogonality of it with respect to p_0, p_1, p_2, p_3.

[1] G. Szego, Orthogonal Polynomials, *Am. Math. Soc.*, vol. 23, Colloquium Publications, 1959, p. 139.

44.4   POWERS OF x

The size of the Vandermonde determinant gives one measure of the linear independence of the powers of x over a set of points.   There are other measures of independence.   Consider again the interval $-1 \leq x \leq 1$ and a linear combination of the powers of x, where to normalize the problem we take the first coefficient equal to one

$$p_n(x) = x^n + a_{n-1}x^{n-1} + \cdots + 1$$

The choice of the coefficients of the nth Chebyshev polynomial (times 2^{n-1}) gives the inequality

$$|p_n(x)| = \left| \frac{T_n}{2^{n-1}} \right| \leq \frac{1}{2^{n-1}} \qquad -1 \leq x \leq 1$$

and shows that the powers of x are practically dependent for large n.   This agrees with our intuitive feelings that the even powers of x look too much like each other to be very independent, and similarly for the odd powers (notice that each Chebyshev polynomial used only one set of these).

Thus the choice of the powers of x (for the interval $-1 \leq x \leq 1$) as the basis for the representation can lead to trouble because they are not very independent. The fact that the powers are easily computed on a computer does not mean that they should be used indiscriminately.   Rather the orthogonal polynomials, discussed again in the next section, provide a better basis, and using the three-term recurrence relation (Sec. 26.6) they are not difficult to compute for moderate n.

44.5   ORTHOGONAL POLYNOMIALS
AND LEAST SQUARES

Orthogonal polynomials and least squares provide an alternate approach to the problem of measuring the linear independence of a set of functions.   Section 26.3 shows that orthogonal functions are linearly independent.

If we rate the linear independence of the powers of x by the least-squares measure

$$\left[\int_{-1}^{1} p_n{}^2(x) \, dx \right]^{1/2}$$

where

$$p_n(x) = x^n + a_{n-1}x^{n-1} + \cdots + a_0$$

we find that corresponding to the Chebyshev polynomial of the previous section we get the Legendre polynomials $P_n(x)$ (Sec. 26.7). The corresponding proof is found by writing

$$p_n(x) = c_n P_n(x) + c_{n-1}P_{n-1}(x) + \cdots + P_0(x)$$

where we know that $c_n = n!/[1 \cdot 3 \cdot 5 \cdots (2n - 1)]$. The integral, due to the orthogonality, is

$$\int_{-1}^{1} p_n^2(x)\, dx = \sum_{k=0}^{n} c_k^2 \cdot \frac{2}{2k + 1}$$

and this is clearly minimized by taking $c_0 = c_1 = \cdots = c_{n-1} = 0$. Thus the least-squares measure is

$$\sqrt{\frac{2}{2n + 1}}\, c_n = \frac{n!\, n!\, 2^n}{(2n)!} \sqrt{\frac{2}{2n + 1}}$$

$$\simeq \frac{(n^n e^{-n})^2\, 2\pi n \cdot 2^n}{(2n)^{2n} e^{-2n} \sqrt{2\pi \cdot 2n} \sqrt{n + 1/2}}$$

$$\simeq \frac{\sqrt{\pi}}{2^n}$$

which is about the same as the Chebyshev measure.

On the other hand consider the linear independence of the orthogonal functions. For purposes of comparison we shall suppose them to be normalized, in the interval $-1 \leq x \leq 1$, that is,

$$\int_{-1}^{1} f_i(x) f_j(x)\, dx = \begin{cases} 0 & i \neq j \\ 1 & i \neq j \end{cases}$$

Consider now the linear combination

$$c_0 f_0(x) + c_1 f_1(x) + \cdots + c_n f_n(x)$$

where all the c_i are not zero. It is natural to normalize the linear combination by requiring that

$$\sum_{i=0}^{n} c_i^2 = 1$$

Thus our measure now becomes

$$\int_{-1}^{1} [c_0 f_0(x) + c_1 f_1(x) + \cdots + c_n f_n(x)]^2\, dx = \sum_{k=0}^{n} c_k^2 = 1$$

for *all* admissible choices of the c_i.

In this framework of ideas we see that the orthogonal functions, including orthogonal polynomials, generally have a nice, uniform amount of linear independence and why they so often form a basis of representation preferable to other mathematically equivalent bases. (See Hilbert matrix arguments in Sec. 25.6.) The equations for the least-squares fit are a nice diagonal set of orthogonal functions and can be very ill-conditioned for the basis $1, x, x^2, \ldots, x^n$.

In still another light, each of the orthogonal polynomials has all its zeros real, distinct, in the interval, and interlacing those of the adjacent polynomials, while the even (odd) powers of x tend to resemble each other more and more as the degree rises. This again gives an indication that for many purposes the orthogonal polynomials form a reasonable basis for representation, and the original powers of x do not. The problem of the basis for representation of our problem is rarely discussed in detail, though we did observe, in Sec. 13.7, the difficulty of having the wrong function get into the basis for the representation and (in Sec. 7.8) the lack of independence in the sinh and cosh representation of functions for large arguments, while the mathematically equivalent e^x and e^{-x} are fine. Near the origin, however, the sinh x and cosh x are more different than the exponentials, so that there is no magic answer for all problems.

The quasi-orthogonal polynomials introduced in Sec. 27.3, provide an example of how to choose a basis for representing functions without paying the high price of trying to make them exactly orthogonal (using perhaps the impractical Schmidt process, Sec. 26.3).

44.6 WHAT SAMPLES?

Two of the fundamental problems in numerical analysis are what samples and what functions, and these are related. We have discussed the effect of sample points on polynomials, and the phenomenon of aliasing (Sec. 31.2) shows the effect on sinusoids. The study of integration formulas using one or more derivatives (Sec. 17.8) shows that a sample of another derivative at a point where the function and the lower derivatives are already known is almost as good as another sample of the function at a new point.

There are versions of the sampling theorem (Sec. 34.6) that use function values and derivatives, and again one sees that the derivative at an old point is about as good as a new function value.

44.7 WHICH BASIS OF FUNCTIONS?

The problem of the basis for representation of the solution is central, yet almost all mathematics is content to affirm that the linear independence is all that matters. In the solution of difference (Chap. 13) and differential equations (Chaps. 23 and

24) this can make a significant difference. In the neighborhood of singularities (Chap. 42) and even in the representation of functions (Chap. 3), for example

$$f(x) = \sqrt{x + 1} - \sqrt{x} \equiv \frac{1}{\sqrt{x + 1} + \sqrt{x}}$$

the form used to represent the function is vital to good computing.

The literature constantly shows that the proper, "natural" set of functions gives spectacular improvements in accuracy and decreases in cost. It is a matter of studying the problem on hand to see which representation is preferable, and many times this depends mainly on the linear independence of the chosen set.

45

EIGENVALUES AND EIGENVECTORS
OF HERMITIAN MATRICES[1]

45.1 WHAT ARE EIGENVALUES AND EIGENVECTORS?

A vector x in Euclidean n-dimensional space has n components x_i, $(i = 1, \ldots, n)$. In calculus we studied (and in Sec. 26.1 we referred to) the product of two n-dimensional vectors x and y, denoted by

$$(x, y) = x_1 y_1 + x_2 y_2 + \cdots + x_n y_n = xy$$

If $(x, y) = 0$, we say (again see Sec. 26.1) that x and y are *orthogonal*. A convenient measure of the "length" of a vector x is the Euclidean norm

$$|x| = \text{norm of } x = (x, x)^{1/2} = \left(\sum x_i^2 \right)^{1/2}$$

When the norm of a vector is 1, it is called a *unit vector*.

A square matrix $A = (a_{i, j})$ multiplied by a column vector x produces a column vector y:

$$Ax = y$$

---

[1] This chapter owes its good points to M. P. Epstein as well as to conversations with P. A. Businger and notes from C. B. Moler.

Thus, matrix multiplication is a linear transformation of the points (vectors) in Euclidean n-dimensional space into points of the same space. The null vector (every component is 0) is usually written as 0 and goes into itself. It is unusual for any other point (vector) to go into itself *except* when $A = I$, the identity matrix with all 1s on the main diagonal and 0s elsewhere, in which case *every* vector goes into itself.

While vectors do not go into themselves, most of the time a square matrix of order n will have n different directions which go into themselves; that is, there are (usually) n different values of a scalar λ for which the following system of equations can be solved:

$$Ax = \lambda x$$

each, of course, giving rise to a different vector x. This is not surprising, since writing the equations in the form

$$(A - \lambda I)x = 0$$

indicates that we can solve the system if and only if the corresponding determinant is zero; namely, if and only if

$$|A - \lambda I| = 0$$

By inspection this determinant is a polynomial of degree n in λ, and in general it will have n distinct zeros, real or complex.

Why should we care about these special values of λ and their corresponding solutions, which, to introduce the jargon, are called, respectively, eigenvalues and eigenvectors? It will turn out that for the cases we are interested in (symmetric and Hermitian matrices) the n eigenvectors are linearly independent and form an orthogonal basis. Thus, an arbitrary vector x can be written in the form

$$x = c_1 x^{(1)} + c_2 x^{(2)} + \cdots + c_n x^{(n)}$$

where the superscript j is used to denote the jth eigenvector. The coefficients c_j are given by

$$c_j = \frac{(x, x^{(j)})}{(x^{(j)}, x^{(j)})} = \frac{(x, x^{(j)})}{|x^{(j)}|^2}$$

When this vector x is multiplied by the matrix A, the individual components of the representation are merely multiplied by their corresponding eigenvalues, that is,

$$Ax = c_1 \lambda_1 x^{(1)} + c_2 \lambda_2 x^{(2)} + \cdots + c_n \lambda_n x^{(n)}$$

Thus, the effect of matrix multiplication by A is particularly easy to follow in the eigenvector representation of an arbitrary vector, since it is a combination of "stretchings" (or "contractions") by factors of λ_j along the axes $x^{(j)}$. This is much of the reason that eigenvalues and eigenvectors play an important role in many applications.

45.2   NOTATION AND HERMITIAN MATRICES

The transpose of a matrix

$$A = (a_{i,j})$$

is written

$$A^T = (a_{j,i})$$

that is, the transpose A^T is obtained from A by writing the rows of A as the columns of A^T (or else the columns as rows) in the same order. Thus, if x is a column vector, then x^T is a row vector, and

$$(x, y) = y^T x = \sum_{i=} y_i x_i$$

$$|x| = (x^T, x)^{1/2}$$

The outer product of two vectors, defined by

$$\langle x, y \rangle = x \cdot y^T$$

is an $n \times n$ matrix (of rank 1) in which the element in the ith row and jth column is $x_i y_j$. This is in contrast with the earlier (inner) product of two vectors, which is a scalar.

It is easy to see from the definition of matrix multiplication that

$$(AB)^T = B^T A^T$$

that is, the transpose of a product is the product of the transposes in reverse order. Finally, it is obvious that

$$(A^T)^T = A$$

All the above can be generalized to matrices with complex elements by making appropriate changes. The central idea is the *conjugate transpose* of a matrix of complex elements in place of the earlier transpose.   The conjugate transpose is

$$A^H = \bar{A}^T = (\bar{a}_{j,i})$$

A matrix for which $A^H = A$ is called Hermitian in honor of the mathematician Hermite.   For such a matrix, if[1]

$$a_{i,j} = p_{i,j} + iq_{i,j}$$

---

[1] Do not confuse the coefficient $i = \sqrt{-1}$ with the subscript i.

then by definition

$$a_{j,\,i} = p_{i,\,j} - iq_{i,\,j}$$

In words, the real part of the Hermitian matrix is symmetric while the imaginary part is "skew-symmetric."

For complex valued vectors the inner product can be defined as

$$(x, y) = y^H x$$

and again the norm is real since

$$|x| = (x, x)^{1/2} = \text{a real scalar}$$

We also have, as before,

$$(AB)^H = B^H A^H \qquad \text{and} \qquad (A^H)^H = A$$

Perhaps the most important property of Hermitian matrices is that all their eigenvalues are real. To see this, let x be any eigenvector with eigenvalue λ. Then

$$Ax = \lambda x$$

Since $A^H = A$,

$$A^H x = \lambda x$$

Therefore, taking the complex transpose of both sides,

$$(A^H x)^H = (\lambda x)^H \qquad \text{or} \qquad x^H A = \bar{\lambda} x^H$$

Multiply this equation on the right by x and the original equation on the left by x^H to get the pair

$$x^H A x = \lambda x^H x$$
$$x^H A x = \bar{\lambda} x^H x$$

Thus $\lambda = \bar{\lambda}$ and λ is real.

We next show that for any two distinct eigenvalues λ_1 and λ_2 of a Hermitian matrix the corresponding eigenvectors are orthogonal. We have

$$Ax^{(1)} = \lambda_1 x^{(1)}$$
$$Ax^{(2)} = \lambda_2 x^{(2)}$$

Multiply the first equation on the left by $x^{(2)H}$ and the second on the left by $x^{(1)H}$:

$$x^{(2)H} A x^{(1)} = \lambda_1 x^{(2)H} x^{(1)}$$
$$x^{(1)H} A x^{(2)} = \lambda_2 x^{(1)H} x^{(2)}$$

Take the conjugate transpose of the first of these equations (recall that $\bar{\lambda}_1 = \lambda_1$, and $A^H = A$), and subtract from the second equation to get

$$(\lambda_2 - \lambda_1)x^{(1)H}x^{(2)} = 0$$

Since we assumed that $\lambda_2 \neq \lambda_1$, we have

$$(x^{(2)}, x^{(1)}) = x^{(1)H}x^{(2)} = 0$$

and the vectors are othogonal.

From this it follows that if a Hermitian matrix has all its eigenvalues distinct then the eigenvectors form an orthogonal basis for Euclidean n-dimensional space. (When the eigenvalues are not all distinct it is still possible to find n independent, orthogonal eigenvectors, but we will not prove this.)

45.3 SIMILARITY REDUCTIONS

The formality of going from symmetric to Hermitian matrices is straightforward. Since it is symmetric matrices that occur most often, and the necessity of using complex conjugate obscures what is going on in the general case, we shall confine the remainder of the chapter to the case of real symmetric matrices.

Suppose that a symmetric matrix A has n eigenvalues $\lambda_1, \lambda_2, \ldots, \lambda_n$ (distinct or not) and let D be the diagonal matrix

$$D = \begin{pmatrix} \lambda_1 & & & 0 \\ & \lambda_2 & & \\ & & \ddots & \\ 0 & & & \lambda_n \end{pmatrix}$$

Let X be the $n \times n$ matrix whose columns are the orthogonal eigenvectors $x^{(1)}$, $x^{(2)}, \ldots, x^{(n)}$. Then the eigenequations

$$Ax^{(i)} = \lambda_i x^{(i)} \qquad i = 1, 2, \ldots, n$$

can be written in the more compact form of matrices

$$AX = XD$$

where D is on the right because it multiplies the columns. The linear independence of the eigenvectors means that the inverse X^{-1} exists and we can multiply on the left to get

$$X^{-1}AX = D$$

Definition: If for any nonsingular matrix S we have

$$B = S^{-1}AS$$

then we say that B is *similar* to A.

Clearly we can suitably multiply on the left and right to get

$$A = SBS^{-1}$$

and A is also similar to B. It is easy to show that similarity is a formal equivalence relation.

The importance of similarity lies in the fact that similar matrices have the same eigenvalues, though the eigenvectors are in general different. To prove this, suppose that

$$Ax = \lambda x \qquad \text{and} \qquad B = S^{-1}AS$$

Multiply the first equation on the left by S^{-1}, and insert $SS^{-1} = I$ between A and x. We get as a result

$$(S^{-1}AS)(S^{-1}x) = \lambda(S^{-1}x)$$

which is the same as

$$B(S^{-1}x) = \lambda(S^{-1}x)$$

Thus λ is an eigenvalue of B with associated eigenvector $(S^{-1}x)$.

The eigenvector matrix X appearing above has an additional interesting property. We have shown that there exist mutually orthogonal eigenvectors in the case of distinct eigenvalues (and asserted that even when they are not distinct there is an orthogonal set of eigenvectors), so that

$$(x^{(i)}, x^{(j)}) = 0 \qquad \text{for} \qquad i \neq j$$

If in addition the set of eigenvectors is orthonormal (Sec. 26.1), then each eigenvector is a unit vector. We can always make them orthonormal, since any non-zero multiple of an eigenvector is still an eigenvector. We then have

$$(x^{(i)}, x^{(i)}) = 1 \qquad \text{for all } i$$

These two sets of inner-product equations are summarized by the matrix equation

$$XX^T = X^TX = I$$

so that we have the important result

$$X^{-1} = X^T$$

Thus, if we normalize our eigenvectors to be unit vectors, we can simplify the similarity relationship between the matrices A and D to

$$X^TAX = D$$

It is natural to call matrices with this property, $X^{-1} = X^T$, *orthogonal* matrices. We will discuss them further in the next section. (In the complex case, a matrix for which $X^H = X^{-1}$ is called *unitary*.)

PROBLEMS

45.3.1 Show that similarity is an equivalence relation (reflexive, symmetric, and transitive).

45.3.2 Carry out the details of the section for the complex case.

45.4 ORTHOGONAL TRANSFORMATIONS

We just defined a matrix Q for which

$$Q^T = Q^{-1}$$

to be orthogonal. To get an idea of why the word orthogonal is used, consider the classic orthogonal transformation

$$\begin{pmatrix} \cos \theta & \sin \theta \\ -\sin \theta & \cos \theta \end{pmatrix}$$

which represents a plane rotation through an angle θ. This matrix transforms any pair of orthogonal axes into another pair of orthogonal axes, from which comes the name. This matrix also has the property that its transpose is its inverse.

The $n \times n$ matrix

$$\begin{pmatrix} 1 & 0 & \cdots & \cdots & \cdots & \cdots & \cdots \\ 0 & 1 & \cdots & \cdots & \cdots & \cdots & \cdots \\ \cdots & \cdots & \cos \theta & \cdots & \cdots & \cdots & \sin \theta \\ \cdots & \cdots & \cdots & 1 & \cdots & \cdots & \cdots \\ \cdots & \cdots & \cdots & \cdots & 1 & \cdots & \cdots \\ \cdots & \cdots & \cdots & \cdots & \cdots & 1 & \cdots \\ \cdots & \cdots & -\sin \theta & \cdots & \cdots & \cdots & \cos \theta \end{pmatrix}$$

also performs a plane rotation, but in Euclidean n-dimensional space. The matrix

$$\begin{pmatrix} \cos \theta & \sin \theta \\ \sin \theta & -\cos \theta \end{pmatrix}$$

is a rotation *plus* a reflection (the determinant is -1). This is also an orthogonal transformation and can likewise be done in n dimensions.

These are the simplest orthogonal transformations since they involve only two coordinates. There are many more orthogonal transformations that are more complicated to write out in detail, and we will look at one particular class of them in the next section.

Orthogonal transformations have the important property that they do not "stretch" the space, that is, if Q is orthogonal, then

$$|Qx| = |x|$$

To prove this, we write the easily verified equation

$$(Ax, y) = (x, A^T y)$$

Replace y by Ay to get

$$(Ax, Ay) = (x, A^T Ay) = (x, y)$$

Now set $y = x$, and we have

$$(Ax, Ax) = (x, x)$$

as required. Thus orthogonal transformations represent only rotations plus reflections but no stretching. Because of this property they seem intuitively more likely to preserve the amount of linear independence of a system fairly well and to hold down roundoff effects, though we shall not try to prove these remarks here.

45.5 HOUSEHOLDER TRANSFORMATIONS

With this background of orthogonal transformations we now turn to the Householder transformation matrices. Let u be a real nonzero vector and let

$$b = \frac{u^T u}{2} = \frac{|u|^2}{2}$$

Then using the outer product $\langle uu^T \rangle = uu^T$, we define a Householder matrix.

$$P = I - \frac{uu^T}{b}$$

For practical computation the above notation is convenient, but for purposes of theory it is simpler to define

$$v = \frac{u}{|u|}$$

so that v is a unit vector and the Householder matrix is

$$P = I - 2vv^T$$

It follows immediately that

$$P^T = I^T - 2(vv^T)^T = I - 2vv^T = P$$

so that P is symmetric.   Furthermore,

$$P^2 = (I - 2vv^T)(I - 2vv^T)$$
$$= I - 4vv^T + 4vv^Tvv^T$$

But $v^Tv = 1$ so the final terms cancel and

$$P^2 = I \qquad \text{or} \qquad P^{-1} = P$$

Since P is symmetric, this is the same as

$$P^{-1} = P^T$$

and the Householder matrix P is orthogonal.

These properties are important because for a symmetric matrix $A, P^{-1}AP = PAP$ and this is similar to A.   Also this is the same as P^TAP preserving symmetry.

We now examine the effect of multiplying a matrix A by a Householder matrix P.   Consider

$$PA = (I - 2vv^T)A = A - 2vv^TA$$

If the components of v are $v_1, v_2, \ldots, v_n$ and the rows of A are denoted by R_1, $R_2 \ldots, R_n$, then v^TA is the row vector

$$v_1R_1 + v_2R_2 + \cdots + v_nR_n$$

Then $2vv^TA$ is the matrix whose kth row is just

$$2v_k(v_1R_1 + \cdots + v_nR_n)$$

and $PA = A - 2vv^TA$ has the kth row equal to

$$R_k - 2v_k(v_1R_1 + \cdots + v_nR_n)$$

In words, PA is obtained from A by subtracting from each row of A a linear combination of the rows of A.   In particular, if $v_k = 0$ then the kth row of PA is the same as the kth row of A.

Similar remarks hold for AP using the columns of A instead of the rows

45.6   TRIDIAGONALIZATION

We will show how to use Householder transformations to reduce a given sym metric matrix to a tridiagonal matrix

$$\begin{pmatrix} \times & \times & 0 & 0 & \ldots & \ldots & 0 \\ \times & \times & \times & 0 & \ldots & \ldots & 0 \\ 0 & \times & \times & \times & \ldots & \ldots & 0 \\ \cdots & \cdots & \cdots & \cdots & \cdots & \cdots & \cdots \\ 0 & 0 & 0 & 0 & \times & \times & \times \\ 0 & 0 & 0 & 0 & 0 & \times & \times \end{pmatrix}$$

with the same eigenvalues as A.

To show what Householder transformations can do, we take an arbitrary column vector x with norm σ, and find a Householder matrix P so that Px has specified components the same as those in x and certain other specified components equal to zero. (Or course analogous statements hold for x and xP when x is a row vector.) For purposes of illustrating the flexibility of the method we will suppose that x and P are six-dimensional, in particular,

$$
x = \begin{pmatrix} x_1 \\ x_2 \\ x_3 \\ x_4 \\ x_5 \\ x_6 \end{pmatrix} \quad \text{and} \quad Px = y = \begin{pmatrix} y_1 \\ y_2 \\ y_3 \\ y_4 \\ y_5 \\ y_6 \end{pmatrix}
$$

and that we want $y_3 = 0 = y_5$. We cannot accomplish this and have all the other y_i equal to the corresponding x_i (unless we started with $x_3 = x_5 = 0$), for we would then have

$$
|y| = \left(\sum y_i^2 \right)^{1/2} < \left(\sum x_i^2 \right)^{1/2} = \sigma
$$

But this contradicts the fact that

$$
|y| = |Px| = |x| = \sigma
$$

We can, however, accomplish what we want by changing only *one* component (other than those we make zero). To see this, we again examine the illustration and suppose that we are willing to change the fourth component. We want. therefore, to get

$$
P \begin{pmatrix} x_1 \\ x_2 \\ x_3 \\ x_4 \\ x_5 \\ x_6 \end{pmatrix} = \begin{pmatrix} x_1 \\ x_2 \\ 0 \\ y_4 \\ 0 \\ x_6 \end{pmatrix}
$$

We have

$$
x_1^2 + x_2^2 + y_4^2 + x_6^2 = \sigma^2
$$

so therefore

$$
y_4 = \pm \left[\sigma^2 - (x_1^2 + x_2^2 + x_6^2) \right]^{1/2}
$$
$$
= \pm \left[x_3^2 + x_4^2 + x_5^2 \right]
$$

To find P we have

$$
y = Px = (I - 2vv^T)x
$$
$$
= x - 2v(v^T x)
$$

or

$$2(v^T x)v = x - y$$

where $v^T x$ is a scalar and v is a unit vector. The direction of v is obviously that of $x - y$. Thus

$$v = \frac{x - y}{|x - y|} = \frac{1}{|x - y|} \begin{pmatrix} 0 \\ 0 \\ x_3 \\ x_4 - y_4 \\ x_5 \\ 0 \end{pmatrix}$$

so we can choose u in the original definition of $P = I - uu^T/b$ as

$$u = \begin{pmatrix} 0 \\ 0 \\ x_3 \\ x_4 - y_4 \\ x_5 \\ 0 \end{pmatrix}$$

with y_4 given as above.

We have a choice of the sign for y_4 so we choose y_4 to have the opposite sign as x_4, thus avoiding possible cancellation and roundoff. In practice, u is easier to use than v, because it avoids the square root necessary to find $|x - y|$.

Notice that y_4 and u depend only on the components of x that will be changed by the transformation. In doing the computation, one can ignore the components that are not changed.

The particular column (row) vectors used are the columns (rows) of A. When we form PA we, in effect, multiply each column of A by P, and when we form AP we multiply each row of A by P.

Returning to the particular six-dimensional case of our example, if

$$\begin{pmatrix} a_{1,1} & a_{1,2} & a_{1,3} & a_{1,4} & a_{1,5} & a_{1,6} \\ a_{2,1} & a_{2,2} & a_{2,3} & a_{2,4} & a_{2,5} & a_{2,6} \\ a_{3,1} & a_{3,2} & a_{3,3} & a_{3,4} & a_{3,5} & a_{3,6} \\ a_{4,1} & a_{4,2} & a_{4,3} & a_{4,4} & a_{4,5} & a_{4,6} \\ a_{5,1} & a_{5,2} & a_{5,3} & a_{5,4} & a_{5,5} & a_{5,6} \\ a_{6,1} & a_{6,2} & a_{6,3} & a_{6,4} & a_{6,5} & a_{6,6} \end{pmatrix}$$

and $a_{i,j} = a_{j,i}$ we can pick an α_1, as described above, and set

$$u_1 = \begin{pmatrix} 0 \\ \alpha_1 \\ a_{3,1} \\ a_{4,1} \\ a_{5,1} \\ a_{6,1} \end{pmatrix}$$

Thus

$$P_1 = I - \frac{u_1 u_1{}^T}{b_1} \quad \text{with} \quad b_1 = \frac{|u_1|^2}{2}$$

so that PA has its first column equal to

$$\begin{pmatrix} a_{1,1} \\ a'_{2,1} \\ 0 \\ 0 \\ 0 \\ 0 \end{pmatrix}$$

for some value of $a'_{2,1}$. As a result $(P_1 A)P_1$ will have its first row equal to

$$(a_{1,1}, a'_{1,2}, 0, 0, 0, 0)$$

with $a'_{1,2} = a'_{2,1}$. Setting $A_2 = P_1 A_1 P_1$ we again have a symmetric matrix similar to A.

$$A_2 = \begin{pmatrix} a_{1,1} & a'_{1,2} & 0 & 0 & 0 & 0 \\ a'_{2,1} & a'_{2,2} & \cdots & \cdots & \cdots & a'_{2,6} \\ 0 & \cdots & \cdots & \cdots & \cdots & \cdots \\ 0 & \cdots & \cdots & \cdots & \cdots & \cdots \\ 0 & \cdots & \cdots & \cdots & \cdots & \cdots \\ 0 & a'_{6,2} & \cdots & \cdots & \cdots & a'_{6,6} \end{pmatrix}$$

If we now choose α_2 appropriately, we have

$$u_2 = \begin{pmatrix} 0 \\ 0 \\ \alpha_2 \\ a'_{4,2} \\ a'_{5,2} \\ a'_{6,2} \end{pmatrix}$$

and $P_2 = I - (u_2 u_2{}^T)/b_2$ as before, and $P_2 A_2$ will have its second column equal to

$$\begin{pmatrix} a'_{1,2} \\ a'_{2,2} \\ a''_{3,2} \\ 0 \\ 0 \\ 0 \end{pmatrix}.$$

P_2 *will not change the first column or row of* A_2. Similarly, $(P_2 A_2)P_2$ will have the second row equal

$$(a'_{2,1}, a'_{2,2}, a''_{2,3}, 0, 0, 0)$$

with $a''_{2,3} = a''_{3,2}$. Then $A_3 = P_2 A_2 P_2$ will also be a symmetric matrix similar to A but with *two* columns and rows reduced to tridiagonal form.

Continuing in this way, we obtain a matrix B which is in tridiagonal form and similar to A. Further, to effect the similarity

$$B = QAQ^{-1}$$

we have chosen Q as the product of the Householder matrices which are orthogonal. This implies that Q is also orthogonal, and

$$B = QAQ^T$$

Note that in the case where A is not symmetric, we can still " zero out " the columns as we did above, but the same Householder transformations will no longer zero out the rows. We will obtain, in the nonsymmetric case, a matrix similar to A and of the form

$$\begin{pmatrix} \times & \times & \times & \times & \times & \times \\ \times & \times & \times & \times & \times & \times \\ 0 & \times & \times & \times & \times & \times \\ 0 & 0 & \times & \times & \times & \times \\ 0 & 0 & 0 & 0 & \times & \times \end{pmatrix}$$

This is called the *Hessenberg* form.

It is worth noting that if we do not require that the final matrix be similar to A (so that the eigenvalues no longer are the same), then we need multiply A on the *left* only by a properly chosen sequence of Householder matrices and will end up with an upper triangular matrix. No requirement on the symmetry of the original matrix would then occur.

Because the Householder matrix is its own inverse, this remark is equivalent to writing the matrix A in the form

$$A = QR.$$

where R is upper triangular and Q (being the product of Householder matrices) is orthogonal. This procedure has application not only in the " QR algorithm " which we discuss in the next section, but also in the solution of systems of linear equations (Chap. 7), particularly the least-squares solution of redundant equations (Chap. 27).

45.7   THE QR ALGORITHM

The QR algorithm is used to compute the eigenvalues of a matrix. The basic idea is expressed above, namely, that we can start with a matrix A (call it A_1) and write it in the form

$$A_1 = Q_1 R_1$$

with Q_1 orthogonal and R_1 upper triangular. We then define

$$A_2 = R_1 Q_1$$

Note that

$$A_2 = Q_1^{-1} A_1 Q_1$$
$$= Q_1^{T} A_1 Q_1$$

so that A_2 is similar to A_1, and is symmetric if A_1 is symmetric. Continuing this process, we write

$$A_2 = Q_2 R_2$$

with Q_2 orthogonal and R_2 upper triangular, and define

$$A_3 = R_2 Q_2$$

In this way we obtain a sequence of matrices $A_1, A_2, A_3 \ldots$, all similar to A. It is an important fact that if A is symmetric and tridiagonal, then so are all the other matrices of this sequence.

This observation is useful in computing the eigenvalues of a symmetric matrix. Given a symmetric matrix we first reduce it to tridiagonal form by Householder transformations and use this result as the starting matrix of the sequence A_i of similar matrices. In this latter part we need to compute only elements on the three main diagonals, and this results in a large savings in computation.

It has been shown by Francis[1] that for any matrix A, the matrices A_n tend "generally" to an upper triangular form.[2]   In particular, for A symmetric the matrices A_n are symmetric and will therefore tend to diagonal form.   The numbers along the diagonal will, of course, be the eigenvalues of A.   Francis also introduced the operations called " shifts " to be performed at each step, whose purpose is to speed convergence.   These shifts are useful but they do not change the general theory, and will not be discussed further here.

45.8   OVERDETERMINED SYSTEMS OF LINEAR EQUATIONS

Suppose that we are given a system of m linear equations in n unknowns with $m > n$,

$$Ax = y$$

where A is an $m \times n$ matrix, and x and y are, respectively, n- and m-dimensional column vectors.   We write the equations in a symbolic block form as in Fig. 45.8.1

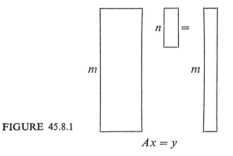

FIGURE 45.8.1

$$Ax = y$$

In general there will be no solution to such an overdetermined system of equations, but we may look for a least-squares solution as in Chaps. 25 to 27.   This means that we want to find a vector x which minimizes the norm

$$|Ax - y|$$

---

[1] J. G. F. Francis, The QR Transformation, *Comp. J.* 4, pts. I and II, pp. 265–271, 332–345, 1961–1962.

[2] If A has nonreal eigenvalues, this statement must be modified because a triangular matrix has its eigenvalues on the diagonal while the process we described uses only real numbers.   In the case in this chapter, when A is symmetric we have proved that the eigenvalues are all real.

Using the methods we have developed, we can find a sequence of House-holder matrices whose product is an orthogonal matrix Q such that QA is in upper triangular form

$$QA = R = \begin{bmatrix} \begin{array}{l} xxx \\ \quad xx \\ \qquad x \\ 0 \end{array} \end{bmatrix}$$

We will assume—what is true most of the time—that the rank of the matrix is n (if not, then the details can be filled in) so that all the diagonal elements are nonzero. If we set $Qy = z$, then we have

$$|Rx - z| = |QAx - Qy| = |Q(Ax - y)| = |Ax - y|$$

since the orthogonal matrix Q does not change the norm. Thus we can equivalently minimize the quantity $|Rx - z|$.

Since R is upper triangular, we can write this equation symbolically in the form

$$R = \begin{array}{c} \overset{n}{} \\ \begin{bmatrix} \hat{R} \\ \hline 0 \end{bmatrix} \begin{array}{l} n \\ m-n \end{array} \end{array} \qquad\qquad Z = \begin{bmatrix} \hat{z} \\ \hline z' \end{bmatrix} \begin{array}{l} n \\ m-n \end{array}$$

Then

$$|Rx - z|^2 = |\hat{R}x - \hat{z}|^2 + |z'|^2$$

Now x is independent of $|z'|^2$, so to minimize $|Rx - z|$ we choose x to minimize $|Rx - z|^2$. Since R is of rank n (by assumption), we can make the latter norm equal to zero by simply solving the triangular system

$$\hat{R}x = \hat{z}$$

exactly. The result is that this solution must be the least-squares solution to the original overdetermined system, and the norm of the residuals is given by $|z'|$.

N + 1

THE ART OF COMPUTING FOR SCIENTISTS AND ENGINEERS

N+1.1 IMPORTANCE OF THE TOPIC

It is unusual[1] in a book on computing to include a chapter on the vague general topic of how to approach and solve problems using computing machines.   The title of the chapter is itself ambiguous; it could refer to the art as practiced by a person computing for scientists and engineers, or it could refer to the art as practiced by scientists and engineers; in fact, it is meant for both.

The subject should not be taken lightly just because at times it is opinion rather than established fact.   I feel that it is more important than many of the specific results discussed in other parts of the book.   At present, the subject is more an art than a science, but this situation is slowly changing because the presence of computers themselves has made possible the mechanization of many processes once believed to require human thought, and active research is now going on in this important area known popularly as artificial intelligence.   The more we learn about how we solve problems, the more work we shall be able to shift from ourselves to the machines.   The art of solving problems with the use of computing machines, is of interest in its own right; it can also help the user in many

---

[1] But see F. S. Acton [1].

situations and greatly increase the value of the machine computations now being done.

Most scientists seem to have been afraid to explore what is generally considered to be the creative process of discovery, but there have been some notable exceptions. Among the mathematicians, " The Method " by Archimedes is one of the earliest examples, and we now have the classic " How to Solve It " by Pólya [48]. Both, however, are concerned with the attack on well-formulated problems, whereas we are concerned with a larger framework of vaguely defined problems from which we hope to extract results to fit an equally vaguely defined situation. The motto of the book,

THE PURPOSE OF COMPUTING IS INSIGHT,
NOT NUMBERS

illustrates our broad frame of reference

Unfortunately, mathematics traditionally emphasizes elegance, which all too often is identified with surprise. Frequently in a mathematical text one will see *consider the following function* ... and from the arguments that follow be compelled to agree with the result, without being given the slightest clue as to why we should consider the function or why it produces the result. Many mathematicians, including the famous C. F. Gauss, pride themselves on removing all the scaffolding that was used to erect the result in the first place and leave only the isolated result to be admired (surprised by). The motto is clearly in direct opposition to this attitude as it applies to numerical methods.

N + 1.2   WHAT ARE WE GOING TO DO WITH THE ANSWER?

Pólya in his book " How to Solve It " emphasizes the importance of understanding the problem. The present author, from many years' experience in computing for others, has found that usually the first question to ask is, What are we going to do with the answers? Will the answers computed actually answer the questions that should have been asked? Frequently one sees proposed studies, and even published papers, that cannot possibly answer the questions they claim will be (or are) answered simply because sufficient attention was not given to whether the data (and computed numbers) are or are not adequate to decide between the possible alternatives.[1]

Again one may ask, Do we need all the answers? Do we need more? Would something else provide a better basis for insight? Many other questions also occur, such as, Are the known conservation laws obeyed by the results?

[1] See, for example, Jeffreys [23].

In order to answer some of these questions, it is well to imagine typical answer sheets and then examine these for their usefulness. More times than one would expect the requested answers will not answer the needs of the research project. The original request may have been to get the answers to a set of simultaneous equations. Sometimes this is all that the computing can give, but many times other items, such as the difficulty of solution, can add to the understanding of the situation being examined. Further, what is to be the measure of accuracy: accuracy in the unknowns, the residuals, or the change in the problem, to name but a few? Were the simultaneous equations necessary? Would an alternative formulation give more insight?

Before going on in this vein, let it also be observed that one cannot expect the problem proposer to know exactly what he wants. In research, and in many stages of development, it is in the nature of the process that we do not know exactly what we are seeking. Indeed, it may be said, "In research, if you know what you are doing, then you shouldn't be doing it." In a sense, if the answer turns out to be exactly what you expected, then you have learned nothing new, although you may have had your confidence increased somewhat.[1]

Trite and obvious as the remark may be, it is important to know what you are seeking. It is less well understood that you should also plan your work to increase the chances that an unusual observation will be found. It is usually worth adding a small amount to the total machine time if in the process many side checks on the model being studied can be included. Thus, for example, in a study of the design of traveling wave tubes the author found that the equations being solved were a result of "linearizing the problem," and he therefore included in the computation some estimates of the probable size of the main nonlinear effects and found (not surprisingly) that in some cases for some ranges of the independent variable this was much too large to be ignored, thus invalidating the modeling, the corresponding computation, and nicely printed sheets of output.

Many of the greatest discoveries have been made by prepared minds making a chance observation and realizing the importance of it. Thus, even though it puts a little extra burden on the output equipment, it is well when planning the output formats to include a few wisely chosen numbers beyond the bare minimum.

In summary, while there are exceptions, it is a good general rule to begin a problem in computation with a searching examination of "What are we going to do with the answers?" An active, imaginative mind can make great contributions to the whole research problem at this stage, and a dull, lazy one can prevent any real insight from emerging from all the hours of machine time used to get the obvious numbers.

---

[1] P. Debye: "If a problem is clearly stated, it has no more interest to the physicist."

One of the most common mistakes in planning a problem is to request too many answers. This is particularly true in problems having many parameters. In such situations, what is generally needed is a good statistician who understands *the design of experiments and optimization theory.* Very often he can plan the search in such a fashion that only a small fraction of the original cases need be solved   The sheer volume of answers can, and very often does, stifle insight.

N+1.3   WHAT DO WE KNOW?

Having decided, tentatively, what we expect to get out of the computation, we next ask What do we know? What information do we have? What is the input? Are we putting in all that we know? For example, if the answer is known to go through the origin, have we included that fact in the input? Do we know some conservation law?

Again we remind the reader that Pólya emphasizes the importance of understanding the problem, but again he is considering mathematically formulated problems and is supposing that there is a complete statement. In the application of mathematics this is, of course, not the situation, nor is it true even in mathematical research. Often further probing of the situation being explored will bring further information to light. In the least-squares fitting of a polynomial in Sec. 27.6 we first ignored the fact that the curve should go through the origin, and as a result we had to do the problem over again. When we used this information, we found a more satisfactory answer.

Sometimes it is difficult to include all the known information in our formulation. Thus, in the above case we knew that the first term had a positive coefficient. We did not include this fact in the input but saved it for an output check. Nevertheless, it is well to get clearly in mind all the relevant information available before going on to the next stage.

Sometimes this exploration of the unknown situation will reveal alternative formulations of the problem, and these in turn may suggest new ideas to be explored. Sometimes we find that unnecessarily restrictive assumptions have been made on the model, and that with little or no extra effort they can be removed. Typically a problem has been needlessly linearized, or viscous damping has been removed. In any case, the role of assumptions should be understood, and checks that will throw light on the validity of some of the assumptions should be included in the computation. Thus the exploration of the input may reveal new or different output requirements.

N+1.4 DESIGNING THE COMPUTATION ROUTINE

Only after we have clearly in mind where we are and where we want to be should we turn to serious work on the question, How do we get from here to there? This is the domain of Pólya's book; all his remarks are relevant, and we assume that the reader either is familiar with them or else will read the book.

Although a problem is posed as one suited to a computing machine, the necessity for using a computer should be questioned. An analytical answer is often very much superior to a numerical result, and sometimes, even when it is more difficult to compute than the original problem, the error estimates may be made much more accurately. Our exploration of what we know about a problem may produce one or more formulations of the problem, some of them merely simple mathematical transformations of the same problem and others, perhaps, quite different. It is expensive to explore a machine-computation plan for all of them, and some initial choices must be made. As a general rule (with many exceptions), the closer the mathematical statement to the fundamental concepts of the field the better, always assuming that scale transformations have been made to make the equations dimensionless. Fancy mathematical transformations usually introduce difficult computational situations.

The plan of computation adopted should make use of as much of the initial information as can be included conveniently. The various mathematical approximations made by the formulas used should be in harmony with the model adopted. The effects of sampling should be examined, as well as the effect of the limited amount of input data.

The plan of computation should include plans for checking both the coding and the results. This point is too often overlooked; thus we recommend, *before* doing any detailed programming, that the questions be asked, How shall I know if the answer is the one that is wanted? What tests shall I apply or have the machine apply? It is essential that some redundant information be computed or found from other sources, so that some checks can be made. It is the experience of the author that a good theoretician can account for almost anything produced, right or wrong, or at least he can waste a lot of his time worrying about whether it is right or wrong.

N+1.5 ITERATION OF THE ABOVE STEPS

We have acted as if these three stages can be completely separated, when in fact they are often interrelated. Nevertheless, it is well to keep the three stages clearly in mind and iterate around them as new information at one stage sheds new light on another.   It is the author's experience that rushing into the stage of arranging the details of the computation is the most common mistake.   This is especially true of the computer expert, because there he feels at home and can show his skill. But all his skill is wasted if he works on the wrong problem or produces numbers that do not answer the real questions.

It is, of course, difficult to be expert enough in a particular field to ask suitable questions about the importance of what is being computed as compared with some other things that might be computed instead.   But there is an art in asking questions.   Socrates claimed not that he knew truth but that he knew how to ask the proper questions of man and draw the truth from him.   He spoke of himself as a midwife.   And it is in somewhat the same way that the consulting computer expert must approach the man with a problem.   In the final analysis, the man with the problem must make the choices and do the research, but well-chosen suggestions from the outsider can help to clarify the nature of the choices and aid in the decisions that must be made at many stages of the work.

N+1.6 A CODE OF ETHICS

Because I have recommended that an active, aggressive role be played by the numerical analyst and that he get into the design stages of the problem and stay in through the interpretation and decision stages, it seems necessary to also suggest some restraints on his role.   The following could serve as commandments.

1   Thou shalt not *impose* your ideas, opinions, and prejudices on the customer, for he is the man with the problem, not you.
2   Thou shalt avoid all unnecessary fancy mathematics, computer jargon, and attempts to raise the "falutin' index" of the problem.
3   Thou shalt not come between the customer and the machine, but shall stand at one side ready and willing to give all the help and aid you can to both.

N+1.7 ESTIMATION OF THE EFFORT NEEDED TO SOLVE THE PROBLEM

In any reasonably mature science, it is necessary to make estimates as to what will happen *before* spending time and money.   In a sense, the more mature the field, the more accurate the estimates.   Judged by this standard, computing is in an elementary state.   Often only the poorest estimates are available.

Some of the estimates that should be made are the following:

1 Will the roundoff errors be serious; if so, to what extent?
2 Will the interval spacing be adequate?
3 If there is an iteration process, how many iterations may reasonably be expected?
4 How much time will it take to code and debug?
5 How will the answers be checked to assure that they are right?
6 How much machine time will be required?
7 When will the answers finally be available?

A glance at these questions will convince the average computing expert that the present state of the art leaves much to be desired.   This is no reason for not trying to make estimates that are as realistic as possible.   Furthermore, proper organization of the computer facility can greatly improve the estimates made for questions 4 and 7.   The provision of monitor systems, automatic coding systems, debugging systems, and a short turnaround time on the computer are musts if these two items are to be estimated accurately.   Stability of the computer center is vital.

This book has tried to approach a number of these questions in various ways. In particular, Part III is believed to provide a basis for many estimates.   However, many of the approaches in the book are merely tentative starts in the direction of making reliable estimates, and much remains to be done.

A person who aspires to become a computer expert in the consulting field should not dismiss these questions as unanswerable but should recognize that reasonably accurate estimates are the mark of a capable craftsman—estimates that are neither too optimistic nor too pessimistic.

N+1.8   LEARNING FROM CHANGES IN THE PLAN

It is almost inevitable that as the computation progresses, new information becomes available, and changes in the plan must be contemplated.   But before adopting a change in plan, an effort should be made to understand why the wrong choice was made in the first place.   Does the change shed any light on the model being used?   Should we still try to get the same kind of answers?   Does the change suggest new or different checks on the validity of the model?   Can some new insights be obtained either from the failure or from a new plan?

A change in plan should not be hastily patched in but should receive the same careful thought and planning as went into the original plan.   As remarked before, if things go as planned, not much can be learned—it is from the unexpected that the great new insights can occasionally arise.   Thus a situation which forces a change in plan should be regarded as an opportunity rather than a failure.   Of course, if the change was due to carelessness or failure to think, then it should be taken as another example of the value of preliminary planning and should be charged to stupidity.

It is all too tempting, when involved in running a problem, to be rushed into making small changes without considering the consequences and the implications, especially if the results have been promised at a certain time.   Yet haste at this time can undo much of the earlier careful work.   It is well to remember that "the man should be the master, not the machine," and by proper organization of the computing facility and one's personal work habits much can be done to keep this clear.

N+1.9   THE OPEN SHOP PHILOSOPHY

If we believe that *The Purpose of Computing Is Insight, Not Numbers,* then it follows that the man who is to get the insight must understand the computing.   If he does not understand what is being done, he is very unlikely to derive much value from the computation.   The bare numbers he can see, but their real meaning may be buried in the computation.   The professional programmer and numerical analyst should not block his view of the problem and of the results of the computation.

Eddington has an illuminating story of a man who went fishing with a certain-sized net.   When he found that the fish caught had a minimum size, he concluded that this was the minimum size of the fish in the sea; he made the mistake of not understanding how the fishing was done.   And so it is with computing; what comes out depends on what goes in *and* on how it is processed.   Without an understanding of the processes used, it is likely that effects due to the model used in the computing will be confused with effects of the model adopted by the user when he formulated the problem.

It has further been found that frequently the process of computing sheds great light on the model being computed.   Computing is a tool that supplies numerical answers, but is also an intellectual tool for examining the world.

It is not likely that great physical insights will arise in the mind of a professional coder who routinely codes problems.   If insights are to arise, and they are what we most want, then it follows that the man with problems must comprehend and follow the computing.   This does not mean that he must do all the detailed work, but without a reasonably thorough understanding of what the computer is doing, it is unlikely that he can either arrange his work to get maximum benefit from the computer or achieve the insights which can and do result from properly arranged computations.

Experience clearly indicates that it is generally easier and better to convert an expert in a given field into a partial expert in computing than it is to try to make a computing expert into an expert in the given field. But if we are to require this, then it falls on the computing experts to make every effort to reduce the burdens of learning and using the computer. Arbitrary rules, special jargon, meaningless forms, changes in the methods and form, and delays in access to the machine should all be reduced to a minimum and carefully monitored to reduce them further when the next machine offers new opportunities to lift from the outsider the burden of the nonessentials of computing.

The study of numerical methods and library routines to be used to increase insight, rather than mere machine efficiency, is still in its infancy, and it is one of the most important areas of future research. To work in this field requires experience in the use of computing in everyday work. It is, however, a field well worth cultivating. Clearly the author does not subscribe to the "black box" theory of library routines. The author does, however, subscribe to the heresy that *some* of the results found in the field of artificial intelligence[1] could profitably be applied to *some* numerical analysis routines.

N+1.10 CLOSING REMARKS

It should not be necessary to remind the reader that most of the above remarks are personal opinions developed by the author while working in a particular laboratory, and they do not necessarily have universal applicability. But in defense of them, they seem to be grounded in common sense as well as in experience. If the reader does not like them, he should not argue the topic but should try to develop his own. It is important for the progress of machine computation that the intuitive methods that we now use become more clearly understood and reduced, when possible, to explicit formulations suitable for use with a computer.

---

[1] See, for example, Nilsson [45] and Slagle [52].

BIBLIOGRAPHY

1   ACTON, F. S.: "Numerical Methods that Work," Harper & Row, Publishers, Incorporated, New York, 1970.

2   AHLBERG, J. H., NILSON, E. N., and WALSH, J. L.: "The Theory of Splines and Their Applications," Academic Press, Inc., New York, 1967.

3   BERNOULLI, JACOB: "Ars Conjectandi," p. 97, 1713.

4   BOOLE, GEORGE: "Treatise on the Calculus of Finite Differences," 4th ed., Chelsea Publishing Company, New York.

5   BROMWICH, T. J. I'a: "An Introduction to the Theory of Infinite Series," 2d ed., Macmillan & Co., Ltd., London, 1947.

6   CAMPBELL, G. A., and FOSTER, R. M.: "Fourier Integrals for Practical Applications," D. Van Nostrand Company, Inc., Princeton, N.J., 1948.

7   CARSLAW, H. S.: "Fourier Series and Integrals," Dover Publications, Inc., New York, 1950.

8   CHAKRAVARTI, P. C.: "Integrals and Sums," p. 61, The Athlone Press, London, 1970.

9   ERDELYI, A., ed.: "Tables of Integral Transforms," vol. 1, McGraw-Hill Book Company, New York, 1954.

10 FEYNMAN, R. P.: "The Character of Physical Law," Chap. 2, The M.I.T. Press, Cambridge, Mass., 1967.

11 FORSYTHE, G., and MOLER, C. B.: "Computer Solution of Linear Algebraic Systems," pp. 11 and 148, Prentice-Hall, Inc., Englewood Cliffs, N.J., 1967.

12 FOX, L.: "Mathematical Tables," vol. 1, Her Majesty's Stationery Office, London, 1956.

13 FOX, L.: "Numerical Solution of Two-point Boundary Problems," Oxford University Press, London, 1957.

14 FOX, L., and PARKER, I. B.: "Chebyshev Polynomials in Numerical Analysis," Oxford Mathematical Handbooks, Oxford University Press, 1968.

15 GASS, S. I.: "Linear Programming," 3d ed., McGraw-Hill Book Company, New York, 1969.

16 GEAR, C. W.: "Numerical Initial-value Problems in Ordinary Differential Equations," Prentice-Hall, Inc., Englewood Cliffs, N.J., 1971.

17 GOLD, B., and RADER, C. M.: "Digital Processing of Signals," McGraw-Hill Book Company, New York, 1969.

18 GRADSHTEYN, I. S., and RYSHIK, I. M.: "Tables of Integrals, Series, and Products," Academic Press, Inc., New York, 1966.

19 HARTREE, D. R.: "Numerical Analysis," 2d ed., p. 116, Oxford University Press, London, 1958.

20 HOESCHLE, D. F.: "Analog-to-digital and Digital-to-analog Conversion Techniques," John Wiley & Sons, Inc., New York, 1968.

21 HOUSEHOLDER, A. S.: "Principles of Numerical Analysis," McGraw-Hill Book Company, New York, 1953.

22 JANSSON, BIRGER: "Random Number Generators," Almqrist and Wiksell, Stockholm, 1966.

23 JEFFREYS, HAROLD: "Theory of Probability," p. 408, 3d ed., Oxford University Press, London, 1967.

24 JOLLEY, L. B. W.: "Summation of Series," Chapman and Hall, Ltd., London, 1925; and Dover Publications, Inc., New York, 1960.

25 JORDAN, CHARLES: "The Calculus of Finite Differences," Chelsea Publishing Company, New York, 1947.

26 JURY, E. I.: "Theory and Applications of the z Transform Method," John Wiley & Sons, Inc., Englewood Cliffs, N.J., 1964.

27 KENDALL, M. G., and STUART, A.: "The Advanced Theory of Statistics," 3d ed., Hafner Publishing Company, Inc., New York, 1969.

28 KNUTH, DONALD E.: "The Art of Computing," vol. 2, "Seminumerical Algorithms," Addison-Wesley Publishing Company, Inc., Reading, Mass., 1969.

29 KOPAL, Z.: "Numerical Analysis," John Wiley & Sons, Inc., New York, 1955.

30 KRYLOV, V. I.: "Approximate Calculation of Integrals," The Macmillan Company, New York, 1962.

31 KRYLOV, V. I., and SKOBLYA, N. S.: "Handbook of Numerical Inversion of Laplace Transforms," Israel Program for Scientific Translation, IPST Press, Jerusalem, Israel.

32 KUO, F. F., and KAISER, J. F.: "System Analysis by Digital Computer," John Wiley & Sons, Inc., New York, 1966.

33 LANCZOS, CORNELIUS: "Applied Analysis," Prentice-Hall, Inc., Englewood Cliffs, N.J., 1956.

34 LANCZOS, CORNELIUS: "Linear Differential Operators," D. Van Nostrand Company, Inc., Princeton, N.J., 1961.

35 LANCZOS, CORNELIUS: "Discourse on Fourier Series," Hafner Publishing Company, Inc., New York, 1966.

36 LEPAGE, W. R.: "Complex Variables and the Laplace Transform for Engineers," McGraw-Hill Book Company, New York, 1961, Dover Publications, Inc., 1980.

37 LIGHTHILL, M. J.: "Fourier Analysis and Generalized Functions," Cambridge University Press, London, 1958.

38 MILNE, W. E.: "Numerical Calculus," Princeton University Press, Princeton, N.J., 1949.

39 MILNE, W. E.: "Numerical Solution of Differential Equations," John Wiley & Sons, Inc., New York, 1953, Dover Publications, Inc., 1970.

40 MILNE-THOMSON, L. M.: "The Calculus of Finite Differences," Macmillan & Co., Ltd., London, 1933.

41 MINEUR, H.: "Techniques de Calcul Numérique," Librairie Polytechnique, Paris, 1952.

42 MOOD, A. M., and GRAYBILL, F. A.: "Introduction to the Theory of Statistics," 2d ed., p. 181, McGraw-Hill Book Company, New York, 1963.

43 MOORE, RAMON E.: "Interval Analysis," pp. 11-145, Prentice-Hall, Inc., Englewood Cliffs, N.J., 1966.

44 MUIR, T.: "A Treatise on the Theory of Determinants," p. 419, Dover Publications, Inc., New York, 1960.

45 NILSSON, NILS J.: "Artificial Intelligence," McGraw-Hill Book Company, New York, 1971.

46 NOBLE, BEN: "Applied Linear Algebra," pp. 16 and 523, Prentice-Hall, Inc., Englewood Cliffs, N.J., 1969.

47 PAIRMAN, ELEANOR: "Tracts for Computers No. 1, Tables of the Digamma and Trigamma Functions," Cambridge University Press, London, 1919.

48 PÓLYA, G.: "How to Solve It," Princeton University Press, Princeton, N.J., 1945.

49 RALSTON, A.: "A First Course in Numerical Analysis," McGraw-Hill Book Company, New York, 1965.

50 SAATY, T. L., and BRAM, J.: "Nonlinear Mathematics," McGraw-Hill Book Company, New York, 1964, Dover Publications, Inc., 1982.

51 SHREIDER, YU. A.: "The Monte Carlo Method," Pergamon Press, New York, 1966.

52 SLAGLE, J. R.: "Artificial Intelligence, The Heuristic Programming Approach," McGraw-Hill Book Company, New York, 1971.

53 SNYDER, MARTIN A.: "Chebyshev Methods in Numerical Approximation," Prentice-Hall Series in Automatic Computation, Prentice-Hall, Inc., Englewood Cliffs, N.J., 1966.

54 SOLODOVNIKOV, V. V.: "Statistical Dynamics of Linear Automatic Control Systems," 2d ed., D. Van Nostrand Company, Inc., Princeton, N.J., 1965.

55 SPIEGEL, M. R.: "Laplace Transforms," Schaum's Outline Series, McGraw-Hill Book Company, New York, 1965.

56 STEFFENSEN, J. F.: "Interpolation," pp. 35–39, Chelsea Publishing Company, New York, 1950.

57 STROUD, A. H., and SEVREST, D.: "Gaussian Quadrature Formulas," Prentice-Hall, Inc., Englewood Cliffs, N.J., 1966.

58 SUSSKIND, A. K., ed.: "Notes on Analog-Digital Conversion Techniques," The M.I.T. Press, Cambridge, Mass., 1957.

59 TRAUB, J. F.: "Iterative Methods for the Solution of Equations," Prentice-Hall, Inc., Englewood Cliffs, N.J., 1964.

60 USPENSKY, J. V.: "Theory of Equations," Chap. 11, McGraw-Hill Book Company, New York, 1948.

61 VAN DER POL, B., and BREMER, H.: "Operational Calculus Based on the Two-sided Laplace Integral," Cambridge University Press, 1959.

62 WATSON, G. N.: "Bessel Functions," 2d ed., Cambridge University Press, London, 1952.

63 WHITTAKER, E. T., and WATSON, G. N.: "Modern Analysis," 4th ed., pp. 127–128, Cambridge University Press, London, 1935.

64 ZYGMUND, A.: "Trigonometric Series," 2d ed., vol. 1, Cambridge University Press, New York, 1959.

INDEX

A CATALOG OF SELECTED
DOVER BOOKS
IN SCIENCE AND MATHEMATICS

A CATALOG OF SELECTED DOVER BOOKS IN SCIENCE AND MATHEMATICS

QUALITATIVE THEORY OF DIFFERENTIAL EQUATIONS, V.V. Nemytskii and V.V. Stepanov. Classic graduate-level text by two prominent Soviet mathematicians covers classical differential equations as well as topological dynamics and ergodic theory. Bibliographies. 523pp. 5⅜ x 8½. 65954-2 Pa. $14.95

MATRICES AND LINEAR ALGEBRA, Hans Schneider and George Phillip Barker. Basic textbook covers theory of matrices and its applications to systems of linear equations and related topics such as determinants, eigenvalues and differential equations. Numerous exercises. 432pp. 5⅜ x 8½. 66014-1 Pa. $10.95

QUANTUM THEORY, David Bohm. This advanced undergraduate-level text presents the quantum theory in terms of qualitative and imaginative concepts, followed by specific applications worked out in mathematical detail. Preface. Index. 655pp. 5⅜ x 8½. 65969-0 Pa. $14.95

ATOMIC PHYSICS (8th edition), Max Born. Nobel laureate's lucid treatment of kinetic theory of gases, elementary particles, nuclear atom, wave-corpuscles, atomic structure and spectral lines, much more. Over 40 appendices, bibliography. 495pp. 5⅜ x 8½. 65984-4 Pa. $13.95

ELECTRONIC STRUCTURE AND THE PROPERTIES OF SOLIDS: The Physics of the Chemical Bond, Walter A. Harrison. Innovative text offers basic understanding of the electronic structure of covalent and ionic solids, simple metals, transition metals and their compounds. Problems. 1980 edition. 582pp. 6⅛ x 9¼. 66021-4 Pa. $16.95

BOUNDARY VALUE PROBLEMS OF HEAT CONDUCTION, M. Necati Özisik. Systematic, comprehensive treatment of modern mathematical methods of solving problems in heat conduction and diffusion. Numerous examples and problems. Selected references. Appendices. 505pp. 5⅜ x 8½. 65990-9 Pa. $12.95

A SHORT HISTORY OF CHEMISTRY (3rd edition), J.R. Partington. Classic exposition explores origins of chemistry, alchemy, early medical chemistry, nature of atmosphere, theory of valency, laws and structure of atomic theory, much more. 428pp. 5⅜ x 8½. (Available in U.S. only) 65977-1 Pa. $11.95

A HISTORY OF ASTRONOMY, A. Pannekoek. Well-balanced, carefully reasoned study covers such topics as Ptolemaic theory, work of Copernicus, Kepler, Newton, Eddington's work on stars, much more. Illustrated. References. 521pp. 5⅜ x 8½. 65994-1 Pa. $12.95

PRINCIPLES OF METEOROLOGICAL ANALYSIS, Walter J. Saucier. Highly respected, abundantly illustrated classic reviews atmospheric variables, hydrostatics, static stability, various analyses (scalar, cross-section, isobaric, isentropic, more). For intermediate meteorology students. 454pp. 6⅛ x 9¼. 65979-8 Pa. $14.95

RELATIVITY, THERMODYNAMICS AND COSMOLOGY, Richard C. Tolman. Landmark study extends thermodynamics to special, general relativity; also applications of relativistic mechanics, thermodynamics to cosmological models. 501pp. 5⅜ x 8½. 65383-8 Pa. $13.95

APPLIED ANALYSIS, Cornelius Lanczos. Classic work on analysis and design of finite processes for approximating solution of analytical problems. Algebraic equations, matrices, harmonic analysis, quadrature methods, much more. 559pp. 5⅜ x 8½. 65656-X Pa. $13.95

INTRODUCTION TO ANALYSIS, Maxwell Rosenlicht. Unusually clear, accessible coverage of set theory, real number system, metric spaces, continuous functions, Riemann integration, multiple integrals, more. Wide range of problems. Undergraduate level. Bibliography. 254pp. 5⅜ x 8½. 65038-3 Pa. $8.95

INTRODUCTION TO QUANTUM MECHANICS With Applications to Chemistry, Linus Pauling & E. Bright Wilson, Jr. Classic undergraduate text by Nobel Prize winner applies quantum mechanics to chemical and physical problems. Numerous tables and figures enhance the text. Chapter bibliographies. Appendices. Index. 468pp. 5⅜ x 8½. 64871-0 Pa. $12.95

ASYMPTOTIC EXPANSIONS OF INTEGRALS, Norman Bleistein & Richard A. Handelsman. Best introduction to important field with applications in a variety of scientific disciplines. New preface. Problems. Diagrams. Tables. Bibliography. Index. 448pp. 5⅜ x 8½. 65082-0 Pa. $12.95

MATHEMATICS APPLIED TO CONTINUUM MECHANICS, Lee A. Segel. Analyzes models of fluid flow and solid deformation. For upper-level math, science and engineering students. 608pp. 5⅜ x 8½. 65369-2 Pa. $14.95

ELEMENTS OF REAL ANALYSIS, David A. Sprecher. Classic text covers fundamental concepts, real number system, point sets, functions of a real variable, Fourier series, much more. Over 500 exercises. 352pp. 5⅜ x 8½. 65385-4 Pa. $11.95

PHYSICAL PRINCIPLES OF THE QUANTUM THEORY, Werner Heisenberg. Nobel Laureate discusses quantum theory, uncertainty, wave mechanics, work of Dirac, Schroedinger, Compton, Wilson, Einstein, etc. 184pp. 5⅜ x 8½. 60113-7 Pa. $6.95

INTRODUCTORY REAL ANALYSIS, A.N. Kolmogorov, S.V. Fomin. Translated by Richard A. Silverman. Self-contained, evenly paced introduction to real and functional analysis. Some 350 problems. 403pp. 5⅜ x 8½. 61226-0 Pa. $10.95

PROBLEMS AND SOLUTIONS IN QUANTUM CHEMISTRY AND PHYSICS, Charles S. Johnson, Jr. and Lee G. Pedersen. Unusually varied problems, detailed solutions in coverage of quantum mechanics, wave mechanics, angular momentum, molecular spectroscopy, scattering theory, more. 280 problems plus 139 supplementary exercises. 430pp. 6½ x 9¼. 65236-X Pa. $13.95

CATALOG OF DOVER BOOKS

ASYMPTOTIC METHODS IN ANALYSIS, N.G. de Bruijn. An inexpensive, comprehensive guide to asymptotic methods—the pioneering work that teaches by explaining worked examples in detail. Index. 224pp. 5⅜ x 8½. 64221-6 Pa. $7.95

OPTICAL RESONANCE AND TWO-LEVEL ATOMS, L. Allen and J. H. Eberly. Clear, comprehensive introduction to basic principles behind all quantum optical resonance phenomena. 53 illustrations. Preface. Index. 256pp. 5⅜ x 8½.
65533-4 Pa. $8.95

COMPLEX VARIABLES, Francis J. Flanigan. Unusual approach, delaying complex algebra till harmonic functions have been analyzed from real variable viewpoint. Includes problems with answers. 364pp. 5⅜ x 8½. 61388-7 Pa. $9.95

ATOMIC SPECTRA AND ATOMIC STRUCTURE, Gerhard Herzberg. One of best introductions; especially for specialist in other fields. Treatment is physical rather than mathematical. 80 illustrations. 257pp. 5⅜ x 8½. 60115-3 Pa. $7.95

APPLIED COMPLEX VARIABLES, John W. Dettman. Step-by-step coverage of fundamentals of analytic function theory—plus lucid exposition of five important applications: Potential Theory; Ordinary Differential Equations; Fourier Transforms; Laplace Transforms; Asymptotic Expansions. 66 figures. Exercises at chapter ends. 512pp. 5⅜ x 8½. 64670-X Pa. $12.95

ULTRASONIC ABSORPTION: An Introduction to the Theory of Sound Absorption and Dispersion in Gases, Liquids and Solids, A.B. Bhatia. Standard reference in the field provides a clear, systematically organized introductory review of fundamental concepts for advanced graduate students, research workers. Numerous diagrams. Bibliography. 440pp. 5⅜ x 8½. 64917-2 Pa. $11.95

UNBOUNDED LINEAR OPERATORS: Theory and Applications, Seymour Goldberg. Classic presents systematic treatment of the theory of unbounded linear operators in normed linear spaces with applications to differential equations. Bibliography. I99pp. 5⅜ x 8½. 64830-3 Pa. $7.95

LIGHT SCATTERING BY SMALL PARTICLES, H.C. van de Hulst. Comprehensive treatment including full range of useful approximation methods for researchers in chemistry, meteorology and astronomy. 44 illustrations. 470pp. 5⅜ x 8½.
64228-3 Pa. $12.95

CONFORMAL MAPPING ON RIEMANN SURFACES, Harvey Cohn. Lucid, insightful book presents ideal coverage of subject. 334 exercises make book perfect for self-study. 55 figures. 352pp. 5⅜ x 8¼. 64025-6 Pa. $11.95

OPTICKS, Sir Isaac Newton. Newton's own experiments with spectroscopy, colors, lenses, reflection, refraction, etc., in language the layman can follow. Foreword by Albert Einstein. 532pp. 5⅜ x 8½. 60205-2 Pa. $12.95

GENERALIZED INTEGRAL TRANSFORMATIONS, A.H. Zemanian. Graduate-level study of recent generalizations of the Laplace, Mellin, Hankel, K. Weierstrass, convolution and other simple transformations. Bibliography. 320pp. 5⅜ x 8¼.
65375-7 Pa. $8.95

THE ELECTROMAGNETIC FIELD, Albert Shadowitz. Comprehensive undergraduate text covers basics of electric and magnetic fields, builds up to electromagnetic theory. Also related topics, including relativity. Over 900 problems. 768pp. 5⅜ x 8¼. 65660-8 Pa. $18.95

FOURIER SERIES, Georgi P. Tolstov. Translated by Richard A. Silverman. A valuable addition to the literature on the subject, moving clearly from subject to subject and theorem to theorem. 107 problems, answers. 336pp. 5⅜ x 8½. 63317-9 Pa. $9.95

THEORY OF ELECTROMAGNETIC WAVE PROPAGATION, Charles Herach Papas. Graduate-level study discusses the Maxwell field equations, radiation from wire antennas, the Doppler effect and more. xiii + 244pp. 5⅜ x 8½. 65678-0 Pa. $6.95

DISTRIBUTION THEORY AND TRANSFORM ANALYSIS: An Introduction to Generalized Functions, with Applications, A.H. Zemanian. Provides basics of distribution theory, describes generalized Fourier and Laplace transformations. Numerous problems. 384pp. 5⅜ x 8½. 65479-6 Pa. $11.95

THE PHYSICS OF WAVES, William C. Elmore and Mark A. Heald. Unique overview of classical wave theory. Acoustics, optics, electromagnetic radiation, more. Ideal as classroom text or for self-study. Problems. 477pp. 5⅜ x 8½. 64926-1 Pa. $13.95

CALCULUS OF VARIATIONS WITH APPLICATIONS, George M. Ewing. Applications-oriented introduction to variational theory develops insight and promotes understanding of specialized books, research papers. Suitable for advanced undergraduate/graduate students as primary, supplementary text. 352pp. 5⅜ x 8½. 64856-7 Pa. $9.95

A TREATISE ON ELECTRICITY AND MAGNETISM, James Clerk Maxwell. Important foundation work of modern physics. Brings to final form Maxwell's theory of electromagnetism and rigorously derives his general equations of field theory. 1,084pp. 5⅜ x 8½. 60636-8, 60637-6 Pa., Two-vol. set $25.90

AN INTRODUCTION TO THE CALCULUS OF VARIATIONS, Charles Fox. Graduate-level text covers variations of an integral, isoperimetrical problems, least action, special relativity, approximations, more. References. 279pp. 5⅜ x 8½. 65499-0 Pa. $8.95

HYDRODYNAMIC AND HYDROMAGNETIC STABILITY, S. Chandrasekhar. Lucid examination of the Rayleigh-Benard problem; clear coverage of the theory of instabilities causing convection. 704pp. 5⅜ x 8¼. 64071-X Pa. $14.95

CALCULUS OF VARIATIONS, Robert Weinstock. Basic introduction covering isoperimetric problems, theory of elasticity, quantum mechanics, electrostatics, etc. Exercises throughout. 326pp. 5⅜ x 8½. 63069-2 Pa. $9.95

DYNAMICS OF FLUIDS IN POROUS MEDIA, Jacob Bear. For advanced students of ground water hydrology, soil mechanics and physics, drainage and irrigation engineering and more. 335 illustrations. Exercises, with answers. 784pp. 6⅛ x 9¼. 65675-6 Pa. $19.95

NUMERICAL METHODS FOR SCIENTISTS AND ENGINEERS, Richard Hamming. Classic text stresses frequency approach in coverage of algorithms, polynomial approximation, Fourier approximation, exponential approximation, other topics. Revised and enlarged 2nd edition. 721pp. 5⅜ x 8½.   65241-6 Pa. $15.95

THEORETICAL SOLID STATE PHYSICS, Vol. 1: Perfect Lattices in Equilibrium; Vol. II: Non-Equilibrium and Disorder, William Jones and Norman H. March. Monumental reference work covers fundamental theory of equilibrium properties of perfect crystalline solids, non-equilibrium properties, defects and disordered systems. Appendices. Problems. Preface. Diagrams. Index. Bibliography. Total of 1,301pp. 5⅜ x 8½. Two volumes.   Vol. I: 65015-4 Pa. $16.95
Vol. II: 65016-2 Pa. $16.95

OPTIMIZATION THEORY WITH APPLICATIONS, Donald A. Pierre. Broad spectrum approach to important topic. Classical theory of minima and maxima, calculus of variations, simplex technique and linear programming, more. Many problems, examples. 640pp. 5⅜ x 8½.   65205-X Pa. $16.95

THE CONTINUUM: A Critical Examination of the Foundation of Analysis, Hermann Weyl. Classic of 20th-century foundational research deals with the conceptual problem posed by the continuum. 156pp. 5⅜ x 8½.   67982-9 Pa. $6.95

ESSAYS ON THE THEORY OF NUMBERS, Richard Dedekind. Two classic essays by great German mathematician: on the theory of irrational numbers; and on transfinite numbers and properties of natural numbers. 115pp. 5⅜ x 8½.
21010-3 Pa. $5.95

THE FUNCTIONS OF MATHEMATICAL PHYSICS, Harry Hochstadt. Comprehensive treatment of orthogonal polynomials, hypergeometric functions, Hill's equation, much more. Bibliography. Index. 322pp. 5⅜ x 8½. 65214-9 Pa. $9.95

NUMBER THEORY AND ITS HISTORY, Oystein Ore. Unusually clear, accessible introduction covers counting, properties of numbers, prime numbers, much more. Bibliography. 380pp. 5⅜ x 8½.   65620-9 Pa. $10.95

THE VARIATIONAL PRINCIPLES OF MECHANICS, Cornelius Lanczos. Graduate level coverage of calculus of variations, equations of motion, relativistic mechanics, more. First inexpensive paperbound edition of classic treatise. Index. Bibliography. 418pp. 5⅜ x 8½.   65067-7 Pa. $12.95

MATHEMATICAL TABLES AND FORMULAS, Robert D. Carmichael and Edwin R. Smith. Logarithms, sines, tangents, trig functions, powers, roots, reciprocals, exponential and hyperbolic functions, formulas and theorems. 269pp. 5⅜ x 8½.
60111-0 Pa. $6.95

THEORETICAL PHYSICS, Georg Joos, with Ira M. Freeman. Classic overview covers essential math, mechanics, electromagnetic theory, thermodynamics, quantum mechanics, nuclear physics, other topics. First paperback edition. xxiii + 885pp. 5⅜ x 8½.   65227-0 Pa. $21.95

CATALOG OF DOVER BOOKS

HANDBOOK OF MATHEMATICAL FUNCTIONS WITH FORMULAS, GRAPHS, AND MATHEMATICAL TABLES, edited by Milton Abramowitz and Irene A. Stegun. Vast compendium: 29 sets of tables, some to as high as 20 places. 1,046pp. 8 x 10½.   61272-4 Pa. $26.95

MATHEMATICAL METHODS IN PHYSICS AND ENGINEERING, John W. Dettman. Algebraically based approach to vectors, mapping, diffraction, other topics in applied math. Also generalized functions, analytic function theory, more. Exercises. 448pp. 5⅜ x 8¼.   65649-7 Pa. $10.95

A SURVEY OF NUMERICAL MATHEMATICS, David M. Young and Robert Todd Gregory. Broad self-contained coverage of computer-oriented numerical algorithms for solving various types of mathematical problems in linear algebra, ordinary and partial, differential equations, much more. Exercises. Total of 1,248pp. 5⅜ x 8½.
Two volumes.   Vol. I: 65691-8 Pa. $16.95
                                                                                Vol. II: 65692-6 Pa. $16.95

TENSOR ANALYSIS FOR PHYSICISTS, J.A. Schouten. Concise exposition of the mathematical basis of tensor analysis, integrated with well-chosen physical examples of the theory. Exercises. Index. Bibliography. 289pp. 5⅜ x 8½.   65582-2 Pa. $8.95

INTRODUCTION TO NUMERICAL ANALYSIS (2nd Edition), F.B. Hildebrand. Classic, fundamental treatment covers computation, approximation, interpolation, numerical differentiation and integration, other topics. 150 new problems. 669pp. 5⅜ x 8½.   65363-3 Pa. $16.95

INVESTIGATIONS ON THE THEORY OF THE BROWNIAN MOVEMENT, Albert Einstein. Five papers (1905–8) investigating dynamics of Brownian motion and evolving elementary theory. Notes by R. Fürth. 122pp. 5⅜ x 8½.
                                                                                60304-0 Pa. $5.95

CATASTROPHE THEORY FOR SCIENTISTS AND ENGINEERS, Robert Gilmore. Advanced-level treatment describes mathematics of theory grounded in the work of Poincaré, R. Thom, other mathematicians. Also important applications to problems in mathematics, physics, chemistry and engineering. 1981 edition. References. 28 tables. 397 black-and-white illustrations. xvii + 666pp. 6⅛ x 9¼.
                                                                                67539-4 Pa. $17.95

AN INTRODUCTION TO STATISTICAL THERMODYNAMICS, Terrell L. Hill. Excellent basic text offers wide-ranging coverage of quantum statistical mechanics, systems of interacting molecules, quantum statistics, more. 523pp. 5⅜ x 8½.
                                                                                65242-4 Pa. $12.95

STATISTICAL PHYSICS, Gregory H. Wannier. Classic text combines thermodynamics, statistical mechanics and kinetic theory in one unified presentation of thermal physics. Problems with solutions. Bibliography. 532pp. 5⅜ x 8½.
                                                                                65401-X Pa. $12.95

CATALOG OF DOVER BOOKS

ORDINARY DIFFERENTIAL EQUATIONS, Morris Tenenbaum and Harry Pollard. Exhaustive survey of ordinary differential equations for undergraduates in mathematics, engineering, science. Thorough analysis of theorems. Diagrams. Bibliography. Index. 818pp. 5⅜ x 8½. 64940-7 Pa. $18.95

STATISTICAL MECHANICS: Principles and Applications, Terrell L. Hill. Standard text covers fundamentals of statistical mechanics, applications to fluctuation theory, imperfect gases, distribution functions, more. 448pp. 5⅜ x 8½. 65390-0 Pa. $11.95

ORDINARY DIFFERENTIAL EQUATIONS AND STABILITY THEORY: An Introduction, David A. Sánchez. Brief, modern treatment. Linear equation, stability theory for autonomous and nonautonomous systems, etc. 164pp. 5⅜ x 8¼. 63828-6 Pa. $6.95

THIRTY YEARS THAT SHOOK PHYSICS: The Story of Quantum Theory, George Gamow. Lucid, accessible introduction to influential theory of energy and matter. Careful explanations of Dirac's anti-particles, Bohr's model of the atom, much more. 12 plates. Numerous drawings. 240pp. 5⅜ x 8½. 24895-X Pa. $7.95

THEORY OF MATRICES, Sam Perlis. Outstanding text covering rank, nonsingularity and inverses in connection with the development of canonical matrices under the relation of equivalence, and without the intervention of determinants. Includes exercises. 237pp. 5⅜ x 8½. 66810-X Pa. $8.95

GREAT EXPERIMENTS IN PHYSICS: Firsthand Accounts from Galileo to Einstein, edited by Morris H. Shamos. 25 crucial discoveries: Newton's laws of motion, Chadwick's study of the neutron, Hertz on electromagnetic waves, more. Original accounts clearly annotated. 370pp. 5⅜ x 8½. 25346-5 Pa. $10.95

INTRODUCTION TO PARTIAL DIFFERENTIAL EQUATIONS WITH APPLICATIONS, E.C. Zachmanoglou and Dale W. Thoe. Essentials of partial differential equations applied to common problems in engineering and the physical sciences. Problems and answers. 416pp. 5⅜ x 8½. 65251-3 Pa. $11.95

BURNHAM'S CELESTIAL HANDBOOK, Robert Burnham, Jr. Thorough guide to the stars beyond our solar system. Exhaustive treatment. Alphabetical by constellation: Andromeda to Cetus in Vol. 1; Chamaeleon to Orion in Vol. 2; and Pavo to Vulpecula in Vol. 3. Hundreds of illustrations. Index in Vol. 3. 2,000pp. 6⅛ x 9¼. 23567-X, 23568-8, 23673-0 Pa., Three-vol. set $44.85

CHEMICAL MAGIC, Leonard A. Ford. Second Edition, Revised by E. Winston Grundmeier. Over 100 unusual stunts demonstrating cold fire, dust explosions, much more. Text explains scientific principles and stresses safety precautions. 128pp. 5⅜ x 8½. 67628-5 Pa. $5.95

AMATEUR ASTRONOMER'S HANDBOOK, J.B. Sidgwick. Timeless, comprehensive coverage of telescopes, mirrors, lenses, mountings, telescope drives, micrometers, spectroscopes, more. 189 illustrations. 576pp. 5⅜ x 8¼. (Available in U.S. only) 24034-7 Pa. $11.95

CATALOG OF DOVER BOOKS

SPECIAL FUNCTIONS, N.N. Lebedev. Translated by Richard Silverman. Famous Russian work treating more important special functions, with applications to specific problems of physics and engineering. 38 figures. 308pp. 5⅜ x 8½. 60624-4 Pa. $9.95

OBSERVATIONAL ASTRONOMY FOR AMATEURS, J.B. Sidgwick. Mine of useful data for observation of sun, moon, planets, asteroids, aurorae, meteors, comets, variables, binaries, etc. 39 illustrations. 384pp. 5⅝ x 8¼. (Available in U.S. only) 24033-9 Pa. $8.95

INTEGRAL EQUATIONS, F.G. Tricomi. Authoritative, well-written treatment of extremely useful mathematical tool with wide applications. Volterra Equations, Fredholm Equations, much more. Advanced undergraduate to graduate level. Exercises. Bibliography. 238pp. 5⅜ x 8½. 64828-1 Pa. $8.95

POPULAR LECTURES ON MATHEMATICAL LOGIC, Hao Wang. Noted logician's lucid treatment of historical developments, set theory, model theory, recursion theory and constructivism, proof theory, more. 3 appendixes. Bibliography. 1981 edition. ix + 283pp. 5⅜ x 8½. 67632-3 Pa. $8.95

MODERN NONLINEAR EQUATIONS, Thomas L. Saaty. Emphasizes practical solution of problems; covers seven types of equations. ". . . a welcome contribution to the existing literature...."–*Math Reviews*. 490pp. 5⅜ x 8½. 64232-1 Pa. $13.95

FUNDAMENTALS OF ASTRODYNAMICS, Roger Bate et al. Modern approach developed by U.S. Air Force Academy. Designed as a first course. Problems, exercises. Numerous illustrations. 455pp. 5⅜ x 8½. 60061-0 Pa. $10.95

INTRODUCTION TO LINEAR ALGEBRA AND DIFFERENTIAL EQUATIONS, John W. Dettman. Excellent text covers complex numbers, determinants, orthonormal bases, Laplace transforms, much more. Exercises with solutions. Undergraduate level. 416pp. 5⅜ x 8½. 65191-6 Pa. $11.95

INCOMPRESSIBLE AERODYNAMICS, edited by Bryan Thwaites. Covers theoretical and experimental treatment of the uniform flow of air and viscous fluids past two-dimensional aerofoils and three-dimensional wings; many other topics. 654pp. 5⅜ x 8½. 65465-6 Pa. $16.95

INTRODUCTION TO DIFFERENCE EQUATIONS, Samuel Goldberg. Exceptionally clear exposition of important discipline with applications to sociology, psychology, economics. Many illustrative examples; over 250 problems. 260pp. 5⅜ x 8½. 65084-7 Pa. $8.95

LAMINAR BOUNDARY LAYERS, edited by L. Rosenhead. Engineering classic covers steady boundary layers in two- and three- dimensional flow, unsteady boundary layers, stability, observational techniques, much more. 708pp. 5⅜ x 8½. 65646-2 Pa. $18 95

LECTURES ON CLASSICAL DIFFERENTIAL GEOMETRY, Second Edition, Dirk J. Struik. Excellent brief introduction covers curves, theory of surfaces, fundamental equations, geometry on a surface, conformal mapping, other topics. Problems. 240pp. 5⅜ x 8½. 65609-8 Pa. $8.95

ROTARY-WING AERODYNAMICS, W.Z. Stepniewski. Clear, concise text covers aerodynamic phenomena of the rotor and offers guidelines for helicopter performance evaluation. Originally prepared for NASA. 537 figures. 640pp. 6⅛ x 9¼.
64647-5 Pa. $16.95

DIFFERENTIAL GEOMETRY, Heinrich W. Guggenheimer. Local differential geometry as an application of advanced calculus and linear algebra. Curvature, transformation groups, surfaces, more. Exercises. 62 figures. 378pp. 5⅜ x 8½.
63433-7 Pa. $9.95

INTRODUCTION TO SPACE DYNAMICS, William Tyrrell Thomson. Comprehensive, classic introduction to space-flight engineering for advanced undergraduate and graduate students. Includes vector algebra, kinematics, transformation of coordinates. Bibliography. Index. 352pp. 5⅜ x 8½. 65113-4 Pa. $9.95

A SURVEY OF MINIMAL SURFACES, Robert Osserman. Up-to-date, in-depth discussion of the field for advanced students. Corrected and enlarged edition covers new developments. Includes numerous problems. 192pp. 5⅜ x 8½. 64998-9 Pa. $8.95

ANALYTICAL MECHANICS OF GEARS, Earle Buckingham. Indispensable reference for modern gear manufacture covers conjugate gear-tooth action, gear-tooth profiles of various gears, many other topics. 263 figures. 102 tables. 546pp. 5⅜ x 8½.
65712-4 Pa. $14.95

SET THEORY AND LOGIC, Robert R. Stoll. Lucid introduction to unified theory of mathematical concepts. Set theory and logic seen as tools for conceptual understanding of real number system. 496pp. 5⅜ x 8¼. 63829-4 Pa. $12.95

A HISTORY OF MECHANICS, René Dugas. Monumental study of mechanical principles from antiquity to quantum mechanics. Contributions of ancient Greeks, Galileo, Leonardo, Kepler, Lagrange, many others. 671pp. 5⅜ x 8½.
65632-2 Pa. $14.95

FAMOUS PROBLEMS OF GEOMETRY AND HOW TO SOLVE THEM, Benjamin Bold. Squaring the circle, trisecting the angle, duplicating the cube: learn their history, why they are impossible to solve, then solve them yourself. 128pp. 5⅜ x 8½. 24297-8 Pa. $4.95

MECHANICAL VIBRATIONS, J.P. Den Hartog. Classic textbook offers lucid explanations and illustrative models, applying theories of vibrations to a variety of practical industrial engineering problems. Numerous figures. 233 problems, solutions. Appendix. Index. Preface. 436pp. 5⅜ x 8½. 64785-4 Pa. $11.95

CURVATURE AND HOMOLOGY, Samuel I. Goldberg. Thorough treatment of specialized branch of differential geometry. Covers Riemannian manifolds, topology of differentiable manifolds, compact Lie groups, other topics. Exercises. 315pp. 5⅜ x 8½. 64314-X Pa. $9.95

HISTORY OF STRENGTH OF MATERIALS, Stephen P. Timoshenko. Excellent historical survey of the strength of materials with many references to the theories of elasticity and structure. 245 figures. 452pp. 5⅜ x 8½. 61187-6 Pa. $12.95

CATALOG OF DOVER BOOKS

GEOMETRY OF COMPLEX NUMBERS, Hans Schwerdtfeger. Illuminating, widely praised book on analytic geometry of circles, the Moebius transformation, and two-dimensional non-Euclidean geometries. 200pp. 5⅜ x 8¼. 63830-8 Pa. $8.95

MECHANICS, J.P. Den Hartog. A classic introductory text or refresher. Hundreds of applications and design problems illuminate fundamentals of trusses, loaded beams and cables, etc. 334 answered problems. 462pp. 5⅜ x 8½. 60754-2 Pa. $11.95

TOPOLOGY, John G. Hocking and Gail S. Young. Superb one-year course in classical topology. Topological spaces and functions, point-set topology, much more. Examples and problems. Bibliography. Index. 384pp. 5⅜ x 8¼. 65676-4 Pa. $10.95

STRENGTH OF MATERIALS, J.P. Den Hartog. Full, clear treatment of basic material (tension, torsion, bending, etc.) plus advanced material on engineering methods, applications. 350 answered problems. 323pp. 5⅜ x 8½. 60755-0 Pa. $9.95

ELEMENTARY CONCEPTS OF TOPOLOGY, Paul Alexandroff. Elegant, intuitive approach to topology from set-theoretic topology to Betti groups; how concepts of topology are useful in math and physics. 25 figures. 57pp. 5⅜ x 8½.
60747-X Pa. $3.95

ADVANCED STRENGTH OF MATERIALS, J.P. Den Hartog. Superbly written advanced text covers torsion, rotating disks, membrane stresses in shells, much more. Many problems and answers. 388pp. 5⅜ x 8½. 65407-9 Pa. $10.95

COMPUTABILITY AND UNSOLVABILITY, Martin Davis. Classic graduate-level introduction to theory of computability, usually referred to as theory of recurrent functions. New preface and appendix. 288pp. 5⅜ x 8½. 61471-9 Pa. $8.95

GENERAL CHEMISTRY, Linus Pauling. Revised 3rd edition of classic first-year text by Nobel laureate. Atomic and molecular structure, quantum mechanics, statistical mechanics, thermodynamics correlated with descriptive chemistry. Problems. 992pp. 5⅜ x 8½. 65622-5 Pa. $19.95

AN INTRODUCTION TO MATRICES, SETS AND GROUPS FOR SCIENCE STUDENTS, G. Stephenson. Concise, readable text introduces sets, groups, and most importantly, matrices to undergraduate students of physics, chemistry, and engineering. Problems. 164pp. 5⅜ x 8½. 65077-4 Pa. $7.95

THE HISTORICAL BACKGROUND OF CHEMISTRY, Henry M. Leicester. Evolution of ideas, not individual biography. Concentrates on formulation of a coherent set of chemical laws. 260pp. 5⅜ x 8½. 61053-5 Pa. $8.95

THE PHILOSOPHY OF MATHEMATICS: An Introductory Essay, Stephan Körner. Surveys the views of Plato, Aristotle, Leibniz & Kant concerning propositions and theories of applied and pure mathematics. Introduction. Two appendices. Index. 198pp. 5⅜ x 8½. 25048-2 Pa. $8.95

THE DEVELOPMENT OF MODERN CHEMISTRY, Aaron J. Ihde. Authoritative history of chemistry from ancient Greek theory to 20th-century innovation. Covers major chemists and their discoveries. 209 illustrations. 14 tables. Bibliographies. Indices. Appendices. 851pp. 5⅜ x 8½. 64235-6 Pa. $18.95

DE RE METALLICA, Georgius Agricola. The famous Hoover translation of greatest treatise on technological chemistry, engineering, geology, mining of early modern times (1556). All 289 original woodcuts. 638pp. 6¾ x 11.   60006-8 Pa. $21.95

SOME THEORY OF SAMPLING, William Edwards Deming. Analysis of the problems, theory and design of sampling techniques for social scientists, industrial managers and others who find statistics increasingly important in their work. 61 tables. 90 figures. xvii + 602pp. 5⅜ x 8½.   64684-X Pa. $16.95

THE VARIOUS AND INGENIOUS MACHINES OF AGOSTINO RAMELLI: A Classic Sixteenth-Century Illustrated Treatise on Technology, Agostino Ramelli. One of the most widely known and copied works on machinery in the 16th century. 194 detailed plates of water pumps, grain mills, cranes, more. 608pp. 9 x 12.
28180-9 Pa. $24.95

LINEAR PROGRAMMING AND ECONOMIC ANALYSIS, Robert Dorfman, Paul A. Samuelson and Robert M. Solow. First comprehensive treatment of linear programming in standard economic analysis. Game theory, modern welfare economics, Leontief input-output, more. 525pp. 5⅜ x 8½.   65491-5 Pa. $14.95

ELEMENTARY DECISION THEORY, Herman Chernoff and Lincoln E. Moses. Clear introduction to statistics and statistical theory covers data processing, probability and random variables, testing hypotheses, much more. Exercises. 364pp. 5⅜ x 8½.   65218-1 Pa. $10.95

THE COMPLEAT STRATEGYST: Being a Primer on the Theory of Games of Strategy, J.D. Williams. Highly entertaining classic describes, with many illustrated examples, how to select best strategies in conflict situations. Prefaces. Appendices. 268pp. 5⅜ x 8½.   25101-2 Pa. $7.95

CONSTRUCTIONS AND COMBINATORIAL PROBLEMS IN DESIGN OF EXPERIMENTS, Damaraju Raghavarao. In-depth reference work examines orthogonal Latin squares, incomplete block designs, tactical configuration, partial geometry, much more. Abundant explanations, examples. 416pp. 5⅜ x 8¼.
65685-3 Pa. $10.95

THE ABSOLUTE DIFFERENTIAL CALCULUS (CALCULUS OF TENSORS), Tullio Levi-Civita. Great 20th-century mathematician's classic work on material necessary for mathematical grasp of theory of relativity. 452pp. 5⅜ x 8½.
63401-9 Pa. $11.95

VECTOR AND TENSOR ANALYSIS WITH APPLICATIONS, A.I. Borisenko and I.E. Tarapov. Concise introduction. Worked-out problems, solutions, exercises. 257pp. 5⅜ x 8¼.   63833-2 Pa. $8.95

THE FOUR-COLOR PROBLEM: Assaults and Conquest, Thomas L. Saaty and Paul G. Kainen. Engrossing, comprehensive account of the century-old combinatorial topological problem, its history and solution. Bibliographies. Index. 110 figures. 228pp. 5⅜ x 8½.   65092-8 Pa. $7.95

CATALYSIS IN CHEMISTRY AND ENZYMOLOGY, William P. Jencks. Exceptionally clear coverage of mechanisms for catalysis, forces in aqueous solution, carbonyl- and acyl-group reactions, practical kinetics, more. 864pp. 5⅜ x 8½.
65460-5 Pa. $19.95

PROBABILITY: An Introduction, Samuel Goldberg. Excellent basic text covers set theory, probability theory for finite sample spaces, binomial theorem, much more. 360 problems. Bibliographies. 322pp. 5⅜ x 8½.
65252-1 Pa. $10.95

LIGHTNING, Martin A. Uman. Revised, updated edition of classic work on the physics of lightning. Phenomena, terminology, measurement, photography, spectroscopy, thunder, more. Reviews recent research. Bibliography. Indices. 320pp. 5⅜ x 8¼.
64575-4 Pa. $8.95

PROBABILITY THEORY: A Concise Course, Y.A. Rozanov. Highly readable, self-contained introduction covers combination of events, dependent events, Bernoulli trials, etc. Translation by Richard Silverman. 148pp. 5⅜ x 8¼.
63544-9 Pa. $7.95

AN INTRODUCTION TO HAMILTONIAN OPTICS, H. A. Buchdahl. Detailed account of the Hamiltonian treatment of aberration theory in geometrical optics. Many classes of optical systems defined in terms of the symmetries they possess. Problems with detailed solutions. 1970 edition. xv + 360pp. 5⅜ x 8½.
67597-1 Pa. $10.95

STATISTICS MANUAL, Edwin L. Crow, et al. Comprehensive, practical collection of classical and modern methods prepared by U.S. Naval Ordnance Test Station. Stress on use. Basics of statistics assumed. 288pp. 5⅜ x 8½.
60599-X Pa. $7.95

DICTIONARY/OUTLINE OF BASIC STATISTICS, John E. Freund and Frank J. Williams. A clear concise dictionary of over 1,000 statistical terms and an outline of statistical formulas covering probability, nonparametric tests, much more. 208pp. 5⅜ x 8½.
66796-0 Pa. $7.95

STATISTICAL METHOD FROM THE VIEWPOINT OF QUALITY CONTROL, Walter A. Shewhart. Important text explains regulation of variables, uses of statistical control to achieve quality control in industry, agriculture, other areas. 192pp. 5⅜ x 8½.
65232-7 Pa. $7.95

METHODS OF THERMODYNAMICS, Howard Reiss. Outstanding text focuses on physical technique of thermodynamics, typical problem areas of understanding, and significance and use of thermodynamic potential. 1965 edition. 238pp. 5⅜ x 8½.
69445-3 Pa. $8.95

STATISTICAL ADJUSTMENT OF DATA, W. Edwards Deming. Introduction to basic concepts of statistics, curve fitting, least squares solution, conditions without parameter, conditions containing parameters. 26 exercises worked out. 271pp. 5⅜ x 8½.
64685-8 Pa. $9.95

TENSOR CALCULUS, J.L. Synge and A. Schild. Widely used introductory text covers spaces and tensors, basic operations in Riemannian space, non-Riemannian spaces, etc. 324pp. 5⅜ x 8¼.
63612-7 Pa. $9.95

CATALOG OF DOVER BOOKS

A CONCISE HISTORY OF MATHEMATICS, Dirk J. Struik. The best brief history of mathematics. Stresses origins and covers every major figure from ancient Near East to 19th century. 41 illustrations. 195pp. 5⅜ x 8½. 60255-9 Pa. $8.95

A SHORT ACCOUNT OF THE HISTORY OF MATHEMATICS, W.W. Rouse Ball. One of clearest, most authoritative surveys from the Egyptians and Phoenicians through 19th-century figures such as Grassman, Galois, Riemann. Fourth edition. 522pp. 5⅜ x 8½. 20630-0 Pa. $11.95

HISTORY OF MATHEMATICS, David E. Smith. Nontechnical survey from ancient Greece and Orient to late 19th century; evolution of arithmetic, geometry, trigonometry, calculating devices, algebra, the calculus. 362 illustrations. 1,355pp. 5⅜ x 8½. 20429-4, 20430-8 Pa., Two-vol. set $26.90

THE GEOMETRY OF RENÉ DESCARTES, René Descartes. The great work founded analytical geometry. Original French text, Descartes' own diagrams, together with definitive Smith-Latham translation. 244pp. 5⅜ x 8½. 60068-8 Pa. $8.95

THE ORIGINS OF THE INFINITESIMAL CALCULUS, Margaret E. Baron. Only fully detailed and documented account of crucial discipline: origins; development by Galileo, Kepler, Cavalieri; contributions of Newton, Leibniz, more. 304pp. 5⅜ x 8½. (Available in U.S. and Canada only) 65371-4 Pa. $9.95

THE HISTORY OF THE CALCULUS AND ITS CONCEPTUAL DEVELOPMENT, Carl B. Boyer. Origins in antiquity, medieval contributions, work of Newton, Leibniz, rigorous formulation. Treatment is verbal. 346pp. 5⅜ x 8½. 60509-4 Pa. $9.95

THE THIRTEEN BOOKS OF EUCLID'S ELEMENTS, translated with introduction and commentary by Sir Thomas L. Heath. Definitive edition. Textual and linguistic notes, mathematical analysis. 2,500 years of critical commentary. Not abridged. 1,414pp. 5⅜ x 8½. 60088-2, 60089-0, 60090-4 Pa., Three-vol. set $32.85

GAMES AND DECISIONS: Introduction and Critical Survey, R. Duncan Luce and Howard Raiffa. Superb nontechnical introduction to game theory, primarily applied to social sciences. Utility theory, zero-sum games, n-person games, decision-making, much more. Bibliography. 509pp. 5⅜ x 8½. 65943-7 Pa. $13.95

THE HISTORICAL ROOTS OF ELEMENTARY MATHEMATICS, Lucas N.H. Bunt, Phillip S. Jones, and Jack D. Bedient. Fundamental underpinnings of modern arithmetic, algebra, geometry and number systems derived from ancient civilizations. 320pp. 5⅜ x 8½. 25563-8 Pa. $8.95

CALCULUS REFRESHER FOR TECHNICAL PEOPLE, A. Albert Klaf. Covers important aspects of integral and differential calculus via 756 questions. 566 problems, most answered. 431pp. 5⅜ x 8½. 20370-0 Pa. $8.95

CATALOG OF DOVER BOOKS

CHALLENGING MATHEMATICAL PROBLEMS WITH ELEMENTARY SOLUTIONS, A.M. Yaglom and I.M. Yaglom. Over 170 challenging problems on probability theory, combinatorial analysis, points and lines, topology, convex polygons, many other topics. Solutions. Total of 445pp. 5⅜ x 8½. Two-vol. set.

Vol. I: 65536-9 Pa. $7.95
Vol. II: 65537-7 Pa. $7.95

FIFTY CHALLENGING PROBLEMS IN PROBABILITY WITH SOLUTIONS, Frederick Mosteller. Remarkable puzzlers, graded in difficulty, illustrate elementary and advanced aspects of probability. Detailed solutions. 88pp. 5⅜ x 8½.

65355-2 Pa. $4.95

EXPERIMENTS IN TOPOLOGY, Stephen Barr. Classic, lively explanation of one of the byways of mathematics. Klein bottles, Moebius strips, projective planes, map coloring, problem of the Koenigsberg bridges, much more, described with clarity and wit. 43 figures. 210pp. 5⅜ x 8½. 25933-1 Pa. $6.95

RELATIVITY IN ILLUSTRATIONS, Jacob T. Schwartz. Clear nontechnical treatment makes relativity more accessible than ever before. Over 60 drawings illustrate concepts more clearly than text alone. Only high school geometry needed. Bibliography. 128pp. 6⅛ x 9¼. 25965-X Pa. $7.95

AN INTRODUCTION TO ORDINARY DIFFERENTIAL EQUATIONS, Earl A. Coddington. A thorough and systematic first course in elementary differential equations for undergraduates in mathematics and science, with many exercises and problems (with answers). Index. 304pp. 5⅜ x 8½. 65942-9 Pa. $8.95

FOURIER SERIES AND ORTHOGONAL FUNCTIONS, Harry F. Davis. An incisive text combining theory and practical example to introduce Fourier series, orthogonal functions and applications of the Fourier method to boundary-value problems. 570 exercises. Answers and notes. 416pp. 5⅜ x 8½. 65973-9 Pa. $11.95

AN INTRODUCTION TO ALGEBRAIC STRUCTURES, Joseph Landin. Superb self-contained text covers "abstract algebra": sets and numbers, theory of groups, theory of rings, much more. Numerous well-chosen examples, exercises. 247pp. 5⅜ x 8½. 65940-2 Pa. $8.95

STARS AND RELATIVITY, Ya. B. Zel'dovich and I. D. Novikov. Vol. 1 of *Relativistic Astrophysics* by famed Russian scientists. General relativity, properties of matter under astrophysical conditions, stars and stellar systems. Deep physical insights, clear presentation. 1971 edition. References. 544pp. 5⅜ x 8½. 69424-0 Pa. $14.95

Prices subject to change without notice.

Available at your book dealer or write for free Mathematics and Science Catalog to Dept. GI, Dover Publications, Inc., 31 East 2nd St., Mineola, N.Y. 11501. Dover publishes more than 250 books each year on science, elementary and advanced mathematics, biology, music, art, literature, history, social sciences and other areas.